THE PHYSICAL CHEMISTRY OF MATERIALS

ENERGY AND ENVIRONMENTAL APPLICATIONS

THE PHYSICAL CHEMISTRY OF MATERIALS

ENERGY AND ENVIRONMENTAL APPLICATIONS

ROLANDO M.A. ROQUE-MALHERBE

CRC Press
Taylor & Francis Group
Boca Raton London New York

CRC Press is an imprint of the
Taylor & Francis Group, an **informa** business

CRC Press
Taylor & Francis Group
6000 Broken Sound Parkway NW, Suite 300
Boca Raton, FL 33487-2742

First issued in paperback 2017

© 2010 by Taylor and Francis Group, LLC
CRC Press is an imprint of Taylor & Francis Group, an Informa business

No claim to original U.S. Government works

ISBN-13: 978-1-4200-8272-2 (hbk)
ISBN-13: 978-1-138-11770-9 (pbk)

Library of Congress Cataloging-in-Publication Data

Roque-Malherbe, Rolando M. A.
  The physical chemistry of materials : energy and environmental applications / Rolando M.A. Roque-Malherbe.
    p. cm.
  Includes bibliographical references and index.
  ISBN 978-1-4200-8272-2 (hardcover : alk. paper)
  1. Materials science. 2. Chemistry, Physical and theoretical. I. Title.

TA403.R567 2010
620.1'1--dc22   2009034795

**Visit the Taylor & Francis Web site at
http://www.taylorandfrancis.com**

**and the CRC Press Web site at
http://www.crcpress.com**

To the loving memory of my mother, Silvia Malherbe;
my father, Rolando Roque;
my grandmothers, Maria Fernandez and Isidra Peña;
my grandfathers, Herminio Roque and Diego Malherbe;
and my favorite pets, Zeolita and Trosia

Contents

Preface

Since ancient times, the development and use of materials has been one of the basic objectives of mankind. Eras, that is, the Stone Age, the Bronze Age, and the Iron Age, have been named after the fundamental material used by mankind to construct their tools. Materials science is the modern activity that provides the raw material for this endless need, demanded by the progress in all fields of industry and technology, of new materials for the development of society.

Metallurgy was one of the first fields where material scientists worked toward developing new alloys for different applications. During the first years, a large number of studies were carried out on the austenite–martensite–cementite phases achieved during the phase transformations of the iron–carbon alloy, which is the foundation for steel production, later the development of stainless steel, and other important alloys for industry, construction, and other fields was produced.

Later, the evolution of the electronic industry initiated the development of an immense variety of materials and devises based, essentially, on the properties of semiconductor, dielectric, ferromagnetic, superconductor, and ferroelectric materials.

In addition, until the second half of the twentieth century, the term ceramic was related to the traditional clays, that is, pottery, bricks, tiles, and cements and glass; however, during the last 50 years, the field of technical ceramics has been rapidly developed, and firmly established.

At the beginning of the twentieth century, the first synthetic polymer, bakelite, was obtained and later, after the First World War, it was proposed that polymers consisted of long chains of atoms held together by covalent bonds. The Second World War gave a huge stimulus to the creation of polymers, which firmly established the field of polymers.

However, important groups of materials cannot be studied in a single volume materials science book. These materials include adsorbents, ion exchangers, ion conductors, catalysts, and permeable materials. Examples of these types of materials are perovskites, zeolites, mesoporous molecular sieves, silica, alumina, active carbons, titanium dioxide, magnesium oxide, clays, pillared clays, hydrotalcites, alkali metal titanates, titanium silicates, polymers, and coordination polymers. These materials have applications in many fields, among others, adsorption, ion conduction, ion exchange, gas separation, membrane reactors, catalysts, catalytic supports, sensors, pollution abatement, detergents, animal nutrition, agriculture, and sustainable energy applications.

The author of this book has been permanently active during his career in the field of materials science, studying diffusion, adsorption, ion exchange, cationic conduction, catalysis and permeation in metals, zeolites, silica, and perovskites. From his experience, the author considers that during the last years, a new field in materials science, that he calls the "physical chemistry of materials," which emphasizes the study of materials for chemical, sustainable energy, and pollution abatement applications, has been developed. With regard to this development, the aim of this book is to teach the methods of syntheses and characterization of adsorbents, ion exchangers, cationic conductors, catalysts, and permeable porous and dense materials and their properties and applications.

Rolando M.A. Roque-Malherbe
Las Piedros, PR, USA
January, 2009

Author

Dr. Rolando M. A. Roque-Malherbe was born in 1948 in Güines, Havana, Cuba. He graduated with a BS in physics from the University of Havana (1970), summa cum laude, specialized (MS equivalent degree) in surface physics at the National Center for Scientific Research, Technical University of Dresden, Germany (1972), magna cum laude, and obtained his PhD in physics (solid state physics) from the Moscow Institute of Steel and Alloys, Russia (1978), magna cum laude. He completed postdoctoral stints at the Technical University of Dresden, Germany; Moscow State University, Russia; the Technical University of Budapest, Hungary; the Institute of Physical Chemistry and Chemical Physics, Russian Academy of Science, Moscow; and the Central Research Institute for Chemistry, Hungarian Academy of Science, Budapest (1978–1984). The group led by him at the National Center for Scientific Research, Higher Pedagogical Institute, Varona, Havana, Cuba (1980–1992), was one of the world leaders in the study and applications of natural zeolites. During this period, he was possibly the only Cuban scientist to receive most awards. In 1993, after a political confrontation with the Cuban regime, he left Cuba with his family as a political refugee. From 1993 to 1999, he worked at various institutions like the Institute of Chemical Technology, Valencia, Spain; at Clark Atlanta University, Atlanta, Georgia; and at Barry University, Miami, Florida. From 1999 to 2004, he was dean and full professor at the School of Sciences in the University of Turabo, Gurabo, Puerto Rico, and currently he is the director of the Institute of Physical and Chemical Applied Research. He has published 121 papers, 5 books, 6 chapters, 30 abstracts, has 15 patents, and made more than 200 presentations at scientific conferences. He is currently an American citizen.

1 Materials Physics

1.1 INTRODUCTION

We discuss briefly some basic topics in materials physics such as crystallography, lattice vibrations, band structure, x-ray diffraction, dielectric relaxation, nuclear magnetic resonance and Mössbauer effects in this chapter. These topics are an important part of the core of this book. Therefore, an initial analysis of these topics is useful, especially for those readers who do not have a solid background in materials physics, to understand some of the different problems that are examined later in the rest of the book.

1.2 CRYSTALLOGRAPHY

1.2.1 Crystalline Structure

An unit cell is a regular repeating pattern that pervades the whole crystal lattice. It is described [1–6] by three vectors: $\bar{a}$, $\bar{b}$, and $\bar{c}$ (Figure 1.1), that outline a parallelepiped, characterized by six parameters. These parameters are the length of the three vectors (a, b, and c) and the angles between them (α, β, and γ). Consequently, all the points that constitute the lattice sites are given by a set of points, which starting from a reference point, are given by

$$\bar{R} = n_1 \bar{a} + n_2 \bar{b} + n_3 \bar{c} \tag{1.1}$$

where n_1, n_2, n_3, are integers running from $-\infty$ to ∞, for a limitless crystal. As a result of this, the lattice is a set of points in space, distinguished by a space periodicity or a translational symmetry. This means that under a translation defined by Equation 1.1, the lattice remains invariant.

If all the lattice points are positioned in the eight corners of a unit cell, then the unit cell is called a primitive unit cell. However, often, for convenience, larger unit cells, which are not primitives, are selected for the description of a particular lattice, as will be explained later.

It is possible, as well, to define the primitive unit cell, by surrounding the lattice points, by planes perpendicularly intersecting the translation vectors between the enclosed lattice point and its nearest neighbors [2,3]. In this case, the lattice point will be included in a primitive unit cell type, which is named the Wigner–Seitz cell (see Figure 1.2).

A concrete building procedure in three dimensions of the Wigner–Seitz cell can be achieved by representing lines from a lattice point to others in the lattice and then drawing planes that cut in half each of the represented lines, and finally taking the minimum polyhedron enclosing the lattice point surrounded by the constructed planes.

Till now, we have only considered a mathematical set of points. However, a material, in reality, is not merely an array of points, but the group of points is a lattice. A real crystalline material is constituted of atoms periodically arranged in the structure, where the condition of periodicity implies a translational invariance with respect to a translation operation, and where a lattice translation operation, $\bar{T}$, is defined as a vector connecting two lattice points, given by Equation 1.1 as

$$\bar{T} = n_1 \bar{a} + n_2 \bar{b} + n_3 \bar{c} \tag{1.2}$$

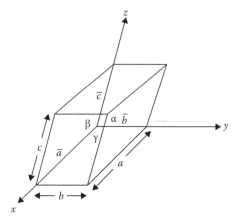

FIGURE 1.1   Unit cell geometrical representation.

Until now, we have considered an infinite lattice, but a real material has limited dimensions, that is, n_1, n_2, n_3 has boundaries. However, an infinite array of unit cells is a good approximation for regions relatively far from the surface, which constitutes the major part of the whole material [5]. At this point, it is necessary to recognize that a real crystal has imperfections, such as vacancies, dislocations, and grain boundaries.

Since a lattice is just a set of points, we will need another entity to describe the real crystal. That is, it is required to locate a set of atoms named "basis" in the vicinity of the lattice sites. Therefore, a crystal will be a combination of a lattice and a basis of atoms. In Figure 1.3, a representation of the operation

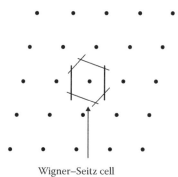

Wigner–Seitz cell

FIGURE 1.2   Wigner–Seitz cell in two dimensions.

$$lattice + basis = crystal$$

is given.

In order to systematize in a logical form the lattices that are compatible with a periodicity condition, the French physicist Auguste Bravais, in 1845, demonstrated that the lattice points in three dimensions, congruent with the periodicity requirement, are the roots of the following trigonometric equation [2]:

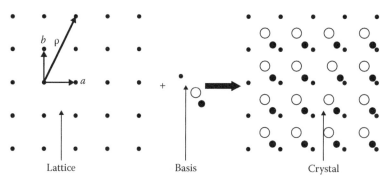

FIGURE 1.3   Representation of the operation: lattice + basis = crystal.

TABLE 1.1

Description of the Seven Crystalline Systems

System	Parameters Describing the Unit Cell
Cubic	$a = b = c$; $\alpha = \beta = \gamma = 90°$
Hexagonal	$a = b \neq c$; $\alpha = \beta = 90°$; $\gamma = 120°$
Rhombohedral or trigonal	$a = b = c$; $\alpha = \beta = \gamma \neq 90°$ and $<120°$
Tetragonal	$a = b \neq c$; $\alpha = \beta = \gamma = 90°$
Orthorhombic	$a \neq b \neq c$; $\alpha = \beta = \gamma = 90°$
Monoclinic	$a \neq b \neq c$; $\alpha = \beta = 90° \neq \gamma$
Triclinic	$a \neq b \neq c$; $\alpha \neq \beta \neq \gamma \neq 90°$

$$\sin^2\left[\frac{\pi\xi}{a}\right] + \sin^2\left[\frac{\pi\eta}{b}\right] + \sin^2\left[\frac{\pi\zeta}{c}\right] = 0 \tag{1.3}$$

where

ξ, η, and ζ are spatial coordinates related with an oblique three-coordinate axis system

$\bar{a}$, $\bar{b}$, and $\bar{c}$ (see Figure 1.1) are the unit vectors of the coordinate system

Bravais then showed that in three dimensions, there are only 14 different lattice types, currently named the Bravais lattices, which are grouped in seven crystal systems [1–3] (see Table 1.1).

Each lattice has an inversion center, a unique set of axes and symmetry planes, and there are possible operations like rotation, reflection, and its combinations [1]. In a case where some symmetry operations leave unchanged a particular point of the fixed lattice, they form a group called the crystallographic point groups. In this regard, there are 32 point groups in three dimensions. Besides, the combination of the point group symmetry operations with the translation symmetry gives rise to the crystallographic space groups. In relation with these operations, there are 230 space groups in three dimensions [1].

Each crystal system is related with a parallelepiped whose vertices are compatible with the sites of the corresponding Bravais lattice (see Figure 1.4) [1–3]. The parallelepiped is described with six parameters, as was previously stated for the unit cell. The most symmetrical crystal system has an essential symmetry, 4 threefold axes, and is named the cubic system. A hexagonal lattice is characterized completely by a regular hexahedral prism, having a sixfold axes as the essential symmetry. This crystal system is named the hexagonal system. The Bravais trigonal lattice is characterized by a geometrical figure that results when a cube is stretched along one of its diagonals (see Figure 1.4). In addition, a rectangular prism with at least one square face has a tetragonal symmetry, that is, a fourfold axes as the essential symmetry, and is the basis of the tetragonal system. Stretching the tetragonal prism along one of the axes produces the orthorhombic prism, having three orthogonal twofold axes as the essential symmetry, and is the origin of the orthorhombic system. To complete the seven crystal systems, it is necessary to include the monoclinic system, which has only a twofold axes as the essential symmetry, and the triclinic system, which has only an inversion center.

Within a given crystal system, a supplementary subdivision is necessary to be made, in order to produce the 14 Bravais lattices. In this regard, it is necessary to make a distinction between the following types of Bravais lattices, that is, primitive (P) or simple (S), base-centered (BC), face-centered (FC), and body-centered (BoC) lattices [1–3].

In Table 1.2, the subtypes corresponding to each crystal system are listed and in Figure 1.4, the 14 Bravais lattices in three dimensions are illustrated.

Among the 14 cells that generate the Bravais lattices (see Figure 1.4), only the P-type cells are considered primitive unit cells. It is possible to generate the other Bravais lattices with primitive unit cells. However, in practice, only unit cells that possess the maximum symmetry are chosen (see Figure 1.4 and Table 1.2) [1–6].

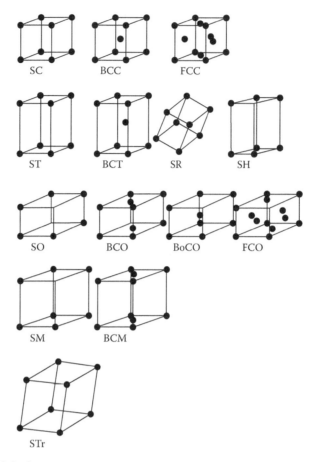

FIGURE 1.4   Bravais lattices.

TABLE 1.2
Subtypes of Lattices in the Seven Crystalline Systems

System	Lattice Types
Cubic	Simple cubic (SC), body-centered cubic (BCC), and face-centered cubic (FCC)
Hexagonal	Simple hexagonal (SH)
Rhombohedral or trigonal	Simple rhombohedral (SR)
Tetragonal	Simple tetragonal (ST) and body-centered tetragonal (BCT)
Orthorhombic	Simple orthorhombic (SO), body-centered orthorhombic (BoCO), face-centered orthorhombic (FCO), and base-centered orthorhombic (BCO)
Monoclinic	Simple monoclinic (SM) and base-centered monoclinic (BCM)
Triclinic	Simple triclinic (STr)

Sources: Schwarzenbach, D., *Crystallography*, John Wiley & Sons, New York, 1997; Kittel, Ch., *Introduction to Solid State Physics*, 8th edn., John Wiley & Sons, New York, 2004; Myers, H.P., *Introduction to Solid State Physics*, 2nd edn., CRC Press, Boca Raton, FL, 1997.

1.2.2 Crystallographic Directions and Planes

The following steps must be followed in order to specify a crystallographic direction:

1. The vector that defines the crystallographic direction should be situated in such a way that it passes through the origin of the lattice coordinate system.
2. The projections of this vector on each of the three axis is determined and measured in terms of the unit cell dimensions, a, b, c, obtaining three integer numbers, n_1, n_2, n_3.
3. These numbers are reduced to smallest integers, u, v, w.
4. These three numbers, enclosed in square brackets and not separated with commas, $[uvw]$, denote the crystallographic direction.

For example, the direction of the positive x-axis is denoted by [100], the direction of the positive y-axis is denoted by [010], and the direction of the positive z-direction is denoted by [001] (see Figure 1.1).

For a crystal having a hexagonal symmetry, a set of four numbers, $[uvtw]$, named the Miller–Bravais coordinate system (see Figure 1.5), is used to describe the crystallographic directions, where the first three numbers, that is, u, v, t, are projections along the axes a_1, a_2, and a_3, describing the basal plane of the hexagonal structure, and w is the projection in the z-direction [2,3].

The following steps should be followed in order to specify a crystallographic plane:

1. The plane ought to be located in such a way that it does not pass through the origin of the lattice coordinate system.
2. After this, the interceptions of the plane on each of the three axis is determined in terms of the unit cell dimensions, a, b, c, and then obtaining three integer numbers p_1, p_2, p_3.
3. The reciprocals of these numbers are then taken and thereafter reduced to smallest integers h, k, l.
4. These three numbers enclosed in parentheses and not separated with commas, that is, (hkl), named the Miller indexes, denote the crystallographic plane.

For example, the plane perpendicular to the x-axis is denoted by (100), the plane perpendicular to the y-axis is denoted by (010), and the plane perpendicular to the positive z-direction is denoted by (001).

For a crystal exhibiting a hexagonal symmetry, a set of four numbers, $(hkil)$, (see Figure 1.5) is used to describe the crystallographic planes, where the first three numbers, that is, h, k, i, are the intercepts of the plane on each of the three axis measured in terms of the unit cell dimensions along the axes a_1, a_2, and a_3, describing the basal plane of the hexagonal structure, and l is the projection in the z-direction.

The position of a point inside the primitive unit cell is determined by a fraction of the axial length, a, b, c. For example, in a body-centered structure, the position of the central point is $\frac{1}{2}\frac{1}{2}\frac{1}{2}$.

1.2.3 Octahedral and Tetrahedral Sites in the FCC Lattice

In the FCC lattice, two types of interstitial sites can be recognized: octahedral sites (O-sites) and tetrahedral sites (T-sites). The O-sites are those which are enclosed by six nearest neighbor atoms at the same distances (see Figure 1.6).

On the other hand, a T-site is the geometric place that is formed when three spheres are in contact with each other, and a fourth sphere is placed in the depression created by the first three. In this case, a tetrahedral site is formed in between the four spheres. That is, if we join three small black spheres located in the centers of the faces (see Figure 1.7), surrounding the diagonal of the cube, we will construct a triangle.

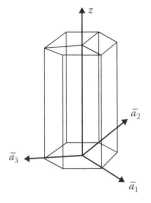

FIGURE 1.5 Miller–Bravais coordinate system.

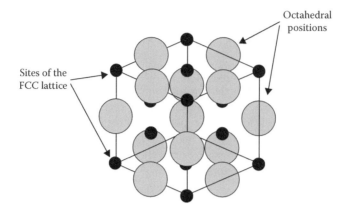

FIGURE 1.6   Octahedral sites.

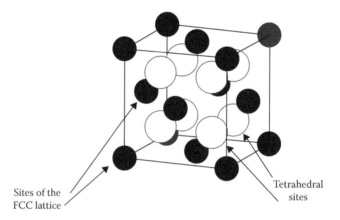

FIGURE 1.7   Tetrahedral sites.

1.2.4   Reciprocal Lattice

A unit cell in the reciprocal lattice is described by the vectors $\overline{a^*}$, $\overline{b^*}$, $\overline{c^*}$, which are defined as follows [2,3,5,6]:

$$\overline{a^*} = 2\pi\frac{\overline{b}\times\overline{c}}{V}, \quad \overline{b^*} = 2\pi\frac{\overline{c}\times\overline{a}}{V}, \quad \text{and} \quad \overline{c^*} = 2\pi\frac{\overline{a}\times\overline{b}}{V} \tag{1.4}$$

where $V = \overline{a} \cdot (\overline{b} \times \overline{c})$
Hence,

$$\overline{a}\cdot\overline{a^*} = \overline{b}\cdot\overline{b^*} = \overline{c}\cdot\overline{c^*} = 2\pi \tag{1.5a}$$

and

$$\overline{a}\cdot\overline{b^*} = \overline{a}\cdot\overline{c^*} = \overline{b}\cdot\overline{a^*} = \overline{b}\cdot\overline{c^*} = \overline{c}\cdot\overline{a^*} = \overline{c}\cdot\overline{b^*} = 0 \tag{1.5b}$$

This means that $\overline{a^*}$ is perpendicular to both $\overline{b}$ and $\overline{c}$, $\overline{b^*}$ is perpendicular to both $\overline{a}$ and $\overline{c}$, and $\overline{c^*}$ is perpendicular to both $\overline{b}$ and $\overline{a}$.

Similar to the direct lattice, all the possible points that lie at the reciprocal lattice can be represented as follows:

$$\overline{G_{hkl}} = h\overline{a^*} + k\overline{b^*} + l\overline{c^*} \tag{1.6}$$

Now, since the Miller indices of a plane implies that the plane intercepts the base vectors at the point $\frac{\overline{a}}{h}, \frac{\overline{b}}{k}, \frac{\overline{c}}{l}$, a triangular portion of the plane has sides

$$\left(\frac{\overline{a}}{h} - \frac{\overline{b}}{k}\right), \left(\frac{\overline{b}}{k} - \frac{\overline{c}}{l}\right), \left(\frac{\overline{c}}{l} - \frac{\overline{a}}{h}\right)$$

Considering Equations 1.5, it is possible to show that

$$\left(\frac{\overline{a}}{h} - \frac{\overline{b}}{k}\right) \bullet \overline{G_{hkl}} = \left(\frac{\overline{b}}{k} - \frac{\overline{c}}{l}\right) \bullet \overline{G_{hkl}} = \left(\frac{\overline{c}}{l} - \frac{\overline{a}}{h}\right) \bullet \overline{G_{hkl}} = 0$$

Consequently, the vector $\overline{G_{hkl}} = \overline{G_{hkl}}$ is perpendicular to the plane (hkl). Then, it is possible to calculate $\left|\overline{G_{hkl}}\right|$, that is, the vector modulus. To perform this calculation, we must define the unit vector in the direction of the vector $\overline{G_{hkl}}$ as follows:

$$\overline{n}_{hkl} = \frac{\overline{G_{hkl}}}{\left|\overline{G_{hkl}}\right|}$$

Subsequently, since by definition the interplanar distance, that is, the distance between the (hkl) planes, is

$$d_{hkl} = \frac{\overline{a}}{h} \bullet \overline{n}_{hkl} = \frac{\overline{a}}{h} \bullet \frac{\overline{G_{hkl}}}{\left|\overline{G_{hkl}}\right|} = \frac{2\pi}{\left|\overline{G_{hkl}}\right|}$$

Consequently,

$$d_{hkl} = \frac{2\pi}{\left|\overline{G_{hkl}}\right|} \tag{1.7}$$

1.3 BLOCH THEOREM

The Bloch theorem is one of the tools that helps us to mathematically deal with solids [5,6]. The mathematical condition behind the Bloch theorem is the fact that the equations which governs the excitations of the crystalline structure such as lattice vibrations, electron states and spin waves are periodic. Then, to solve the Schrödinger equation for a crystalline solid where the potential is periodic, $\{V(\overline{r} + \overline{R}) = V(\overline{r})\}$, this theorem is applied [5,6].

If $V(\bar{r})$ is the potential "seen" by an electron belonging to the solid, then the one electron wave function, $\psi(\bar{r})$, satisfies the Schrödinger equation:

$$-\frac{\hbar^2}{2m}\nabla^2\psi(\bar{r})+V(\bar{r})\psi(\bar{r})=E\psi(\bar{r}) \qquad (1.8)$$

In the case of lattice waves and spin waves, the procedure is different but the principle is the same. The periodic potential is represented with the help of a Fourier series

$$V(\bar{r})=\sum_{\bar{G}_{hkl}} V_{\bar{G}_{hkl}}\, e^{i\bar{G}_{hkl}\cdot\bar{r}}$$

where $\overline{G_{hkl}}=\overline{d^*_{hkl}}=h\overline{a^*}+k\overline{b^*}+l\overline{c^*}$ is the reciprocal lattice vector. Since $V(\bar{r})$ is a real function, it is necessary that

$$V^*_{\bar{G}_{hkl}}=V_{-\bar{G}_{hkl}}$$

since

$$V^*(\bar{r})=\sum_{\bar{G}_{hkl}} V_{\bar{G}_{hkl}}\, e^{-i\bar{G}_{hkl}\cdot\bar{r}}=\sum_{\bar{G}_{hkl}} V_{-\bar{G}_{hkl}}\, e^{i\bar{G}_{hkl}\cdot\bar{r}}=V(\bar{r})$$

Given that the Schrödinger equation

$$\left(-\frac{\hbar^2}{2m}\nabla^2+V(\bar{r})-E\right)\psi(\bar{r})=\left(\hat{H}(r)-E\right)\psi(\bar{r})$$

is periodic, that is,

$$\left(\hat{H}(r)-E\right)\psi(\bar{r})=\left(\hat{H}(r+\bar{R})-E\right)\psi(\bar{r}+\bar{R})$$

Then, the wave function $\psi(\bar{r})$ and the wave function $\psi(\bar{r}+\bar{R})$ must differ only in a constant, then

$$\psi(\bar{r}+\bar{R})=\vartheta_{\bar{R}}\psi(\bar{r})$$

where the condition of normalization required by all the wave functions requires that

$$\left|\vartheta_{\bar{R}}\right|^2=1$$

Consequently,

$$\vartheta_{\bar{R}}=e^{-i.\alpha(R)}$$

where $\alpha(\bar{R})$ is a real number. Besides, since

$$\vartheta_{\bar{R}_1}\vartheta_{\bar{R}_2}=\vartheta_{\bar{R}_1+\bar{R}_2}$$

we will then have that

$$\alpha(\bar{R}_1) + \alpha(\bar{R}_2) = \alpha(\bar{R}_1 + \bar{R}_2)$$

Subsequently,

$$\alpha(\bar{R}) = \bar{k} \cdot \bar{R}$$

and

$$\vartheta_{\bar{R}} = e^{-i.\bar{k} \cdot \bar{R}}$$

$$\psi(\bar{r} + \bar{R}) = e^{-i\bar{k} \cdot \bar{R}} \psi(\bar{r})$$

Therefore, the periodic function

$$u(\bar{r}) = e^{-i\bar{k} \cdot \bar{r}} \psi(\bar{r})$$

Have the correct form to be a solution of Equation 1.8. As a result, the Bloch theorem affirms that the solution to the Schrödinger equation may be a plane wave multiplied by a periodic function, that is [5,6],

$$\psi_{\bar{k}}(\bar{r}) = e^{i\bar{k} \cdot \bar{r}} u_{\bar{k}}(\bar{r}) \tag{1.9a}$$

where the periodic function is given by

$$u_{\bar{k}}(\bar{r}) = \sum_{\bar{G}_{hkl}} u_{\bar{G}_{hkl}}(\bar{k}) e^{i\bar{G}_{hkl} \cdot \bar{r}} \tag{1.9b}$$

It is necessary to state now that the rigorous fulfillment of the Bloch theorem needs an infinity lattice. In order to calculate the number of states in a finite crystal, a mathematical requirement named the Born–Karman cyclic boundary condition is introduced. That is, if we consider that a crystal with dimensions $N_1\bar{a}$, $N_2\bar{b}$, $N_3\bar{c}$ is cyclic in three dimensions, then [5]

$$\psi(\bar{r} + N_1\bar{a}) = \psi(\bar{r}), \quad \psi(\bar{r} + N_2\bar{b}) = \psi(\bar{r}), \quad \text{and} \quad \psi(\bar{r} + N_3\bar{c}) = \psi(\bar{r})$$

For a Bloch state, the above conditions mean that

$$e^{-i.\bar{k} \cdot N_1\bar{a}} = e^{-i.\bar{k} \cdot N_2\bar{b}} = e^{-i.\bar{k} \cdot N_3\bar{c}}$$

This condition can be satisfied only if

$$\bar{k} = \frac{2\pi m_1}{N_1} \bar{a}^* + \frac{2\pi m_2}{N_2} \bar{b}^* + \frac{2\pi m_3}{N_3} \bar{c}^*$$

where

m_1, m_2, and m_3 are integers
$\overline{a^*}$, $\overline{b^*}$, $\overline{c^*}$ are the reciprocal lattice vectors

The allowed values of m_1, m_2, and m_3 must run through the values:

$$0 \leq m_1 \leq N_1, \quad 0 \leq m_2 \leq N_2, \quad \text{and} \quad 0 \leq m_3 \leq N_3$$

However, this is not the proper range, and the appropriate extent is

$$-\frac{N_1}{2} \leq m_1 \leq \frac{N_1}{2}, \quad -\frac{N_2}{2} \leq m_2 \leq \frac{N_2}{2}, \quad \text{and} \quad -\frac{N_3}{2} \leq m_3 \leq \frac{N_3}{2}$$

which will give a cell centered in origin, as was previously observed for the Wigner–Seitz in real space, but now in the $\overline{k}$ space. This cell is named the Brilloin zone, which is the Wigner–Seitz cell in the $\overline{k}$ space or inverse space.

The number of allowed states is then $N_1 \times N_2 \times N_3 = M$, which is the number of cells in a real macroscopic finite crystal. That is, the number of allowed wave vectors in a Brilloin zone is exactly the number of unit cells in the crystal under consideration.

1.4 LATTICE VIBRATIONS

1.4.1 PHONONS

Lattice vibrations are fundamental for the understanding of several phenomena in solids, such as heat capacity, heat conduction, thermal expansion, and the Debye–Waller factor. To mathematically deal with lattice vibrations, the following procedure will be undertaken [7]: the solid will be considered as a crystal lattice of atoms, behaving as a system of coupled harmonic oscillators. Thereafter, the normal oscillations of this system can be found, where the normal modes behave as uncoupled harmonic oscillators, and the number of normal vibration modes will be equal to the degrees of freedom of the crystal, that is, $3nM$, where n is the number of atoms in the unit cell and M is the number of units cell in the crystal [8].

In order to solve this problem, it is possible to use the Hamiltonian procedure of classical mechanics [8]. Hence, the classical Hamiltonian of a system of coupled harmonic oscillators can be written as follows [7]:

$$H = \sum_i \frac{(p_i')^2}{2m_i} + \sum_{i,j} \frac{1}{2} C_{i,j}' q_i' q_j' \tag{1.10}$$

where

q_i' are the coordinates of displacement from the equilibrium position
$p_i' = m_i \dfrac{dq_i'}{dt}$ are the impulses
$C_{i,j}' = C_{j,i}'$ are constants

The Hamiltonian can be simplified if we made the following substitutions in order to eliminate the constant

$$q_i = q_i' \sqrt{m_i}$$

and

$$C_{i,j} = \frac{C'_{i,j}}{\sqrt{m_i m_j}}$$

And finally,

$$p_i = \frac{\partial L}{\partial \dot{q}_i} = \frac{p'_i}{\sqrt{m_i}}$$

where L is the Lagrangian function. Consequently, the Hamiltonian can be written as follows:

$$H = \sum_i \frac{(p_i)^2}{2} + \sum_{i,j} \frac{1}{2} C_{i,j} q_i q_j \qquad (1.11)$$

Following the rules of the Hamiltonian method, the equations of motion can be written as follows:

$$\dot{p}_i = -\frac{\partial H}{\partial q_i} = -\sum_i C_{i,j} q_j \quad \text{and} \quad \dot{q}_i = \frac{\partial H}{\partial p_i} = p_i \qquad (1.12)$$

Equation 1.12 is a system of linear differential equations with constant coefficients. Then, following the rules for solving this type of an equation, its solution can be written in the following form [7]:

$$q_i^\beta = e^{-i\omega\beta t} c_i^\beta$$

where
  $\omega_\beta = 2\pi\, \nu_\beta$ are the angular frequencies
  ν_β are frequencies

The condition for solving this system is [9]

$$|C_{i,j} - \omega^2 \delta_{i,j}| = 0 \qquad (1.13)$$

which gives an equation that allows us to get the values of ω_β and the corresponding orthogonal vectors c_i^β

$$\sum_i c_i^\beta c_i^\delta = \delta_{\beta\delta}$$

where the general solution for q_i has the following form:

$$q_i = \sum_\beta L_\beta q_i^\beta \qquad (1.14)$$

where L_β are constants. In essence, during the previous procedure we have separated the motion of the system in normal vibration modes, where each one has a frequency ω_β. Thereafter, the motion of the system is described as a sum of normal vibration modes.

Now making the following substitution [7]

$$Q_\beta = L_\beta e^{-i\omega_\beta t}$$

it is then possible to make the following variable substitution:

$$q_i = \sum_\beta Q_\beta c_i^\beta$$

And then get [10]

$$H = \sum_i h_\beta \tag{1.15}$$

where

$$h_\beta = \frac{1}{2} p_\beta^2 + \frac{1}{2} \omega_\beta^2 Q_\beta^2 \tag{1.16}$$

If we now change the coordinates and the momentum by their quantum mechanical corresponding operators, we will get

$$\hat{H} = \sum_i \hat{h}_\beta$$

in which

$$\hat{h}_\beta = -\frac{\hbar^2}{2} \frac{\partial^2}{\partial Q_\beta^2} + \frac{1}{2} \omega_\beta^2 Q_\beta^2$$

where

$$\hbar = \frac{h}{2\pi}$$

and h is the Planck's constant. This is the Schrödinger equation for a quantum harmonic oscillator of frequency ω_β. Therefore, the energy of the system will be

$$E = \sum_\beta \left(N_\beta + \frac{1}{2} \right) \hbar \omega_\beta \tag{1.17}$$

where

$$E_n = \left(n + \frac{1}{2} \right) \hbar \omega \tag{1.18}$$

are the energy levels of a quantum harmonic oscillator. Consequently, we have reduced the lattice energy to the summation of the energy of different noncoupled harmonic oscillators.

It is very well known that Einstein, developing Planck's ideas, quantized the electromagnetic field by introducing a quantum particle named the photon. Consequently, each mode or state of a classical electromagnetic field is characterized by an angular frequency, ω, and a wave vector, $\bar{k} = \dfrac{2\pi}{\lambda}\bar{s}$, in which $\bar{s}$ is a unit vector normal to the wave fronts. Then, the modes or states are replaced by the photon that carries energy

$$E = \hbar\omega$$

and momentum

$$\bar{p} = \hbar\bar{k}$$

where

$\hbar = \dfrac{h}{2\pi}$

$\omega = 2\pi\nu$ is the angular frequency

ν is the frequency of the electromagnetic radiation

λ is the wavelength of the electromagnetic radiation

Similarly, during their effort to understand the thermal energy of solids, Einstein and Debye quantized the lattice waves and the resulting quantum was named phonon. Consequently, it is possible to consider the lattice waves as a gas of noninteracting quasiparticles named phonons, which carries energy, $E = \hbar\omega$, and momentum, $\bar{p} = \hbar\bar{k}$. That is, each normal mode of oscillation, which is a one-dimensional harmonic oscillator, can be considered as a one-phonon state.

1.4.2 Bose–Einstein Distribution

It is possible to calculate the average energy for a single oscillation mode, following the canonical ensemble methodology [6,11] as

$$\langle E \rangle = \frac{\sum_{0}^{\infty}\left(n+\frac{1}{2}\right)\hbar\omega e^{-\frac{\left(n+\frac{1}{2}\right)\hbar\omega}{kT}}}{\sum_{0}^{\infty}e^{-\frac{\left(n+\frac{1}{2}\right)\hbar\omega}{kT}}} = \frac{\hbar\omega}{2} + \frac{\sum_{0}^{\infty}n\hbar\omega e^{-\frac{\left(n+\frac{1}{2}\right)\hbar\omega}{kT}}}{\sum_{0}^{\infty}e^{-\frac{\left(n+\frac{1}{2}\right)\hbar\omega}{kT}}}$$

where k is the Boltzmann constant. It is easy to show that

$$\langle E \rangle = \frac{\hbar\omega}{2} - \frac{\partial}{\partial\left(\frac{1}{kT}\right)}\ln\left(\sum_{0}^{\infty}e^{-\frac{\hbar\omega}{kT}}\right) = \left(\frac{e^{-\frac{\hbar\omega}{kT}}}{1-e^{-\frac{\hbar\omega}{kT}}} + \frac{1}{2}\right)\hbar\omega = \left(n(\omega,T) + \frac{1}{2}\right)\hbar\omega$$

Consequently,

$$n(\omega, T) = \frac{1}{e^{\frac{\hbar\omega}{kT}} - 1} \tag{1.19}$$

which is the Bose–Einstein distribution function. Consequently, phonons behave as bosons [12]. If we use Equation 1.19 to describe each vibration mode, then

$$n_\beta(\omega_\beta, T) = \frac{1}{e^{\frac{\hbar\omega_\beta}{kT}} - 1} \tag{1.20}$$

Then, Equation 1.20 tells us that there are on average $n_\beta(\omega_\beta, T)$ phonons in the β mode, where this mode contributes energy

$$\langle E_\beta \rangle = \left(n_\beta(\omega_\beta, T) + \frac{1}{2} \right) \hbar\omega_\beta$$

1.4.3 Heat Capacity of Solids

The average energy in the canonical ensemble of the whole system is

$$U = \langle E_T \rangle = E_0 + \sum_\beta \hbar\omega_\beta \left(\frac{1}{e^{\beta\hbar\omega_\beta} - 1} \right) \tag{1.21}$$

Besides, the canonical partition function [11] of the system of oscillators is [13]

$$Z = e^{-\frac{E_0}{kT}} \prod_\beta \frac{1}{1 - e^{-\frac{\hbar\omega_\beta}{kT}}}$$

Then,

$$\ln Z = \frac{E_0}{kT} - \sum_\beta \ln\left(1 - e^{-\frac{\hbar\omega_\beta}{kT}} \right) \tag{1.22}$$

We will now attempt an analysis of Equation 1.21 for n mol of a metallic, ionic, or covalent crystal, with 1 ion per lattice site, that is, for an Avogadro number, N_A, of ions at a high temperature. At these conditions, $kT \gg \hbar\omega_\beta$, and, consequently,

$$\langle E_T \rangle = E_0 + \sum_\beta \hbar\omega_\alpha \left(\frac{1}{e^{\frac{\hbar\omega_\beta}{kT}} - 1} \right) = E_0 + \sum_\alpha kT = E_0 + 3NkT = E_0 + 3nRT \tag{1.23}$$

where n is the number of moles.

Since the heat capacity at constant volume is defined as

$$C_V = \left(\frac{\partial U}{\partial T}\right)_V = \left(\frac{\partial \langle E_T \rangle}{\partial T}\right)_V \qquad (1.24)$$

then with the help of Equations 1.23 and 1.24, we can obtain, for $n = 1$

$$C_V = 3R$$

which is the Dulong–Petit law, where $R = kN_A$ is the ideal gas constant. The same result can as well be obtained with the following argument: a classical harmonic oscillator included in a system of harmonic oscillators (as is the proposed model of a solid) in thermal equilibrium at a temperature T has an average energy equal to kT, since the number of normal modes is $3N$, where $N = nN_A$ is the number of atoms in the solid, N_A is the Avogadro number, and n, the number of moles. Then, the average classical internal energy of a solid for $n = 1$ is $3RT$ and $C_V = 3R$.

However, we need to know the behavior of solids at all temperatures. Einstein, in 1907, to deal with the problem, assumed that all the normal vibration modes have the same angular frequency ω_E. As a result, Equation 1.21 will take the following form [12]:

$$\langle E_T \rangle = E_0 + \frac{3N_A\hbar\omega_E}{e^{\frac{\hbar\omega_E}{kT}} - 1} = E_0 + \frac{3N_A k\Theta_E}{e^{\frac{\Theta_E}{T}} - 1}$$

where
$k\Theta_E = \hbar\omega_E$
Θ_E is a characteristic temperature of the system

Consequently, the heat capacity at a constant volume will be

$$C_V = 3N_A k \left(\frac{\Theta_E}{T}\right)^2 \frac{e^{\frac{\Theta_E}{T}}}{\left(e^{\frac{\Theta_E}{T}} - 1\right)^2}$$

where the limit for the high temperature is $C_V = 3R$

Debye, in 1912, made more realistic assumptions in order to deal with the lattice vibration problem. He considered that because of the large number of atoms in the crystal the number of normal vibration modes is very high, and it is possible to consider that the vibrations are continuously distributed over a specified range of frequencies, $0 < \nu < \nu_m$, where the distribution is such that the number of normal vibration modes in the interval from ν to $\nu + d\nu$ is $g(\nu)d\nu$. Consequently, in Equation 1.22, it is possible to substitute the summation for the integration. Therefore [13],

$$\ln Z = -\frac{E_0}{kT} - \int_0^{\nu_m} \ln\left(1 - e^{-\frac{h\nu_\alpha}{kT}}\right) g(\nu)d\nu \qquad (1.25)$$

The density of elastic standing waves in a continuous solid is given by [14]

$$g(\nu) = \frac{12\pi V \nu^2}{V_s^3} \qquad (1.26a)$$

where

V_s is the average speed of sound waves in the solid

ν is the frequency of the standing wave

V is the volume of the solid

The derivation of Equation 1.26a is carried out by calculating the number of standing waves in a cubic cavity of volume V, and follows a process similar to that applied in Section 1.5.3 for calculating the density of states for an electron gas [14].

Now, since

$$\int_0^{\nu_m} g(\nu)\,d\nu = \int_0^{\nu_m} \frac{12\pi\nu^2}{V_s^3}\,d\nu = 3N_A$$

then

$$\nu_m = \left(\frac{3N_A V_s^3}{4\pi V}\right)^{\frac{1}{3}}$$

and

$$g(\nu) = \frac{9N_A}{\nu_m^3}\nu^2 \qquad (1.26b)$$

Substituting Equation 1.26 in Equation 1.25, we will get

$$\ln Z = -\frac{E_0}{kT} - \frac{9N_A}{\nu_m^3}\int_0^{\omega_m} \nu^2 \ln\left(1 - e^{-\frac{h\nu}{kT}}\right)d\nu$$

Then, using [11]

$$U = kT^2\left(\frac{\partial \ln Z}{\partial V}\right) \quad \text{and} \quad C_V = \left(\frac{\partial U}{\partial T}\right)_V = \left(\frac{\partial \langle E_T\rangle}{\partial T}\right)_V$$

we will get (Figure 1.8).

$$C_V = 9N_A k\left(\frac{T}{\Theta_D}\right)^{\frac{\Theta_D}{T}}\int_0^{\frac{\Theta_D}{T}} \frac{y^4 e^y}{e^y - 1}\,dy \qquad (1.27)$$

where

$k\Theta_D = h\nu_m$ defines the Debye temperature, Θ_D

$y = \dfrac{h\nu}{kT}$ is an integration variable

The integral in Equation 1.27 cannot be analytically solved; however, for a high temperature, $\dfrac{T}{\Theta_D} \gg 1$,

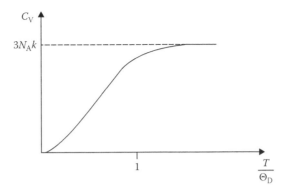

FIGURE 1.8 Graphic representation of the Debye law of specific heat.

$$C_V = 9N_A k \left(\frac{T}{\Theta_D}\right)^{\frac{\Theta_D}{T}} \int_0^{\frac{\Theta_D}{T}} y^4 dy = 9N_A k \left(\frac{T}{\Theta_D}\right)^3 \left(\frac{1}{3}\right)\left(\frac{\Theta_D}{T}\right)^3 = 3N_A k$$

On the other hand, the integral in Equation 1.27 for a low temperature, $\frac{T}{\Theta_D} \ll 1$, can be written as follows:

$$C_V = 9N_A k \left(\frac{T}{\Theta_D}\right) \int_0^{\frac{\Theta_D}{T}} \frac{y^4 e^y}{e^y - 1} dy \approx 9N_A k \left(\frac{T}{\Theta_D}\right) \int_0^{\infty} \frac{y^4 e^y}{e^y - 1} dy \tag{1.28}$$

Then, the integral in the right of Equation 1.28 can be integrated as follows:

$$C_V = \frac{12\pi^4}{5} N_A k \left(\frac{T}{\Theta_D}\right)^3$$

1.5 ELECTRONS IN CRYSTALLINE SOLID MATERIALS

1.5.1 ELECTRON GAS

In a free atom of a metallic element, the valence electron moves in an orbital around the ion formed by the nucleus and the core electrons. When a solid metal is formed, these external orbitals overlap and interact. Subsequently, the outer electrons do not belong anymore to the atom. In this case, the wave function describing the state of these electrons is a solution of the Schrödinger equation for the motion in the potential of all the ions. As a consequence, in a metal, the bonding is carried out by the conduction electrons that form a cloud of electrons, which fills the space between the metal ions and mutually joins the ions throughout the Coulombic attraction between the electron gas and positive metal ions [14–16]. In this regard, the metallic crystal is held together by electrostatic forces of attraction between the positively charged metal ions and the nonlocalized, negatively charged electrons, that is, the electron gas. In the framework of the electron gas model or the Drude model, the system is formed by the cations plus a free electron gas. The premises behind the Drude model are [14–16]

- Electrons collide with positive ions.
- Collisions are instantaneous events.
- Electrons lose all extra energy gained from the external electric field during a collision.
- Between collisions the electrons moves freely.
- Mutual repulsion between electrons is ignored.
- Finally, it is possible to state that the electron is confined to an energy band, named the conduction band, as will be explained later.

Now, if free electrons are influenced by an external electric field, $\bar{E}_x$, then a net electron drift in the x-direction is produced (see Figure 1.9). This net drift, along the force, which is created by the electric field, is superimposed on the chaotic motion of the electron gas. The end result of this process is that, following numerous scattering episodes, the electron has moved by a net distance, Δx, from its initial position in the direction of the positive terminal.

Following these assumptions, the Newton motion equation, along the x-axis, for the electrons in the free electron gas is given by

$$m_e \frac{dv_x}{dt} = e\, E_x - m_e \frac{v_x}{\tau}$$

where
τ is the time between collisions
m_e and e are the mass and charge of the electron

Then, the steady-state solution of the Newton equation for the electron in the electron gas under the influence of an external electric field is given by

$$v_x^{drift} = \frac{e\tau}{m_e} E_x$$

Now,

$$J_x = \sigma E_x$$

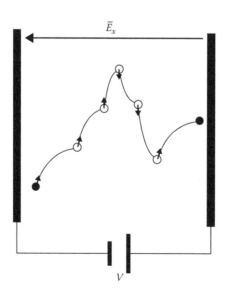

FIGURE 1.9   Electron trajectories in the electron gas or Drude model.

where

J_x is the current density

σ is the conductivity

And with the help of the definition of mobility, M,

$$v_x^{\text{drift}} = ME_x$$

It is possible to show that

$$\sigma = \frac{ne^2\tau}{m_e} = neM \tag{1.29}$$

The previously described theory in its original form assumes that the classical kinetic theory of gases is applicable to the electron gas, that is, electrons are expected to have velocities that are temperature dependent according to the Maxwell–Boltzmann distribution law. But, the Maxwell–Boltzmann energy distribution has no restrictions to the number of species allowed to have exactly the same energy. However, in the case of electrons, there are restrictions to the number of electrons with identical energy, that is, the Pauli exclusion principle; consequently, we have to apply a different form of statistics, the Fermi–Dirac statistics.

1.5.2 FERMI–DIRAC DISTRIBUTION

One of the simplest procedures to get the expression for the Fermi–Dirac (F–D) and the Bose–Einstein (B–E) distributions, is to apply the grand canonical ensemble methodology for a system of noninteracting indistinguishable particles, that is, fermions for the Fermi–Dirac distribution and bosons for the Bose–Einstein distribution. For these systems, the grand canonical partition function can be expressed as follows [12]:

$$\Theta = \sum_{N=0}^{\infty} \lambda^N \sum_{\{N_k\}} e^{\frac{-\sum_k N_k \varepsilon_k}{kT}} \tag{1.30}$$

where

ε_k are the energy states of the individual particles is the number of particles in the system

$\lambda = e^{-\frac{\mu}{kT}}$, in which μ is the chemical potential of the system of N indistinguishable noninteracting particles

The summation over $\{N_k\}$ means that we are summing the particle distributions in the energy states accessible to the system where

$$N = \sum_k N_k$$

and

$$E_j = \sum_k N_k \varepsilon_k$$

is the energy of the particle system; then, rearranging Equation 1.30 leads to

$$\Theta = \sum_{N=0}^{\infty} \lambda^N \sum_{\{N_k\}} e^{-\frac{\sum_k N_k \varepsilon_k}{kT}} = \sum_{N=0}^{\infty} \sum_{\{N_k\}} \lambda^{\sum_i N_i} e^{-\frac{\sum_k N_k \varepsilon_k}{kT}} = \sum_{N=0}^{\infty} \sum_{\{N_k\}} \prod_k \left(\lambda e^{-\frac{\varepsilon_k}{kT}} \right)^{N_k} \qquad (1.31)$$

And continuing with the rearrangement of Equation 1.31, we will get

$$\Theta = \sum_{N_1}^{N_1^{max}} \sum_{N_2}^{N_2^{max}} \cdots \prod_k \left(\lambda e^{-\frac{\varepsilon_k}{kT}} \right)^{N_k} = \sum_{N_1=0}^{N_1^{max}} \lambda e^{-\frac{\varepsilon_1}{kT}} \sum_{N_2=0}^{N_2^{max}} \lambda e^{-\frac{\varepsilon_2}{kT}} \cdots = \prod_k \sum_{N_k=0}^{N_k^{max}} \left(\lambda e^{-\frac{\varepsilon_k}{kT}} \right)^{N_k} \qquad (1.32)$$

We know from the Pauli principle that for fermions $N_k = 0$ and $N_k = 1$. Consequently,

$$\Theta = \prod_k \left(1 + \lambda e^{-\frac{\varepsilon_k}{kT}} \right)^{N_k} = \prod_k \left(1 + e^{-\frac{\varepsilon_k - \mu}{kT}} \right)^{N_k}$$

Since [11,12]

$$\overline{N} = kT \left(\frac{\partial \ln \Theta(V,T,\mu)}{\partial \mu} \right)_{V,T} = \sum_k \frac{\lambda e^{-\frac{\varepsilon_k}{kT}}}{1 + \lambda e^{-\frac{\varepsilon_k}{kT}}} \qquad (1.33)$$

the average number of particles in the state k in the Fermi–Dirac distribution is

$$\overline{N}_k = \frac{\lambda e^{-\frac{\varepsilon_k}{kT}}}{1 + \lambda e^{-\frac{\varepsilon_k}{kT}}} \qquad (1.34)$$

As a corollary, in the case of bosons, since $N_k = 0, 1, 2, 3,\ldots, \infty$, then

$$\overline{N}_k = \frac{\lambda e^{-\frac{\varepsilon_k}{kT}}}{1 - \lambda e^{-\frac{\varepsilon_k}{kT}}} \qquad (1.35)$$

which is equivalent to the previously obtained Bose–Einstein distribution, since in the case of bosons, there is no restriction on the total number of particles, and $\mu = 0$ [17].

In this regard, the probability of finding an electron in a state with energy E is given by the Fermi–Dirac distribution function, $f(E)$, which is expressed as follows (Figure 1.10):

$$f_{FD}(E) = \frac{1}{e^{\frac{E-\mu}{kT}} + 1} = \frac{1}{e^{\frac{E-E_F}{kT}} + 1}$$

where
    E is the state energy
    $\mu = E_F$ is the Fermi energy level

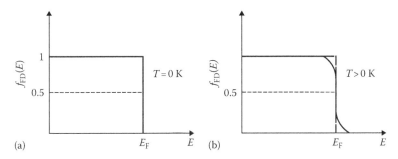

FIGURE 1.10 (a) Fermi–Dirac distribution for $T = 0\,\mathrm{K}$ and (b) Fermi–Dirac distribution for $T > 0\,\mathrm{K}$.

k is the Boltzmann constant
T is the absolute temperature

The Fermi–Dirac distribution describes the statistics of electrons in the conduction band of a solid when the electrons interact with each other and the environment, so that they obey the Pauli exclusion principle.

In Figure 1.10, it is shown that the Fermi level is the energy of the highest occupied quantum state in a system of fermions at $0\,\mathrm{K}$, and that above $0\,\mathrm{K}$, because of thermal excitation, some of the electrons are at energies above E_F.

1.5.3 DENSITY OF STATES FOR THE ELECTRON GAS

We will now calculate the density of electron states in the case of the electron gas. In this model, the core electrons are considered as nearly localized, and must be distinguished from the conduction electrons, which are supposed to freely move in Bloch states throughout the whole crystal [5]. Because of the fact that the potential is constant, the single-particle Hamiltonian is merely the kinetic energy of the electron, that is,

$$\hat{H} = -\frac{\hbar^2}{2m}\nabla^2 \tag{1.36}$$

Then, the conduction electron states are plane waves, that is,

$$\psi_{\bar{k}} = e^{i\bar{k}\cdot\bar{r}} \tag{1.37}$$

But, the real wave function must include the spin coordinate, then [6]

$$\psi_{\bar{k},s} = e^{i\bar{k}\cdot\bar{r}}\chi(s) \tag{1.38}$$

where

$$\chi\left(\frac{1}{2}\right) = \begin{pmatrix} 1 \\ 0 \end{pmatrix}$$

and

$$\chi\left(-\frac{1}{2}\right) = \begin{pmatrix} 0 \\ 1 \end{pmatrix}$$

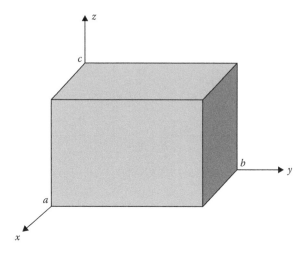

FIGURE 1.11   Box of volume $V = abc$ where the electrons are confined.

Substituting Equation 1.38 in Equation 1.36, we will get the energy of the electrons that is independent of the spin state [12,15]

$$E^0(k) = \frac{\hbar^2 k^2}{2m_e} \tag{1.39}$$

where $k = |\bar{k}|$. Then, the system in consideration is equivalent to a quantum system of noninteracting electrons in the three-dimensional potential box (see Figure 1.11) [11,17]. In this case, the possible energies for electrons confined in a cubic box of volume, $V = abc$, are given by

$$E(n_1, n_2, n_3) = \frac{h^2}{8m_e}\left(\frac{n_1^2}{a^2} + \frac{n_2^2}{b^2} + \frac{n_3^2}{c^2}\right)$$

where n_1, n_2, and n_3 are quantum numbers, each of which can be any integer number except 0. For a square box, where $a = b = c = L$, we will have

$$E(n_1, n_2, n_3) = \frac{h^2}{8L^2 m_e}(n_1^2 + n_2^2 + n_3^2) = \frac{h^2 R^2}{8m_e L^2} \tag{1.40}$$

where we have defined the sphere of radius

$$R^2 = (n_1^2 + n_2^2 + n_3^2) = \frac{E}{A}$$

in which

$$A = \frac{h^2}{8L^2 m_e}$$

Consequently, the number of states that can be accommodated in the space defined by $\bar{n} = n_1\bar{i} + n_2\bar{j} + n_3\bar{k}$ (see Figure 1.12) is

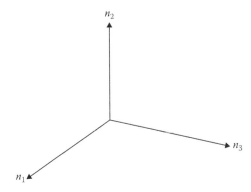

FIGURE 1.12 $\bar{n} = n_1\bar{i} + n_2\bar{j} + n_3\bar{k}$, space.

$$\eta = 2\left(\frac{1}{8}\right)\left(\frac{4}{3}\pi R^3\right) = \frac{1}{3}\pi\left(\frac{E}{A}\right)^{\frac{3}{2}} \tag{1.41}$$

where the factor 2 is due to the two spin states and the factor 1/8 is because only positive numbers of the quantum states are allowed. Then the density of states can be defined as follows:

$$g(E) = \frac{d\eta}{dE} = \frac{\pi}{2}\left(\frac{\sqrt{E}}{A^{3/2}}\right) \tag{1.42}$$

In this regard, if the probability of occupancy of a state at an energy E is $f_{FD}(E)$, in agreement with the Fermi–Dirac distribution, we are dealing with electrons, which are fermions. Then, the product $f_{FD}(E)g(E)$ is the number of electrons per unit energy per unit volume. Consequently, the area under the curve with the energy axis gives

$$N = \int_0^\infty g(E)f_{FD}(E)dE \tag{1.43}$$

which is the number of free electrons in volume V.

We can now calculate the value of the Fermi energy level, because as the electrons fulfill the Pauli exclusion principle, only two electrons can occupy one energy state thereafter, since at $T = 0$ [K], $f_{FD}(E) = 1$, for $E < E_F(0)$ and $f_{FD}(E) = 0$; for $E > E_F(0)$, then

$$N = \int_0^{E_F(0)} g(E)\,dE = \frac{\pi}{2A^{3/2}}\int_0^{E_F(0)} \sqrt{E}\,dE = \frac{\pi}{3A^{3/2}}\left[E_F(0)\right]^{3/2}$$

And as a result

$$E_F(0) = \frac{h^2}{8m_e}\left(\frac{3N}{\pi L^3}\right)^{2/3} = \frac{h^2}{8m_e}\left(\frac{3n}{\pi}\right)^{2/} \tag{1.44}$$

where $n = \left(\dfrac{N}{V}\right)$ and $V = L^3$

It is easy now to calculate the mean energy of an electron in a solid, $\bar{\varepsilon}_{average}$, at $T = 0$ [K], as follows:

$$\bar{\varepsilon}_{average}(0) = \frac{1}{N}\int_0^\infty Eg(E)f_{FD}(E)dE = \frac{1}{N}\int_0^{E_F(0)} Eg(E)dE = \left(\frac{3}{5}\right)E_F(0)$$

Above absolute zero, the average energy is approximately [2,15]

$$\bar{\varepsilon}_{average}(T) = \left(\frac{3}{5}\right)E_F(0)\left[1 + \frac{5\pi^2}{12}\left(\frac{kT}{E_F(0)}\right)^2\right]$$

Since $E_F(0) \gg kT$

$$\bar{\varepsilon}_{average}(T) \approx \bar{\varepsilon}_{average}(0) = \frac{1}{2}m(\bar{v}_F)^2$$

where $\bar{v}_F$ is the root mean-square speed of the electrons in the valence band of a solid around the Fermi level. Then

$$\bar{v}_F = \left(\frac{6E_F(0)}{5m}\right)^{1/2} \tag{1.45}$$

This velocity of the electron is independent of temperature, in contradiction to the Maxwell–Boltzmann statistic, which states that

$$\left(\frac{1}{2}\right)m\langle v_e^2\rangle = \frac{3}{2}kT$$

1.5.4 Energy Band Model

The electron gas model adequately describes the conduction of electrons in metals; however, it has a problem, that is, the electrons with energy near the Fermi level have wavelength values comparable to the lattice parameters of the crystal. Consequently, strong diffraction effects must be present (see below the diffraction condition (Equation 1.47). A more realistic description of the state of the electrons inside solids is necessary. This more accurate description is carried out with the help of the Bloch and Wilson band model [18].

If the problem is mathematically treated as a perturbation of the free-electron gas energy states caused by the presence of the periodic potential, $V(\bar{r})$, in the Schrödinger equation, then

$$-\frac{\hbar^2}{2m}\nabla^2\psi(\bar{r}) + V(\bar{r})\psi(\bar{r}) = E\psi(\bar{r})$$

Then [5],

$$E(\bar{k}) = E^0(k) + \int \psi_{\bar{k}}(\bar{r})V(\bar{r})\psi_{\bar{k}}(\bar{r})d^3\bar{r} + \sum_{\bar{k}'}\frac{\int \psi_{\bar{k}}(\bar{r})V(\bar{r})\psi_{\bar{k}'}(\bar{r})d^3\bar{r}}{E^0(\bar{k}) - E^0(\bar{k}')} \tag{1.46}$$

where

$$E^0(k) = \frac{\hbar^2 k^2}{2m}$$

Since diffraction is an effect linked to scattering, if a beam of fast electrons is being directed into a crystal, its scattering process will be described by the Born approximation where the rate of transition between the initial state, $\Psi_{\bar{k}}$, and the final state, $\Psi_{\bar{k}'}$, is given by [10]

$$P_{\bar{k},\bar{k}'} = \int \Psi_{\bar{k}} V(\bar{r}) \Psi_{\bar{k}'} d^3\bar{r}$$

Given that [5,6]

$$V(\bar{r}) = \sum_{\bar{G}_{hkl}} V_{\bar{G}_{hkl}} e^{i\bar{G}_{hkl} \cdot \bar{r}}$$

and

$$\Psi_{\bar{k}} = e^{i\bar{k}.\bar{r}}$$

then

$$P_{\bar{k},\bar{k}'} = \sum_{\bar{G}_{hkl}} \int e^{i(\bar{k}+\bar{G}_{hkl}-\bar{k}')} d^3\bar{r}$$

where

$$P_{\bar{k},\bar{k}'} = V_{\bar{G}_{hkl}}$$

If the diffraction condition for electrons in a crystal (Equation 1.47)

$$\bar{k} - \bar{k}' = \bar{G}_{hkl} \tag{1.47}$$

is fulfilled, then

$$P_{\bar{k},\bar{k}'} = 0$$

Subsequently, introducing the diffraction condition in Equation 1.46, we will get [5]

$$E(\bar{k}) = \frac{\hbar^2 k^2}{2m} + V_0 + \sum_{\bar{G}_{hkl} \neq 0} \frac{\left| V_{\bar{G}_{hkl}} \right|^2}{E^0(\bar{k}) - E^0(\bar{k}' - \bar{G}_{hkl})}$$

Consequently, the periodicity condition of the potential produces the segmentation in the energy bands.

A more exact treatment is made using the Bloch theorem. In this sense, the solution of the Schrödinger equation may be a plane wave multiplied by a periodic function, that is,

$$\psi_{\bar{k}}(\bar{r}) = e^{-\bar{k}\bullet\bar{r}} u_{\bar{k}}(\bar{r})$$

where

$$u_{\bar{k}}(\bar{r}) = \sum_{\bar{G}_{hkl}} u_{\bar{G}_{hkl}}(\bar{k}) e^{i\bar{G}_{hkl}\bullet\bar{r}} \tag{1.48}$$

Due to the periodicity of $u_{\bar{k}}(\bar{r})$, if we insert Equation 1.48 in the Schrödinger equation [6]

$$\left(-\frac{\hbar^2}{2m}\nabla^2 + \sum_{\bar{G}'_{hkl}} V_{\bar{G}'_{hkl}} e^{i\bar{G}'_{hkl}\bullet\bar{r}} - E \right) \sum_{\bar{G}_{hkl}} u_{\bar{G}_{hkl}} e^{i\bar{G}_{hkl}\bullet\bar{r}} = 0$$

Then

$$\left(\frac{\hbar^2}{2m}(\bar{k}+\bar{G})^2 - E \right) u_{\bar{G}_{hkl}}(\bar{k}) + \sum_{\bar{G}'_{hkl}} V_{\bar{G}'_{hkl}} u_{\bar{G}_{hkl}-\bar{G}'_{hkl}} = 0 \tag{1.49}$$

This equation is named the Bloch difference equation and is a set of coupled linear equations whose nontrivial solution conditions are

$$\left| \left(\frac{\hbar^2}{2m}(\bar{k}+\bar{G})^2 - E \right) \delta_{\bar{G}_{hkl},\bar{G}''_{hkl}} + V_{\bar{G}_{hkl}-\bar{G}''_{hkl}} \right| = 0 \tag{1.50}$$

This is named the Hill determinant. After solving, the resulting secular determinant for the root of $E_n(\bar{k})$ provides a more accurate method for calculating the band structure of solids, where $n = 1$ refers to the first band, $n = 2$ to the second, and so on.

1.5.5 MOLECULAR ORBITAL APPROACH FOR THE FORMATION OF ENERGY BANDS

A crystalline solid can be considered as a huge, single molecule; subsequently, the electronic wave functions of this giant molecule can be constructed with the help of the molecular orbital (MO) methodology [19]. That is, the electrons are introduced into crystal orbitals, which are extended along the entire crystal, where each crystal orbital can accommodate two electrons with opposite spins. A good approximation for the construction of a crystal MO is the linear combination of atomic orbitals (LCAO) method, where the MOs are constructed as a LCAO of the atoms composing the crystal [19].

For example, in metals, because of their large electrical conductivity, it seems that at least some of the electrons can move freely through the bulk of the metal, while the core electrons remain in their atomic orbital, similar to the isolated atoms forming the metal. For example, let us take into account the formation of a linear array of lithium atoms from individual lithium atoms: Li–Li; Li–Li–Li; Li–Li–Li–Li…. Then, the first stage is the formation of a lithium molecule, Li_2. This molecule is analogous to the hydrogen molecule, H_2 [15,19]. In the formation of the H_2 molecule, two MOs are formed, that is, the bonding MO

$$\Psi_\sigma = \psi_{1s}(\bar{r}_A) + \psi_{1s}(\bar{r}_B)$$

and the antibonding MO

$$\Psi_\sigma = \psi_{1s}(\bar{r}_A) - \psi_{1s}(\bar{r}_B)$$

where the two electrons pair their spins and occupy the bonding orbital. Then the two lithium atoms are bound together by a pair of valence electrons, where each lithium atom supplies its 2s electron to form a covalent molecular bond (see Figure 1.13). In this case, the molecule formed occurs in lithium vapor.

We will now take into account the hypothetical linear molecule, Li_3. The valence electron cloud is spherical; then, in the course of the linear combination of atomic orbitals, the three atomic valence electron clouds overlap to form one continuous distribution, and two distributions with nodes, that is, three MOs (see Figure 1.14). While the length of the chain is augmented, the number of electronic states, into which the atomic 2s state splits during the linear combination of atomic orbitals, increases. In this regard, the number of states equals the number of atoms.

A similar situation takes place when lithium chains are placed side by side or stacked on top of each other, so that finally the space lattice of the lithium crystal is obtained. In this case, the electronic states have energies that are bounded by an upper and lower limiting value, forming an energy band of closely spaced values (see Figure 1.14). Similarly, energy bands can also result from overlapping p and d orbitals.

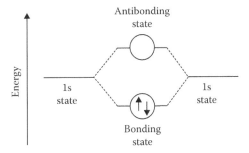

FIGURE 1.13   Energy of the states formed during the establishment of a Li_2 molecule.

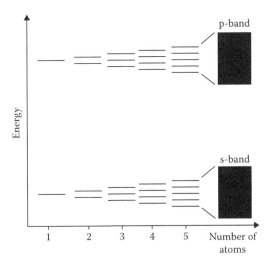

FIGURE 1.14   Band formation process.

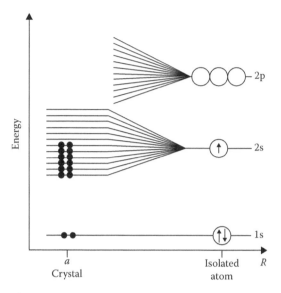

FIGURE 1.15 Band formation process for a Li crystalline solid.

The electronic states within an energy band are filled progressively by pairs of electrons in the same way that the orbitals of an atom are filled in accordance with the Pauli principle. This means that for lithium, the electronic states of the 2s band will be exactly half filled (Figure 1.15).

To summarize, the formation of a 2s-energy band from the 2s orbitals when N Li atoms are gathered together to form the Li crystal is shown in Figure 1.15. There are, N 2s-electrons but there are $2N$ states in the band, therefore the 2s band is only half full. Besides, the atomic 1s orbital, which is close to the Li nucleus, that is, is the two 1s electrons which are the core electrons, remains undisturbed in the solid, that is, each Li atom has a closed K-shell, specifically a full 1s orbital. Consequently, in general, when a solid metal is formed, the external orbitals overlap. As a consequence of this process, the outer electrons move without restraint through the metal, while the core electrons remains in their atomic orbital.

On the other hand, in covalently bonded materials like carbon, silicon, and germanium, the formation of energy bands first involves the hybridization of the outer s- and p-orbitals to form four identical orbitals, ψ_{hyb}, which form an angle of 109.5° with each other, that is, each C, Si, and Ge atom is tetrahedrally coordinated with the other C, Si, and Ge atom, respectively (Figure 1.16), resulting in a diamond-type structure.

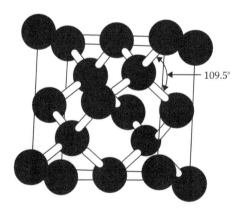

FIGURE 1.16 Tetrahedral bonding of atoms in a diamond-type structure of C, Si, and Ge crystals.

When these atoms are close enough, the ψ_{hyb} orbitals on two neighboring atoms can overlap to form a bonding orbital and an antibonding orbital [13,15]. In the crystal, the bonding orbital overlap to give the valence band, which is full of electrons, while the antibonding orbital overlap to give the conduction band, which is empty (see Figure 1.17). Since the conduction band is empty in the case of intrinsic semiconductors and insulators, these materials only conduct by the thermal excitation of electrons to the conduction band and by the formation of holes in the valence band (see Figure 1.18).

This excitation process is an activated process of electron jumps through the band gap, E_g. If the energy gap is low as in the case of semiconductors, the conductivity is low but noticeable. However, in the case of insulators, since the energy gap is high, the conductivity is very low.

Similarly, the covalent compound ZnS (zinc blende) is a semiconductor that has a structure similar to diamond, where the Zn atoms occupy the FCC lattice sites, and the S atoms occupy four of the eight tetrahedral sites of the FCC lattice (see Section 1.2.2). Analogous semiconducting properties are obtained when elements from the IIIA and VA columns of the periodic table are formed, for example, InAs, GaAs, and InP and also in the case when elements from the IIB and VIA columns of the periodic table are created, for instance, ZnTe and ZnSe.

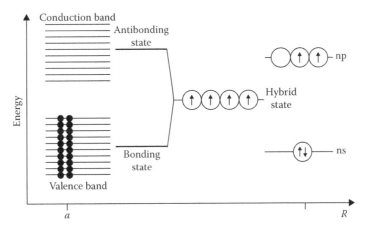

FIGURE 1.17   Band formation process for a C, Si, Ge, or α-Sn crystal.

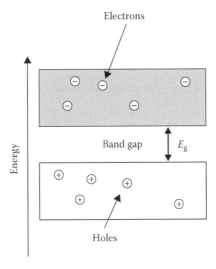

FIGURE 1.18   Formation of holes in the valence band by thermal excitation of electrons to the conduction band.

1.6 X-RAY DIFFRACTION

1.6.1 GENERAL INTRODUCTION

X-ray diffraction [20–26] is the most powerful method for the study of crystalline materials. The effect of x-ray generation during a glow discharge was casually discovered in 1895 by Wilhelm Röntgen at the University of Würtzburg in Germany. Some years later, in 1912, at the University of Munich, Max von Laue and collaborators carried out one of the most important experiments of modern physics, the Laue–Knipping–Friedrich experiment, which established that x-radiation consisted of electromagnetic waves. Additionally, the experiment clearly showed that the crystals were composed of atoms arranged on a space lattice, since the electromagnetic x-ray radiation was interfering during its scattering by the crystal atoms.

To generate an x-ray beam, a vacuum tube is needed where an electron beam, produced by a heated filament, is collimated and accelerated by an electric potential of several kilovolts, that is, from 20 to 45 kV (Figure 1.19). This beam is directed to a metallic anode (Figure 1.19). The electrons hitting the anode will convey a fraction of their energy to the electrons of the target material, a process resulting in the electronic excitation of the atoms composing the metallic anode. The x-ray tube has to be evacuated to allow electron movement. Finally, in order to dissipate the heat produced by this process in the metallic anode, it is normally water cooled.

The x-ray tube produces two kinds of radiations: the continuous spectrum (Figure 1.20) and the characteristic spectrum (Figure 1.21). The continuous spectrum is a plot of the intensity of the x-ray

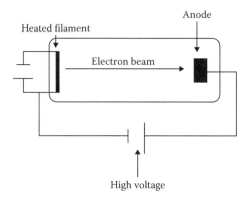

FIGURE 1.19 Schematic representation of an x-ray tube.

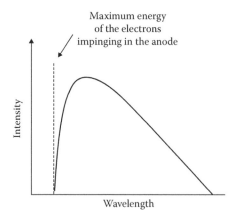

FIGURE 1.20 Schematic representation of a continuous spectrum.

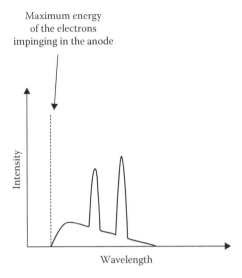

Maximum energy
of the electrons
impinging in the anode

Intensity

Wavelength

FIGURE 1.21 Schematic representation of a characteristic spectrum.

emission of the tube, which is measured in counts per second, and which is contingent on the anode material and on the high voltage imposed versus the wavelengths of the emitted x-rays. The mechanism of production of this radiation is by the deceleration of the electrons in the beam by the atoms that compose the metallic anode. Explicitly, this is a braking radiation, in German, Bremsstrahlung, which is the name normally given to this radiation. The Bremsstrahlung is then generated when photons are emitted, when the electrons in the beam lose kinetic energy.

The second type of spectrum (Figure 1.21), called the characteristic spectrum, is produced as a result of specific electronic transitions that take place within individual atoms of the anode material as will be explained below.

The discovery of x-rays provided crystallographers a powerful tool for the thorough determination of crystal structures and unit cell sizes [20–26]. X-rays have wavelengths between 0.2 and 10 nm. As x-rays possess dimensions comparable to the interplanar distances in crystals, x-ray crystallography is an ideal nondestructive method for material characterization, since nanometer parameters as well as macroscopic properties of the tested samples can be determined from x-ray diffraction data.

The basic properties of a wave significant in diffraction are wavelength, λ, that is, the distance between two adjacent peaks of the wave; wave amplitude, $|A|$, specifically, half the difference between peak and depression; intensity, $I \propto |A|^2$, and phase, φ, which is the location of a peak relative to other waves, measured as a fraction of the wavelength or as an angle in the range $0°$ to $360°$.

1.6.2 X-Ray Scattering

Diffraction is a process composed of elastic or coherent scattering of the x-ray radiation with the dispersion centers of the material, and thereafter these scattered rays interfering between them.

To understand the scattering process, we will distinguish it from other processes, starting with electron scattering, which was studied in 1906 by J.J. Thomson. He found that the intensity scattered by an electron interacting with an x-ray radiation is given by following equation [20,26]:

$$I = I_0 \frac{K}{r^2} \left(\frac{1 + \cos^2(2\theta)}{2} \right) \tag{1.51}$$

with

$$K = \frac{e^4}{16\pi^2\varepsilon_0^2 m_e^2 c^4}$$

where
  I_0 is the intensity of the incident beam
  r is the distance from the scattering electron to the detector
  e is the electron charge
  m_e is the mass of the electron
  ε_0 is the permittivity in free space
  c is the speed of light

The expression given by Equation 1.51 is normally named the polarization factor.

The next step is the scattering by an atom. This effect is basically the addition of the scattering of the electron cloud around the nucleus, since each electron in the atom scatters part of the incident radiation in a coherent form in agreement with the Thomson equation. Owing to the fact that the electrons in an atom are located at different points within the atom, and the fact that the x-ray wavelength is of the same order as the atomic dimensions, there will be path differences between waves scattered by different electrons; if these path differences are less than one wavelength, then the interference will be partially destructive [20,22,26]. To describe this effect, the parameter f is defined, also called the atomic scattering factor, which is the ratio of the amplitude scattered by an atom, A_a, to the amplitude scattered by an electron, A_e, that is [21]

$$f = \frac{A_a}{A_e}$$

Then, $f = Z$, for atomic dispersions in the forward direction, that is, for an angle $\theta = 0$. At higher scattering angles, f will decrease proportionally to the function $\frac{\sin\theta}{\lambda}$ [22].

The real form of the f function is calculated by integrating scattering over the electron distribution around an atom.

Equation 1.52 shows the calculation of the scattered intensity by a set of atoms, since, the lattice of a crystalline material is combined with a complicated motif or basis (see Figure 1.3). In this regard, the amplitude for the wave scattered from a set of atoms can be expressed as the superposition of scatterings from the individual atoms as follows [6,20,22]:

$$F(\overline{S}) = \sum_{\overline{\alpha}} f_{\overline{\rho}}(\overline{S}) e^{i\overline{S}\bullet\overline{\alpha}} \tag{1.52}$$

where
  $f_{\overline{\rho}}(\overline{S})$ is the atomic scattering factor for a given atom
  $\overline{S} = \Delta\overline{k}$ is the wave vector change during the scattering
  $\overline{\alpha}$ indicates the position of every atom in the set of scattering atoms

For a crystalline solid,

$$\overline{\rho} = \overline{R} + \overline{\alpha}_i \tag{1.53}$$

where
  $\bar{R} = n_1\,\bar{a} + n_2\,\bar{b} + n_3\,\bar{c}$ is a vector that indicates the position of the lattice points
  $\bar{\alpha}_i$ is a vector indicating the position of atoms inside the unit cell

Then, in the case of crystalline solids, the scattering amplitude is given by [2,6,22]

$$F(\bar{S}) = \sum_{\bar{R}}\sum_{i} f_j(\bar{S})e^{i\bar{S}\bullet(\bar{R}+\bar{\alpha}_i)} = \left(\sum_i f(\bar{S})e^{i\bar{S}\bullet\bar{\alpha}_i}\right)\left(\sum_{\bar{R}} e^{i\bar{S}\bullet\bar{R}}\right) = \Phi(\bar{S})\Lambda \tag{1.54}$$

where the first term is a summation over the basis of atoms

$$\Phi(\bar{S}) = \left(\sum_i f(\bar{S})e^{i\bar{S}\bullet\bar{\alpha}_i}\right) \tag{1.55}$$

which is named the structure factor, and the second term, that is, a summation over the lattice points,

$$\Lambda = \left[\sum_{\bar{R}} e^{i\bar{S}\bullet\bar{R}}\right]$$

is called the lattice sum term, because of the translation symmetry of the crystalline structure is only different from zero when

$$\bar{S} = \overline{G_{hkl}} = h\overline{a^*} + k\overline{b^*} + l\overline{c^*}$$

which, as was previously shown for electrons, (Equation 1.47) is equivalent to the constructive interference condition.

1.6.3 DIFFRACTION CONDITIONS

The fact that the scattering is elastic, during x-ray diffraction, implies the conservation of energy and momentum. Then,

$$\omega_i = \omega_s$$

where
  ω_i is the frequency of the incident beam
  ω_s is the frequency of the scattered radiation

and

$$|\bar{p}_i| = |\bar{p}_s|$$

where
  $|\bar{p}_i|$ is the magnitude of the momentum of the incident radiation
  $|\bar{p}_s|$ is the magnitude of the scattered momentum

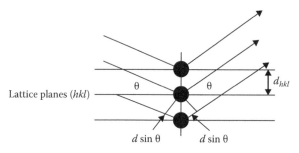

FIGURE 1.22 Bragg's law.

Since, the scattered ray interfere; subsequently, for rays reflected by two adjacent planes, the condition for constructive interference is given by (see Figure 1.22) [22]

$$\text{Path difference} = 2d_{hkl}\sin\theta = n\lambda \tag{1.56}$$

This expression is known as the Bragg law. In this regard, we know $\bar{k} = \dfrac{2\pi}{\lambda}\bar{s}$, in which $\bar{s}$ is a unit vector normal to the wave fronts. Then, if $\bar{s}_s$, and $\bar{s}_i$ are unit vectors along the directions of the diffracted and incident beams (see Figure 1.23), respectively, it is possible to show that the Bragg condition is equivalent to [2,26]:

$$2\pi\left(\frac{\bar{s}_s}{\lambda} - \frac{\bar{s}_i}{\lambda}\right) = \overline{G_{hkl}} = h\overline{a^*} + k\overline{b^*} + l\overline{c^*} \tag{1.57}$$

since (see Figure 1.23)

$$2\pi\left|\frac{\bar{s}_s}{\lambda} - \frac{\bar{s}_i}{\lambda}\right| = \frac{4\pi}{\lambda}\sin\theta = |\overline{G_{hkl}}| = \frac{2\pi}{d_{hkl}}$$

Consequently,

$$2d_{hkl}\sin\theta = \lambda$$

which is the Bragg constructive interference condition. Therefore, the condition expressed in Equation 1.57 can be written as

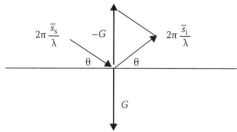

FIGURE 1.23 Graphical representation of the condition $2\pi\left(\dfrac{\bar{s}_s}{\lambda} - \dfrac{\bar{s}_i}{\lambda}\right) = \overline{G_{hkl}}$.

$$\overline{a} \bullet (\overline{s}_s - \overline{s}_i) = h\lambda, \quad \overline{b} \bullet (\overline{s}_s - \overline{s}_i) = k\lambda, \quad \text{and} \quad \overline{c} \bullet (\overline{s}_s - \overline{s}_i) = l\lambda$$

where h, k, and l are integers, which are the Laue diffraction conditions. Finally, the conditions expressed in Equations 1.56 and 1.57 are equivalent to the scattering condition given by

$$\Delta \overline{k} = G_{hkl}$$

which was as well found in the case of electrons (see Equation 1.47).

1.6.4 POWDER DIFFRACTION METHOD

The great majority of the applications of the x-ray diffraction methodology in material characterizations are carried out with the help of diffractometers, which use the Bragg–Brentano geometry. The principal characteristics of the Bragg–Brentano geometry are shown in Figure 1.24.

A specimen, located at the sample plane, is supported on a flat support bench that is free to rotate about its perpendicular axis, which is located at the origin (see Figures 1.24 and 1.25). The rotation is such that the angle of the incident x-ray beam with respect to the sample plane is θ, and the angle between the diffracted beam and the incident beam is 2θ.

The Bragg–Brentano type of diffractometer is composed of an x-ray tube with a metallic anode that supplies x-rays that are scattered from the sample and focused at the slit before hitting the detector. In some cases, a monochromator capable of yielding a monochromatic x-ray beam is added. The sample is rotated, relative to the x-ray at angles from 0° to 90° with the help of a goniometer, where the powdered sample is placed on the sample holder. Electronic equipment is used to amplify and filter signal pulses from the detector.

The powdered material sample to be tested is generally further ground in order to get a very fine powder, where the crystalline grains have arbitrary orientations. With the help of these random grain

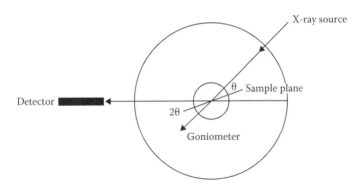

FIGURE 1.24   Bragg–Brentano geometry.

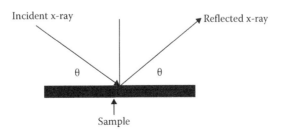

FIGURE 1.25   Sample irradiation in a Bragg–Brentano diffractometer.

directions, it is predicted that by rotating the sample relative to the incident x-ray (Figure 1.25), we can find all angles where diffraction take place.

Now, we will calculate the intensity diffracted by a powdered material that consists of a set of randomly oriented and spaced crystallites of the material under test. In this case, the position of a scattering center can be given as follows [6]:

$$\bar{\rho} = \bar{P} + \overline{R_{\bar{P}}} + \overline{\alpha_{\bar{P}_i}}$$

where

$\bar{P}$ is the position of a particular crystallite

$R_{\bar{P}}$ is the position vector of the lattice points in the crystallite located at $\bar{P}$

$\overline{\alpha_{\bar{P}_i}}$ is the vector that gives the position of the atoms inside a unit cell of the crystallite located at $\bar{P}$

Then, the scattering amplitude in this case will be

$$F(\bar{S}) = \sum_P \sum_{R_{\bar{P}}} \sum_i f_j(\bar{S}) e^{i\bar{S}\bullet(\bar{P}+\bar{R}_P+\overline{\alpha_{P_i}})} = \left(\sum_{\bar{P}} e^{i\bar{S}\bullet\bar{P}}\right)\left(\sum_i f(\bar{S}) e^{i\bar{S}\bullet\overline{\alpha_{\bar{P}_i}}}\right)\left(\sum_{R_{\bar{P}}} e^{i\bar{S}\bullet\overline{R\bar{P}}}\right) \tag{1.58}$$

where the term $\sum_{R_{\bar{P}}} e^{i\bar{S}\bullet\overline{R_{\bar{P}}}}$, for a three-dimensional crystallite with a parallelepiped form, implies broadening of the diffraction peak. Since [22,26]

$$\sum_{R_{\bar{P}}} e^{i\bar{S}\bullet\overline{R_{\bar{P}}}} = \left(\frac{\sin\left(\frac{N_a}{2}\right)\bar{S}\bullet\bar{a}}{\sin\bar{S}\bullet\frac{\bar{a}}{2}}\right)\left(\frac{\sin\left(\frac{N_b}{2}\right)\bar{S}\bullet\bar{b}}{\sin\bar{S}\bullet\frac{\bar{b}}{2}}\right)\left(\frac{\sin\left(\frac{N_c}{2}\right)\bar{S}\bullet\bar{c}}{\sin\bar{S}\bullet\frac{\bar{c}}{2}}\right) = B_F(\theta) \tag{1.59}$$

where $M = N_a \bullet N_b \bullet N_c$ is the number of cells in the crystallite.

The term $B_F(\theta)$ indicates that for a crystalline powder the diffraction peak will be broadened, with the smaller crystallites generating a larger spread.

To conclude this section, it is necessary to state that the other two terms have the meaning previously explained.

1.6.5 OTHER FACTORS AFFECTING THE SCATTERING INTENSITY OF A POWDERED SAMPLE

There are other factors affecting the intensity of the peaks on a x-ray diffraction profile of a powdered sample. We have analyzed the structure factor, the polarization factor, and the broadening of the lines because of the dimensions of the crystallites. Now, we will analyze the multiplicity factor, the Lorentz factor, the absorption factor, the temperature factor, and the texture factor [21,22,24,26].

1.6.5.1 Multiplicity Factor

The multiplicity factor, m, specifies the number of equivalent lattice planes that may all cause reflection at the same Bragg angle position, that is, the number of equally spaced planes cutting a unit cell in a particular, (hkl), crystalline plane family. In the case of low symmetry systems, the multiplicity factor will be low every time. On the other hand, for high symmetry systems, a single family of planes might be duplicated various times, and each replica will add to the intensity of the diffraction peak.

In this regard, for the (100) reflection of a powdered cubic material, some crystallites will be oriented in such a way that the (100) reflection will occur, and for others in such a position the (010) or (001) reflection will occur, but since $d_{100} = d_{010} = d_{001}$, these reflections contribute to the same peak [21].

In conclusion, the multiplicity factor identifies the number of equivalent lattice planes that might all cause reflection at the same Bragg angle. The multiplicity of the different structures has been tabulated (see for example Ref. [24]).

1.6.5.2 Lorentz Factor

The Lorentz factor is a geometrical factor that influences the intensity of the diffracted beam and operates as follows: when a crystal is rotated for unequal lengths of time, the different planes are passed through the Bragg positions then, the total amount of radiation in every reflection is proportional to this time chance to reflect, and therefore the Lorentz factor is essentially this time factor [21,22,24,26]. It is inversely proportional to the velocity with which the plane passes through the condition of reflection, that is, the diffractometer moves at a constant rate, and the extent of time at every point in the diffracting condition will be a function of the diffraction angle. As the angle augments, more time is spent in the diffracting condition [21,22,24,26]. Consequently, this effect must be rectified by incorporating the calculation of intensity, a term called the Lorentz factor [20–22,25]:

$$L = \frac{1}{4\sin^2\theta\cos\theta}$$

This term is normally combined with the atomic scattering polarization term:

$$P = \frac{1}{2}(1 + \cos^2 2\theta)$$

To get the so-called Lorentz polarization (LP) factor

$$LP = \frac{1 + \cos^2 2\theta}{\sin^2\theta\cos\theta} \tag{1.60}$$

where the constant factor, $\frac{1}{8}$, is omitted [21].

1.6.5.3 Absorption Factor

Throughout the passage through materials, x-rays undergo an attenuation of intensity as a result of their absorption. In this regard, the Lambert–Beer law [20,26]

$$I = I_0 e^{-\mu x} = I_0 e^{-\mu_m \rho x} \tag{1.61}$$

is able to describe the absorption effect, where the intensity I_0 that enters into a sample will be exponentially reduced to a quantity $I_0\, e^{-\mu_m \rho x}$ after a distance x, in which the parameter $\mu = \mu_m \rho$ is named the linear absorption coefficient where μ_m is the mass absorption coefficient, and ρ is the density of the sample. The linear absorption coefficient depends on the wavelength of the radiation used, the chemical composition of the sample, and its density.

In the case of testing a powdered sample in a Bragg–Brentano diffractometer, the sample has the shape of a flat plate located parallel to the reflecting plane, making equal angles with the incident and diffracted beams; then, if we have a single phase in the sample, the absorption factor is given by [21]

$$A = \frac{1}{2\mu} \tag{1.62}$$

It must be noted that this factor is independent of the diffraction angle θ. This fact is caused by the precise equilibrium between two opposite effects, that is, when θ is small, the sample area irradiated by the incident beam of fixed cross section is large, but the penetration depth of the beam is small. On the contrary when θ is large, the area irradiated is small but the penetration depth is high, since volume is area multiplied by depth, then the effective volume irradiated is almost constant and, therefore, independent of θ.

1.6.5.4 Temperature Factor

The change in the intensity with temperature is calculated with the temperature factor. This change is produced by the crystal lattice vibrations, that is, the scattering atoms or ions vibrate around their standard positions as was previously explained (see Section 1.4); consequently, as the crystal temperature increases, the intensity of the Bragg-reflected beams decreases without affecting the peak positions [25]. Debye and Waller were the first to study the effect of thermal vibration on the intensities of the diffraction maxima. They showed that thermal vibrations do not break up the coherent diffraction; this effect merely reduces the intensity of the peaks by an exponential correction factor, named the temperature factor, $D(\theta)$ [2,26], given by

$$D(\theta) = e^{-M(T)} \tag{1.63}$$

where

$$M(T) = 2\pi^2 \left(\frac{\langle u^2 \rangle}{d^2} \right) = 8\pi^2 \langle u^2 \rangle \left(\frac{\sin\theta}{\lambda} \right)^2 = B \left(\frac{\sin\theta}{\lambda} \right)^2$$

in which

$\langle u^2 \rangle$ is the mean-square displacement of the vibrating atom or ion in the direction normal to the diffracting planes

d is the interplanar distance

B is the Debye–Waller factor equal to $B = 8\pi^2 \langle u^2 \rangle$

Then, as T increases, B will increase, because at a high temperature this parameter is directly related to the thermal energy, kT.

1.6.6 Intensity of a Diffraction Peak

The factors that are included when calculating the intensity of a powder diffraction peak in a Bragg–Brentano geometry for a pure sample, composed of three-dimensional crystallites with a parallelepiped form, are the structure factor $|F_{hkl}|^2 = |F(S)|^2$, the multiplicity factor, m_{hkl}, the Lorentz polarization factor, $LP(\theta)$, the absorption factor, A, the temperature factor, $D(\theta)$, and the particle-size broadening factor, $B_F(\theta)$. Then, the line intensity of a powder x-ray diffraction pattern is given by [20–22,24–26]

$$I_{\mathrm{d}}(\theta) = I_0 K_{\mathrm{e}} \mathrm{LP}(\theta) F_{hkl}^2 m_{hkl} \left(\frac{1}{V_{\mathrm{c}}}\right)^2 B_{\mathrm{F}}(\theta) D(\theta) A \tag{1.64}$$

where

$I_{\mathrm{d}}(\theta)$ is the diffracted intensity
I_0 is the intensity of the incident beam
K_{e} is a constant for a particular experiment
V_{c} is the volume of the unit cell of the crystal

and the meaning of the other factors have been previously explained.

1.7   DIELECTRIC PHENOMENA IN MATERIALS

1.7.1   INTRODUCTION

As soon as a nonpolar material is exposed to a static electric field, $\bar{E}$, dipoles become excited. That is, a local charge difference is induced within the neutral species, as the centers of "gravity" for the equal amount of positive and negative charges, $\pm Q$, turn out to be separated by a small distance, $\bar{d}$, generating a dipole with a dipole moment

$$\bar{p} = Q\bar{d} \tag{1.65}$$

which is related to the microscopic or local electric field, $\bar{E}_{\mathrm{Local}}$, but different from the macroscopic electric field [27].

In this section, a simple description of the dielectric polarization process is provided, and later to describe dielectric relaxation processes, the polarization mechanisms of materials produced by macroscopic static electric fields are analyzed. The relation between the macroscopic electric response and microscopic properties such as electronic, ionic, orientational, and hopping charge polarizabilities is very complex and is out of the scope of this book. This problem was successfully treated by Lorentz. He established that a remarkable improvement of the obtained results can be obtained at all frequencies by proposing the existence of a local field, which diverges from the macroscopic electric field by a correction factor, the Lorentz local-field factor [27].

1.7.1.1   Electronic Polarization

The mechanism of electronic polarization operates in all atoms and molecules, since the centre of "gravity" of the electrons surrounding the positive cores are displaced by the electric field. This effect is extremely fast, and thus effective up to optical frequencies. The dipole moment for this polarization mechanism can be written as follows:

$$\bar{p}_{\mathrm{e}} = \alpha_{\mathrm{e}} \bar{E} \tag{1.66}$$

where α_{e} is the electronic polarizability.

1.7.1.2   Ionic Polarization

The ionic polarization mechanism refers to a polarization mechanism in materials that contain ions. The dipole moment in this case is produced by the separation of the positive and negative ions included in the structure of these materials. The dipole moment for the ionic polarization mechanism can be expressed as

$$\bar{p}_{\mathrm{i}} = \alpha_{\mathrm{i}} \bar{E} \tag{1.67}$$

where α_{i} is the ionic polarizability.

1.7.1.3 Dipolar (or Orientation) Polarization

The dipolar polarization mechanism is present in substances containing molecules with permanent dipole moments. In these compounds, the bulk material is neutral, since normally the orientation of the polar molecules constituting the material is randomly distributed due to the action of thermal energy. In this situation, under the influence of the electric field, the dipoles will be partially oriented in the direction of the electric field. The dipole moment can be written as

$$\overline{p_d} = \alpha_d \overline{E} \tag{1.68}$$

where α_d is the dipolar polarizability.

1.7.1.4 Hopping of Charge Carriers' Polarization

In the case of materials that contain mobile positive ions included in a fixed anionic framework, for example, zeolites [28], jumps of mobile charge carriers between localized sites in a framework of charges of different signs occurs, creating polarization. In general, this is a slow process.

1.7.1.5 Interfacial Polarization

The interfacial polarization mechanism is generally present in heterogeneous systems, that is, materials composed of different dielectric substances. In this case, the disparity of dielectric properties induces the mobile positive and negative charges to be deposited on the interfaces of different insulating materials, thus forming dipoles. Then, grain boundaries and interfaces between different materials frequently give rise to interfacial polarization. This phenomenon is often very slow [28–30].

1.7.2 Susceptibility and Dielectric Constant

The macroscopic polarization vector, $\overline{P}$, and the electric field, $\overline{E}$, which generally are oriented in the same direction, are interrelated by the following expression:

$$\overline{P} = \chi \varepsilon_0 \overline{E} \tag{1.69}$$

where
   χ is the susceptibility of the dielectric, a dimensionless number which is 1 for ideal vacuum
   ε_0 is the permittivity of vacuum

The capacity of an empty parallel plate capacitor, that is, without a dielectric material between the parallel plates (see Figure 1.26) is

$$C_0 = \frac{Q_0}{V} = \varepsilon_0 \frac{A}{d}$$

And the capacity of a capacitor filled with a dielectric slab is

$$C = \frac{Q}{V} = \varepsilon_r \varepsilon_0 \frac{A}{d}$$

where $\varepsilon_r = C/C_0$ is the relative permittivity or dielectric constant.
The vector of electric displacement, $\overline{D}$, is defined as follows:

$$\overline{D} = \varepsilon_0 \overline{E} + \overline{P}$$

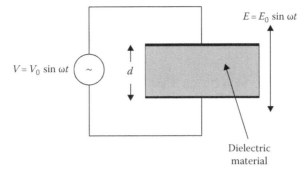

$$V = V_0 \sin \omega t$$

$$E = E_0 \sin \omega t$$

Dielectric
material

FIGURE 1.26 Time-varying electric field applied across a parallel plate capacitor.

Consequently, it is easy to show that

$$\overline{D} = \varepsilon_0 \overline{E} + \chi \varepsilon_0 \overline{E} = (1 + \chi)\varepsilon_0 \overline{E} = \varepsilon_r \varepsilon_0 \overline{E}$$

Then, susceptibility can be written as follows:

$$\chi = \varepsilon_r - 1 \tag{1.70}$$

1.7.3 Complex Permittivity

In the case of a time-varying electric field applied across a parallel plate capacitor of unit area having a separation between plates, d (Figure 1.26), the total current is given by [29–32]

$$J_T(t) = J + \frac{\partial D(t)}{\partial t} = \sigma_0 E(t) + \tilde{\varepsilon}\frac{\partial E}{\partial t} \tag{1.71}$$

where
  J is the conduction current
  $\dfrac{\partial D(t)}{\partial t}$ is the displacement current density

  σ_0 is the conductivity
  $\tilde{\varepsilon}$ is the complex permittivity, which is a complex variable defined as $\tilde{\varepsilon} = \varepsilon' - i\varepsilon''$, where ε' and ε''
    are the real and imaginary parts of the permittivity, respectively.

The complex permittivity can be written as

$$\tilde{\varepsilon} = \left(\varepsilon_r' - i\varepsilon_r''\right)\varepsilon_0$$

where
  $\varepsilon_r' = \varepsilon'/\varepsilon_0$ is the real part of the relative permittivity
  $\varepsilon_r'' = \varepsilon''/\varepsilon_0$ is the imaginary part of the relative permittivity

The real term, ε_r', physically means the dielectric constant and the imaginary term, ε_r'', means the loss factor. Complex permittivity is introduced to take into account the dielectric losses due to resistance, or friction due to the polarization and orientation of the electric dipoles. If it is applied, a sinusoidal field due to a sinusoidal voltage (see Figure 1.26) can be expressed as follows:

$$E = E_0 e^{i\omega t} \tag{1.72}$$

Then introducing Equation 1.72 in Equation 1.71, we get

$$J_T(t) = (\sigma_0 + \omega\varepsilon'')E(t) + i\omega\varepsilon' E(t) \tag{1.73}$$

where the first term represents loss of energy, which is, in turn, composed of the term represented by conductivity characterized by a loss of energy due to charge scattering during cationic migration, and the other term characterized by the imaginary term, ε'', due to resistance to the polarization process.

1.7.4 DIELECTRIC RELAXATION

We now analyze a case where we have an instantaneous increase or a reduction of the electric field, $\bar{E}$. This will lead to a polarization or depolarization process, which will follow with some delay or retardation due to the increase or reduction of the electric field, respectively. Consequently, in relation with a time-dependent variation of the electric field, $\bar{E} = \bar{E}(t)$, the dielectric properties of the materials become dynamic events. In this regard, the time dependency of $\bar{P} = \bar{P}(t)$ will not be the same as that of $\bar{E} = \bar{E}(t)$, since the different polarization processes have different time delays, with respect to the appearance of the electric field. This delay is obviously related to the time-dependent behavior of the susceptibility $\chi = \chi(t)$.

The time, τ, required by the different polarization processes previously analyzed is different. Specifically, the time needed for the electronic polarization, τ_e, is in the range 10^{-16} s $< \tau_e < 10^{-15}$ s. The time required for the ionic polarization, τ_i, is in the range 10^{-13} s $< \tau_i < 10^{-12}$ s; additionally, the time required for the dipolar polarization, τ_d, is in the range 10^{-7} s $< \tau_d < 10^{-5}$ s. As well, the time required for the charge-hopping polarization process, τ_d, is in the range 10^{-5} s $< \tau_{ch} < 10^{-4}$ s. Finally, the time required for the interfacial or space charge polarization process, τ_{sc}, is in the range 10^{-1} s $< \tau_{sc} < 10^1$ s.

The whole polarization of an arbitrary dielectric material is given by

$$P = P_e + P_i + P_d + P_{ch} + P_{sc} \tag{1.74}$$

where P_e, P_i, P_d, P_{ch}, and P_{sc} are related to the electronic, ionic, dipolar, charge-hopping, and space charge polarization mechanisms. Subsequently, since the dielectric response time of the electronic and ionic polarization is very small, it is possible for frequencies, f, greater than, 10^{12} Hz, the external electric field can be considered static. This fact means that the electronic and ionic susceptibilities can be lumped in a constant, susceptibility, $\chi(\infty)$, explicitly:

$$P_e + P_i = \varepsilon_0\chi(\infty)E(t) = P_\infty \tag{1.75}$$

Let us assume now, for example, that a step-like constant electric field of magnitude E_0 is applied within a dielectric at any time t_0, and remains constant for $t \geq t_0$ (see Figure 1.27). Then, $P_\infty = P_e + P_i$ is almost instantaneously established. Thereafter, the acting relaxation processes (i.e., dipolar and/or charge-hopping and/or space charge polarization mechanisms) provoke that the polarization is not instantaneously established.

That is, the polarization process represented by $P_T(t)$ will be established throughout a time evolution. To mathematically express this process, the following time convolution integral (normally named the Duhamel's integral) [32] is used:

$$P_T(t) = \varepsilon_0\chi(\infty)E(t) + \varepsilon_0\int_{-\infty}^{\infty} f(t-\tau)E(\tau)d\tau = P_\infty + P \tag{1.76a}$$

where $f(t) = 0$ for $t < 0$

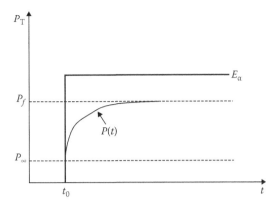

FIGURE 1.27 Step-like constant electric field of magnitude $E = E_\alpha \, 1(t - t_0)$ and the corresponding polarization process.

where $f(t)$ is the so-called dielectric response function and

$$P(t) = \varepsilon_0 \int\limits_{-\infty}^{\infty} f(t - \tau)E(\tau)\mathrm{d}\tau \tag{1.76b}$$

represents polarization where the large-relaxation time dielectric processes are involved, that is, dipolar, and/or charge-hopping and/or space charge polarization mechanisms. If we apply the Fourier transform and the convolution theorem [9] to Equation 1.76b, then we can write Equation 1.76b in the frequency domain as follows

$$P(\omega) = \varepsilon_0 \chi(\omega)E(\omega)$$

in which

$$\tilde{\chi}(\omega) = \int\limits_{-\infty}^{\infty} f(t)\mathrm{e}^{i\omega t}\mathrm{d}t$$

where

$$\tilde{\chi}(\omega) = \tilde{\varepsilon}_r(\omega) - \varepsilon_r(\infty) = \tilde{\varepsilon}_r(\omega) - \left(1 + \chi(\infty)\right)$$

is the dielectric susceptibility and $\tilde{\varepsilon}_r(\omega)$ is the complex dielectric function.

The electric displacement for a time-varying external voltage is given by [31]

$$D(t) = \varepsilon_0 \varepsilon_r(\infty)E(t) + P(t)$$

Thereafter, introducing the previous expression into Equation 1.71, we get

$$J_T(t) = \sigma_0 E(t) + \varepsilon_0 \varepsilon_r(\infty)\frac{\partial E(t)}{\partial t} + \frac{\partial P(t)}{\partial t} \tag{1.77}$$

Now it is possible to state that the external potential described in Figure 1.27 can be mathematically expressed as follows:

$$E(t) = E_\alpha \left[1(t - t_0) \right] \tag{1.78}$$

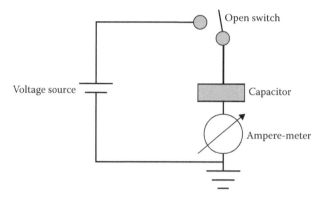

FIGURE 1.28 Polarization test circuit.

where the factor $1(t - t_0)$ is the Heaviside function, which is used to indicate the unit step. Subsequently, introducing Equation 1.78 into the convolution integral [9] (Equation 1.76), we get [32]

$$J_T(t) = \sigma_0 E_\alpha 1(t - t_0) + \varepsilon_0 [\varepsilon_r(\infty)\delta(t) + f(t)]E_\alpha \qquad (1.79)$$

where [9]

$$\int_{-\infty}^{\infty} \delta(t)\,dt = 1$$

in which $\delta(t)$ is the Dirac's delta function.

Equation 1.79 is the basis for a measurement method of the dielectric response function $f(t)$. Upon connecting the switch of the circuit shown in Figure 1.28, a polarization current, $i_{pol}(t)$, through the capacitor can be recorded, according to the following equation

$$i_p(t) = C_0 U_0 \left[\frac{\sigma_0}{\varepsilon_0} + \varepsilon_r(\infty)\delta(t) + f(t) \right]$$

which is obtained from Equation 1.79 when applied to the circuit represented in Figure 1.28, where $C_0 = Q_0/U_0 = \varepsilon_0 (A/d)$ is the vacuum capacitance of the test object, $i_p(t) = J_T(t) A$, and $E_0 = U_0/d$.

1.7.5 DEBYE RELAXATION MODEL FOR THE DIPOLAR MECHANISM

We will suppose now that our system is composed of polar molecules. In this case, the polarization can be expressed as follows:

$$P = P_e + P_i + P_d$$

since: $P_e + P_i = \varepsilon_0 \chi(\infty)E(t)$. Subsequently, we will only need to consider the dipolar polarization relaxation.

If the constant electric field is suddenly changed from E_0 to 0 at time t_0 (see Figure 1.29) [15,31] for a system of polar molecules, then the induced polarization, P, has to decrease from $P_\infty + P_d$ to

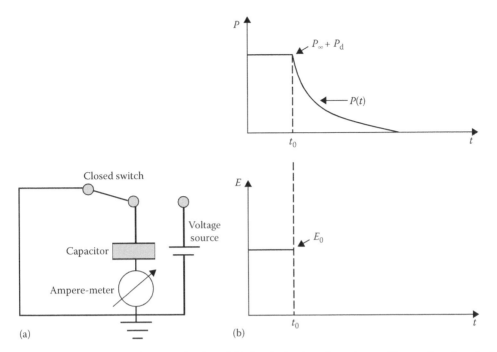

FIGURE 1.29 (a) Depolarization test circuit and (b) depolarization after a heavy side-step-function electric field.

a final value of 0. The steady decrease component is achieved by random collisions of the dipolar molecules [33].

In the frame of the Debye model, the dipolar polarization decay is given by

$$\frac{dP_d}{dt} = -\frac{P_d}{\tau} \tag{1.80}$$

where τ is the relaxation time of the process.

The methodology for the calculation of the complex relative permittivity for the dipolar relaxation mechanism is founded on the calculation of the dielectric response function, $f(t)$, for a depolarization produced by the discharge of a previously charged capacitor. In Figure 1.29a, a circuit is shown where a capacitor is inserted in which a dipolar dielectric material is enclosed in the parallel plate capacitor of area, A, and thickness, d, with empty capacitance $C_0 = Q_0/U_0 = \varepsilon_0(A/d)$, and $E_0 = U_0/d$. In Figure 1.29b, the corresponding depolarization process is shown.

The depolarization current generated in the circuit is mathematically expressed by [29]

$$i_d = A\frac{dP}{dt} = C_0 U_0 f(t) \tag{1.81}$$

which is as well obtained by applying Equation 1.77 to the depolarization case. Consequently, solving Equation 1.80 and substituting the solution in Equation 1.81 allows us to get

$$f(t) = f_0 e^{-\frac{t}{\tau}} \tag{1.82}$$

Now,

$$\tilde{\chi}(\omega) = \int_0^\infty f(t) e^{-i\omega t} dt$$

Then, calculating the previous integral we get

$$\tilde{\chi}(\omega) = \frac{\chi(0)}{1 + i\omega\tau}$$

As was previously established,

$$\tilde{\chi}(\omega) = \tilde{\varepsilon}_r(\omega) - \varepsilon_r(\infty)$$

Subsequently, the complex dielectric constant is given by

$$\tilde{\varepsilon}_r(\omega) = \varepsilon_r(\infty) + \frac{\varepsilon_r(0) - \varepsilon_r(\infty)}{1 + i\omega\tau}$$

Consequently,

$$\varepsilon_r'(\omega) = \varepsilon_r(\infty) + \frac{\varepsilon_r(0) - \varepsilon_r(\infty)}{[1 + (\omega\tau)^2]} \quad \text{and} \quad \varepsilon_r''(\omega) = \frac{(\varepsilon_r(0) - \varepsilon_r(\infty))\omega\tau}{[1 + (\omega\tau)^2]}$$

In Figure 1.30, the plots of $\varepsilon_r'(\omega)$ are shown, that is, the real part of the complex relative permittivity, and $\varepsilon_r''(\omega)$, that is, the imaginary part of the complex relative permittivity.

1.7.6 MODEL TO DESCRIBE DIELECTRIC RELAXATION FOR A CHARGE-HOPPING PROCESS

We now develop a model using a dehydrated zeolite as the test material, since in this case the polarization takes place by the charge-hopping mechanism [28,34]. We consider that during cationic migration in zeolites, the charge compensating cations jump from one site to the other [34]. Therefore, since the zeolite framework is negatively charged, a microscopic change in the polarization occurs [34] (see Figure 1.31).

The methodology for the calculation of the complex relative permittivity for the cation-hopping relaxation mechanism is very similar to those applied in the previous (Section 1.7.5). It is also based on

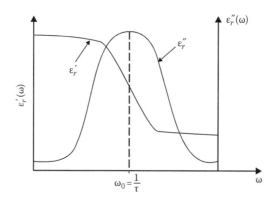

FIGURE 1.30 Graph of $\varepsilon'(\omega)$, that is, the real part of the complex relative permittivity, and $\varepsilon''(\omega)$, that is, the imaginary part of the complex relative permittivity.

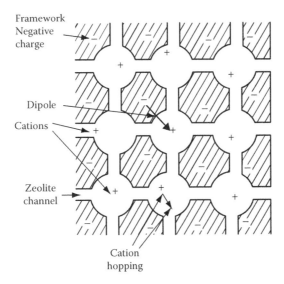

FIGURE 1.31   Schematic representation of the channels and cavities of a dehydrated zeolite.

the use of the dielectric response function, $f(t)$, under a transient excitation, specifically created in a circuit similar to that reported in Figure 1.29a, where the parallel plate capacitor of area, A, and thickness, d, with empty capacitance $C_0 = Q_0/U_0 = \varepsilon_0(A/d)$, and $E_0 = U_0/d$, is filled with a homoionic aluminosilicate zeolite (see Figure 1.32) [34].

The depolarization current (see Figure 1.29b) is also given by Equation 1.81 [29]. Now, we consider the time dependence of the cationic concentration, $C(x,t)$, in the zeolite filling the gap of the condenser after the step-like constant electric field fulfills the condition (see Figure 1.33):

$$\frac{dC(x,t)}{dt} = R(t)$$

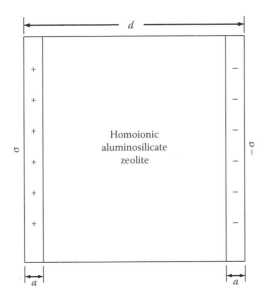

FIGURE 1.32   Schematic representation of polarized homoionic aluminosilicate zeolite filling the gap of a parallel plate capacitor.

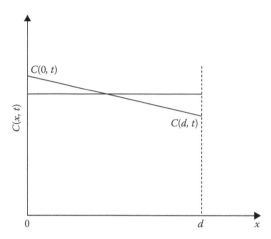

FIGURE 1.33 Profile of the cationic concentration in the zeolite after the application of a step-like constant electric field.

The volumetric charge in the vicinity of the plates of the capacitor, on the side of the zeolite (see Figure 1.32), is [34]

$$\rho_0(t) = (C(0,t) - C_0)Ze = \rho_0'(t)Ze \tag{1.83a}$$

and

$$\rho_d(t) = -\left(C(d,t) - C_0\right)Ze = \rho_d'(t)Ze \tag{1.83b}$$

where
  Ze is the charge of the cation present in the zeolite
  C_0 is the equilibrium volumetric concentration of cations in the zeolite

and because of charge conservation

$$\rho_0(t) = -\rho_d(t)$$

Now, it is possible to state that the surface charge concentration is given by [34]

$$\sigma = \rho_0(t)a$$

where a is the width of the surface charge area (see Figure 1.32). Now, it is easy to calculate the depolarization field [34]

$$E_d = \frac{P_d}{\chi \varepsilon_0} = \frac{\rho_0(t)a}{\chi \varepsilon_0} \tag{1.84}$$

since

$$P_d = \sigma = \rho_0(t)a$$

As it is well known, the transport equation for cations after the step-like constant electric field is given by [35]

$$J = MC(x,t)E_d - D\frac{\partial C(x,t)}{\partial t} \tag{1.85a}$$

where

$$M = \frac{ZeD}{kT} \tag{1.85b}$$

is the mobility and D is the self-diffusion coefficient for cationic migration.

The solution of Equation 1.85, after the transient field is applied, for the geometry of the present system (see Figure 1.32) in the steady state is [35]

$$J = \frac{EM\left[C(0,t) - C(d,t)e^{\eta}\right]}{1 - e^{\eta}}$$

where

$$\eta = \frac{MEd}{D}$$

For the system under study, $\eta \approx 10^5$; consequently, the diffusion term will be insignificant with respect to the field effect over the cationic transport [34]. Then, the expression for the flux is

$$J \approx MC(d,t)E_d \tag{1.86}$$

Now applying the continuity equation, taking into account the geometric conditions of the system under study [34], we get

$$\frac{1}{\Delta t}\left(\frac{\Delta N}{Aa}\right) \approx \frac{\partial \rho_0'}{\partial t} = -\frac{\partial J}{\partial x} \approx \frac{1}{a}\left(\frac{\Delta N}{A\Delta t}\right) = -\frac{J}{a} \tag{1.87}$$

where
   ΔN is the number of cations flowing out of the charge formed in the edges of the zeolite, during the time interval, Δt (see Figure 1.32)
   $V = aA$ is the volume where this charge is developed.

It is now easy to show, taking into consideration Equations 1.83, 1.86, and 1.87, that [34]

$$-\frac{\partial \rho_0'}{\partial t} = \frac{MC(d,t)E_d}{a} = \frac{MZeC(d,t)\rho_0'(t)}{\chi \varepsilon_0} = \zeta C(d,t)\rho_0'(t) \tag{1.88a}$$

where

$$\zeta = \frac{MZe}{\chi \varepsilon_0} \tag{1.88b}$$

Now, it is possible to make another approximation

$$C(d,t) \approx C_0$$

Then, Equation 1.88a reduces to

$$-\frac{1}{\rho_0'}\left(\frac{\partial \rho_0'}{\partial t}\right) = \zeta C_0 \tag{1.89}$$

Then, applying Equations 1.85 and 1.89, and taking into account Equation 1.88b, gives

$$\frac{1}{\rho_0'}\left(\frac{\partial \rho_0'}{\partial t}\right) = -\frac{(Ze)^2 C_0}{\chi \varepsilon_0 kT} D = -QD \tag{1.90a}$$

where

$$Q = \frac{(Ze)^2 C_0}{\chi \varepsilon_0 kT} \tag{1.90b}$$

Then, the solution of Equation 1.90a is [34]

$$\rho_0'(t) = \rho_0'(0)e^{-QDt} \tag{1.91}$$

Now, from Equation 1.80, it is easy to get

$$f(t) = KQDe^{-QDt}$$

where K is a constant.
Consequently,

$$\tilde{\chi}(\omega) = \int_0^\infty f(t)e^{-i\omega t}dt$$

and

$$KQD\int_0^\infty e^{-QDt}e^{-i\omega t}dt = \frac{\chi(0)}{1+i\dfrac{\omega}{QD}}$$

Subsequently,

$$\tilde{\chi}(\omega) = \frac{\chi(0)}{1+i\dfrac{\omega}{QD}} = \frac{\chi(0)}{1+i\dfrac{\omega}{\omega_0}} = \frac{\chi(0)}{1+\left(\dfrac{\omega}{\omega_0}\right)^2} - i\left(\frac{\chi(0)}{1+\left(\dfrac{\omega}{\omega_0}\right)^2}\right)\left(\frac{\omega}{\omega_0}\right) \tag{1.92}$$

where

$$\omega_0 = QD$$

It is now possible to consider that [34]

$$D = D_0 e^{-\frac{E_a}{kT}}$$

Afterward,

$$\omega_0 = QD_0 e^{-\frac{E_a}{kT}} \tag{1.93}$$

Then, the real part of $\tilde{\chi}(\omega)$ is

$$\chi'(\omega) = \frac{\chi(0)}{1 + \left(\dfrac{\omega}{\omega_0}\right)^2} \tag{1.94}$$

Equation 1.92 fulfills all the conditions imposed on a generalized susceptibility, that is [36], it is a complex function of frequency, where

$$\chi'(-\omega) = \chi'(\omega) \quad \text{and} \quad \chi''(-\omega) = -\chi''(\omega)$$

and when

$$\omega \to \infty: \chi'(\omega) \to \chi'(\infty)$$

where $\chi'(\infty)$ is a real finite term. Finally, it is necessary to state that a relation between the real and imaginary parts, named the Kramers–Kronig relation, exists. With respect to the fulfillment of these relations, it is necessary to affirm that they are obtained by applying the Cauchy integral formula to $\tilde{\chi}(\omega)$, that is [9],

$$\chi(\omega) = \frac{i}{\pi} P \int_{-\infty}^{\infty} \frac{\chi(y)}{y - \omega} dy$$

where P indicates the Cauchy principal value of the integral. Now, separating the real and the imaginary parts

$$\chi'(\omega) = \frac{1}{\pi} P \int_{-\infty}^{\infty} \frac{\chi''(y)}{y - \omega} dy \quad \text{and} \quad \chi''(\omega) = \frac{1}{\pi} P \int_{-\infty}^{\infty} \frac{\chi'(y)}{y - \omega} dy$$

We get the Kramers–Kronig relation between the real χ' and the imaginary χ'' parts of the electric susceptibility.

In Figure 1.34, an example of the fitting of Equation 1.94 to experimental data to obtain a least-square approximation of the experimental data to the theoretical expression for the real part of the dielectric susceptibility is given [34].

In addition, in Table 1.3, the activation energies for the self-diffusion coefficient for some dehydrated homoionic zeolites calculated by the fitting of Equation 1.90 to experimental data in order to get ω_0 at different temperatures are given; then, with the help of Equation 1.93, the activation energy, E_a, is obtained.

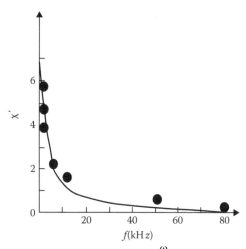

FIGURE 1.34 Fitting of the experimental data χ' vs. $f = \dfrac{\omega}{2\pi}$ corresponding to the sample Na-MOR, at 393 K. (From Roque-Malherbe, R. and Hernández Vélez, M., *J. Thermal Anal.*, 36, 1025, 1990; 36, 2455, 1990.)

TABLE 1.3
Activation Energies for Cationic Self-Diffusion for Some Dehydrated Homoionic Zeolites

Zeolite	$E\left[\dfrac{kJ}{mol}\right]$
Na-HEU	27
Na-MOR	32
Na-FAU	24
Ca-HEU	37

Source: Roque-Malherbe, R. and Hernández Vélez, M., *J. Thermal Anal.*, 36, 1025, 1990.

1.8 NUCLEAR MAGNETIC RESONANCE

1.8.1 INTRODUCTION

The effect of nuclear magnetic resonance (NMR) was at first experimentally demonstrated and explained by Purcell and Bloch in 1945 [37,38]. They received the Nobel Prize in 1952 for this work. Since then, NMR has become a powerful tool in the analysis of the chemical composition and structure of matter, and is widely applied in physics, chemistry, materials science, and biology [39–42].

Atomic nuclei have spin angular momentum, $\overline{I}$, also called nuclear spin [41,42]. In general, the spin is a quantum property of elementary particles with no classical equivalent. It is quantized, and, consequently, can only possess definite discrete values. The nuclei spin angular momentum is a consequence of the particles composing the nucleus, that is, the protons and neutrons which both have spin quantum number 1/2. Consequently, a nucleus with an even mass number, A, has an integral value of the nuclear spin quantum number, I. Conversely, a nucleus with an odd mass number, A, has a half integral value of the nuclear spin quantum number, I. In this sense, nuclei like ^1H, ^{13}C, ^{15}N, ^{19}F, ^{31}P, ^{29}Si, and ^{129}Xe, having $I = 1/2$, are very significant in NMR [39–41].

The quantity I characterizes the scalar magnitude of the spin angular momentum vector, which is given by

$$\left|\overline{I}\right| = \left(I(I+1)\right)^{1/2}\left(\frac{h}{2\pi}\right) = \left(I(I+1)\right)^{1/2}\hbar$$

Nuclei with a nonzero nuclear spin quantum number also have a magnetic moment. The magnitude of the magnetic dipole moment vector is

$$\overline{\mu} = \gamma \overline{I}$$

in which the proportionality constant, γ, is named the magnetogyric ratio, or the gyromagnetic ratio where

$$\gamma = \frac{e}{2m_p}g_N$$

and

  e is the charge of the electron
  m_p is the mass of the proton
  g_N is the nuclear factor of g

Additionally, the scalar magnitude of the magnetic dipole moment vector is given by [43]

$$\left|\overline{\mu}\right| = \gamma \left(I(I+1)\right)^{1/2}\hbar$$

On the other hand, the z-component quantum number is then denoted by m_I, where

$$I_z = m_I\hbar$$

in which $I < m_I < I$.

Owing to the fact that the proton mass is higher than the electron mass, the nuclear magnetic moments are about 2000 times smaller than the electron spin magnetic moments.

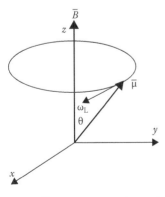

FIGURE 1.35 Precession of nuclear dipole magnetic moment.

1.8.2 Nuclear Zeeman Effect

In the absence of a magnetic field, the nuclear spins are arbitrarily oriented. However, when the spins are placed in an external magnetic field, $\overline{B}$, oriented in the z-direction, the magnetic moments associated with the nuclear spins will interact with the magnetic field (see Figure 1.35) [44]. In this regard, the precession of the nuclear spin effect is very similar to the spinning top in a gravitational field, where the top has a movement of precession about the gravitational field direction.

The energy of a nuclear dipole moment in an external magnetic field, $\overline{B}$, is given by

$$E = -\overline{\mu}\bullet\overline{B} = -\gamma\overline{I}\bullet\overline{B} = -\gamma\left|\overline{I}\right|\bullet\left|\overline{B}\right|\cos\theta = -\gamma\left|\overline{B}\right|m_I\hbar$$

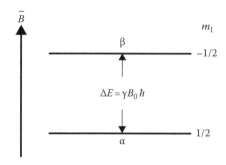

FIGURE 1.36 Energy separation between nuclear spin states for a system immersed in an external magnetic field.

where

θ is the angle between $\bar{I}$ and $\bar{B}$

$|\bar{I}| \cos \theta = m_I$ and $\hbar = \dfrac{h}{2\pi}$

Here, we only analyze nuclei with $I = 1/2$. In Figure 1.36, the diagram of the energy separation between nuclear spin states, for a nuclear spin ($I = 1/2$) in an external magnetic field, is shown. The energy between both states is

$$\Delta E = \gamma \left|\bar{B}\right| \left|\Delta m_I\right| \hbar = \gamma B_0 \hbar$$

where $|\bar{B}| = B_0$ and $|\Delta m_I| = 1$.

This is called the Zeeman splitting, and is shown as an energy level diagram (Figure 1.36). These two states are given a variety of labels, but are commonly referred as "spin down," or state α, and "spin up" or state β, where the spin-down state has a higher energy than the spin-up state. Transitions between the two states can be induced by absorption or emission of a photon, such that $\Delta E = \gamma B_0 \hbar$.

Transitions between the two energy states, spin up and spin down, can occur by absorption or emission of electromagnetic radiation of frequency, ν_L, which is given by the Larmor equation [44]:

$$\nu_L = \frac{\gamma}{2\pi} B_0 \quad \text{or} \quad \omega_L = 2\pi\nu_L = \gamma B_0$$

This frequency depends, for a given species of nuclei, purely on the applied magnetic field. It is the strength of the field experienced by the nucleus that enables the determination of the structure in spectroscopy experiments, and the position in imaging experiments.

1.8.3 Magnetization and Time Evolution of the Magnetization

In a real system, there is not just one isolated nucleus. In reality, there are many nuclei, and all them can occupy a particular spin state. After a certain time after the application of the external magnetic field, the spin system will reach the state of thermal equilibrium with a thermostat. This means that we should consider an ensemble of spins; consequently, applying the canonical ensemble methodology, it is easy to calculate the ratio of the populations of the two spin energy states [45]:

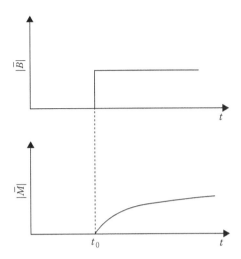

FIGURE 1.37 Development of longitudinal magnetization.

$$\frac{N_{-1/2}}{N_{1/2}} = \exp\left(-\frac{\Delta E}{kT}\right) = \exp\left(-\frac{\hbar\gamma B_0}{RT}\right)$$

Now, since $\hbar\gamma B_0 < RT$, it is possible then to make the following approximation:

$$\frac{N_{-1/2}}{N_{1/2}} = 1 - \left(\frac{\hbar\gamma B_0}{RT}\right)$$

This equation means that the spin populates the two Zeeman energy levels according to the Boltzmann distribution.

The difference in occupation of both states provokes a net magnetization in the system of spins (Figure 1.37) [46].

The net magnetization is developed because the number of spins filling the lower energy level, $N_{1/2}$, outnumbers the amount occupying the upper level, $N_{-1/2}$. Consequently, in equilibrium, a net quantity of spins, $N = N_{1/2} - N_{-1/2}$, will be pointing in the direction of the external field, and then the spin system is magnetized as a paramagnetic material. Then, the z-component of the equilibrium magnetization of the system of spins, which is called the longitudinal magnetization, M_z^0, is given by

$$M_z^0 = (N_{1/2} - N_{-1/2})\mu$$

The spin magnetization vector, $\bar{M}$, will experience a torque, since it is placed in a magnetic field, $\bar{B}$, and will then establish a precession movement. The equation of motion for $\bar{M}$ can subsequently be written as follows (Figure 1.38):

$$\frac{d\bar{M}}{dt} = \gamma\bar{M} \times \bar{B}$$

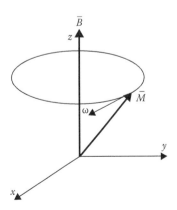

FIGURE 1.38 Precession of the magnetization vector in a static magnetic field aligned along the z-axis.

Now, if in addition to the static magnetic field, $\bar{B} = B_0\bar{k}$, a time varying field is applied along z [47], then

$$\overline{B}_1 = B_1(\cos \omega_0 t)\overline{i} - B_1(\sin \omega_0 t)\overline{j}$$

is applied perpendicularly to $\overline{B} = B_0\overline{k}$, and oscillating at ω_0, then the total magnetic field applied to the magnetization field will be

$$\overline{B}_T = B_1(\cos \omega_0 t)\overline{i} - B_1(\sin \omega_0 t)\overline{j} + B_0\overline{k}$$

Consequently, the equation of motion of the magnetization field will be now

$$\frac{d\overline{M}}{dt} = \gamma\overline{M} \times \overline{B}_T = \gamma\overline{M} \times \left(B_1(\cos \omega_0 t)\overline{i} - B_1(\sin \omega_0 t)\overline{j} + B_0\overline{k}\right)$$

1.8.4 NUCLEAR MAGNETIC RESONANCE EXPERIMENT

In order to detect the NMR signal, it is necessary to have a radio frequency (r.f.) coil in the transverse plane, that is, perpendicular to the static magnetic field, $\overline{B} = B_0\overline{k}$, which runs through the z-axis, and with the help of this coil, an electromagnetic field is induced (Figure 1.39) [42].

A coil located around the x-axis will produce a magnetic field, which alternates in direction when an alternating r.f. current is passed through the coil (see Figure 1.39) [42]. In a frame of reference turning about the z-axis at a frequency identical to that of the alternating current, the magnetic field along the x'-axis will be constant, just as in the direct current case in the laboratory frame [47,48].

Now, once the alternating current passing through the coil is connected or disconnected, it produces a pulsed $\overline{B}_1$ magnetic field along the x'-axis. Therefore, the system of spins included in the atoms composing the sample under test in the NMR experiment reacts to this pulse causing the net magnetization vector to rotate about the direction of the applied magnetic field $\overline{B}_1$. As a result, the rotation angle is reliant on the pulse time-span, τ, and its magnitude, B_1, as follows [42,47]:

$$\theta = 2\pi\gamma\tau B_1$$

A 90° pulse is one which rotates the magnetization vector clockwise by 90° about the x'-axis, that is, a 90° pulse rotates the equilibrium magnetization down to the y'-axis.

1.8.5 SPIN-LATTICE RELAXATION TIME (T_1), SPIN–SPIN RELAXATION TIME (T_2), AND THE BLOCH EQUATIONS

As was previously stated,

$$M_z^0 = \left(N_{1/2} - N_{-1/2}\right)\mu$$

where $\dfrac{N_{-1/2}}{N_{1/2}} = \exp\left(-\dfrac{\hbar\gamma B_0}{RT}\right)$, in which $N_{1/2}$ and $N_{-1/2}$ are referred

as unit volumes, and $\mu = |\overline{\mu}| = \gamma|\overline{I}|$.

A 90° pulse excites enough low-energy nuclei to equalize the populations of the spin-up and spin-down states, that is, $N_{1/2} \approx N_{-1/2}$ [42]. Consequently, the equilibrium state will be disrupted. Subsequently, since the application of a resonant r.f. pulse will disturb the spin system, there must be a process of coming back

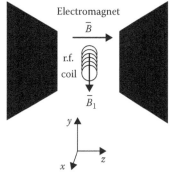

FIGURE 1.39 Schematic representation of a nuclear magnetic resonance equipment.

to equilibrium. This relaxation process involves the exchange of energy between the spin system and its surroundings.

Such a process is called spin-lattice relaxation, in which the rate at which equilibrium is restored is characterized by the spin-lattice or longitudinal relaxation time, T_1, and the rate of approximation to equilibrium is governed by the following equation [2]

$$\frac{dM_z}{dt} = \frac{M_z^0 - M_z}{T_1}$$

where T_1 is the spin-lattice relaxation time. The solution of the previous equation gives the dependence of time, t, on the longitudinal magnetization:

$$M_z = M_z^0 \left(1 - \exp\left[-\frac{t}{T_1} \right] \right)$$

On the other hand, the components of the magnetization in the x-axis, M_x, and the y-axis, M_y, are called the transverse magnetization [48].

Now, it is necessary to assert that in addition to the rotation, the magnetization de-phases, since each of the spin packets experience a somewhat dissimilar magnetic field. This effect is produced because the spin systems do not only exchange energy with the surrounding lattice, but also among themselves, and consequently rotate at slightly different Larmor frequencies [2,47]. Therefore, the longer the elapsed time, the bigger the phase difference.

Consequently, the spin systems not only exchange energy with the surrounding lattice, but also among them. This exchange of energy between them is generally a faster process than the spin-lattice relaxation process, and it is characterized by the spin–spin relaxation time, T_2. Then, the rate of approximation to equilibrium, in the case of the spin–spin relaxation, is governed by the following equations [2,47]:

$$\frac{dM_x}{dt} = -\frac{M_x}{T_2} \quad \text{and} \quad \frac{dM_y}{dt} = -\frac{M_y}{T_2}$$

where

T_2 is the spin–spin relaxation time
M_x and M_y, are the components of the magnetization in the, x- and y-axis

and the solution of the previous equation gives the dependence of time, t, of the transversal magnetization

$$M_x = M_x^0 \exp\left[-\frac{t}{T_2} \right] \quad \text{and} \quad M_y = M_y^0 \exp\left[-\frac{t}{T_2} \right]$$

Then, the magnetization in the x- and y-planes moves to zero, and subsequently the longitudinal magnetization increases until we get M_z^0 along the z-axis.

Now, in order to combine the spin-lattice and spin–spin natural processes occurring in a spin system with the conditions imposed in an NMR experiment, the Bloch equations are suggested:

$$\frac{dM_x}{dt} = (\gamma \overline{M} \times \overline{B})_x - \frac{M_x}{T_2}$$

$$\frac{dM_y}{dt} = (\gamma \overline{M} \times \overline{B})_y - \frac{M_y}{T_2}$$

and

$$\frac{dM_z}{dt} = (\gamma \overline{M} \times \overline{B})_z + \frac{M_0 - M_z}{T_1}$$

These are a set of coupled differential equations, and when these set of equations are correctly integrated, they will yield the components of the magnetization as a function of time.

1.9 MÖSSBAUER EFFECT

1.9.1 INTRODUCTION

In 1957, Rudolph Mössbauer, during his graduate studies, discovered an outstanding effect [49] that has generated an entire field in physics, that is, a very high-resolution spectroscopy in the γ-ray region of the spectrum named Mössbauer spectroscopy [50–56]. The effect consists of the fact that a γ photon emitted by an excited nucleus can be resonantly absorbed by another nucleus [50–56]. This means that a recoilless emission and absorption has occurred.

This methodology has been applied in many areas, such as the measurement of lifetimes of excited nuclear states and nuclear magnetic moments, the investigation of electric and magnetic fields in atoms and crystals, in the analysis of special relativity, the equivalence principle, and also in other applications [50–57].

1.9.2 MÖSSBAUER EFFECT

As is very well known, during a transition involving an excited and a ground or less excited energy levels, the system emits a photon. The energy distribution in the frequency space of the photon has a natural line width, Γ, ensuing from the finite lifetime, τ, of the excited state (Figure 1.40).

If the emitted photon hits on a system of the same nature as the one emitting, but in its ground state; after that, exists a certain probability that it will be absorbed, then lifting up the system to an excited state, a phenomenon called resonant absorption of photons.

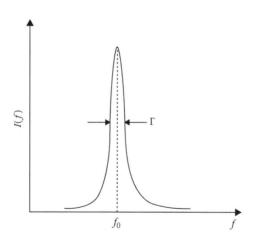

FIGURE 1.40 Emission spectrum.

$$p' = mv \qquad p_\gamma^e = \frac{E_\gamma^e}{c}$$

FIGURE 1.41 Recoil of the emitter during an emission transition.

However, in a transition, not the entire energy change during the emission transition gets to the photon, as energy and momentum are conserved during the transition, and a certain quantity of energy goes to the recoil of the emitter system (Figure 1.41).

For a system originally at rest, the conservation of momentum implies

$$p' = mv = p_\gamma^e = \frac{E_\gamma^e}{c}$$

where
E_γ^e is the energy of the emitted photon
m is the mass of the emitter
c is the speed of light

Besides, the conservation of the energy means that the energy of the emitted photon will be given by [56]

$$E_\gamma^e = E_\gamma - E_R$$

where E_R, the recoil energy, is

$$E_R = \frac{(E_\gamma^e)^2}{2mc^2} \approx \frac{(E_\gamma)^2}{2mc^2}$$

The emitter particle recoils with a momentum E_γ^e/c.

Similarly, for an absorbing particle initially at rest, when the photon is absorbed, the absorber will recoil. Then, the distributions of the emission and absorption energies are, consequently, separated by two times the recoil energy, $2E_R$ (see Figure 1.42). In this case, the chance of resonant absorption is proportional to the overlap of both distributions.

For atomic transitions, the recoil energies are small in comparison with the natural line widths. Therefore, in this case there is a high chance of resonant absorption during the emission and absorption processes. In contrast, for nuclear transitions in the case of gases, the emitting and absorbing particles are atoms; subsequently, the mass is somewhat small, resulting in a large recoil energy. Accordingly, there is generally no resonant absorption for these nuclear transitions, since the recoil energies are by far larger than the natural line width and there is a small or no overlap of transition energy probabilities (see Figure 1.42).

Mössbauer's discovery [49] consisted in the fact that when the nuclei of the emitter and the absorber are included in a solid matrix, they vibrate in a crystal lattice [49,54,56]. Therefore, owing to the essential quantum character of solid vibrations (see Section 1.4), the atoms located in a solid matrix are limited to a certain collection of quantized lattice vibration energies [54]. Consequently, if the recoil energy is smaller than the lowest quantized lattice vibration energy, E_V, then $E_V = k\Theta_D$, in which, k is the Boltzmann constant and Θ_D is the Debye temperature of the solid. In this case, this

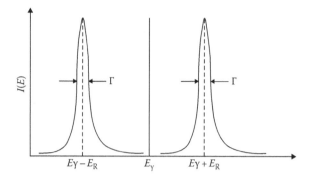

FIGURE 1.42 Emission and absorption with recoil.

energy is not enough to excite the crystal lattice to the next vibration state. Therefore, a fraction of the nuclear transitions occurs, as if the entire crystal acts as the recoiling particle, sharing with the entire crystal the momentum of the γ photon [54,56].

Since the mass of the solid, M, is particularly large in comparison to that of a solitary atom, in this case

$$E_E = \frac{(E_\gamma^e)^2}{2Mc^2} \ll \frac{(E_\gamma)^2}{2mc^2}$$

Consequently, resonance emission and absorption can take place (see Figure 1.42).

The probability, f, of a recoil-free process in a solid is given by the previously studied Debye–Waller factor [3,56]

$$f(T) = \left(\frac{<x^2>}{(\bar{\lambda})^2} \right)$$

where $<x^2>$ is the mean-square deviation of the atoms from the equilibrium positions in the crystal lattice and $(\bar{\lambda})^{-2} = (2\pi\nu/c)^2$ is the inverse of the square of the wavenumber of the emitted γ photon. The probability, f, of a recoil-free process at $T = 0\,K$ is given by [56]

$$f(0) = \left(\frac{3E_R}{2k\Theta_D} \right)$$

where
  k is the Boltzmann constant
  Θ_D is the Debye temperature

As a matter of fact, Mössbauer isotopes are those that exhibit an appreciable probability, f, for a recoil-free process when they are included in a solid. The elements that have Mössbauer isotopes are Fe, Ru, Sn, Sb, Te, I, W, Ir, Au, Eu, Gd, Dy, Er, Yb, and Np [54]. However, the most extensively used isotope is ^{57}Fe.

REFERENCES

1. D. Schwarzenbach, *Crystallography*, John Wiley & Sons, New York, 1997.
2. Ch. Kittel, *Introduction to Solid State Physics* (8th edition), John Wiley & Sons, New York, 2004.

3. H.P. Myers, *Introduction to Solid State Physics* (2nd edition), CRC Press, Boca Raton, FL, 1997.

4. R. Asokami, *Solid State Physics*, Anshan Publishing, Tunbridge Wells, U.K., 2006.

5. J.M. Ziman, *Principles of the Theory of Solids* (2nd edition), Cambridge University Press, Cambridge, UK, 1972.

6. J.I. Gersten and F.W. Smith, *The Physics and Chemistry of Materials*, John Wiley & Sons, New York, 2001.

7. R.P. Feynman, *Statistical Mechanics. A Set of Lectures*, W.A. Benjamin Inc., Reading, MA, 1972.

8. H. Goldstein, C.P. Poole, and J.L. Safko, *Classical Mechanics* (3rd edition), Addison-Wesley, New York, 2001.

9. G.B. Arfken and H.J. Weber, *Mathematical Methods for Physicists* (5th edition), Harcourt/Academic Press, San Diego, CA, 2001.

10. A. Messiah, *Quantum Mechanics*, Dover Publications, New York, 1999.

11. R. Roque-Malherbe, *Adsorption and Diffusion in Nanoporous Materials*, CRC Press, Boca Raton, FL, 2007.

12. D.A. McQuarrie, *Statistical Mechanics*, University Science Books, Sausalito, CA, 2000.

13. I.N. Levine, *Physical Chemistry* (5th edition), McGraw-Hill, Boston, MA, 2002.

14. A. Beiser, *Concepts of Modern Physics* (6th edition), McGraw-Hill, Boston, MA, 2003.

15. S.O. Kasap, *Principles of Electronic Materials* (2nd edition), McGraw-Hill Higher Education, New York, 2002.

16. N.W. Ashcroft and N.D. Mermin, *Solid State Physics*, Brooks/Cole, Belmont, MA, 1976.

17. F. Reif, *Fundamentals of Statistical and Thermal Physics*, McGraw-Hill, Boston, MA, 1965.

18. Ch. Kittel, *Quantum Theory of Solids* (2nd edition), John Wiley & Sons, New York, 2004.

19. I.N. Levine, *Quantum Chemistry* (5th edition), Prentice Hall, Upper Saddle River, NJ, 2001.

20. A. Guinier, *X-Ray Diffraction in Crystals, Imperfect Crystals and Amorphous Bodies*, Dover Publications Inc., New York, 1994.

21. B.D. Cullity and S.R. Stock, *Elements of X-Ray Diffraction* (3rd edition), Prentice Hall, Upper Saddle River, NJ, 2001.

22. R.C. Reynolds, in *Modern Powder Diffraction*, D.L. Bish and J.E. Post, (editors), *Reviews in Mineralogy*, Vol. 20, The Mineralogical Society of America, BookCrafters Inc., Chelsea, MI, 1989, p. 1.

23. L.H. Schwartz and J.B. Cohen, *Diffraction from Materials* (2nd edition), Springer-Verlag, New York, 1987.

24. R. Jenkins and R.L. Snyder, *Introduction to X-Ray Powder Diffractometry*, John Wiley & Sons, New York, 1996.

25. B.E. Warren, *X-ray Diffraction*, Addison-Wesley, Reading, MA, 1969.

26. M. Birkholz, *Thin Film Analysis by X-Ray Scattering*, John Wiley & Sons, New York, 2006.

27. A. Lagendijk, B. Nienhuis, B.A. van Tiggelen, and P. de Vries, *Phys. Rev. Lett.*, 79, 657 (1997).

28. R. Roque-Malherbe, in *Handbook of Surfaces and Interfaces of Materials*, Vol. 2, H.S. Nalwa, (editor), Academic Press, New York, Chapter 13, 2001, p. 509.

29. A.K. Jonscher, *Dielectric Relaxation in Solids*, Chelsea Dielectric Press, London, UK, 1983.

30. A.K. Jonscher, *The Universal Dielectric Response. A Review of Data and Their New Interpretation*, Chelsea Dielectric Group, Pulton Place, London, UK, 1978.

31. K.Ch. Kao, *Dielectric Phenomena in Solids*, Elsevier, Amsterdam, the Netherlands, 2004.

32. W.S. Zaengl, 12th International Symposium on High Voltage Engineering, Bangalore, India, 20–24 August, 2001 Keynote Speeches, Session 9.

33. P. Debye, *Polar Molecules*, Chemical Catalog Co., New York, 1929.

34. R. Roque-Malherbe and M. Hernández Vélez, *J. Thermal Anal.*, 36, 1025 (1990); 36, 2455 (1990).

35. A.T. Fromhold, *Theory of Metal Oxidation*, North Holland Pub. Co., New York, 1976.

36. L.D. Landau and E.M. Lifshitz, *Statistical Physics*, Pergamon Press, New York, 1958.

37. E.M. Purcell, H.C. Torrey, and R.V. Pound, *Phys. Rev.*, 69, 127 (1946).

38. F. Bloch, W.W. Hansen, and M. Packard, *Phys. Rev.*, 69, 37 (1946).

39. P.J. Hore, *Nuclear Magnetic Resonance*, Oxford University, Oxford, UK, 1995.

40. J.H.H. Nelson, *Nuclear Magnetic Resonance Spectroscopy*, Prentice Hall, Upper Saddle River, NJ, 2002.

41. B. Cowan, *Nuclear Magnetic Resonance and Relaxation*, Cambridge University Press, Cambridge, UK, 2005.

42. H. Günther, *NMR Spectroscopy* (2nd edition), John Wiley & Sons, New York, 2001.

43. P.W. Atkins, *Physical Chemistry* (6th edition), W.H. Freeman and Co., New York, 1998.

44. S.T. Thornton and A. Rex, *Modern Physics* (2nd edition), Thomson Learning, Singapore, 2002.

45. T.L. Hill, *An Introduction to Statistical Thermodynamics*, Dover Publications, New York, 1986.

46. A.C. Larsson, A nuclear magnetic resonance study of dialkyldithiophosphate complexes, PhD thesis, Lulea University of Technology, Department of Chemical Engineering and Geosciences, Division of Chemistry, Porsön, Lulea, Sweden, 2004.

47. J.P. Hornak, *The Basics of Nuclear Magnetic Resonance*, www.cis.rit.edu/htbooks/nmr/inside.htm, 2002.

48. D.A. Skoog, F.J. Holler, and T.A. Nieman, *Principles of Instrumental Analysis* (5th edition), Saunders College Publishing, Philadelphia, PA, 1998.

49. R. Mössbauer, *Z. Physik*, 151, 124 (1958).

50. H. Frauenfelder, *The Mössbauer Effect*, W. A. Benjamin, New York, 1962.

51. G.K. Wertheim, *Mössbauer Effect Principles and Applications*, Academic Press, New York, 1964.

52. U. Gonser (editor), *Mössbauer Spectroscopy*, Springer-Verlag, New York, 1975.

53. T.C. Gibb, *Principles of Mössbauer Spectroscopy*, Chapman & Hall, London, 1977.

54. D.P. Dickson and F.J. Berry, (editors) *Mössbauer Spectroscopy*, Cambridge University Press, Cambridge, UK, 1986, p. 1.

55. A.G. Maddock, *Mössbauer Spectroscopy: Principles and Applications of the Techniques*, Horwood Publishing Limited, Westergate, Chichester, UK, 1997.

56. A.C. Melissinos and J. Napolitano, *Experimental Modern Physics* (2nd edition), Academic Press, New York, 2003.

57. R.V. Parish, in *Mössbauer Spectroscopy*, D.P. Dickson and F.J. Berry, (editors), Cambridge University Press, Cambridge, UK, 1986, p. 17.

2 Structure of Adsorbents, Ion Exchangers, Ion Conductors, Catalysts, and Permeable Materials

2.1 INTRODUCTION

The creation and manipulation of materials is an essential goal of mankind. Even, complete epoch has been designed by the basic material used by humanity to construct his tools and weapons. In this regard, new materials are essential objectives of materials science research [1–22]. In order to develop new materials, modern materials science has become interdisciplinary, that is, physics, chemistry, and engineering are in some way blended in order to get a final product, which requires the following fundamental components: synthesis, characterization, study of the obtained material's properties, and applications. However, to scientifically understand and master the preceding operations, it is necessary to thoroughly study the structure of the materials of our interest. To continue, the description of physical chemistry of materials, here, we study the structures of the materials of interest, that is, adsorbents, ion exchangers, ion conductors, catalysts, and permeable materials. These materials have many applications in numerous areas that are significant for chemical, sustainable energy, and pollution abatement applications, such as ion conduction, ion exchange, gas separation, membrane reactors, coatings, catalysts, catalyst supports, and sensors.

Among the materials belonging to these groups, we concentrate on those of an inorganic nature. However, at the end of this chapter, we review the fundamental aspects of polymer structure.

2.2 TRANSITION METAL CATALYSTS

2.2.1 METALLIC CATALYSTS' PERFORMANCE

These catalysts are composed of one or several metallic active components, deposited on a high surface area support, whose purpose is the dispersion of the catalytically active component or components and their stabilization [23–27]. The most important metallic catalysts are transition metals, since they possess a relatively high reactivity, exhibit different oxidation states, and have different crystalline structures. In this regard, highly dispersed transition clusters of metals, such as Fe, Ru, Pt, Pd, Ni, Ag, Cu, W, Mn, and Cr and some alloys, and intermetallic compounds, such as Pt–Ir, Pt–Re, and Pt–Sn, normally dispersed on high surface area supports are applied as catalysts.

The dispersion of the catalytic metal is necessary, because these metals are often expensive. Thereafter, if it is applied in a finely dispersed form as particles on a high surface area support, a large fraction of the metal atoms are exposed to the reactant molecules [26]. Then, we save in metal amount and increase the efficiency of the catalyst.

In catalysis, one of the key concepts to understand catalytic action is the so-called active sites [28]. The essential concept behind this term is the fact that catalytic activity in solids is restricted to specific sites in the catalyst surface. Another factor influencing catalytic activity is the geometric factor, that is, a properly spaced array of atoms on the solid surface, named Balandin multiplets, exists with which the reactant interacts [29]. Another factor is the electronic factor, which deals with the correlation between catalytic activity and bulk electronic properties of the catalysts.

Transition metals are active as catalysts because of their capacity to chemisorb atoms, given that the main role of transition metals as catalysts is to atomize gaseous molecules, such as H_2, O_2, N_2, and CO, thereby providing atoms to other reactants and reaction intermediates [27].

Transition metal nanoparticles supported on different substrates are used as catalysts for different reactions, such as hydrogenations and enantioselective-synthesis of organic compounds, oxidations and epoxidations, reduction, and decomposition [24,25]. Among the supports that have been applied in the preparation of supported transition metal nanoparticles are active carbon, silica, titanium dioxide, and alumina.

Chemisorption is vital in catalysis. Transition metals such as Fe, Pd, Pt, Ir, Ni, Co, Cr, Mn, Ti, Hf, Zr, V, Nb, Mo, W, Ru, and Os have the ability to chemisorb simple molecules such as O_2, CO, H_2, CO_2, C_2H_4, and N_2 [24,25,27]. However, if chemisorption is very strong, the catalytic sites are blocked. Therefore, it is necessary that an intermediate between weak chemisorption, when there is no reaction, and strong chemisorption, when the sites are blocked [24,25], is available. In this sense, the first d-block metals form especially stable surface bonds, while the noble metals form weak bonds. These properties are unfavorable to catalysis. Hence, the best metallic catalysts are in between these two groups [27].

2.2.2 BAND STRUCTURE OF TRANSITION METALS

The elevated catalytic activities of transition metals have been attributed to their partially unoccupied d-electron states. Then, information regarding the band structure of these metals is obviously of interest for understanding the catalytic action of these metals [30–34].

In general terms, transition metals are those which have incompletely filled d-bands. The progression in the filling of the d-band in the first long-transition metal series is as follows: Ti(HCP), V(BCC), Cr(BCC), Fe(BCC), Co(FCC), Ni(FCC), Cu(FCC), Zn(HCP), and is not highly influenced by the structural difference between the body-centered cubic (BCC) and face-centered cubic (FCC) lattices. However, this is not the case for the hexagonal close-packed (HCP) lattice [10]. An analogous pattern is expected for the second and third series.

The band system is schematically represented in Figure 2.1 [35]. The difference in the band structures are the position of the Fermi level, which increases in energy from Ti to Zn, and the band width; in addition, these d-electrons have a mixed character, that is, they can move as free electrons, as well as exhibit a localized character [10].

The heat of adsorption increases from Zn to Ti, that is, from right to left in the periodic table [24], since the interaction between the metal and the chemisorbed molecule is principally determined by the so-called one-electron energy term [36]. That is, in a transition metal surface, the d-electrons lie in a band around the Fermi level (see Figure 2.1) where they can interact with the adsorbate states forming bonding and antibonding states [35]. In this situation, the d-electron contribution to the surface chemical bond is proportional to $(1 - f_d)$, where f_d is the degree of filling of the d-band. Therefore, the first transition elements with less d-electrons, that is, with lower degree of filling, form stronger chemical bonds [24]. Both strong and weak chemical bonds are unfavorable for the catalytic action of a metal.

2.2.3 BODY-CENTERED CUBIC IRON AS A CATALYST

In the Haber–Bosch process, ammonia is formed from the reaction between N_2 and H_2, using a Fe_3O_4 (magnetite) catalyst promoted with Al_2O_3, CaO, K_2O and a moderately small amount of iron and other elements [25]. In this mixture, the catalytically active

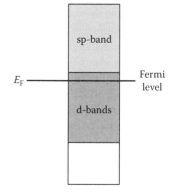

FIGURE 2.1 Schematic representation of d-bands and the sp-band.

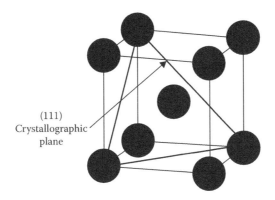

FIGURE 2.2　BCC α-Fe structure.

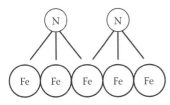

FIGURE 2.3　Chemisorbed nitrogen on an iron atom surface.

phase is iron, which is formed from the magnetite through a reduction produced by the reactant mixture. Since the mechanism of the reaction is related to the velocity of dissociative nitrogen chemisorption, the reaction is related with the structure of iron and the band structure of this metal.

The iron structure is BCC up to 912°C, and is called α-Fe (see Figure 2.2). Besides, the electronic arrangement of iron is ${[Ar]3d^6 4s^2}$. The catalytic properties of this metal are closely related to its electronic configuration and consequently to its band structure. It is a well-known fact that d-electrons contribute to the bonding of atoms and molecules to the iron surface. That is, the chemisorption of molecules on the iron surface (see Figure 2.3) is one crucial step in catalysis with metals since the chemisorbed molecules are atomized, and consequently become more reactive. The dissociative chemisorption of nitrogen on iron atoms, shown in Figure 2.3 [27], is the crucial step in ammonia synthesis and ammonia dissociation.

Another important factor in the catalytic properties of Fe and other metallic catalysts is the structure of these catalysts. It is a well-known fact that during ammonia synthesis, when it is exposed on the surface of a Fe catalyst's (111) plane, the catalyst is more active [14].

In general, during the preparation of metallic catalysts, more than one type of crystallographic planes are uncovered, each of which exposes a characteristic pattern of atoms, for example, the crystallographic planes (111), (110), (100), and (010) [27].

Besides, the geometric factor is important in the catalytic activity of iron, that is, the catalyst surface has irregularities such as steps and kinks that expose atoms with low coordination numbers which appears to be particularly reactive [22].

2.2.4　FACE-CENTERED CUBIC PLATINUM AS A CATALYST

One of the most extensively studied reactions in the past years, and whose elementary steps appears to be best understood, is the catalytic oxidation of carbon monoxide on a Pt catalyst [23]:

$$2CO + O_2 \leftrightarrow 2CO_2$$

This reaction is a very important process in automobile catalytic converters.

The platinum structure is FCC (see Figure 2.4) and the electronic arrangement of Pt is ${[Xe]5d^9 6s^1}$.

The band structure of platinum is also the cause of its catalytic properties. Platinum does not form very strong, nor weak bonds, and is, consequently, one of the most active among the transition metals in several reactions. In fact, platinum is a multipurpose, heterogeneous metallic catalyst,

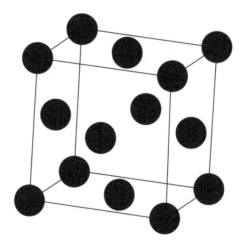

FIGURE 2.4 FCC Pt structure.

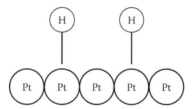

FIGURE 2.5 Chemisorbed hydrogen on a surface composed of platinum atoms.

possibly the most versatile metallic catalyst. It is used in reducing and oxidizing conditions, for example, for dehydrocyclization, isomerization, and hydrogenation of hydrocarbons and oxidation of NH_3 and CO, as was previously noted [14,22].

Platinum is active as a catalyst because of its capacity to chemisorb atoms, that is, in some case its role as catalyst is to atomize gaseous molecules, such as H_2, O_2, N_2, and CO, giving atoms to other reactants and reaction intermediates (see Figure 2.5) [14,27]. Nickel and palladium, which have the same position as platinum in the first and second series of transition elements and the same FCC structure, have catalytic properties very similar to those of platinum.

2.2.5 Hexagonal Close-Packed Cobalt as a Catalyst

Cobalt is a good catalyst for hydrogenation reactions; it is used in the synthesis of low molecular weight hydrocarbons and for the conversion of nitriles in primary amines [37]. It can as well be applied in Fisher–Tropsch synthesis [30]. The structure of cobalt is HCP (see Figure 2.6) and the electronic arrangement of Co is $\{[Ar]3d^74s^2\}$.

2.2.6 Balandin Volcano Plot

The Balandin volcano plot illustrates a relation between a certain type of catalytic activity index, A_C^X, versus some enthalpic function, H_C^X, related with the heat of chemisorption [25]. In Figure 2.7

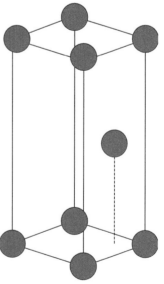

FIGURE 2.6 HCP structure of Co.

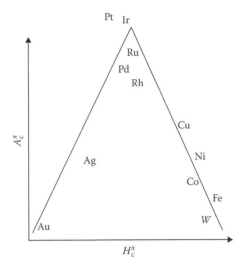

FIGURE 2.7 Balandin volcano plot of A_C^X vs. H_C^X.

(volcano plot), a peak of maximum catalytic activity for transition metals located in the middle of the periodic table, that is, Pt(FCC), Pd(FCC), Rh(FCC), Ir(FCC), Ru(HCP) [22,24,25,27], is evident. This fact is not closely related with the crystalline structure of the transition metal.

2.3 NONMETALLIC CATALYSTS

2.3.1 SIMPLE OXIDES

The theory of catalysis on oxides has been developed by Hauffe [38], Wolkenstein [39], and others. Normally, the oxides applied in catalysis are semiconductors with chemical formulas AO, AO_2, A_2O_3, A_3O_4, etc., where A denotes the metallic element.

The composition of these oxides normally departs from the precise stoichiometry, expressed in their chemical formulae. For example, in the case of a stoichiometric oxide, such as AO_δ, where $\delta = 0$, we will have only thermal disorder, where the concentration of vacancies, and interstitials will be determined by the Schottky, Frenkel, and anti-Frenkel mechanisms [40–42] (these defects are explained in more detail in Chapter 5). In the case of the Schotky mechanism, the following equilibrium, described with the help of the Kroger–Vink notation, [43] develops [40]

$$\text{nil} \leftrightarrow V_A'' + V_O^{\bullet\bullet}$$

which describes the formation of a cation, V_A'', and an anion, $V_O^{\bullet\bullet}$, vacancies, from the neutral lattice, nil. The Frenkel equilibrium is expressed as follows [41]

$$A_A^x + V_i^x \leftrightarrow V_A'' + A_i^{\bullet\bullet}$$

which explains the formation of a cation, V_A'', vacancy and a cation, $A_i^{\bullet\bullet}$, interstitial, from a neutral lattice A_i^x cation position and a neutral lattice V_i^x interstitial site. These cation vacancies are twice negatively charged, and these anion interstitial vacancies are twice positively charged with respect to the neutral lattice. The anti-Frenkel equilibrium is described by [41]

$$O_O^x + V_i^x \leftrightarrow V_O^{\bullet\bullet} + O_i''$$

which gives a description of the formation of an anion, $V_O^{\bullet\bullet}$, vacancy and an anion, $O_i^{''}$, interstitial, from a neutral lattice O_i^x anion position and a neutral lattice O_i^x interstitial site.

The catalytic activity of these simple oxides has been correlated [38,39] with its ability to chemisorb simple molecules such as CO and N_2O via an electron transfer, which results in a change in the electron transport properties in the semiconductor solid oxide. In this sense, the catalytic activity of simple oxides has been correlated with the band gap, and the number of d-electrons for 3d metal oxides [22].

2.3.2 ROCK-SALT-STRUCTURE CATALYSTS

Magnesium oxide (MgO) is a good example of a catalyst with rock salt structure, that is, NaCl-type structure (Figure 2.8).

The NaCl-type structure of MgO is characterized by an FCC cubic close-packed (CCP) anionic framework of O^{2-}, with all the octahedral positions filled with Mg^{2+}. In the CCP structure, there are four octahedral sites per unit cell and eight tetrahedral sites (see Figures 1.6 and 1.7).

Magnesium oxide doped with Li is used as a catalyst for the oxidative conversion of methane to higher hydrocarbons [44] and for the selective oxidation of methane [25].

2.3.3 RUTILE-TYPE CATALYST

There are three TiO_2 polymorphs, namely, rutile, anatase, and brookite; an additional synthetic phase called TiO_2 (B); and some high-pressure polymorphs have been also reported [45,46]. Both the rutile and anatase polymorphs have photocatalytic properties; however, these properties are more developed in the anatase polymorph [46].

The rutile structure is tetragonal (see Figure 2.9). That is, it is formed by a cationic framework of a tetragonal body-centered structure with the Ti^{4+} cation, located in the lattice sites and the O^{2-} anions arranged as a TiO_6 octahedron connected through opposite edges alongside the c-axis.

The n-type semiconductor properties of TiO_2-based materials are basic in realizing the photocatalytic functions [45]. Rutile is a semiconductor with a band gap of 3.10 eV [47]. By introducing particular impurities of C and/or N in the TiO_2 structure, the band gap can be changed [45,47].

Heterogeneous photocatalysis using titanium dioxide as the photocatalyst is initiated by the production of electron–hole pairs, as follows (Figure 2.10) [48]:

$$TiO_2 \xrightarrow{h\nu} TiO_2[e^-(CB) + h^+(VB)]$$

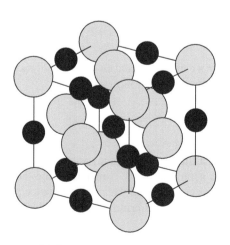

FIGURE 2.8 NaCl-type structure of MgO.

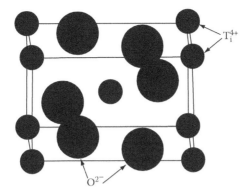

FIGURE 2.9   Rutile titanium dioxide.

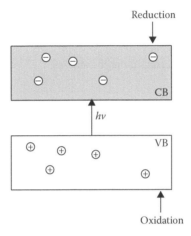

FIGURE 2.10   Oxidation–reduction by a photocatalyst.

In the case of water splitting, this process takes place, in most cases, through oxidation by hydroxyl radicals formed when the photogenerated holes interact with adsorbed water in the following way (Figure 2.11) [48,49]:

$$4h^+(\text{VB}) + 2\text{H}_2\text{O} \rightarrow \text{O}_2 + 4\text{H}^+$$

That is, photocatalytic water splitting is an oxidation–reduction process, where the oxidation process takes place in the valence band and the reduction process in the conduction band, in the subsequent manner (see Figure 2.11):

$$4e^-(\text{CB}) + 4\text{H}^+ \rightarrow 2\text{H}_2$$

Pt-loaded TiO_2 has been known to decompose water to O_2 and H_2, by the UV-light irradiation of H_2O vapor (see Figure 2.11) [50].

2.3.4   CORUNDUM-TYPE CATALYSTS

Corundum is one of the many polimorphs of alumina (Al_2O_3). The corundum-type structure is the structural shape of hematite ($\alpha\text{Fe}_2\text{O}_3$). In Figure 2.12, layers A and B of a close-packed arrangement of spheres, and the formation of the corresponding octahedral and tetrahedral sites are shown [51,52].

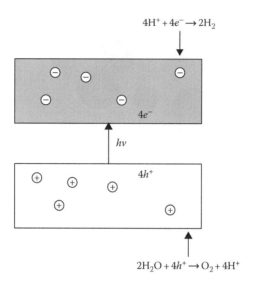

$$4H^+ + 4e^- \rightarrow 2H_2$$

$4e^-$

hv

$4h^+$

$$2H_2O + 4h^+ \rightarrow O_2 + 4H^+$$

FIGURE 2.11 Water splitting by semiconductor photocatalysts.

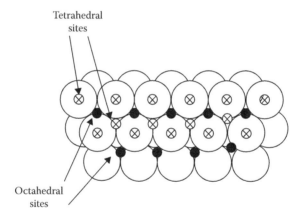

Tetrahedral sites

Octahedral sites

FIGURE 2.12 Tetrahedral and octahedral sites in the HCP structure.

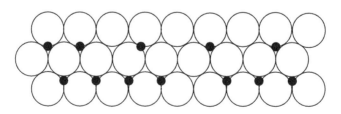

FIGURE 2.13 A Layer of O^{2-} with 2/3 of the octahedral sites occupied by Fe^{3+}.

The corundum-type structure of αFe_2O_3 is formed by a HCP array of O^{2-} anions with Fe^{3+} cations filling 2/3 of the octahedral sites located between the contiguous A, B sequence of layers (see Figure 2.13).

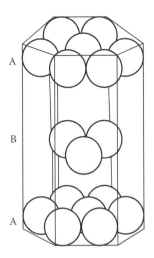

FIGURE 2.14 Three-dimensional HCP structure.

In the HCP crystal structure, similar to the CCP, the coordination number of the lattice sites is 12, and two octahedral sites per basic hexagonal prism, that is, six per each HCP cell are present (see Figure 2.14).

Iron oxide with a corundum-type structure, that is, αFe_2O_3, is a good catalyst for reactions such as the catalytic oxidation of SO_2 [53]. Studies carried out on this reaction show that SO_2 adsorption is the rate-determining step; it contributes electrons to the conduction band and the adsorption of O_2 withdraws the conduction electrons from an oxygen vacancy [53].

2.3.5 WURTZITE-TYPE CATALYSTS

Zinc oxide (ZnO) is a versatile catalyst that crystallizes to the wurtzite-type structure [54–56]. In the structure of ZnO, each one of the zinc and oxygen ions is tetrahedrally coordinated. To be precise, in the wurtzite-type structure, a HCP framework of O^{2-} anions is formed, where half, 1/2, of the 12 tetrahedral sites present in the HCP structure (see Figure 2.15) are occupied by Zn^{2+} cations [52].

ZnO with wurtzite-type structure is a good photocatalyst [54,55], besides in combination with Cu, it is used as a reforming catalyst [56]. As a photocatalyst, ZnO is efficient for the photodecomposition of organic compounds, but in comparison with TiO_2, ZnO shows appreciable instability during irradiation [54].

2.3.6 FLUORITE-TYPE CATALYSTS

Cerium oxide, or ceria (CeO_2), is a component of an autoexhaust catalyst that crystallizes to the fluorite structure [45]. In the fluorite structure of CeO_2, a CCP framework of Ce^{4+} ions is formed, where the eight tetrahedral sites present in the FCC structure are occupied by O^{2-} ions (see Figure 2.16, where the links between atoms to make the tetrahedral positions clear are shown) [52].

CeO_2 with fluorite-type structure is a component in NO_2 dissociation catalysts [57] and biomass gasification catalysts [58].

2.3.7 SPINEL-TYPE CATALYSTS

Magnetite (Fe_3O_4) is a component of the water–gas shift reaction that crystallizes to the inverse spinel structure. The general formula of the oxides known as spinel is AB_2O_4. In the normal spinel-type structure, A is an A^{2+} metal, and B is a B^{3+} metal. O^{2-} forms a CCP anionic framework, where the A atoms occupy 1/8 of the tetrahedral sites and the B atoms occupy 1/2 of the octahedral sites (see Figures 1.6 and 1.7). An example of a normal spinel is $MgAl_2O_4$.

Ferrites are spinels with a general formula AFe_2O_4. But, bulk Fe_3O_4 has an inverse spinel structure, where Fe^{3+} is located in the tetrahedral (T) sites and Fe^{2+} and Fe^{3+} are located in the octahedral (O) sites. The inverse spinel structure consists of a FCC oxygen anionic framework (CCP), with the Fe^{3+} ions filling 1/8 of the tetrahedral A sites, and equal amounts of Fe^{2+} and Fe^{3+} ions filling 1/2 of the octahedral B sites.

Electron hopping between Fe^{2+} and Fe^{3+} ions occurs at the B sites and results in a high conductivity and an average charge of $Fe^{2.5+}$ (see Figure 2.17).

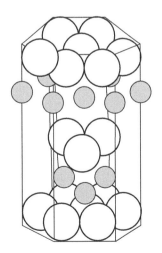

FIGURE 2.15 Wurtzite-type structure.

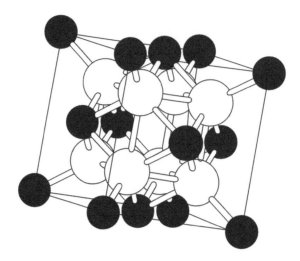

FIGURE 2.16   Fluorite-type structure.

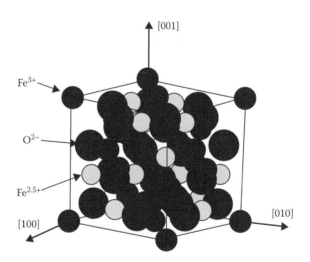

FIGURE 2.17   Spinel-type structure.

2.3.8   ZINC BLENDE–TYPE STRUCTURE

In Section 1.5.5, the tetrahedral bonding of the atoms in the C, Si, Ge, and α-Sn crystals was discussed. All these materials crystallize to a diamond-type structure. This structure can be geometrically represented as two FCC lattices displaced with respect to each other through the diagonal by 1/4 of the length of the diagonal (see Figure 1.16). The diamond-type structure could be also visualized as an FCC lattice with half the tetrahedral sites occupied by carbon atoms. In this regard, the zinc blende (ZnS)–type structure is closely related to the diamond-type structure. But in this case the Zn atoms occupy half the tetrahedral sites, and the S atoms occupy the lattice points of the FCC Bravais lattice.

The zinc blende–type of structure is found in several compounds, for example, SiC, BP, BN, AlP, AlAs, GaP, CdSe, and GaSb. The majority of these compounds are semiconductors and some of them are applied as photocatalysts.

2.4 PERMEABLE MATERIALS

2.4.1 INTRODUCTION

Permeable materials for the production of membranes have been employed for the treatment of a diversity of fluids [16,59–64]. A membrane is a perm-selective barrier between two phases capable of being permeated owing to a driving force, such as pressure, concentration, or electric field gradient [16,59,60].

Membranes are classified as organic or inorganic, taking into account the material used for their syntheses; porous or dense, based on the porosity of the material applied; and symmetric and asymmetric for a membrane made of a single porous or dense material or for a membrane made of a porous support and a dense end, respectively [16,64]. We are fundamentally interested here in asymmetric inorganic membranes made of a porous end to bring mechanical stability to the membrane and made of alumina, silica, carbon, zeolites, and other materials, and a dense end to give selectivity to the membrane (see Chapter 10). However, we also analyze the performance of porous polymers.

Inorganic membranes, which are described in detail in Chapter 10, are employed in gas separation [16,59,60,62–64], catalytic reactors [62], gasification of coal [60], water decomposition [60], and other applications [16,59–64].

Polymeric membranes also have vast applications in several processes, such as desalination using reverse osmosis membranes. Filtration, in a wide sense, with polymeric membranes can be applied in gas separation processes, biochemical processing, wastewater treatment, food and beverage production, and pharmaceutical applications [59–61].

Membrane polymeric materials for separation applications are made of polyamide, polypropylene, polyvinylidene fluoride, polysulfone, polyethersulfone, cellulose acetate, cellulose diacetate, polystyrene resins cross-linked with divinylbenzene, and others (see Section 2.9) [59–61]. The use of polyamide membrane filters is suggested for particle-removing filtration of water, aqueous solutions and solvents, as well as for the sterile filtration of liquids. The polysulfone and polyethersulfone membranes are widely applied in the biotechnological and pharmaceutical industries for the purification of enzymes and peptides. Cellulose acetate membrane filters are hydrophilic, and consequently, are suitable as a filtering membrane for aqueous and alcoholic media.

2.4.2 PALLADIUM: A HYDROGEN PERMEABLE MATERIAL

Hydrogen produces interstitial solid solutions with the majority of metals, particularly with those belonging to the transition metals of groups IV and V. Particularly, Pd, which exhibits a negative enthalpy of hydrogen solution, possesses, consequently, high hydrogen solubility, even forming hydrides at high concentration. Others, for example, the transition metals of groups VI and VII, and Fe, possess positive enthalpy of hydrogen solution and, therefore, low hydrogen solubility [65].

In metals, the reaction of hydrogen with a material is represented as follows [66]:

$$\left(\frac{1}{2}\right)H_2(g) \rightarrow H_i^{\bullet'} + e'$$

where
$H_i^{\bullet'}$ is an interstitial proton
e' is a conduction electron

This is possible because the conduction electrons are delocalized in the whole crystal, as was previously analyzed, and the protons are surrounded by conduction band electrons [67]. Consequently, in metals, the proton can have a high coordination number, and can be interstitially located in tetrahedral and/or octahedral sites [66].

Palladium, which shows an FCC structure, is a good absorbent of hydrogen. In this regard, permeation of hydrogen through Pd membranes, for its purification, is a well-known process.

2.4.3 Yttrium Oxide (Y_2O_3)-Stabilized Zirconium Oxide (ZrO_2)

As early as 1899, Nernst observed an effect of ionic conductivity in a $ZrO_2 + 9\%$ Y_2O_3 system. In this regard, in 1937, the first solid oxide–electrolyte fuel cell using this material was constructed [68].

Pure, bulk zirconia exhibits three structures in different ranges of temperatures at atmospheric pressure [69]. The most stable thermodynamic form is monoclinic, and it transforms to unquenchable tetragonal and cubic (fluorite) [45,70]. That is, zirconium oxide (ZrO_2), also named zirconia, is monoclinic up to 1370 K, and at this temperature has a phase transformation to a tetragonal structure [68]. To use zirconia, it is necessary to stabilize it. To do that, different oxides such as CaO, MgO, Y_2O_3, and others are used, in order to make solid solutions which have a stable cubic structure. For example, adding 16–26 wt % of CaO to zirconia produces a cubic phase at all temperatures [70]. The tetragonal–cubic transformation of ZrO_2 also takes place by adding Y_2O_3. Accumulated data indicate that the minimum amount of Y_2O_3 needed to fully stabilize ZrO_2 is 7–8 wt %, and the obtained material is a well-known oxygen-permeable material [45,70–72].

Yttria-stabilized zirconia ($[Zr_{1-x} Y_x]O_{2-x/2}$) is known in the literature as YSZ and has a fluorite-type structure [67] (see Figure 2.16). This material has a high oxygen ion conductivity and is, therefore, applied as a high-temperature electrolyte material, for example, in high-temperature fuel cells [68,73].

2.4.4 Hydrogen-Permeable Perovskites

The perovskite group of materials are oxides possessing analogous structures with a general formula ABO_3, where A is a cation larger in size than B. Even though the most numerous and most interesting compounds with a perovskite structure are oxides, some carbides, nitrides, halides, and hydrides also crystallize to this structure [74]. In the ABO_3 perovskite structure, the A site can be M^+ as Na or K; M^{2+} as Ca, Sr, or Ba; or M^{3+} as La or Fe; the B site can be occupied by M^{5+} as Nb or W; M^{4+} as Ce or Ti; or M^{3+} as Mn or Fe [75].

The extensive variety of properties that these compounds show is derived from the fact that around 90% of the metallic natural elements of the periodic table are known to be stable in a perovskite-type oxide structure [74]. Besides, the possibility of synthesizing multicomponent perovskites by partial substitution of cations in positions A and B gives rise to substituted compounds with a formula $A_{1-x}A'_xB_{1-y}B'_yO_{3-\delta}$. The resulting materials can be catalysts, insulators, semiconductors, superconductors, or ionic conductors.

In Figure 2.18a, the corner-sharing octahedra that form the skeleton of the ideal close-packed perovskite cubic structure is shown, in which the center of the cube is occupied by the A cation. Alternatively, this structure can be viewed with the B cation placed in the center the cube (see Figure 2.18b) [8].

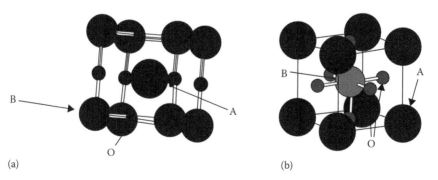

(a)   (b)

FIGURE 2.18 ABO_3 ideal perovskite structure. (a) The center of the cube is occupied by the A cation and (b) the B cation placed in the center of the cube.

In the ideal structure, where the atoms are in contact, the B–O distance is equal to $a/2$, where a is the cubic unit-cell parameter; likewise, the A–O distance is $a/\sqrt{2}$, then the following correlation between the ionic radii holds [76]:

$$(r_A + r_O) = \sqrt{2}(r_B + r_O)$$

To measure the departure from the ideal condition, Goldschmidt introduced a tolerance factor, t, defined by the equation:

$$t = \frac{(r_A + r_O)}{\sqrt{2}(r_B + r_O)}$$

However, a large number of perovskite structures are distorted to orthorhombic, rhombohedral, or tetragonal, which can be approximated as cubic with t deviated from 1. In most cases, t varies between 0.75 and 1 [77].

In relation to the permeability of perovskites, in the 1980s, Iwahara and collaborators [78,79] investigated protonic conductivity in $SrCeO_3$ and $BaCeO_3$ doped with trivalent cations such as Y, Yb, Gd, and Eu, and identified these materials as good, high-temperature proton conductors [78–85]. Thereafter, corresponding applications in solid oxide fuel cells (SOFCs) and other uses were developed. Subsequently, numerous studies have been carried out on the structure, transport properties, and chemical stability of doped $SrCeO_3$ and $BaCeO_3$ and other materials [86–96].

2.4.5   SILVER IODIDE: A FAST ION CONDUCTOR

Alpha silver iodide (α-AgI), a fast ion conductor, is one of the different polymorphic structures of AgI showing a cubic structure [51], where I^- occupies anionic positions, that is, the Cl^- sites in the CsCl-type structure (see Figure 2.19). On the other hand, the low temperature phase, that is, β-AgI, exhibits a hexagonal wurtzite-type structure.

The fast, ion conduction properties of α-AgI is a result of the large anionic radius of I^- and the low cationic radius of Ag^+, immersed in a cubic structure containing 42 vacant sites between octahedral and tetrahedral sites. Because of this fact, Ag^+ has a high mobility in this cubic structure [22].

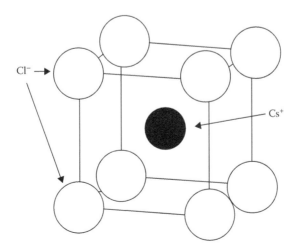

FIGURE 2.19   Cesium chloride–type structure.

2.5 CRYSTALLINE AND ORDERED NANOPOROUS ADSORBENTS AND CATALYSTS

2.5.1 ZEOLITE ADSORBENTS

The study of zeolites as adsorbent materials began in 1938, when Barrer published a series of papers on the adsorptive properties of zeolites [97]. In the last 50 years, zeolites, both natural and synthetic, have become one of the most important materials in modern technology [97–107]. Today, the production and application of zeolites for industrial processes is a multimillion dollar industry.

Aluminosilicate zeolites are three-dimensional microporous crystalline solids (see an example in Figure 2.20). These materials are manufactured from $(AlO_2)^-$ and (SiO_2) tetrahedral structures [97,98,107–109]. The (TO_2) tetrahedra are linked by oxygen atoms, which are shared between two tetrahedra [16,107–109]. To indicate the tetrahedral structure the generic chemical formula TO_2, is used and not TO_4. The presence of tetra-coordinated Al(III) generates a negative charge in the framework. This negative charge must be balanced by extra-framework cations [one per Al(III)]; consequently, the chemical composition of aluminosilicate zeolites can be expressed as [98,104]:

$$M_{x/n}\left\{(AlO_2)_x(SiO_2)\right\}zH_2O$$

where
   M are the balancing cations (charge, $+n$) compensating the charge from the Al(III)
   zH_2O is the water contained in the voids of the zeolite

The balancing cations can be a metal or another species, for example, ammonium. Since this cation is located within the zeolite channel or cavity, it can be exchanged, giving the zeolite its ion-exchange property [98,104].

In Figure 2.20, the FAU-type framework, which is congruent with the structure of the natural zeolite, faujasite, and the synthetic zeolites, X, Y, LZ-210, and SAPO-37, is shown [108].

The FAU framework structure is composed of sodalite cages, linked by 6–6 secondary building units (SBUs) like the carbon atom in the structure of diamond. The formed cube has an axis, $a \approx 24.3$ Å, and the framework produced by the union of these cubes contains windows, which lead to an approximately spherical cavity with a radius, $R \approx 6.9$ Å, known as the supercage or β-cage [108,109]. These supercages have tetrahedral symmetry, and are opened through four 12-member ring (MR) windows, each with a diameter of $d \approx 7.4$ Å [108,109].

Si and Al are the T atoms in aluminosilicate zeolites. However, other elements, such as P, Ge, Ga, Fe, B, Be, Cr, V, Zn, Zr, Co, Mn, and other metals can as well be T atoms [16,107,108]. These elements are tetrahedrally combined to form a zeolite, with or without charge compensating cations, in such a way that the electroneutrality principle is fulfilled [16,97,98].

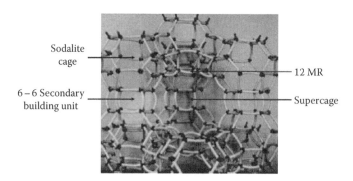

FIGURE 2.20　Structure of the zeolite-type framework FAU.

Most of the existing zeolites are synthetic, that is, more than 1000 different materials, and this number is constantly increasing [108]. Besides, about 40 natural zeolites have been found in the earth's crust [104].

In zeolites, cages, channels, and pores of different sizes and shapes, depending on the zeolitic structure, are found [109]. These dimensions depend on the arrangement of the $(AlO_2)^-$ and (SiO_2) or, in general, the (TO_2) tetrahedra to form substructures.

The classification of pores in zeolite adsorbents can be simplified by using the term "primary porosity" to refer to the cavities and channels, which constitute the zeolite framework (Figure 2.21), and "secondary porosity" (Figure 2.22) in order to designate the macroporosity and the mesoporosity developed between the zeolite crystals (Figure 2.22) [16,101,103]. Now, it is necessary to recognize that the illustration of the micropore represented in Figure 2.21 is only a simple scheme to show the topological relation between the molecules inside the pore and the pore walls. Because the slit geometry clearly shows this relation, however, zeolites do not exhibit slit pore geometry (see Chapter 5).

Zeolites have been shown to be good adsorbents for H_2O, NH_3, H_2S, NO, NO_2, SO_2, and CO_2, linear and branched hydrocarbons, aromatic hydrocarbons, alcohols, ketones, and other molecules [16,103,104].

With respect to Figure 2.22, it should be acknowledged that the design of the secondary porosity represented in Figure 2.22 is simply a diagram to show the spatial relations between the molecules outside the microporosity and the outer surface of the zeolite crystallites.

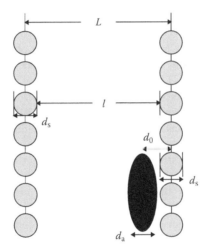

FIGURE 2.21 Schematic representation of a pore of diameter (L), where a molecule is adsorbed at a distance, d_0, on the surface.

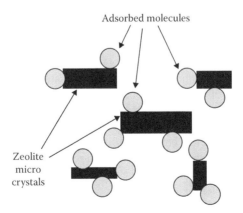

FIGURE 2.22 Schematic representation of adsorption in the secondary porosity of a zeolite.

2.5.2 Mesoporous Molecular Sieve Adsorbents

The use of zeolites and related materials is limited to small molecules, since the dimensions and ease of access of pores in these materials are restrained to the microporous scale. Consequently, during the last years, a significant effort has been focused on obtaining materials showing larger pore size [110–116]. As a significant result of this endeavor, in 1992, researchers at the Mobil Corporation discovered the M41S family of mesoporous molecular sieves [110,111]. These materials possess exceptionally large uniform pore structures.

This new family of mesoporous silica and aluminosilicate compounds were obtained by the introduction of supramolecular assemblies. Micellar aggregates, rather than molecular species, were used as structure-directing agents. Then, the growth of inorganic or hybrid networks templated by structured surfactant assemblies permitted the construction of novel types of nanostructured materials in the mesoscopic scale (2–100 nm) [110,113,117].

This supramolecular directing concept has led to a family of materials whose structure, composition, and pore size can be tailored during synthesis by variation of the reactant stoichiometry, nature of the surfactant molecule, or by postsynthesis functionalization techniques [113].

The obtained solid phases are characterized by an ordered, not crystalline, pore wall structure, presenting sharp mesopore-size dispersions. These mesoporous materials are ordered, but not crystalline, because of the lack of precise atomic positioning in the pore wall structure as was shown by MAS-NMR and Raman spectroscopy [115]. This gave rise to inorganic solids with enormous differences in morphology and structure [113].

The original members of the M41S family consisted of MCM-41 (hexagonal phase) (Figure 2.23), MCM-48 (cubic Ia3d phase), and MCM-50 (a stabilized lamellar phase) [110,111,117]. The structure of MCM-41 has a hexagonal stacking of uniform diameter porous tubes, whose size can vary from about 15 Å to more than 100 Å [111,117].

MCM-48, a cubic material, exhibits an x-ray diffraction pattern consisting of several peaks that can be assigned to the Ia3d space group [117]. The structure of MCM-48 has been proposed to be bicontinuous with a simplified representation of two, infinite, three-dimensional, mutually intertwined, unconnected network of rods [118].

MCM-50, a stabilized lamellar structure, exhibits an x-ray diffraction pattern consisting of several low-angle peaks that can be indexed to (h00) reflections, that is, the material is pillared-layered consisting of two-dimensional sheets [117].

Other mesoporous molecular sieve materials are SBA-1 (cubic Pm3n phase) [115] and SBA-2 (cubic p63/mmc phase) [119].

Subsequent synthesis efforts have produced new materials such as SBA-15. This mesoporous silica presents a well-defined hexagonal structure, large surface area, and high hydrothermal stability, which gives them high potential for a variety of applications [120,121].

All of these mesoporous materials are characterized by having narrow pore size distributions comparable to microporous materials, and extraordinary hydrocarbon sorption capacities. These materials are important mostly because of their exceptional ordered mesopore structure, which is not common for other adsorbent materials. As a result, mesoporous-ordered silica has been proposed as a reference material for the study of adsorption processes in mesopores [16]. In addition to the pore structure, the most striking features of these novel materials are the large BET surface area and pore volume, which make this family of materials excellent adsorbents in many applications [16].

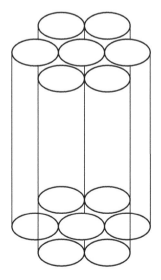

FIGURE 2.23 Graphic representation of the MCM-41 mesoporous molecular sieve.

2.5.3 Zeolite Catalysts

Zeolites are, perhaps, the most commonly used catalysts in industry for oil refining, petrochemistry, and in organic synthesis for the manufacture of fine chemicals [122–126].

In order to get acidic zeolite catalysts, zeolites are ion exchanged with ammonium cations:

$$\text{Na-Zeolite} + NH_4^+ \leftrightarrow NH_4\text{-Zeolite} + Na^+$$

And then they are transformed into acidic zeolites by the decomposition of the exchanged ammonium cations [104,125].

$$NH_4\text{-Zeolite} \xrightarrow{\text{Heat}} H\text{-Zeolite} + NH_3$$

This procedure can be succeeded by the ultrastabilization process, which is one of the basic operations in the industrial production of acid catalysts, consisting of a controlled dealumination process produced by thermal treatment in a water vapor atmosphere, which increases the thermal stability of the zeolite.

Acidic zeolite materials are the main catalysts in the cracking process, which is the most important process among industrial chemical processes. Broad studies of heterogeneous cracking catalysts, started in the 1950s, discovered that the basic nature of cracking catalysts is acidic, and generation of acidic sites on solids has been extensively studied.

Another important zeolite catalyst is the so-called bifunctional catalyst. The thermal reduction of zeolites previously exchanged with metals is the method currently used for the preparation of bifunctional catalysts for hydrocarbon conversion. The bifunctional zeolite catalysts are composed of both acidic sites and metal clusters. The preparation methods of these catalysts encompass three steps: ion exchange, calcinations, and reduction, (Section 3.2.1.4) [123,127].

In addition to the Brönsted acidity in zeolites, in these materials the Lewis acidity is present as well. According to Lewis, an acid is an electron pair acceptor, a definition which is broader than that given by Brönsted, since a proton is a particular case of an electron pair acceptor. Then, the definition of Lewis covers practically all acid–base processes, whereas the definition of Brönsted represents only a particular type of process [128]. The Lewis acidity is related to the existence of an extra-framework Al (EFAL) species formed during the zeolite dealumination process [128].

Additionally, basic catalytic activity in zeolites has been observed. The side chain alkylation of toluene effectively catalyzed by alkali ion-exchanged X- and Y-type zeolites has been described [129]. This is a typical base-catalyzed reaction, and the activity fluctuated with the type of the exchanged alkali cation and the type of zeolite, revealing that basic properties can be controlled by selecting the exchanged cation and the type of zeolite [129]. The most significant basic site in zeolites is the framework negative oxygen, which is a Lewis basic site [130].

2.5.4 Pillared Clay Catalysts

Clay minerals are hydrous aluminum phyllosilicates made of sheets or layers composed of tetrahedra and octahedra. This mineral type includes the following groups: kaolinite, smectite, illite, and chlorite. In the case of smectite, each layer comprises two sublayers of tetrahedra with an inserted octahedral layer, where, between layers, an interlayer space where the exchangeable cations are located is formed [131–133]. In Figure 2.24

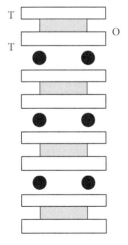

FIGURE 2.24 Schematic representation of the structure of montmorillonite.

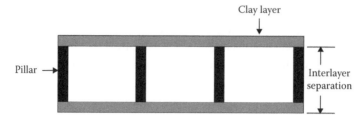

FIGURE 2.25 Schematic representation of a pillared clay.

[122,131], the schematic representation of the structure of montmorillonite is shown; in this figure, the white rectangles represent tetrahedral coordinated layers, the gray rectangles represent octahedral layers, and the black spheres represent exchangeable cations.

The principal class of swelling clays is smectite. This type of mineral comprises beidellite, hectorite, fluorhectorite, saponite, sauconite, montmorillonite, and nontronite. Smectites can be described merely on the basis of layers enclosing two sheets of silica sandwiching a layer of octahedral Al or Mg, that is, 2:1 layered clays (see Figure 2.24) where the layer of alumina octahedra is inserted into the two layers of silica tetrahedra.

Starting from clays, it is possible to develop an additional group of materials with large pores, whose framework is composed of layered structures with pillars in the interlamellar region [122]. These materials are the so-called pillared clays (PILCs).

The intercalation of alkylammonium cations in clays, specifically tetraalkylammonium, was carried out by Barrer and MacLeod. The idea of maintaining the clays permanently expanded by intercalating strong inorganic pillars was advanced later [132]. In Figure 2.25, a schematic representation of a PILC is shown.

The creation of Brönsted acidic sites in a di-octahedral clay such as beidellite, or a tri-octahedral smectite, such as saponite, is carried out using the same methodology previously explained for zeolites:

$$Na\text{-}PILCS + NH_4^+ \leftrightarrow NH_4\text{-}PILCS + Na^+$$

And after heating in an air flow, the acidic form of the PILCs is obtained as follows:

$$NH_4\text{-}PILCS \xrightarrow{\text{Heat}} H\text{-}PILCS + NH_3$$

Besides the Brönsted acidic sites created by the above explained methodology, in PILCs that are ion exchanged with multivalent cations and are partially dehydrated, Lewis acidic sites are produced [132].

The main interest in developing acid PILCs was due to their potential applications as cracking catalysts. Certainly the prospect of making PILCs in which the big gaseous oil molecules can diffuse and meet the active acidic sites was a motivation for the development of these catalysts [122].

2.6 ION-EXCHANGE CRYSTALLINE MATERIALS

We discuss crystalline inorganic ion exchangers, such as [97,98,104,134–146] zeolites, hydrotalcites, alkali metal titanates, titanium silicates, and zirconium phosphates, and also organic ionic exchangers such as the ion-exchange resins. However, details are given only in the case of the archetypal crystalline ionic exchangers, that is, aluminosilicate zeolites [97,98,104,134].

2.6.1 ZEOLITES

In ion exchange, the solid, that is, the ion exchanger, consisting of a matrix of fixed ions, called the coions, forms a charged framework, whose charge is balanced by mobile ions. These are located in definite sites of the channels or cavities which conform the solid matrix, and are called the counter-ions. The ion-exchange reaction in aluminosilicate zeolites is represented by:

$$z_B A\{z_A^+\} + BZ \leftrightarrow z_A B\{z_B^+\} + AZ \tag{2.1}$$

where
$z_A e^+$ and $z_B e^+$ are the charges of cations A and B, respectively
$A\{z_A^+\}$ and $B\{z_B^+\}$ describe the cations A and B in solution
AZ and BZ are the cations A and B in the zeolite

The total exchange capacity (TEC) of a aluminosilicate zeolite is a function of the framework, that is, the Si/Al ratio. It is easy to get a numerical relation between the TEC in mequiv/g, C, and the number of Al atoms per framework unit cell, N^{Al} [104]:

$$C = \frac{N^{Al}}{N_{Av} \rho V_c} \tag{2.2}$$

where
N_{Av} is the Avogadro number
ρ is the zeolite density
V_c is the volume of the framework unit cell

The relation follows from N^{Al} / N_{Av}, that is, the total number of equivalents of exchangeable cations per unit cell, where ρV_c is the mass of the unit cell.

Aluminosilicate zeolites are a key group of crystalline inorganic ion exchangers that are useful in industrial, agricultural, and environmental applications, extending from wastewater cleaning, agriculture, aquaculture, to detergency and the removal of radioactive nuclei from nuclear wastewater [104]. On the other hand, ionic exchange is the most important method currently used to modify zeolites [134]. The presence of Al(III) in the zeolite structure in tetrahedral sites generates a negative charge in the zeolite framework. In order to compensate this negative charge, subsequently, extraframework cations, one per Al(III), are included in the zeolite cavities and channels. These extraframework charge-compensating cations are able to be exchanged during ion-exchange reactions.

Aluminosilicate zeolites are recognized as excellent ionic exchangers [98,101,104,134,142–145]. Using zeolites, mainly clinoptilolite, chabazite, and phillipsite, different process have been developed for the treatment of municipal wastewater, purification of radioactive wastewater, for the processing of agricultural wastewater with high chemical oxygen demand, animal nourishment, aquaculture, and the removal of heavy metals from wastewater [104].

However, the major use of ionic exchange in zeolites today is the application of the synthetic zeolite A in detergents for the softening of laundry water.

2.6.2 HYDROTALCITES

Hydrotalcites are layered double hydroxides with the general formula [137]:

$$\left[M^{(II)}_{1-x} M^{(III)}_x (OH)_2 \right]^{x+} [A]^-_x \bullet m H_2 O$$

where
$M^{(II)}$ can be Ca^{2+}, Mg^{2+}, Mn^{2+}, Fe^{2+}, Co^{2+}, N^{2+}, and Zn^{2+}
$M^{(III)}$ can be Al^{3+}, Cr^{3+}, Mn^{3+}, Fe^{3+}, Co^{3+}, and Ga^{3+}
A can be Cl^-, Br^-, I^-, NO_3^-, CO_3^{2-}, and SO_4^{2-}, silicate-, polyoxometalate-, and/or organic anions

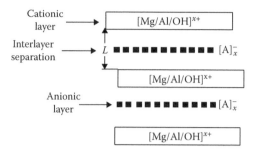

FIGURE 2.26 Schematic representation of the hydrotalcite structure.

The structure of hydrotalcites is linked to that of the natural mineral brucite, $Mg(OH)_2$ [136,137], and is based on a stacking of hydroxide layers, where part of the $M^{(II)}$ ions are substituted by $M^{(III)}$; this substitution produces positively charged $[M^{(II)}/M^{(III)}/OH]$ layers, which are balanced by anions placed between them. Consequently, these materials are anion exchangers, where the negative ions are located in layers alternating with the positive layers all along the c-direction of the rhombohedral or hexagonal unit cells (see Figure 2.26) [137].

2.6.3 Titanates

2.6.3.1 Alkali Metal Titanates

These materials have been obtained in compliance with any of the following two series: $M_2Ti_nO_{2n+1}$ with $n = 1$–9 and $Na_4Ti_nO_{2n+2}$ with $n = 1, 3, 5$ [136]. For example, $K_2Ti_2O_5$, $Na_2Ti_3O_7$, $Na_2Ti_4O_9$, and $Na_4Ti_5O_{12}$, have layered structures, built from TiO_6-octahedra sharing edges, which are joined together through the corners to other similar chains of the octahedra, thus forming layers. The alkaline metal ions are located in between the layers [139].

Among these titanates, ion-exchange behavior has been demonstrated in $K_2Ti_2O_5$, $Na_2Ti_3O_7$, and $Na_2Ti_4O_9$.

2.6.3.2 Titanium Silicates

ETS-4 and ETS-10 are microporous titanosilicates discovered in 1989 [138]. Unlike conventional zeolites, the framework of these titanium silicates are constituted from SiO_4 tetrahedra and TiO_6 octahedra by corner-sharing oxygen atoms. The presence of each tetravalent Ti atom in an octahedron generates two negative charges, which are balanced by exchangeable cations Na^+ and K^+. Such unique framework properties imply that both ETS-4 and ETS-10 are very promising ion exchangers [139].

From a structural point of view, ETS-4 is an analogue of the mineral zorite, while ETS-10 is comparable topologically to zeolite β. As a consequence of the high amount of disorder in these materials, the determination of their structures represents a challenging task [138,146]. ETS-4 is, as was previously stated, essentially a synthetic analogue of the mineral zorite, where, even though larger openings are present in its structure, faulting ensures that access to the crystal interior of ETS-4 occurs through relatively narrow eight-membered rings [139]. The disorder in ETS-10 is described in terms of diverse stacking sequences of the same titanosilicate unit $[Si_{40}Ti_8O_{104}]^{16-}$, giving rise to numerous probable polymorphs [138]. The frequently described are polymorphs A and B; both of them include a three-dimensional 12-ring pore system, where, in polymorph A, the 12-ring pores [146] are assembled in a zigzag mode, forming a spiral channel and a tetragonal lattice, while the stacking in polymorph B leads to a diagonal array of the 12-ring pores with a monoclinic lattice [138]. The Ti(IV) atoms are observed to be octahedrally coordinated by oxygen atoms to the four Si tetrahedra and connected to each other by O–Ti–O chains; as a result, in the anionic structure, there is a −2 charge related to every Ti site, where this charge is compensated by extraframework

Na⁺ and K⁺ ions [138,147]. Then, the pore structure of ETS-10 contains 12 rings in all three dimensions; these are straight along the crystallographic directions: [100] and [010] and twisted along the direction of disorder. A small number of microporous zeolitic materials with a three-dimensional 12-ring pore system are known, and in this aspect ETS-10 has excellent diffusion characteristics [139].

2.6.4 ZIRCONIUM PHOSPHATES

About 40 years ago, Clearfield and coworkers started a fertile activity in the synthesis and characterization of crystalline zirconium phosphates [136,140,148]. After the initial findings, an enormous research effort directed toward the syntheses and characterization and investigation of the properties of the many phases of crystalline zirconium phosphate have been carried out [140]. In this regard, it has been shown that, in general, all of the zirconium phosphate phases have excellent ion-exchange properties.

Zirconium phosphates consist of phosphates of both group IVA (group 14) and group IVB (group 4) elements [136]. Zirconium bis (monohydrogen orthophosphate) monohydrate, that is, α-$Zr(O_3POH)_2 \cdot H_2O$, called α-zirconium phosphate (α-ZrP), is the most extensively characterized zirconium phosphate [135,148] (see Figure 2.27). These phosphates have a layered structure where the metal atoms are arranged practically in a plane and are bridged by phosphate groups as shown in Figure 2.27 [135,148]. In this way, three phosphate–oxygen bonds to metal atoms are formed, so that the remaining phosphate oxygen, which is bonded to a proton, points into the interlayer space. Consequently, in this α-layer, the metal atoms are six-coordinated and the phosphates are located in a tetrahedral coordination, forming T–O–T-type layers as in smectite clays [135,136] (see Figure 2.24). The main dissimilarity is that the phosphate groups are in inverted positions relative to the silicate groups in clays; then, as a result, the negative layer charge is concentrated on the proton-bearing oxygen rather than spread over the many silicate oxygens in the clay interlayer boundary [136].

Even though the layers are held together merely by van der Waals forces, the interlayer distance is small and the bottlenecks offer just enough space for an unhydrated K⁺ to diffuse into the interlamellar spacing [136].

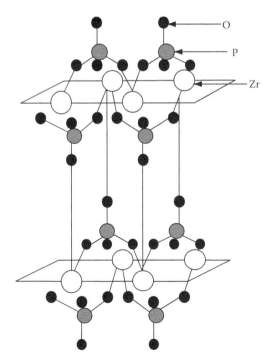

FIGURE 2.27 Idealized representation of α-ZrP.

2.7 AMORPHOUS SILICA ADSORBENTS AND CATALYTIC SUPPORTS

2.7.1 AMORPHOUS SILICA

Silica is one of the most abundant chemical substances on earth. It can be both crystalline or amorphous. The crystalline forms of silica are quartz, cristobalite, and tridymite [51,52]. The amorphous forms, which are normally porous [149] are; precipitated silica, silica gel, colloidal silica sols, and pyrogenic silica [150–156]. According to the definition of the International Union of Pure and Applied Chemistry (IUPAC), porous materials can be classified as follows: microporous materials are those with pore diameters from 3 to 20 Å; mesoporous materials are those that have pore diameters between 20 and 500 Å; and macroporous materials are those with pores bigger than 500 Å [149].

Porous silica is one of the different forms of amorphous silica. It can be prepared by acidification of basic aqueous silicate solutions, and when reaction conditions are properly adjusted, porous silica gels are obtained [150]. If water is evaporated from the pores of silica hydrogels prepared in this fashion, porous xerogels are obtained [153].

Other forms of silica, such as pyrogenic silica and mineral opals are normally nonporous. Pyrogenic silica is composed of silica particles with a very narrow particle size distribution. This material is obtained by vaporizing SiO_2 in an arc or a plasma jet, or by the oxidation of silicon compounds [152]. Artificial opals are materials characterized by the presence of SiO_2 microspheres [16].

Solgel processing specifies a type of solid material synthesis method performed in a liquid and at low temperatures [152]. The produced inorganic solids, mostly oxides or hydroxides, are formed by chemical transformation of chemical solutes termed precursors. The solid is formed as the result of a polymerization process that involves the establishment of M–OH–M or M–O–M bridges between the metallic atoms M of the precursor molecules [152]. The drying process, after the gel formation, is carried out at a relatively low temperature to produce a xerogel, or by a supercritical drying process.

During thermal drying, or room temperature evaporation, capillary forces provoke stresses on the gel. This effect raises the coordination numbers of the particles, and produces collapse of the network, that is, particle agglomeration [153] (see Figure 2.28).

In contrast, aerogels (Figure 2.28) are materials synthesized by traditional solgel chemistry, but are dried basically by a supercritical drying process, where as a result, the dry samples keep the porous texture, which they had in the wet stage [152]. The supercritical extraction of the solvent from a gel does not induce capillary stresses due to the lack of solvent–vapor interfaces. Therefore, the compressive forces exerted on the gel network are significantly diminished relative to those created during the formation of a xerogel. Aerogels, consequently, retain a much stronger resemblance

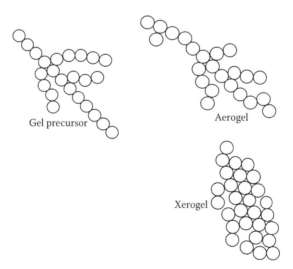

FIGURE 2.28 Formation of the aerogel and xerogel during the drying process.

FIGURE 2.29   SEM micrograph of sample 81 (bar = 1000 nm).

to their original gel network structure than xerogels do, are materials with lower apparent densities, and have larger specific surface areas than xerogels [152,153].

In the case of xerogels, there is no doubt that the amorphous framework is made up of very small globular units of size of 10–20 Å; commonly, these particles are densely packed. In the case of aerogels, the structure is more open, and the primary particles are not densely packed [152].

Synthetic silica microspheres (Figure 2.29) can easily be obtained by using the Stobe–Fink–Bohn (SFB) method [154–157], as explained in Section 3.2.3.

Amorphous silicas play an important role in many different fields, since siliceous materials are used as adsorbents, catalysts, nanomaterial supports, chromatographic stationary phases, in ultrafiltration membrane synthesis, and other large-surface, and porosity-related applications [16,150–156].

The common factor linking the different forms of silica are the tetrahedral silicon–oxygen blocks; if the tetrahedra are randomly packed, with a nonperiodic structure, various forms of amorphous silica result [16]. This random association of tetrahedra shapes the complexity of the nanoscale and mesoscale morphologies of amorphous silica pore systems. Any porous medium can be described as a three-dimensional arrangement of matter and empty space where matter and empty space are divided by an interface, which in the case of amorphous silica have a virtually unlimited complexity [158].

In the author's opinion, the better approach to experimentally study the morphology of the silica surface is with the help of physical adsorption (see Chapter 6). Then, with the obtained, adsorption data, some well-defined parameters can be calculated, such as surface area, pore volume, and pore size distribution. This line of attack (see Chapter 4) should be complemented with a study of the morphology of these materials by scanning electron microscopy (SEM), transmission electron microscopy (TEM), scanning probe microscopy (SPM), or atomic force microscopy (AFM), and the characterization of their molecular and supramolecular structure by Fourier transform infrared (FTIR) spectrometry, nuclear magnetic resonance (NMR) spectrometry, thermal methods, and possibly with other methodologies.

The molecular properties of silica are strongly affected by the nature of their surface sites [16,159–161]. In solgel-synthesized porous silica surfaces, the unsaturated surface valencies are satisfied by surface hydroxyl functionalities, which, depending on the calcination temperature, exist as (a) vicinal (hydrogen-bonded silanols), (b) geminal (two hydroxyl groups attached to the same silicon atom), or (c) isolated (no hydrogen bonds possible) silanol sites (see Figure 2.30) [159,161].

The hydrogen bond interaction of OH groups at the surface is determined by the Si–O–Si ring size, and its opening degree, the number of hydroxyls per silicon site, and the surface curvature. The concentration of OH groups at the surface is approximately $2–5 \times 10^{18}$ OH/m^2, and it is found to be almost independent of the synthesis conditions of porous silica [162].

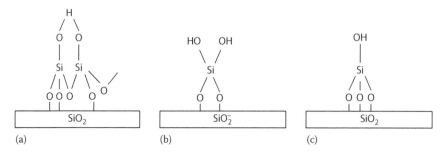

(a)   (b)   (c)

FIGURE 2.30 Schematic presentation of three types of silanol groups occurring on silica surfaces: (a) vicinal silanols, (b) geminal silanols, and (c) isolated silanols.

These silanols are preferential adsorption sites for different molecules. As a result, surface hydroxyls are, indeed, particularly reactive with H_2O and other polar molecules, such as NH_3, which is likely to be physically adsorbed to form a multiple hydrogen-bonded layer [16].

2.7.2 AMORPHOUS SILICA AS ADSORBENTS AND CATALYTIC SUPPORTS

As can be inferred from the description of the surface chemistry of silica, one of the major adsorption applications of silica is the adsorption of polar molecules (see Chapter 6) [16]. On the other hand, many practical catalysts consist of one or several active components deposited on a high surface area support. In heterogeneous catalysis, transition metal nanoparticles are supported on different substrates and utilized as catalysts for different reactions, such as hydrogenations and enantioselective synthesis of organic compounds [163], oxidations and epoxidations [164], and reduction and decomposition [151]. For example, a porous silica xerogel is a good support for Co in the Fischer–Tropsch catalysts for the diesel fuel conversion of syngas [165]. The surface decoration of partially agglomerated amorphous silica nanospheres with SnO_x nanocrystals of diameter 3–6 nm to produce extremely small, tin oxide nanostructures on a silica nanosurface to produce improved catalysts was as well demonstrated [166]. Besides, in various organic syntheses, catalytic hydrogenation is carried out in a liquid phase using a metal, such as Pd, Pt, or Rh supported, in silica as the catalyst [167–170].

2.8 ACTIVE CARBON AND OTHER CARBON FORMS AS ADSORBENTS AND CATALYTIC SUPPORTS

Activated carbons [171–182] are amorphous materials showing highly developed adsorbent properties. These materials can be produced from approximately all carbon-rich materials, including wood, fruit stones, peat, lignite, anthracite, shells, and other raw materials. The properties of the produced adsorbent materials will depend not merely on the preparation technique but as well on the carbonaceous raw material used for their production. Actually, lignocellulosic materials account for 47% of the total raw materials used for active carbon production [178].

The method normally applied for the conversion of the organic raw material into activated carbon is carbonization, that is, pyrolysis under inert atmosphere. This procedure is followed by activation, explicit heat treatment with an oxidizing agent, or by simultaneous carbonization and activation with a dehydrating compound [171,178].

The structure of activated carbon consists of carbon atoms that are ordered in parallel stacks of hexagonal layers similar to graphite [51,52]. To be precise, graphite is formed of layer planes of sp^2 (see Figure 2.31), carbon atoms forming regular hexagons. That is, each carbon is bonded to other three carbons by a σ-bond, and the p_z orbitals containing one electron form a delocalized π-bond (see Figure 2.32) [180,181]. The different layers are linked by van der Waals forces [51].

Active carbon has a high density of carbon atoms in graphite-like sheets [178]. These sheets are spatially organized in such a way that small slit-shaped pores are present in the bulk material [16].

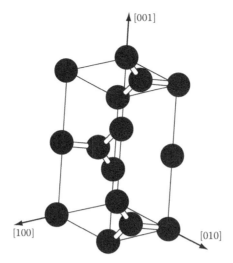

FIGURE 2.31 Structure of hexagonal graphite.

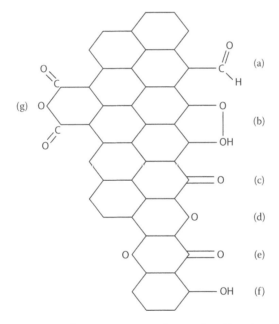

FIGURE 2.32 Significant oxygen surface groups on a carbon surface (a) carboxyl, (b) lactone, (c) carbonyl, (d) ether, (e) pyrone, (f) phenol, and (g) carboxyl anhydride.

All of these features considerably improve the sorption capacity of carbons, and their capacity to remove contaminants and pollutants interacting with the surface of carbons in a dispersive way [177]. In addition, active carbon contains heteroatoms such as oxygen, and, to a smaller degree, nitrogen and sulfur. These atoms are bound to the activated carbon surface in the form of functional groups, which are acidic or basic, giving the activated carbon surface an acidic or basic character, respectively [173,178]. It is as well necessary to state that the chemical heterogeneity of the carbon surface is mostly the result of the presence of heteroatoms [175].

In addition to the developed pore structure, small pore sizes, and large surface area, surface hydrophobicity is an extremely helpful property of active carbon [170–175]. This property is particularly useful for the adsorption of organic species [176].

Various heteroatoms, including oxygen, hydrogen, nitrogen, phosphorous, and sulfur, can be located in the carbon matrix, in the form of single atoms and/or functional groups [180]. These atoms are chemically connected to the carbon atoms with unsaturated valences that are located at the edges of graphite basal planes [51].

Oxygen is the main heteroatom in the carbon matrix, and the occurrence of functional groups, such as carboxyl, carbonyl, phenols, enols, lactones, and quinones, has been suggested [176,181]. These surface groups can be produced during the activation procedure and can as well be introduced subsequent to preparation by an oxidation treatment [178] (see Figure 2.32).

The diverse types of oxygen functional groups determine the acidic and basic character of the carbon surfaces [175,182]. The acidic character is typically linked with surface complexes like carboxyl, lactone, and phenol, while the basic nature is regularly assigned to surface groups, such as pyrone, ether, and carbonyl [173,178] (see Figure 2.32).

Our current comprehension of the adsorption of organic compounds by active carbon reveals that this phenomenon is controlled by two major interactions [180,183]: physical interactions, which include size exclusion and microporosity effects, and chemical interactions, which depends on the chemical nature of the adsorbate surface and the solvent.

For instance, in liquid-phase adsorption, it has been established that the adsorption capacity of an activated carbon depends on the adsorbent's pore structure, ash content, functional groups [184–186], the nature of the adsorbate, its pK_a, functional groups present, polarity, molecular weight, and size [187], and, finally, the solution conditions, such as pH, ionic strength, and the adsorbate concentration [188].

Because surface functional groups influence the adsorption properties and the reactivity of activated carbons, many methods, including heat treatment, oxidation, amination, and impregnation with various inorganic compounds, have been developed in order to modify activated carbons [183]. These modifications can alter the surface reactivity, as well the structural and chemical properties of the carbon, which can be characterized using various methods, as described in detail elsewhere [176].

Consequently, the exceptional adsorption properties of active carbon result from its high surface area, adequate pore size distribution, broad range of surface functional groups, and relatively high mechanical strength [179]. As a result, porous carbonaceous materials are routinely applied in several industrial processes for the removal of impurities from gases and liquids, including gas separation and purification, vehicle exhaust emission control, solvent recovery, in environmental technologies, or for high-grade products [172,179]. Besides, it can be used as a catalyst support [178].

Another carbon-related material very promising for diverse applications is carbon nanotubes (CNT) [189,190]. It is known that carbon, because of the different hybridizations assumed by it, can bond in diverse modes to produced dissimilar bulk structures, for example, the two solid phases of pure carbon, that is, graphite and diamond, fullerenes, and active carbon and others [189]. In this regard, CNTs consists of graphitic sheets that have been rolled up into a cylindrical shape. CNTs have exceptional properties, such as small diameters, specifically, up to 100 nm, relatively long length in the order of micrometers, high mechanical strength, excellent thermal conduction, and high thermal and chemical stabilities [190]. When a single layer of graphene sheet is rolled to produce a cylindrical shape, a single-walled carbon nanotube (SWCNT) is obtained; on the other hand, if multiple layers of graphene sheets are rolled to shape a hollow cylindrical shape, a multiwalled carbon nanotube (MWCNT) is obtained [189].

2.9 POLYMERS

2.9.1 INTRODUCTION

The term polymer was introduced by Berzelius in 1833. These are materials comprised of huge molecules that have very high molecular weight, and are composed of a large number of repeating units named "mers." A proper comprehension of the structure of polymers did not come into view until the second decade of the twentieth century. Previously, these compounds were considered molecular clusters held together by an unidentified force. However, in 1922, Staudinger proposed that polymers consist of long chains of atoms held together by covalent bonds.

As a matter of fact, mankind knows polymers from ancient times, due to the existence of naturally occurring polymers such as latex, starches, cotton, wool, leather, silk, amber, proteins, enzymes, starches, cellulose, lignin, and others. The other type of polymers are synthetic polymers. Braconnot, in 1811, perhaps made the first significant contribution to polymer science by developing compounds derived from cellulose. Later, cellulose nitrate was obtained in 1846 by Schönbein, afterward in 1872, its industrial production was established. Besides, in 1839, Goodyear found out by accident that by heating latex with sulfur its properties were altered creating a flexible and temperature-stable rubber. This process is named vulcanization.

Carothers, in 1928, showed that polymers can be produced from their component monomers by logical chemical means. This researcher made, in 1935, the first significant application of polymers when he invented nylon in the Du Pont laboratories. World War II brought a big stimulus to the creation of polymers, firmly establishing the field of polymer science and technology. After the war, the production of nylon, polyethylene, polyvinylchloride, polytetrafluoroethylene, polypropylene, polystyrene, and other polymers formed the basis of a fast growing industry. Currently, synthetic polymers are produced commercially on a very large scale and have a wide range of properties and uses. To mention only some applications, it can be pointed out that nylon is used in stockings, tires, ropes, fabrics and other items, and polycarbonates and acrylonitrile–butadiene–styrene are applied as roofing materials [13].

We are interested in the application of polymers as adsorbents, ion exchangers, fuel cells, and permeable materials. In this regard, the first resins with some of these properties were obtained by D'Aleleio in 1944 based on the copolymerization of styrene and divinylbenzene. Unfunctionalized polystyrene resins cross-linked with divinylbenzene (Amberlite) are widely applied as adsorbents [191,192]. In addition, the polystyrene–divinylbenzene resins functionalized with sulfuric acid (sulfonation) to create negatively charged sulfonic sites are applied as cation exchangers, and treated by chloromethylation followed by amination produce anionic resins [193,194].

2.9.2 Polymer Structure

It is possible to classify polymers by their structure as linear, branched, cross-linked, and network polymers. In some polymers, called homopolymers, merely one monomer (α) is used for the formation of the chains, while in others two or more diverse monomers ($\alpha,\beta,\gamma,\ldots$) can be combined to get different structures forming copolymers of linear, branched, cross-linked, and network polymeric molecular structures. Besides, on the basis of their properties, polymers are categorized as thermoplastics, elastomers, and thermosets. Thermoplastics are the majority of the polymers in use. They are linear or branched polymers characterized by the fact that they soften or melt, reversibly, when heated. Elastomers are cross-linked polymers that are highly elastic, that is, they can be lengthened or compressed to a considerable extent reversibly. Finally, thermosets are network polymers that are normally rigid and when heated do not soften or melt reversibly.

2.9.2.1 Linear or Chain Polymers

The linear polymeric structure is the simple conformation for a polymer molecule (see Figure 2.33) in which the monomers are consecutively assembled in a single chain, of the same monomer, to form a homopolymer $\alpha - \alpha - \alpha - \alpha - \alpha - \alpha - \cdots$. In the case of copolymers, the linear structure can have a random distribution of two or more monomers to form a random copolymer $\alpha - \alpha - \beta - \alpha - \beta - \beta - \alpha - \alpha - \beta - \cdots$, that is, when the sequence has a chaotic order; in other cases, when the sequence is not chaotic, alternating copolymers $\alpha - \beta - \alpha - \beta - \alpha - \beta - \alpha - \beta \cdots$ or block copolymers $\alpha - \alpha - \beta - \beta - \alpha - \alpha - \beta - \beta - \cdots$ are formed where a periodic order in the sequence of monomers is observed [8,13].

FIGURE 2.33   Schematic representation of a linear polymer.

TABLE 2.1
Some Monomers which Produce Linear Polymers

Polymer	Monomer
Polyethylene	H H | | —C — C— | | H H
Polyvinylchloride	H H | | —C — C— | | H Cl
Polypropylene	H H | | —C — C— | | H CH$_3$

Some of the most important linear or chain polymers belong to the category of homopolymers, for example, polyethylene, polyvinylchloride, and polypropylene (see Table 2.1) where the vinyl-type monomer (see Figure 2.34) with different side groups, R, is repeated [8,195].

Polyethylene is the most widespread polymer; it is used in bags, containers, and wire insulation. Polyvinylchloride is used in the production of containers, plumbing accessories, synthetic leather, and floor tiles, and polypropylene is applied in the fabrication of rugs and containers [196].

Other important chain polymers, where the vinyl monomer with a side group unit (Figure 2.34) is repeated, are polystyrene and poly (vinyl acetate) (see Table Monomer 2.2) [8].

TABLE 2.2
Two More Examples of Chain Polymers

Polymer	Side group
Polystyrene	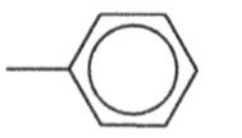 (benzene ring)
Poly (vinyl acetate)	O ‖ —O —C — CH$_3$

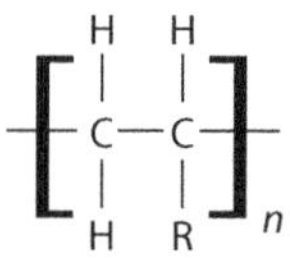

$$\left[\begin{array}{cc} H & H \\ | & | \\ C & C \\ | & | \\ H & R \end{array}\right]_n$$

FIGURE 2.34   Vinyl-type repeating unit with a side group.

TABLE 2.3
Acrylate and Methacrylate Polymers

Polymer	Side Group X	Side Group R
Poly(ethyl acrylate)	–H	$-C_2H_5$
Poly(methyl methacrylate)	$-CH_3$	$-CH_3$
Poly(ethyl methacrylate)	$-CH_3$	$-C_2H_5$

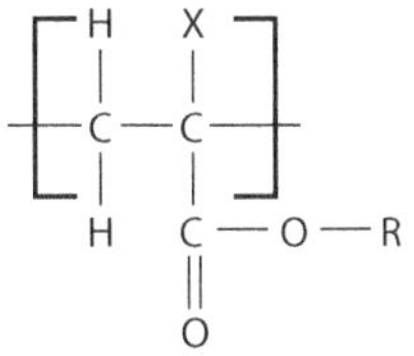

FIGURE 2.35 Repeating unit with two side groups for acrylate and methacrylate polymers.

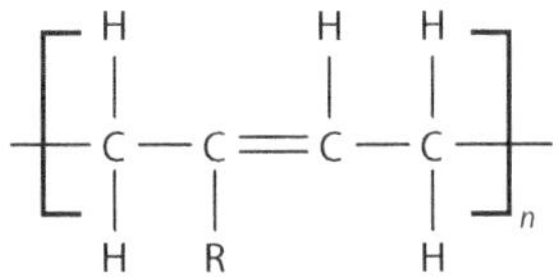

FIGURE 2.36 Repeating unit with one side group for diene polymers.

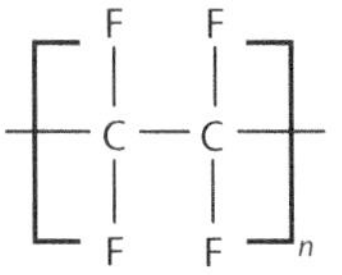

FIGURE 2.37 PTFE.

Polystyrene is applied fundamentally in the fabrication of molded objects and poly (vinyl acetate) is used mainly in the preparation of paints and adhesives. These polymers are obtained by polymerizing monomers, by the chain kinetic scheme, which involves the opening of the double bonds to form a linear molecule [196].

Another type of important linear polymers are the acrylates and methacrylates. In Figure 2.35, the repeating unit with two side groups from which these polymer types are derived is shown [195,197]. In Table 2.3, three examples of this type of polymers, specifically, poly (ethyl acrylate), which is applied in paints; poly (methyl methacrylate), which is applied fundamentally as a construction material in place of glass; and poly (ethyl methacrylate), which is applied as adhesives, are reported.

A supplementary form of linear polymers is those of the "diene" class. In this case, the repeating unit with one side group from which these polymer types are derived is shown in Figure 2.36 [197].

In this class, polybutadiene (R = –H), polyisoprene (R = $-CH_3$), and polychloroprene (R = –Cl) are included [195].

A very important chain polymer from the vinylidene class [195,197] is polytetrafluoroethylene (PTFE), which is widely applied in cooking pots, as a sealing material, and also has other uses. In Figure 2.37, the repeating unit in the case of PTFE is shown [8].

An additional, important, type of chain polymers is the polyamides, polyesters, and polyurethanes [195–197]. Among the polyamides, nylon is a very important member. In Figure 2.38, the repeating unit of one of the member of the nylon family, that is, the repeating unit of polyhexamethylene adipamide, or nylon 6,6 is represented [8,195]. As was previously commented, nylon is applied in stockings, ropes, fabrics, and has other uses.

In addition, among the polyesters, poly(ethylene terephtalate) or Dacron, is another important member. In Figure 2.39, the repeating unit of poly(ethylene terephtalate) is shown [8,195]. This polymer is applied in the production of fabrics.

In regard to linear copolymers, the most common ones are ethylene–propylene, styrene–butadiene, acrylonitrile–butadiene, and acrylonitrile–butadiene–styrene [195]. These materials are

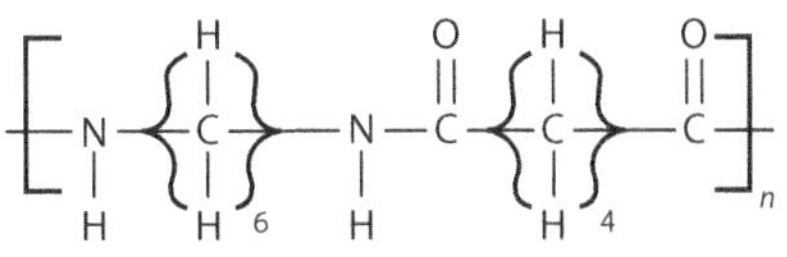

FIGURE 2.38 Polyhexamethylene adipamide.

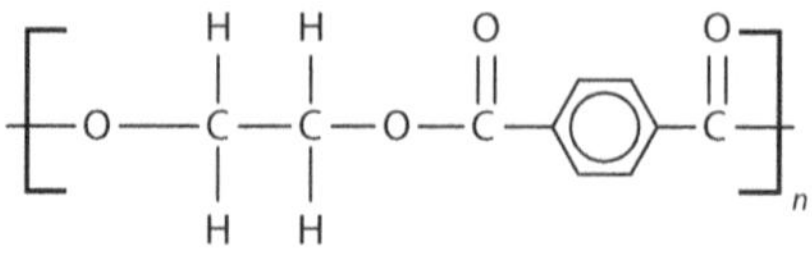

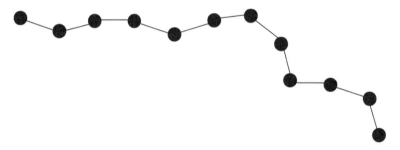

FIGURE 2.39   Poly(ethylene terephtalate).

rubbers and elastomers, for example, the random copolymer styrene–butadiene is a rubber applied for the production of automobile tires. Besides, the acrylonitrile–butadiene rubber is another random highly elastic copolymer that is applied for the production of gasoline pipes [8]. The acrylonitrile–butadiene–styrene rubber is applied as roofing materials [13].

In general, the polymer chain of organic polymers, like those analyzed here, are formed by a carbon chain backbone, where the polymer chain of molecules is generally not straight, since bending and twisting of the chain is possible. The angle between single bonded carbon atoms is approximately 109°, consequently the chain has a zigzag pattern (see Figure 2.40), where while the single bond keeps the angle near 109°, the polymer chain can rotate around the single bond [8]. On the other hand, double and triple bonds are rigid.

Consequently, a molecular polymer chain bends forming coils and curves (see Figure 2.41), where neighboring chains interlace and entangle. The massive entanglement of the chain molecules in elastic materials is the origin of the big expansion exhibited by elastic materials [8].

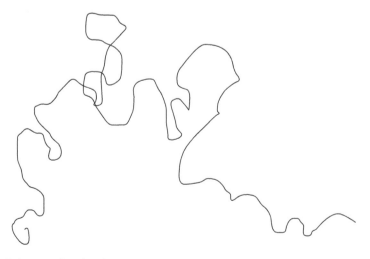

FIGURE 2.40   Zigzag pattern of a chain polymer molecule.

FIGURE 2.41   Polymer coil molecule.

2.9.2.2 Branched Polymers

The branched polymeric structure shows a conformation for the polymer molecule, where branches or side chains radiate to from a linear polymer. In this case, the monomer, α, is consecutively assembled in the linear chain, and the monomer, β, forms the branches (see Figure 2.42), where both monomers α and β can be identical as in the case of low-density polyethylene.

The branches, formed during the synthesis process in this type of polymers, are regarded as part of the main linear chain. These branches diminish the packing efficacy of the polymer (see Figure 2.43) and reduce the density of the polymer [8], since the intermolecular forces that bond the different polymer molecules are weaker, given the distance between the main chains is bigger for the branched polymer. This is the situation for low-density polyethylene.

Dendrimers, a very important set of materials, are a particular case of branched polymers in which every monomer unit is branched, that is, the polymer consists of a repeating unit that comprises a branching group. In this regard, the amount of branches augments, in agreement to a power law expression, with the number of generations of the polymer. Thereafter, dendrimers are macromolecules with a roughly spherical shape.

2.9.2.3 Cross-Linked Polymers

We are interested in the application of porous polymers for the preparation of adsorbents [191,192], ion exchangers [193,194], and membranes [60,198,199]. For these applications, cross-linked polymers are fundamentally applied.

FIGURE 2.42 Branched polymer molecule.

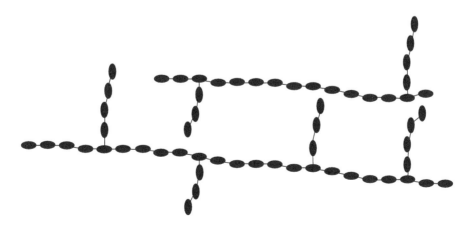

FIGURE 2.43 Representation of a branched polymer.

During cross-linking, the different chains bond together to form a certain type of network (see Figure 2.44), where the formed material is named a thermoset, since it does not flow during heating. A typical cross-linked polymer is obtained by the copolymerization of styrene and divinylbenzene.

The resulting material contains small pores, a consequence of the fact that divinylbenzene binds simultaneously linear chains of styrene at various points (see Figure 2.45); consequently, these materials can be good adsorbents [200–202]. In fact, as was previously stated, polystyrene resins cross-linked with divinylbenzene (Amberlite) are extensively used as adsorbents.

More concretely, we are interested in highly cross-linked, permanently porous polymers. These materials, which have a permanent porous structure produced during their synthesis and preserved in the dry state [203], are employed in a broad variety of applications [204] as adsorbents and ion exchangers [191–194].

The inner porous morphology of these resins is distinguished by interconnected channels that form a porous network, which pervades the rigid, significantly cross-linked polymer matrix [205]. These materials are often synthesized by suspension polymerization [206], where a polymerization mixture which includes a cross-linking monomer, a functional comonomer or comonomer, an initiator, and a porogenic agent is polymerized.

As common examples of monomers used for the production of resins, methacrylate and styrene can be mentioned, where the cross-linkers are ethylene dimethacrylate (EDMA) or divinyl benzene (DVB), respectively [205]. The most important resins of this type are the poly(styrene–divinyl benzene) type or Amberlite resins, as well as the Wofatit and Lewatit [192] types. These materials have been widely applied as adsorbents [191,192,207] and as ion exchangers [193,194,208].

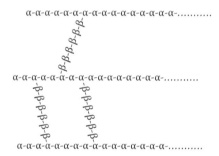

FIGURE 2.44   Representation of a cross-linked polymer.

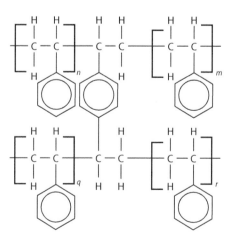

FIGURE 2.45   Poly(styrene–divinylbenzene) copolymer.

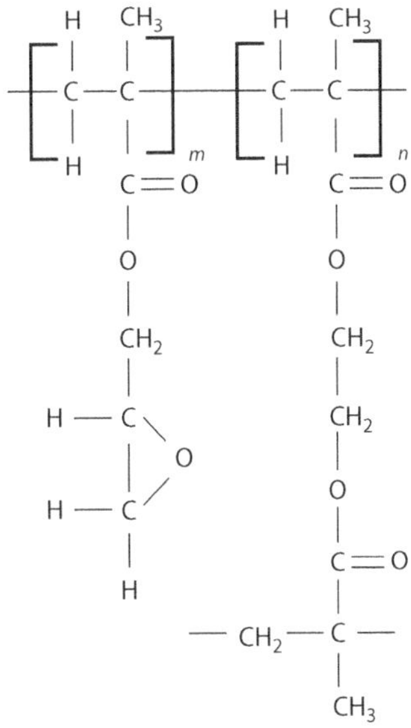

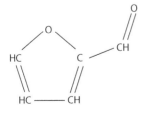

FIGURE 2.46 Poly(glycidyl methacrylate–ethyleneglycol dimethacrylate) copolymer.

An additional porous polymer is poly(glycidyl methacrylate–ethyleneglycol dimethacrylate) (see Figure 2.46) that is synthesized by suspension polymerization in the presence of an inert porogen in the polymerization reaction, obtaining a material with an internal macroporous morphology characterized by an interconnected pore network, which permeates the extensively cross-linked polymer matrix [209].

Another method applied to produce porous polymers is based on the addition of an inorganic matrix of well-known porosity, for example, silica gel or aluminum oxide, to the reacting mixture [210–212]. Subsequent to the polymerization process, the inorganic pattern is eliminated by dissolution without destruction of the produced polymer [211]. These materials develop a complex pore system [211–213].

2.9.3 FURFURAL RESINS

Furfural ($C_5H_4O_2$) is a chemical obtained from agricultural products such as sugar cane, corncob, oat, wheat fiber, and others. Chemically, it is an aromatic aldehyde (see Figure 2.47), and takes part in the same type of reactions as other aldehydes.

It is a nonstable and highly reactive compound. In particular, furfural can react through the carbonyl group (=CO) and the furanic ring, that is, the carbonyl group can participate in addition reactions and the furanic ring can be broken [214]. In this regard, when furfural is heated in the presence of acids, it solidifies into a solid, thermosetting resin [215–219]. The polymerization process, in this case, follows an acidic mechanism where the aldolic condensation of furfural is produced [220]. With the help of the polymerization of furfural, in the presence of a Brönsted acid catalyst, it is possible to get porous polymers that are good adsorbents [217,218].

Besides, furfural polymerizes in the presence of an acid catalyst with phenol, urea, and acetone. In this regard, the phenol–furfural and urea–furfural resins are important [221–223]. Furfural reacts usually as does all α-substituted aldehydes; in this regard, with phenol it condenses in the presence of either alkali or acid to form synthetic resins in a reaction that is very similar to that of phenol with formaldehyde or acetaldehyde [223].

However, the most important furan resins are those produced with 2-furfuryl alcohol, for example, the 2-furfuryl alcohol–formaldehyde-based resins, which are normally synthesized by a condensation reaction catalyzed by acidic sites and promoted by heat [224]; or the poly(furfuryl alcohol) thermosetting resin that is usually synthesized by the cationic condensation of its monomer 2-furfuryl alcohol, which polymerizes exothermically in the presence of a catalyst such as acid and iodine in methylene chloride, producing black, amorphous, and branched and/or cross-linked structures [225].

FIGURE 2.47 Furfural structure.

2.9.4 COORDINATION POLYMERS

In recent years, great attention has been given to the application of coordination polymers in the development of new porous materials to be applied in gas adsorption and other fields [226–236]. Some authors consider that in comparison with usual porous materials, for example, zeolites and activated carbons, these materials are more promissory, as a consequence of the fact that their framework is designable and more flexible because of a diversity of coordination geometries of metal centers and multifunctionality of bridging organic parts [228]. In this regard, the synthesis and characterization of a coordination polymer having a three-dimensional network structure bridging a two-dimensional layer of porous copper(II) terephthalate with triethylenediamine as a pillar ligand has been reported [229]. This methodology has been expanded to obtain a series of coordination polymers having a three-dimensional network structure bridging a two-dimensional layer of porous copper(II) dicarboxylate 1,2 with triethylenediamine as a pillar ligand [228]. This novel family of compounds are also known as metal–organic frameworks (MOFs), because they form zeolitic-like frameworks [230–236]. The design and synthesis of MOFs has produced a great amount of structures that have been shown to possess effective gas and liquid adsorption properties [230–236]. MOFs are crystalline inorganic–organic hybrid materials consisting of metal ions and organic molecules linked in space to create infinite one-, two-, or three-dimensional frameworks [232]. The modularity of MOFs, particularly, the capability to alter the organic and/or inorganic components, provides a mean to change and control the properties of such materials; in this regard, the inner empty spaces can be functionalized with organic groups by applying the proper synthesis design [233]. Specially, porous structures assembled from discrete metalcarboxylate clusters and organic links have been revealed to be open to a methodical change in pore size and functionality, a feature that has led to the synthesis of MOFs capable of outstanding hydrogen and methane storage properties [232].

The stable porosity of such MOFs is attributable to the structural properties of the metal-carboxylate clusters, where each metal ion is locked into place by the carboxylates, to create rigid units of simple geometry, referred to as SBUs [232], similarly to the case of inorganic zeolites.

Another particularly interesting example is the zeolitic imidazolate frameworks (ZIFs) [230]. This is a new class of nanoporous compounds that consists of tetrahedral clusters of MN_4, where M = Co, Cu, Zn, etc. are linked by simple imidazolate ligands [232]. As a subclass of the group of MOFs, the ZIFs display adjustable pore size and chemical functionality of standard MOFs; simultaneously, they have the outstanding chemical stability and the abounding structural diversity of inorganic zeolites [230]. Other similar structures, named isoreticular metal–organic frameworks (IRMOFs), are members of a new group of crystalline materials shaped from metal-oxide clusters bridged by functionalized organic links [231].

An additional family of organometallic materials is the cyanometallates, which are Prussian blue analogues. These are microporous materials, similar to zeolites, with relatively large adsorption space and small access windows [237–241]. These Prussian blue analogues develop zeolite-like structures based upon a simple cubic $(T[M(CN)_6])$ framework, in which octahedral $[M(CN)_6]^{n-}$ complexes are linked via octahedrally coordinated, nitrogen-bound T^{m+} ions [237]. In the prototypic compound, that is, Prussian blue, specifically $(Fe_4[Fe(CN)_6]_3 \cdot 14H_2O)$, charge balance with the Fe^{3+} ions conducts to vacancies at one-quarter of the $[Fe(CN)_6]^{4-}$ complexes [242].

The porous character of the cyanometallates structure is linked to the coordination taken up by the metal (M) in the hexacyanometallate anion, $[M(CN)_6]^{n-}$, that shapes the salt of the cyanometallate anion [238]. For example, hexacyanometallates can be considered as a three-dimensional framework of a molecular block, the hexacyanometallate anion, $[M(CN)_6]^{n-}$, through a transition metal cation T^{m+}, which connects the N ends of the adjacent blocks [239].

These compounds are lightweight microporous crystalline materials, with comparatively ample adsorption space, small access windows (0.5–1.0 nm), and intermediate to low adsorption interaction with the adsorbate [237,238]. The relatively low adsorbent–adsorbate interactions seem suitable for use where constant cycles of adsorption and regeneration are needed [239]. These are characteristics that are appealing for the separation and storage of small-sized molecules and make these materials as promising as many of the metal organic frameworks currently being evaluated as gas storage materials [239,240].

REFERENCES

1. F. Schutz and K. Unger, in *Preparation of Solid Catalysts*, G. Ertl, H. Knozinger, and J. Weitkamp, (editors), Wiley-VCH, Weinheim, Germany, 1999, p. 60.
2. I.K. Pokhodnya and E.O. Paton, (editors), *Advanced Materials Science: 21st Century*, Vol. 1, Cambridge International Science Publishers, Cambridge, UK, 1999.
3. Ch. Kittel, *Quantum Theory of Solids* (2nd edition), John Wiley & Sons, New York, 2004.
4. E.N. Kaufmann, (editor), *Characterization of Materials*, Vols. I and II, John Wiley & Sons, New York, 2003.
5. R. Asokami, *Solid State Physics*, Anshan Publishing, Tunbridge Wells, Kent, UK, 2006.
6. S.O. Kasap, *Principles of Electronic Materials* (2nd edition), McGraw-Hill Higher Education, New York, 2002.
7. W.F. Smith, *Principles of Materials Science and Engineering* (3rd edition), McGraw-Hill Companies Inc., New York, 1999.
8. W.D. Callister Jr., *Materials Science and Engineering: An Introduction* (7th edition), John Wiley & Sons, New York, 2006.
9. Ch. Kittel, *Introduction to Solid State Physics*, John Wiley & Sons, New York, 1971.
10. H.P. Myers, *Introduction to Solid State Physics* (2nd edition), CRC Press, Boca Raton, FL, 1997.
11. J.M. Ziman, *Principles of the Theory of Solids* (2nd edition), Cambridge University Press, Cambridge, UK, 1972.
12. N.W. Ashcroft and N.D. Mermin, *Solid State Physics*, Brooks/Cole, Belmont, MA, 1976.
13. J.I. Gersten and F.W. Smith, *The Physics and Chemistry of Materials*, John Wiley & Sons, New York, 2001.
14. G.A. Somorjai, *Chem. Rev.*, 96, 1223 (1996).
15. B.S. Bosktein, M.I. Mendelev, and D.J. Srolovitz, *Thermodynamics and Kinetics in Materials Science*, Oxford University Press, Oxford, UK, 2005.
16. R. Roque-Malherbe, *Adsorption and Diffusion in Nanoporous Materials*, CRC Press, Boca Raton, FL, 2007.
17. I.N. Levine, *Physical Chemistry* (5th edition), McGraw-Hill, Boston, MA, 2002.
18. A. Beiser, *Concepts of Modern Physics* (6th edition), McGraw-Hill, Boston, MA, 2003.
19. D.A. McQuarrie, *Statistical Mechanics*, University Science Books, Sausalito, CA, 2000.
20. S.T. Thornton and A. Rex, *Modern Physics* (2nd edition), Thomson Learning, Singapore, 2002.
21. P.W. Atkins, *Physical Chemistry* (6th edition), W.H. Freeman & Co., New York, 1998.
22. C.N.R. Rao and J. Gopalakrishnan, *New Directions in Solid State Chemistry* (2nd edition), Cambridge University Press, Cambridge, UK, 1997.
23. A.E. Van Diepen, M. Makkee, and J.A. Moulin, in *Environmental Catalysis*, F.J.J.G. Janssen and R.A. van Santen, (editors), Imperial College Press, London, UK, 1999.
24. G.A. Somorjai, *Introduction to Surface Chemistry and Catalysis*, John Wiley & Sons, New York, 1994.
25. J.M. Thomas and W.J. Thomas, *Principle and Practice of Heterogeneous Catalysis*, VCH Publishers, New York, 1997.
26. M. Che, O. Clause, and Ch. Marcilly, in *Preparation of Solid Catalysts*, G. Ertl, H. Knozinger, and J. Weitkamp, (editors), Wiley-VCH, Weinheim, Germany, 1999, p. 315.
27. D.V. Schriver and A.T. Atkins, *Inorganic Chemistry* (3rd edition), W.H. Freeman & Co. New York, 1999.
28. H.S. Taylor, *Proc. Roy. Soc. Lond.*, A108, 105 (1952).
29. A.A. Balandin, *Adv. Catal.*, 19, 1 (1969).
30. J.H. Sinfelt, *Prog. Solid State Chem.*, 10, 55 (1975).
31. J.H. Sinfelt and G.D. Meitzner, *Acc. Chem. Res.*, 26, 1 (1993).
32. A. Mackintosh and O.K. Andersen, in *Electrons at the Fermi Surface*, M. Soringford, (editor), Cambridge University Press, Cambridge, UK, 1980.
33. N.F. Mott, *Adv. Phys.*, 13, 325 (1964).
34. L.F. Mattheis, *Phys. Rev. A*, 134, A970 (1964).
35. P.H.T. Philipsen and E.J. Baerends, *J. Phys. Chem. B*, 110, 12470 (2006).
36. J.K. Nørskov, *Phys. Rev. B*, 26, 2875 (1982).
37. M.S. Wainwright, in *Preparation of Solid Catalysts*, G. Ertl, H. Knozinger, and J. Weitkamp, (editors), Wiley-VCH, Weinheim, Germany, 1999. p. 28.
38. K. Hauffe, *Reactionen in und Festenstoffen* (2nd edition), Springer-Verlag, Berlin, 1966.
39. Th. Wolkenstein, *Adv. Catal.*, 12, 189 (1960).
40. M. Martin, in *Diffusion in Condensed Matter*, P. Heithans and J. Karger, (editors), Springer-Verlag, Berlin, 2005, p. 209.

41. J. Frenkel, *Z. Physik*, 53, 652 (1926).
42. C. Wagner and W. Schottky, *Z. Phys. Chem.*, B11, 163 (1930).
43. F. Kroger and V. Vink, *Solid State Phys.*, 3, 307 (1956).
44. V.R. Choudhary, S.T. Chaudhari, and M.Y. Pandit, *J. Chem. Soc., Chem. Commun.*, 1158 (1991).
45. M. Fernandez-Garcia, A. Martinez-Arias, J.C. Hanson, and J.A. Rodríguez, *Chem. Rev.*, 104, 4063 (2004).
46. T.L. Thompson and J.T. Yates Jr., *Chem. Rev.*, 106, 4428 (2006).
47. A.L. Linsebigler, G. Lu, and J.T. Yates Jr., *Chem. Rev.*, 95, 735 (1995).
48. A. Hameed and M.A. Gondal, *J. Mol. Catal. A*, 219, 109 (2004).
49. C. Chen, P. Lei, H. Ji, W. Ma, and J. Zhao, *Environ. Sci. Technol.*, 38, 329 (2004).
50. M. Yagi and M. Kaneko, *Chem. Rev.*, 101, 21 (2001).
51. C.E. Housecroft and A.G. Sharpe, *Inorganic Chemsitry* (2nd edition), Pearson Education Limited, Essex, UK, 2005.
52. W. Porterfield, *Inorganic Chemistry. A Unified Approach* (2nd edition), Academic Press, San Diego, CA, 1993.
53. K.H. Kim and J.S. Chol, *J. Phys. Chem.*, 85, 2447 (1981).
54. M.A. Fox and M.T. Dulay, *Chem. Rev.*, 93, 341 (1993).
55. A. Maldotti, A. Molinari, and R. Amadelli, *Chem. Rev.*, 102, 3811 (2002).
56. J.D. Holladay, Y. Wang, and E. Jones, *Chem. Rev.*, 104, 4767 (2004).
57. L.L. Wikstrom and K.E. Nobe, *Ind. Eng. Chem. Proc. Des. Dev.*, 4, 191 (1965).
58. M. Asadullah, T. Miyazawa, S-i. Ito, K. Kunimori, and K. Tomishige, *Energ. Fuel.*, 17, 842 (2003).
59. M. Mulder, *Basic Principles of Membrane Technology*, Kluwer Academic Publishers, Dordrecht, the Netherlands, 1996.
60. R.W. Baker, *Membrane Technology and Applications*, John Wiley & Sons, New York, 2004.
61. M.F. Goosen, S.S. Sablani, and R. Roque-Malherbe, in *Handbook of Membrane Separations: Chemical, Pharmaceutical, and Biotechnological Applications*, A.K. Pabby, A.N. Sastre, and S.S. Rizvi, (editors), CRC Press, Boca Raton, FL, 2008, p. 325.
62. G. Saracco, H.W.J.P. Neomagus, G.F. Versteeg, and W.P.M. Swaaij, *Chem. Eng. Sci.*, 54, 1997 (1999).
63. H.P. Hsieh, *Inorganic Membranes for Separation, and Reaction, Membrane Science and Technology Series 3*, Elsevier, Amsterdam, the Netherlands, 1996.
64. S. Morooka and K. Kusakabe, *MRS Bull.*, March, 25, 1999.
65. H. Mehrer, in *Diffusion in Condensed Matter*, P. Heithans and J. Karger, (editors), Springer-Verlag, Berlin, 2005, p. 3.
66. R. Oesten and R.A. Higgins, *Ionics*, 1, 427 (1995).
67. K.-D. Kreuer, *Chem. Mater.*, 8, 610 (1996).
68. T.H. Etsell and S.N. Flengas, *Chem. Rev.*, 70, 339 (1970).
69. C.J. Howard, E. Kisi, and O. Ohtaka, *J. Am. Ceram. Soc.*, 74, 2321 (1991).
70. D.R. Askeland, *The Science and Engineering of Materials* (3rd edition), PWS Publishing Company, Boston, MA, 1994.
71. J.B. Goodenough, in *Mixed Ionic Electronic Conducting Perovskites for Advanced Energy Systems*, N. Orlovskaya and N. Browning, (editors), *NATO Science Series*, Vol. 173, Kluwer Academic Publishers, Dordrecht, the Netherlands, 2004, p. 1.
72. G. Adachi, N. Imanaka, and S. Tamura, *Chem. Rev.*, 102, 2405 (2002).
73. C.-Y. Wang, *Chem. Rev.*, 104, 4727 (2004).
74. M.A. Peña and J.L.G. Fierro, *Chem. Rev.*, 101, 1981 (2001).
75. J. Twu and P.K. Gallagher, in *Properties and Applications of Perovskite Type Oxides*, L.G. Tejuca and J.L.G. Fierro, (editors), Marcel Dekker, New York, 1993, p. 1.
76. V.M. Goldschmidt, *Skr. Nor. Viedenk.-Akad., Kl. I: Mater.-Naturvidensk. Kl.*, No. 8 (1926).
77. K.S. Knight, *Solid State Ionics*, 74, 109 (1994).
78. H. Iwahara, H. Uchida, and K. Morimoto, *J. Electrochem. Soc.*, 137, 462 (1990).
79. H. Iwahara, K. Uchida, and K. Ogaki, *J. Electrochem. Soc.*, 135, 529 (1988).
80. K.D. Kreuer, *Solid State Ionics*, 97, 1 (1997).
81. N. Bonanos, B. Ellis, K.S. Knight, and M.N. Mahmood, *Solid State Ionics*, 35, 179 (1989).
82. N. Bonanos, *Solid State Ionics*, 53–56, 967 (1992).
83. N. Bonanos, *J. Phys. Chem. Solids*, 54, 867 (1993).
84. H. Iwahara, H. Uchida, and K. Morimoto, *J. Electrochem. Soc.*, 137, 462 (1990).
85. D. Shima and S. Haile, *Solid State Ionics*, 97, 443 (1997).
86. T. Yajima and H. Iwahara, *Solid State Ionics*, 50, 281 (1992).
87. K. Liang and A. Nowick, *Solid State Ionics*, 61, 77 (1993).

88. T. Norby, *Solid State Ionics*, 40–41, 857 (1990).
89. S. Nieto, R. Polanco, and R. Roque-Malherbe, *J. Phys. Chem. C*, 111, 2809 (2007).
90. C.N.R. Rao, J. Gopalakrishnan, and K. Vidyasagar, *Indian J. Chem. Sect.*, 23A, 265 (1984).
91. D.M. Smyth, *Annu. Rev. Mater. Sci.*, 15, 329 (1985).
92. D.M. Smyth, in *Properties and Applications of Perovskite-Type Oxides*, L. Tejuca, and J.L.G., Fierro, (editors), Marcel Dekker, New York, 1993, p. 47.
93. T. Norby and R. Hausgrud, in *Non-Porous Inorganic Membranes*, A.F. Sammells and M.V. Mundschau, (editors), Wiley-VCH Verlag GmbH & Co., Weiheim, Germany, 2006, p. 1.
94. J. Wu, L.P. Li, W.T.P. Espinosa, and S.M. Haile, *J. Mater. Res.*, 19, 2366 (2004).
95. J. Wu, R.A. Davies, M.S. Islam, and S.M. Haile, *Chem. Mater.*, 17, 846 (2005).
96. L. Li, A. Li, and E. Iglesia, *Stud. Surf. Sci. Catal.*, 136, 357 (2001).
97. R.M. Barrer, *Zeolites and Clay Minerals as Sorbents and Molecular Sieves*, Academic Press, London, 1978.
98. D.W. Breck, *Zeolite Molecular Sieves*, John Wiley & Sons, New York, 1974.
99. E.F. Vansant, *Pore Size Engineering in Zeolites*, John Wiley & Sons, New York, 1990.
100. R. Szostak, *Handbook of Molecular Sieves*, Van Nostrand-Reinhold, New York, 1992.
101. G.V. Tsitsisvili, T.G. Andronikashvili, G.N. Kirov, and L.D. Filizova, *Natural Zeolites*, Ellis Horwood, New York, 1992.
102. R. Roque-Malherbe, L. Lemes-Fernandez, L. Lopez-Colado, C. de las Pozas, and A. Montes-Caraballal, in *Natural Zeolites '93 Conference Volume International Committee on Natural Zeolites*, D.W. Ming and F.A. Mumpton, (editors), Brockport, NY, 1995, p. 299.
103. R. Roque-Malherbe, *Mic. Mes. Mat.*, 41, 227 (2000).
104. R. Roque-Malherbe, in *Handbook of Surfaces and Interfaces of Materials*, Vol. 5, H.S. Nalwa, (editor), Academic Press, New York, Chapter 12, 2001, p. 495.
105. M. Guisnet and J.-P. Gilson, (editors), *Zeolites for Cleaner Technologies*, Imperial College Press, London, 2002.
106. H.G. Karge, M. Hunger, and H.K. Beyer, in *Catalysis and Zeolites. Fundamentals and Applications*, J. Weitkamp and L. Puppe, (editors), Springer-Verlag, Berlin, 1999, p. 198.
107. F. Marquez-Linares and R. Roque-Malherbe, *Facets-IUMRS, J.*, 2(4), 14 (2003); 3(1), 8 (2004).
108. Ch. Baerlocher, W.M. Meier, and D.M. Olson, *Atlas of Zeolite Framework Types* (5th edition), Elsevier, Amsterdam, the Netherlands, 2001.
109. J. Smith, *Chem. Rev.*, 88, 149 (1988).
110. C.T. Kresge, M.E. Leonowicz, W.J. Roth, J.C. Vartuli, and J.S. Beck, *Nature*, 359, 710 (1992).
111. J.S. Beck, J.C. Vartuli, W.J. Roth, M.E. Leonowicz, C.T. Kresge, K.D. Schmitt, C.T.-W. Chu et al., *J. Am. Chem. Soc.*, 114, 10834 (1992).
112. X.S. Zhao, G.Q. Lu, and G.J. Millar, *Ind. Eng. Chem. Res.*, 35, 2075 (1996).
113. G.J.A.A. Soler-Illia, C. Sanchez, B. Lebeau, and J. Patarin, *Chem. Rev.*, 102, 4093 (2002).
114. C.S. Cundy and P.A. Cox, *Chem. Rev.*, 103, 663 (2003).
115. M.E. Davies, *Nature*, 417, 813 (2002).
116. J.M. Garces, A. Kuperman, D.M. Millar, M. Olken, A. Pyzik, and W. Rafaniello, *Adv. Mat.*, 12, 1725, (2000).
117. T.J. Barton, L.M. Bull, G. Klemperer, D.A. Loy, B. McEnaney, M. Misono, P.A. Monson et al., *Chem. Mater.*, 11, 2633 (1999).
118. A. Monnier, F. Schüth, Q. Huo, D. Kumar, D. Margolese, R.S. Maxwell, G.D. Stucky et al., *Science*, 261, 1299 (1993).
119. Q. Huo, R. Leon, P. Petroff, and G.D. Stucky, *Science*, 268, 1324 (1995).
120. D. Zhao, J. Feng, Q. Huo, N. Melosh, G.H. Fredrickson, B.F. Chmelka, and G.D. Stucky, *Science*, 279, 548 (1998).
121. A. Nossov, E. Haddad, F. Guenneau, A. Galarneau, F. Di Renzo, F. Fajula, and A. Gedeon, *J. Phys. Chem. B*, 107, 12456 (2003).
122. A. Corma, *Chem. Rev.*, 97, 2373 (1997).
123. C. de las Pozas, R. López-Cordero, C. Díaz-Aguilas, M. Cora, and R. Roque-Malherbe, *J. Solid State Chem.*, 114, 108 (1995).
124. J.A. Martens and P.A. Jacobs, in *Catalysis and Zeolites: Fundamentals and Applications*, J. Weitkamp and L. Puppe, (editors), Springer-Verlag, Berlin, 1999, p. 53.
125. G.H. Kühl, in *Catalysis and Zeolites: Fundamentals and Applications*, J. Weitkamp and L. Puppe, (editors), Springer-Verlag, Berlin, 1999, p. 81.
126. A. Corma, *Chem. Rev.*, 95, 559 (1995).

127. W.M.H. Sachtler, in *Preparation of Solid Catalysts*, G. Ertl, H. Knozinger, and J. Weitkamp, (editors), Wiley-VCH, New York, 1997, p. 388.

128. A. Corma and H. Garcia, *Chem. Rev.*, 103, 4307 (2003).

129. H. Hattori, *Chem. Rev.*, 95, 537 (1995).

130. D. Barthomeuf, *J. Phys. Chem.*, 88, 42 (1984).

131. S.L. Suib, *Chem Rev.*, 93, 803 (1993).

132. A. Clearfield and M. Kuchenmeister, in *Supramolecular Architecture*, T. Bein, (editor), American Chemical Society Symposium Series, Washington, DC, 1992, p. 128.

133. D.E. Vaughan, *Catal. Today*, 2, 187 (1988).

134. H.S. Sherry, in *Handbook of Zeolite Science and Technology*, S.M. Auerbach, K.A. Corrado, and P.K. Dutta, (editors), Marcel Dekker, New York, 2003, p. 1007.

135. A. Marti and J. Colon, *Inorg. Chem.*, 42, 2830 (2003).

136. A. Clearfield, *Chem. Rev.*, 88, 125 (1988).

137. R.P. Bontchev, S. Liu, J.L. Krumhansl, J. Voigt, and T.M. Nenoff, *Chem. Mater.*, 15, 3669 (2003).

138. M.E. Grillo, and J. Carrazza, *J. Phys. Chem.*, 100, 12261 (1996).

139. J. Rocha and Z. Lin, *Rev. Min. and Geochem.*, 57, 173 (2005).

140. A.E. Gash, P.K. Dorhout, and S.H. Strauss, *Inorg. Chem.*, 39, 5538 (2000).

141. P.G. Menon and B. Delmon, in *Preparation of Solid Catalysts*, G. Ertl, H. Knozinger, and J. Weitkamp, (editors), Wiley-VCH, Weinheim, Germany, 1999, p. 109.

142. R. Roque-Malherbe, W. del Valle, N. Planas, K. Gómez, D. Ledes, L. Garay, and J. Ducongé, in *Zeolites '02 Book of Abstracts*, P. Misaelides, (editor), 7th International Conference on the Occurrence, Properties and Use of Natural Zeolites, Thessaloniki, Greece, June 3–7, 2002, p. 316.

143. C. Colella, in *Natural Zeolites 93 Conference Volume*, D.W. Ming and F.A. Mumpton, (editors), International Committee on Natural Zeolites, Brockport, NY, 1995, p. 363.

144. R. Roque-Malherbe, A. Picart, C. Diaz, and G. Rodriguez, Patent Certificate 21055, ONIITEM, C Street and 15 Avenue, Vedado, Havana, Cuba, (1986).

145. R. Roque-Malherbe, W. del Valle, J. Duongé, and E. Toledo, *Int. J. Environ. Pollut.*, 31, 292 (2007).

146. I.R. Agger, M.W. Anderson, J. Rocha, D.P. Luigi, M. Naderi, and A.K. Baggaley, in *Proceedings of the 12th International Zeolite Conference*, M.M.J. Treacy, B.K. Marcus, M.E. Bisher, and J.B. Higgins, (editors), Materials Research Society, Warrendale, PA, 1999, p. IV-2457.

147. I. Delevoye, S. Ganapathy, T. Kumar, C. Fernadez, and J.-P. Amoreux, in *Proceedings of the 12th International Zeolite Conference*, M.M.J. Treacy, B.K. Marcus, M.E. Bisher, and J.B. Higgins, (editors), Materials Research Society, Warrendale, PA, 1999, p. IV-2985.

148. A. Clearfield and Y. Ortiz-Avila, in *Supramolecular Architecture*, T. Bein, (editor), American Chemical Society Symposium Series, Washington, DC, 1992. p. 176.

149. K.S.W. Sing, D.H. Everett, R.A.W. Haul, L. Moscou, R.A. Pirotti, J. Rouquerol, and T. Siemieniewska, *Pure App. Chem.*, 57, 603 (1985).

150. R.K. Iler, *The Chemistry of Silica*, John Wiley & Sons, New York, 1979.

151. C. Burda, X. Chen, R. Narayanan, and M.A. El-Sayed, *Chem. Rev.*, 105, 1025 (2005).

152. A.C. Pierre and G.M. Pajonk, *Chem. Rev.*, 102, 4243 (2002).

153. B.L. Cushing, V.L. Kolesnichenko, and C.J. O'Connor, *Chem. Rev.*, 104, 3893 (2004).

154. R. Roque-Malherbe and F. Marquez-Linares, *Mat. Sci. Semicond. Proc.*, 7, 467 (2004).

155. R. Roque-Malherbe and F. Marquez-Linares, *Surf. Interf. Anal.*, 37, 393 (2005).

156. F. Marquez-Linares and R. Roque-Malherbe, *J. Nanosc. Nanotech.*, 6, 1114 (2006).

157. W. Stobe, A. Fink, and E. Bohn, *J. Colloids Interface Sci.*, 26, 62 (1968).

158. H. van Damme, in *Adsorption on Silica Surfaces*, E. Papirer, (editor), Marcel Dekker, New York, 2000, p. 119.

159. B.A. Morrow and I.D. Gay, in *Adsorption on Silica Surfaces*, E. Papirer, (editor), Marcel Dekker, New York, 2000, p. 9.

160. J. Goworek, in *Adsorption on Silica Surfaces*, E. Papirer, (editor), Marcel Dekker, New York, 2000, p. 167.

161. R. Duchateau, *Chem. Rev.*, 102, 3525 (2002).

162. C.J. Brinker and G.W. Scherer, *Sol-Gel Science*, Academic Press, New York, 1990.

163. B.F.G. Johnson, *Top. Catal.*, 24, 147 (2003).

164. M. Haruta, *Chem. Rec.*, 3, 75 (2003).

165. B.C. Dunn, D.J. Covington, P. Cole, R.J. Pugmire, H.L.C. Meuzelaar, R.D. Ernst, E.C. Heider, and E.M. Eyring, *Energ. Fuel.*, 18, 1519 (2004).

166. J.L. Gole and Z.L. Wand, *Nano Lett.*, 1, 449 (2001).

167. L. Guczi, A. Horvath, A. Beck, and A. Sarkany, *Stud. Surf. Sci. Catal.*, 145, 351 (2003).

168. O. Dominguez-Quintero, S. Martinez, Y. Henriquez, L. D'Ornelas, H. Krentzien, and J. Osuna, *J. Mol. Catal. A*, 197, 185 (2003).
169. T. Lopez, M. Asomoza, P. Bosch, E. Garcia-Figueroa, and R. Gomez, *J. Catal.*, 138, 463 (1992).
170. J. Panpranot, K. Pattamakomsan, P. Praserthdam, and J.G. Goodwin Jr., *Eng. Chem. Res.*, 43, 6014 (2004).
171. J.W. Patrick, (editor), *Porosity in Carbons*, Edward Arnold, London, 1995.
172. F. Derbyshire, M. Jagtoyen, R. Andrews, A. Rao, I. Martin-Guillon, and E.A. Grulke, in *Chemistry and Physics of Carbon*, Vol. 27, L.R. Radovic, (editor), Marcel Dekker, New York,, 2001, p. 1.
173. L.R. Radovic, C. Moreno-Castilla, and J. Rivera-Utrilla, in *Chemistry and Physics of Carbon*, Vol. 27, L.R. Radovic, (editor), Marcel Dekker, New York, 2001, p. 227.
174. M.M. Dubinin, *Carbon*, 18, 355 (1980).
175. H.P. Boehm, *Adv. Catal.*, 16 (1966) 179; *Carbon*, 32, 759 (1994).
176. C.A. Leon-Leon and L.R. Radovic, in *Chemistry and Physics of Carbon*, Vol. 24, P.A. Thrower, (editor), Marcel Dekker, New York, 1992, p. 213.
177. N.N. Avgul and A.V. Kiselev, in *Chemistry and Physics of Carbon*, Vol. 6, P.J. Walker Jr., (editor), Marcel Dekker, New York, 1970, p. 1.
178. F. Rodriguez-Reinoso and A. Sepulveda-Escribano, in *Handbook of Surfaces and Interfaces of Materials*, Vol. 5, H.S. Nalwa, (editor), Academic Press, New York, 2001, p. 309.
179. D. Lozano-Castello, D. Cazorla-Amoros, A. Linares-Solano, and D.F. Quinn, *J. Phys. Chem. B*, 106, 9372 (2002).
180. V.V. Turov and R. Leboda, in *Chemistry and Physics of Carbon*, L.R. Radovic, (editor), Marcel Dekker, New York, Vol. 27, 2001, p. 67.
181. S. Ege, *Organic Chemistry* (4th edition), Houghton-Mifflin Co., Boston, MA, 1999.
182. I.I. Salame and T.J. Bandosz, *Langmuir*, 16, 5435 (2000).
183. T. Karanfil and J.E. Kilduff, *Environ. Sci. Technol.*, 33, 3217 (1999).
184. C. Hsieh and H. Teng, *J. Colloid Interface Sci.*, 230, 171 (2000).
185. C.C. Leng and N.G. Pinto, *Carbon*, 35, 1375 (1997).
186. K. Laszlo and L.G. Nagy, *Carbon*, 35, 593 (1997).
187. H. Tamon and M. Okazaki, *Carbon*, 34, 741 (1996).
188. L.R. Radovic, I.F. Silva, J.I. Ume, J.A. Menendez, Y. Leon, C.A. Leon, and A.W. Scaroni, *Carbon*, 35, 1399 (1997).
189. P.M. Ajayan, *Chem. Rev.*, 99, 1787 (1999).
190. D. Tasis, N. Tagmatarchis, A. Bianco, and M. Prato, *Chem. Rev.*, 106, 1105 (2006).
191. K. Wagner and S. Schulz, *J. Chem. Eng. Data*, 46, 322 (2001).
192. J. Germain, J. Hradil, J.M.J. Frechet, and F. Svec, *Chem. Mater.*, 18, 4430 (2006).
193. F.J. DeSilva, 25th Annual Water Quality Association Conference, Fort Worth, TX, March 1999.
194. M. Vilensky, B. Berkowitz, and A. Warshawsky, *Environ. Sci. Technol.*, 36, 1851 (2002).
195. L.H. Sperling, *Introduction to Physical Polymer Science*, John Wiley & Sons, New York, 2006.
196. R.O. Ebedele, *Polymer Science and Technology*, CRC Press, Boca Raton, FL, 2000.
197. D.I. Bower, *Introduction to Polymer Physics*, Cambridge University Press, Cambridge, UK, 2002.
198. R.W. Baker, in *Membrane Separation Systems. Recent Developments and Future Directions*, Vol. II, R.W. Baker, E.L. Cussler, W. Eykamp, W.J. Koros, R.L. Riley, and H. Strathmann, (editors), SciTech Publishing Inc., Raleigh, NC, 1991. p. 100.
199. H. Strathmann, in *Handbook of Industrial Membrane Technology*, M. Porter, (editor), Noyes Publications, Park Ridge, NJ, 1990, p. 1.
200. J.R. Benson, *Am. Lab.*, April, 2003.
201. M.I. Abrams and J.R. Millar, *React. Funct. Polym.*, 35, 7 (1997).
202. S.M. Howdle, K. Jerábek, V. Leocorbo, P.C. Marr, and D.C. Sherrington, *Polymer*, 41, 7273 (2000).
203. D.C. Sherrington, *Chem. Commun.*, 2275 (1998).
204. F. Schuth, K.S.W. Sing, and J. Weitkamp, (editors), *Handbook of Porous Solids*, John Wiley & Sons, New York, 2002.
205. A.K. Hebb, K. Senoo, R. Bhat, and A.I. Cooper, *Chem. Mater.*, 15, 2061 (2003); *Composites Sci. Tech.*, 63, 2379 (2003).
206. E. Vivaldo-Lima, P.E. Wood, A.E. Hamielec, and A. Penlidis, *Ind. Eng. Chem. Res.*, 36, 939 (1997).
207. L. Shen, Y. Duan Lei, and F. Wania, *J. Chem. Eng. Data*, 47, 944 (2002).
208. G. Schwachula and G. Popov, *Pure Appl. Chem.*, 54, 2103 (1982).
209. N.S. Pujari, A.R. Vishwakarma, T.S. Pathak, A.M. Kotha, and S. Ponrathnam, *Bull. Mater. Sci.*, 27, 529 (2004).
210. B. Feibush and N.-H. Li, US Patent No. 4,933,372 (1990).

211. A. Deryło-Marczewska, J. Goworek, and W. Zgrajka, *Langmuir*, 17, 6518 (2001).
212. A. Deryło-Marczewska, J. Goworek, S. Pikus, E. Kobylas, and W. Zgrajka, *Langmuir*, 18, 7538 (2002).
213. A. Deryło-Marczewska, J. Goworek, R. Kusak, and W. Zgrajka, *Applied Surface Science*, 195, 117 (2002).
214. A. Dunlop and F.N. Peters, *The Furans*, American Chemical Society Monograph Series, Washington, DC, 1958.
215. R. Sanchez, PhD dissertation, National Center for Scientific Research, Havana, Cuba, 1988.
216. R. Roque-Malherbe, J. Onate-Martinez, and E. Navarro, *J. Mat. Sci. Lett.*, 12, 1037 (1993).
217. R. Sanchez and C. Hernandez, *Eur. Polym. J.*, 30, 51 (1994).
218. R. Sanchez, C. Hernandez, R. Roque-Malherbe, and H. Campaña, Cuban Patent Certificate No. 21644 A1, Cuban Office for the Industrial Property, Havana, 1987.
219. T. Budinova, D. Savova, N. Petrov, M. Razvigorova, V. Minkova, N. Ciliz, E. Apak, and E. Ekinci, *Ind. Eng. Chem. Res.*, 42, 2223 (2003).
220. I.V. Kaminskii, N.V. Ungurean, and V.I. Ilinskii, *Plat. Massy*, 12, 9 (1960).
221. P. Trickey, C.S. Miner, and H.J. Brownlee, *Ind. Eng. Chem.*, 15, 65 (1923).
222. L. Brown and D. Watson, *Ind. Eng. Chem.*, 51, 683 (1959).
223. A.J. Norton, *Ind. Eng. Chem.*, 40, 236 (1948).
224. Z. Laszlo-Hedvig and M. Szeszlay, *Concise Polymeric Materials Encyclopedia*, CRC Press, Boca Raton, FL, 1998, p. 548.
225. H. Wang and J. Yao, *Ind. Eng. Chem. Res.*, 45, 6393 (2006).
226. W. Mori, F. Inoue, K. Yoshida, H. Nakayama, and S. Takamizawa, *Chem. Lett.*, 1219 (1997).
227. K. Seki, S. Takamizawa, and W. Mori, *Chem. Lett.*, 122 (2001).
228. K. Seki and W. Mori, *J. Phys. Chem. B*, 106, 1389 (2002).
229. K. Seki, S. Takamizawa, and W. Mori, *Chem. Lett.*, 332 (2001).
230. H. Wu, W. Zhou, and T. Yildirim, *J. Amer. Chem. Soc.*, 129, 5314 (2007).
231. B. Chen, M. Eddaoudi, S.T. Hyde, M. O'Keeffe, and O.M. Yaghi, *Science*, 291, 1021 (2001).
232. N.L. Rosi, J. Kim, M. Eddaoudi, B. Chem, M. O'Keeffe, and O.M. Yaghi, *J. Am. Chem. Soc.*, 127, 1504 (2005).
233. N. Zabukovec-Logar and V. Kaučič, *Acta Chim. Slov.*, 53, 117 (2006).
234. M. Jacoby, *Chem. Eng. News*, 83, 43 (2005).
235. B. Chen, M. Eddaoudi, T. Reineke, J.W. Kampf, M. O'Keeffe, and O.M. Yaghi, *J. Am. Chem. Soc.*, 122, 11559 (2000).
236. M. Eddaoudi, H. Li, and O.M. Yaghi, *J. Am. Chem. Soc.*, 122, 1391 (2000).
237. S.S. Kaye and J.R. Long, *J. Am. Chem. Soc.*, 127, 6506 (2005).
238. S. Natesakhawat, J.T. Culp, C. Matranga, and B. Bockrath, *J. Phys. Chem. B.*, 111, 1055 (2007).
239. J. Roque, E. Regueraa, J. Balmaseda, J. Rodríguez-Hernández, L. Reguerad, and L.F. del Castillo, *Mic. Mes. Mat.*, 103, 57 (2007).
240. E. Lima, J. Balmaseda, and E. Reguera, *Langmuir*, 23, 5752 (2007).
241. L. Reguera, J. Balmaseda, C.P. Krap, M. Avila, and E. Reguera, *J. Phys. Chem. C*, 112, 17443 (2008).
242. H.J. Buser, D. Schwarzenbach, W. Petter, and A. Ludi, *Inorg. Chem.*, 16, 2704 (1977).

3 Synthesis Methods of Catalyst Adsorbents, Ion Exchangers, and Permeable Materials

3.1 INTRODUCTION

Synthesizing, or producing, new materials or modifying existing ones is as old as our civilization. The concept of synthesis describes a set of chemical and physical procedures that result in a material with a group of properties, which allow its various applications. In this regard, obtaining new materials or modifying existing materials is only possible through a superior comprehension of the fundamental chemical and physical principles that govern the synthesis procedures, the structure of the material, and the desired properties. Behind this effort is a growing demand for better materials, and only during the last 100 years, these efforts have grown into a science—materials science.

Society requires new materials for sustainable development through a sustainable energy economy, and consequently pollution abatement policies have increased the demand for new specific materials. Materials science for sustainable energy and pollution abatement is the scientific endeavor that provides these new materials in order maintain the development of a sustainable society.

The first steps in the development of a new material are its synthesis, preparation, and manufacture, or whatever concept we use means the method applied in order to get the material. In this chapter, we explain some of the most important methods of material synthesis.

3.1.1 Nucleation and Growth: Johnson–Mehl–Avrami Equation

The majority of the existing procedures to synthesize materials includes phase transformation, which in one form or the other, in the majority of cases goes through a nucleation and growth process. It is impossible to give a complete exposition of this subject here. However, a general description of the nucleation and growth process [1–5] is given, using the example of solidification.

Normally, the first and crucial step in any phase transformation is the nucleation process. In the case of solidification, the archetypal phase transition, this process involves the ordering of groups of atoms in the liquid to form very small solid clusters (see Figure 3.1). The ordering process occurs though fluctuations, which take place at temperatures both above and below the melting point. In these conditions, clusters produced above the melting temperature go back to the liquid because it is the stable phase. Conversely, clusters created below the melting temperature can grow to solid nuclei, given their sizes are satisfactorily large to be stable.

From a thermodynamic point of view, the barrier for nucleation is related to the large surface energy of the solid–liquid interface in relation to the increase in energy between the solid and the liquid phases for a small cluster. In a general case of nucleation, this process takes place, as a rule, on heterogeneities which exist in every sample. During solidification, the inclusion of supplementary substrates in the melt gives nucleation sites a diminished energy barrier for nucleation that increases the nucleation rate. This process is recognized as heterogeneous nucleation.

Generally, nucleation is a kinetic process where a small number of atoms form a stable cluster of atoms arranged closely to the structure of the new phase or phases, named nucleus, within the old phase. This nucleus subsequently operates as the primary construction block for a growing grain (see Figure 3.1).

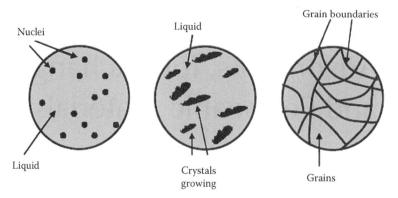

FIGURE 3.1   Nucleation and growth process in the case of solidification.

When a grain has nucleated, it is energetically favorable to increase its size. In the solidification of pure metals, for example, the growth rate of the grain is chiefly managed by the subtraction of the latent heat liberated owing to the phase transformation. In the general case, following a nucleation event, grains grow throughout the old phase until the transformation is complete. Then, the overall transformation is the result of the simultaneously occurring grain nucleation and grain growth (see Figure 3.1).

The experimental kinetic data, plotted as transformed fraction, f, versus time, t, normally present a sigmoid shape (see Figure 3.2). These curves are explained in terms of the nucleation growth mechanism, which takes into account that the new phase forms nuclei randomly in the bulk of the old phase and then these nuclei grow. The growth occurs in the interfacial region between the old and new phases. While the reaction advances, the interface increases constantly, making the reaction accelerate up to an inflection point; ahead of this point the growing nuclei overlaps and the reaction decreases its rate up to the completion of the reaction at $f \rightarrow 1$.

The sigmoid form can be analyzed by splitting the curve into four regions (see Figure 3.2); the first region is the induction period, which ends at $t = \tau$. Afterward, there occurs an acceleration of the transformation, which ends approximately at $t = t_{1/2}$. Thereafter, a deceleration period starts, which runs in the range $0.5 < f < 1$. Finally, we have transformation completion when $f = 1$.

The induction process is normally dominated by the nucleation process. The acceleration step tends to be dominated by growth, while the deceleration is a consequence of the extinction of growth, because of the impingement of diverse growth regions at grain boundaries [5].

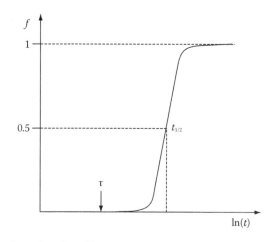

FIGURE 3.2   Phase transformation sigmoid curve.

The kinetic theory presented by Johnson and Mehl [2] and Avrami [3] predicts the volume fraction transformed, f, as a function of time, t, during an isothermal phase transformation. The derivation of the Johnson–Mehl–Avrami kinetics is based on the grouping of the three individual partial processes, that is, nucleation, growth, and impingement of growing particles [5].

The two most important nucleation processes are continuous nucleation, that is, when the nucleation rate is temperature dependent according to an Arrhenius equation, and the site saturation process, that is, when all nuclei are present before the growth starts. The two growth processes normally considered are volume diffusion controlled and interface controlled. Finally, the process that interferes with growth is the hard impingement of homogeneously dispersed growing particles.

The different combinations of nucleation, growth, and impingement processes give rise to the Johnson–Mehl–Avrami kinetic model [4], which results in the following equation

$$f(t) = 1 - \exp\left(-g\int_0^t \dot{P}(x)\left[\phi(t-x)\right]^q dx\right) \qquad (3.1)$$

where
  g is a geometry factor
  $\dot{P}(t)$ is the rate of the nucleation process
  $\phi(t)$ is the growth rate
  q is the dimensionality of the growth, that is, for a three-dimensional grain $q = 3$ and $g = 4\pi/3$.

For a solid-state reaction, one of the solutions of Equation 3.1 is the Avrami–Erofeev equation [3]. The phase transition model that derives this equation supposes that the germ nuclei of the new phase are distributed randomly within the solid; following a nucleation event, grains grow throughout the old phase until the transformation is complete. Then, the Avrami–Erofeev equation is [3]

$$f(t) = 1 - \exp-(kt)^n \qquad (3.2)$$

where
  $3 < n < 4$ for three-dimensional growth
  $2 < n < 3$ for two-dimensional growth or $1 < n < 2$ growth

The Avrami–Erofeev scheme is one of the most widely used for studying solid-state kinetics; it has been effectively applied to a diversity of solid-state processes including decompositions.

3.2 METHODS FOR THE PREPARATION OF METALLIC-SUPPORTED CATALYSTS

In this section, some of the most important methods for catalyst preparation including impregnation, grafting, precipitation, and chemical vapor deposition (CVD) are discussed [6–18].

3.2.1 DEPOSITION OF THE ACTIVE COMPONENT

3.2.1.1 Impregnation

This is a process designed to cover a catalyst support, such as silica, alumina, mesoporous molecular sieves, or other supports with a metallic catalyst, or other catalytically active materials. The process is carried out by contacting the solid support, for a precise time, with a solution containing the active elements, to introduce a solution of the precursor into the pores of the support. During the impregnation process, the support can be completely free of the solvent when the precursor is dissolved. In this case, the impregnation process is termed capillary impregnation, and if already filled with the solvent, then the process is called diffusional impregnation [8].

Throughout the impregnation process, either one process or additional parallel processes, such as adsorption, selective reaction, or even ion exchange, can occur. Occasionally, the process of impregnation is carried out by percolation of the impregnating solution through the support bed, or by sinking the support in the impregnation solution [7]. Two or more active elements may be hosted both in a sole step by coimpregnation, or one after the other in successive impregnations.

Washing, drying, and frequent calcination occurs between and after impregnations [7,8] and in order to prepare metallic-supported catalysts, it is as well necessary to carry out a reduction process [9]. The most common reducing agent is H_2.

To conclude, this method for metal catalyst preparation consists of the impregnation of a preformed support with metal precursors and subsequent calcination and reduction.

3.2.1.2 Grafting

Another deposition method is that involving the creation of a strong bond, for example, a covalent bond, between the support and the active element [7,8]. This process is typically called grafting or anchoring. This procedure is carried out by a chemical reaction between functional hydroxyl groups on the surface of the support and a properly chosen inorganic or organometallic compound of the active element.

For example, as was previously explained in Chapter 2, silica samples have a high concentration of silanols in the surface. These silanols (see Figure 2.30) can be functionalized via simple elimination reactions. Attaching an organic functional group to a silica surface by means of a covalent bond is the most reliable method of modification of the silica surface. The covalent bond used for binding is generally the Si–O–Si bond, where one of the silicon atoms is on the silica surface, and the other comes from organosilicon compounds, such as a trialkoxyorganosilane (see Figure 3.3) [6]. Explicitly, the Si–O–Si bond is formed by the reaction of a Si–OH group previously located on the silica surface with the applied organosilicon compound containing a leaving group of high reactivity on the silicon atom.

Another functionalization process is carried out with a metal carbonyl, as follows [12]:

$$[s{-}OH] + M_3CO_{12} \rightarrow [s{-}O]_y M_3HCO_{10} + 2CO$$

The next step is the elimination of the physisorbed metal complexes by purging with an inert gas and then the decomposition of the metal carbonyl at a higher temperature.

In summary, the first step to functionalize a support is to thermally treat it, in order to remove physisorbed water and dehydroxylate the surface [11]. The second step is the anchoring process, which is the chemical reaction of condensation between the chemical precursor and the –OH surface group [s–OH] [12].

3.2.1.3 Precipitation

This is a technique frequently used for the preparation of both support precursors and catalyst precursors and takes place when two or more solutions are mixed by a suitable method [7]. The growth of nanometer-sized materials, in some cases, involves the process of precipitation of a solid phase from the solution. In this regard, for nanoparticle formation, the solution has to be supersaturated by directly dissolving the solute at a higher temperature and then cooling to low temperatures, or by the addition of the needed reactants to generate a supersaturated solution during the reaction.

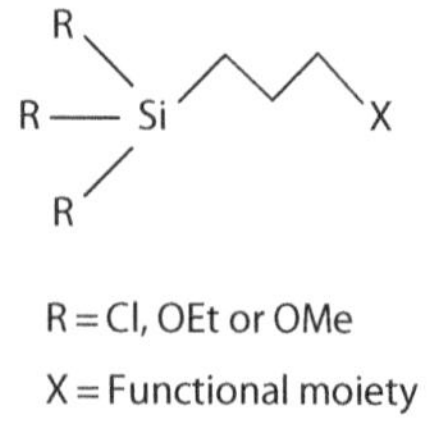

For a particular solvent, there is a definite solubility for a solute, whereby the supplement of any surplus solute will result in the precipitation and

R = Cl, OEt or OMe

X = Functional moiety

FIGURE 3.3 Trialkoxy-organosilane.

formation of nanocrystals. Then, precipitation consists of nucleation followed by growth, where the kinetics of nucleation and growth in homogeneous solutions can be adjusted by controlled chemical release and by controlling the kinetics of precipitation; thus, it is essential to control temperature, pH, and the concentration of the reactants and ions [13].

3.2.1.4 Bifunctional Zeolite Catalysts

The bifunctional zeolite catalysts are composed of both acid sites and metal clusters. The preparation methods of these catalysts encompasses three steps: ion exchange, calcination, and reduction. The ion-exchange process is carried out with aqueous solutions of salts or, more commonly, of complexes of the metals that will be incorporated into the zeolite cavities and channels.

For the case of metals with high catalytic activity such as Pt, Pd, and Rh, the complexed ions $[Pt(NH_3)_4]^{2+}$, $[Pd(NH_3)_4]^{2+}$, and $[Rh(NH_3)_5(H_2O)]^{3+}$, in aqueous solution are exchanged with sodium zeolite [14]. After the ion exchange, the produced solid is calcined in a flow of air or oxygen, in order to remove water and the ligands of the exchanged cations. Finally, the solid produced after calcination is reduced in a hydrogen flow, by passing typically a stream of a mixture of 4 wt % H_2 + 96 wt % of Ar or N_2. The reduction process is described by the following reaction [10,14]

$$M^{n+} + \frac{n}{2}H_2 \rightarrow M^0 + nH^+$$

which produces metallic clusters and Brönsted acid sites.

3.2.1.5 Chemical Vapor Deposition

Chemical vapor deposition (CVD) is a process of chemical nature applied to produce materials with high purity and special properties. It consists of the deposition of a material by chemical reaction from the gas phase [7].

CVD is a flexible and simple process to produce supported catalysts [16]. In a typical CVD process (see Figure 3.4), a substrate is attacked by one or more gaseous precursors that react and/or decompose on the substrate surface to produce the chosen deposit.

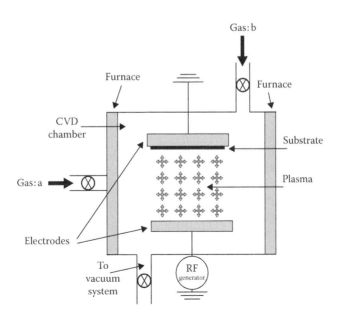

FIGURE 3.4 Chemical vapor deposition facility.

The chemical reactions used in CVD are pyrolysis, hydrolysis, disproportionation, reduction, oxidation, carburization, and nitridization [16]. The selection of the precursors is regulated by general features that can be summarized as follows [17]: stability at room temperature, enough volatility at low temperature, high purity, ability to react plainly on or with the support, and ability to react without the production of side or parasitic reactions.

The most important type of precursors are [17] metal carbonyls of Co, Cr, Fe, Mn, Mo, Ni, Os, Re, Ru, V, W, Ir, Os, and Pt; chlorides of Si, B, Mo, Nb, Ti, V, and Zr; and other compounds. Oxygen, air, or steam are used as oxidizing agents.

Supported metallic Ni, Cr, Fe, Mo, Co, V, Ti, Sn, Ru, and Pd catalysts have been prepared with the help of the CVD method fundamentally using carbonyl and chloride precursors [16].

3.2.1.6 Case Study: Preparation of Ni Bifunctional Catalysts Supported on Homoionic: Na, K, Ca, and Mg Clinoptilolite

A Ni bifunctional catalyst supported in a homoionic natural clinoptilolite will be used, as an example, to further explain the methodology of preparation of this type of catalysts [18]. As was previously commented, the thermal reduction of zeolites previously exchanged with metals is the method currently used for the preparation of bifunctional catalysts for hydrocarbon conversion. To produce these catalysts, synthetic [19,20] and natural zeolites [18,21–23] are used. During this procedure, the zeolite is exchanged with the cationic form of the metal that will be used as the catalyst and afterward the obtained exchanged zeolite is reduced in a H_2 atmosphere at about 450°C and maintained at this temperature for about 2 h [18,19–23].

In general, bifunctional catalysts are applied in hydrocarbon hydrogenation. In this regard, clinoptilolite [20,21] exchanged with Ni^{2+}, and then thermally reduced, was used for the hydrogenation of ethylbenzene [21] and 1-hexene [24,25]. A natural erionite ore exchanged with NH_4^+ and Ni^{2+} and then calcined was tested for the hydrocracking of n-paraffin; the catalyst was tested in a pilot plant and it was demonstrated that a catalytic life of more than 1 year is possible [22,23].

Other examples of the use of natural zeolites in catalysis has been the use of Cd-exchanged chabazite, clinoptilolite, erionite, and mordenite as catalysts in the hydration of acetylene to acetaldehyde [26], and a Cu-exchanged natural mordenite, which was studied as a catalyst for the selective reduction of NO with NH_3 [27].

In this section, the natural zeolite labeled HC, which was used for the preparation of the Ni bifunctional catalysts, was obtained from the deposits of Castillas, La Habana, Cuba, with the following elemental chemical composition, in oxide wt %: SiO_2: 66.8; Al_2O_3: 13.1; Fe_2O_3: 1.3; CaO: 3.2; MgO: 1.2; Na_2O: 0.6; K_2O: 1.9; H_2O: 12.1 [18,25] and a mineralogical composition, in wt %: clinoptilolite 85 and others 15, where the others include montmorillonite (2–10 wt %), quartz (1–5 wt %), calcite (1–6 wt %), feldspars (0–1 wt %), magnetite (0–1 wt %), and volcanic glass (3–6 wt %) [18,25] (see Table 4.1).

The sample HC was refluxed five times during 2 h in a 3 M solution of NaCl, KCl, CaCl, and MgCl. Later, the obtained samples, that is, NaHC, KHC, CaHC, and MgHC, were refluxed five times during 2 h in a 1 M solution of $Ni(NO_3)_2$ to obtain the following samples: NiNaHC, NiKHC, NiCaHC, and NiMgHC [18].

In Table 3.1, the ionic compositions of the homoionic samples Na–HC, K–HC, Ca–HC, Mg–HC, and the original HC sample are displayed [18,28]. The reported data reveal that Na and K are selectively exchanged in the clinoptilolite-containing zeolite rock, HC. On the other hand, the degree of exchange for the bivalent cations, Ca and Mg, is inferior than those reported for the monovalent cations.

In Table 3.2, [18] the data corresponding to the ionic exchange of Ni^{2+} in the samples NaHC, KHC, CaHC, and MgHC are reported. It is evident that in all cases, Ni^{2+} exchanges with the appropriate ion, that is, with Na^+ in the case of NiNaHC, K^+ in the case of NiKHC, Ca^{2+} in the case of NiCaHC, and Mg^{2+} in the case of NiMgHC [18].

TABLE 3.1
Cationic Composition of the Samples HC, NaHC, KHC, CaHC, and MgHC

Sample	$Na^+ \left[mequiv/g \right]$	$K^+ \left[mequiv/g \right]$	$Ca^{2+} \left[mequiv/g \right]$	$Mg^{2+} \left[mequiv/g \right]$
HC	0.2	0.4	0.8	0.7
NaHC	1.6	0.3	0.1	0.0
KHC	0.2	1.7	0.0	0.1
CaHC	0.2	0.3	1.0	0.5
MgHC	0.1	0.3	0.2	1.4

Note: The error in the determination of the cationic composition is 0.1 mequiv/g.

TABLE 3.2
Cationic Composition of the Samples NiNaHC, NiKHC, NiCaHC, and NiMgHC

Sample	$Na^+ \left[mequiv/g \right]$	$K^+ \left[mequiv/g \right]$	$Ca^{2+} \left[mequiv/g \right]$	$Mg^{2+} \left[mequiv/g \right]$	$Ni^{2+} \left[mequiv/g \right]$
NiNaHC	0.4	0.2	0.2	0.1	1.1
NiKHC	0.0	1.2	0.0	0.1	0.8
NiCaHC	0.1	0.2	0.8	0.4	0.6
NiMgHC	0.1	0.4	0.3	0.7	0.7

Note: The error in the determination of the cationic composition is 0.1 mequiv/g.

The reduction process [18]

$$Ni^{2+} + H_2 \rightarrow M^0 + 2H^+$$

was carried out in the reduction cell of a temperature-programmed reduction (TPR) equipment (see Section 4.7.4). In a quartz U-tube flow reactor, of a TPR equipment, used as reduction cell, 0.25 mg of the sample was placed [18,29]. The samples were heated at a rate of 10 K/min from a temperature of 300 K up to 823 K, and maintained at this temperature for 4 h in a flow of 20 cm^3/min (at STP) of dry air. This process was followed by cooling to 300 K. Then, the flow was switched to a reductive hydrogen–argon mixture of molar ratio $H_2/Ar = 7/3$, at a flow rate of 25 cm^3/min (at STP), and the sample heated at a heating rate of 10 K/min from a temperature of 300 K up to 1273 K.

The consumption of hydrogen was monitored with a thermal conductivity detector (TCD) held constantly at 100°C and recorded at a signal rate of 1 point/s [18,29]. The hydrogen consumption was quantified by means of calibration with pure CuO (Merck, >99.9%). The temperature of the sample bed was measured by means of a thermocouple inside the U-tube (with the tip within the catalyst bed) that followed an exact linear ramp throughout the TPR run. A cold trap was used to prevent water passing through the TCD.

The TPR experimental conditions were selected according to criteria reported elsewhere [18,29]. It was shown that Ni^{2+} was reduced in all cases (see Figure 3.5) [18].

The TPR profiles corresponding to samples NiNaHC and NiKHC show only one peak, indicating the existence of only one reduction process (see Figure 3.4) [18]. The obtained catalysts were tested by a hydrogenation reaction with 1-hexene [24,25].

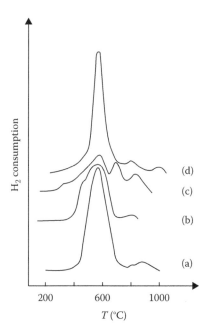

FIGURE 3.5 Thermoreduction profiles of (a) NiNaHC, (b) NiMgHC, (c) NiCaHC, and (d) NiKHC.

3.3 SYNTHESIS OF INORGANIC SOLIDS

3.3.1 Solid-State Reaction Method

Powders of different inorganic solids can be prepared using the solid-state reaction method. A solid-state reaction is a reaction between solids, that is, both the starting materials and the products are solids. This is a well-known and established method for the preparation of inorganic powdered materials where oxides, carbonates, nitrates, and other inorganic compounds are mixed, following stoichiometric proportions. After that, the powders are, in general, thoroughly milled, and finally thermally treated to get the required product. In this case, keeping the mixture in the solid state during the whole transformation allows good control of the composition, allowing homogeneity at the atomic level and a high degree of dispersion of the obtained powder [30].

In the solid-state reaction, nucleation and growth have a fundamental role, because, in essence, the solid-state reaction is a phase transformation. In this type of reactions, nucleation and growth follow similar principles as those previously analyzed in Section 3.1; the principal difference being the increased role of diffusion in solid-state reactions [30].

In order to better understand the methodology of solid-state reactions, we now analyze the production of a typical material. The synthesis of perovskites by a solid-state reaction is a good example [31–35]. To carry out this process, carbonates and oxides of the A- and B-type elements, corresponding to the ABO_3 perovskite formula in the desired molar ratios, to get the required composition of the final product, are thoroughly ball milled in acetone or isopropanol as the milling media [31]. The obtained product is dried at about $100°C$ and subsequently calcined in air at around $600°C$, for 4–8 h using both heating and cooling rates of $2°C/min$. Subsequently, the calcined samples are again briefly ground and then sieved. Thereafter, a second calcination step is carried out at $1300°C–1600°C$ for another 5–15 h, using both heating and cooling rates of $2°C/min$, to ensure that a single perovskite phase will be formed. Finally, this calcined sample will also be ground and sieved [31,32]. After this, it is possible to prepare dense compacts by uniaxially pressing powders at about $250 MPa$, followed by sintering at $1400°C–1600°C$ in air for 5–10 h, using both heating and cooling rates of $2°C/min$ [32]. Now, to explain the details related to the solid-state reaction process, it is better to use a specific reaction, for example, the synthesis of $BaCeO_3$-based proton conductor perovskites [32], in order to obtain a sample with composition $BaCe_{0.95}Yb_{0.05}O_{3-\delta}$ [31].

FIGURE 3.6   SEM micrograph of a powder of the synthesized $BaCe_{0.95}Yb_{0.05}O_{3-\delta}$ perovskite at a magnification of $2000\times$ (bar = 1 μm).

In this case, $BaCO_3$, CeO_2, and Yb_2O_3 in the desired molar ratios, to get the required composition of the final product, are ball milled for 48 h using isopropanol as the milling media. The obtained product is dried at $100°C$, cured for 24 h, and subsequently calcined in air at $600°C$ for 6 h, using both heating and cooling rates of $2°C/min$. The calcined samples are ground and sieved, and a second calcination step is carried out at $1300°C$ for another 10 h, using both heating and cooling rates of $2°C/min$, to ensure that a single perovskite phase is formed. This calcined sample is also ground and sieved. In Figure 3.6, a SEM micrograph is shown, which clearly characterizes the morphology of the crystals which constitute a powder of the obtained $BaCe_{0.95}Yb_{0.05}O_{3-\delta}$ perovskite [31].

An additional example is the solid-state reaction between bulk samples of copper oxide and alumina [36]. This reaction between CuO powder and single-crystal alumina substrates in air at $1100°C$ produces both $CuAl_2O_4$ and $CuAlO_2$; [36]. Another example is the synthesis of the ferrites $MgAl_2O_4$ and $MgFe_2O_4$ by solid-state reactions, which occurs by counter-diffusion of the Mg^{2+}, Fe^{3+}, and Al^{3+} ions through the relatively rigid oxygen lattice of the spinel or ferrite [37].

3.3.2 SOLGEL METHODOLOGIES

3.3.2.1 Introduction

The solgel methodology is a basic procedure for the synthesis of different materials [38–46]. The solgel process in essence means the synthesis of a solid compound by a chemical reaction in solution at a low temperature, [38–40]. Depending on the different precursors used, the solgel procedures can be fundamentally divided into different types [40,41], such as the Pechini method [42,43]; the solgel route based upon hydrolysis–condensation of metal alkoxides [43–49]; and the gelation route based upon concentration of aqueous solutions involving metal chelates, often called as "chelate gel" route [40].

The solgel methodologies are frequently applied since it presents some benefits, such as superior control of the stoichiometry and purity, bigger flexibility for the development of thin films, the possibility to get new compositions, and finally an improved capacity to control the material powder particle size [41].

Apart from tetraethylorthosilicate (TEOS), the majority of the metal alkoxides have the following setbacks [40]: high cost, unavailability, toxicity, and fast hydrolysis rate. Thus, in the synthesis of multicomponent oxide materials, including more than one kind of metal ion, the Pechini and gelation routes are very frequently used as the alternatives to the metal alkoxides–based solgel process [40].

3.3.2.2 Pechini Method

The Pechini method is a chemical solution method named after its inventor, Maggio Pechini, in 1967 [42]. Pechini developed a modified solgel process for metals that are not suitable for traditional solgel type reactions due to their unfavorable hydrolysis equilibria. This method includes a combined process of metal complex formation and in situ polymerization of organics. The benefit of the Pechini method is based on the elimination of the prerequisite that the metals involved form suitable hydroxo complexes. Since, chelating agents are apt in developing stable complexes with a diversity of metals over wide pH ranges; allowing the synthesis of oxides of considerable complexity [40].

This method is well-known and is used for the synthesis of homogeneous multicomponent metal oxide materials; it includes a combined process of metal complex formation and in situ polymerization of organics. It relies on the development of complexes of alkali metals, alkaline earths, transition metals, or even nonmetals with bi- and tridentate organic chelating agents such as citric acid [40].

To carry out the process, normally, an R-hydroxycarboxylic acid, such as citric acid, is used to form stable metal complexes; then, their polyesterification with a polyhydroxy alcohol such as ethylene glycol (EG) or polyethylene glycol forms a polymeric resin [39,42]. That is, a polyalcohol, such as EG, is added to create connections between the chelates by a polyesterification reaction, resulting in the gelation of the reaction mixture [39,42]. The immobilization of metal complexes in such rigid organic polymer networks reduces segregation of particular metal ions, ensuring compositional homogeneity. Following the drying process, the gel is heated to start the pyrolysis of the organic species, resulting in agglomerated submicron oxide particles; that is, the calcination of the polymeric resin at a moderate temperature (500°C–1000°C) generates a pure phase multicomponent metal oxide [40].

The method is normally performed as follows [32,39]: an aqueous solution of suitable oxides or salts is mixed with an α-hydroxycarboxylic acid, such as citric acid, and EG is added to the solution. At heating, an esterification process runs in the system leading to the formation of a stable gel in which the metal ions remain fixed. In the modified Pechini process, ethylenediaminetetraacetic acid (EDTA) has been used to replace citric acid due to its stronger chelating power.

To better comprehend the Pechini procedure, the synthesis of perovskites could be a good example, more specifically the synthesis of $Ba_xCe_{0.85}M_{0.15}O_3$, where M = Nd, Gd, Yb, and, $x =$ 0.95–1.05, by a modification of the Pechini process [43]. To carry out the synthesis, the precursors $Ba(NO_3)_2$, $Ce(NO_3)_3 \cdot 6H_2O$, $Nd(NO_3)_3 6H_2O$, $Yb(NO_3)_3 \cdot 4.44H_2O$, and $Gd(NO_3)_3 \cdot 5.45H_2O$ are dissolved in water, alongside with EDTA and EG, which serve as polymerization/complexation agents [32]. The molar ratios were set at $EDTA/\Sigma$ Metal = 2 and EDTA/EG = 1/3, correspondingly [32]. Thereafter, the evaporation of water and polymerization of the EG take place upon mild heating. Finally, the resulting product was calcined at 1300°C for 10 h.

3.3.3 Solgel Route Based on the Hydrolysis–Condensation of Metal Alkoxides

Another route for the production of materials involves the reaction of hydrolysis–condensation of metal alkoxides with water. We study here the important case of amorphous silica synthesis. In this case [38,39,44], silicic acid is first produced by the hydrolysis of a silicon alkoxide, formally a silicic acid ether. The silicic acids consequently formed can either undergo self-condensation, or condensation with the alkoxide. The global reaction continues as a condensation polymerization to form high molecular weight polysilicates. These polysilicates then connect together to form a network, whose pores are filled with solvent molecules, that is, a gel is formed [45].

The solgel procedure, as was previously recognized, is a method for solid materials synthesis, carried out in a liquid at a low temperature (typically $T < 100°C$). The synthesized inorganic solids are formed by chemical transformation of chemical solutes termed precursors [39]. More specifically for silica synthesis, solgel processing refers to the hydrolysis and condensation of alkoxide-based precursors, such as, TEOS, that is, $Si(OC_2H_5)_4$ [11].

To better understand this solgel procedure, the synthesis of silica microspheres using the Stobe–Fink–Bohn (SFB) method [46] and a modification of the SFB method [47–49] are excellent examples. The SFB method consists of the hydrolysis of TEOS ($Si(C_2H_5O)_4$) in ethanol, methanol, n-propanol, or n-butanol in the presence of ammonia as a catalyst [46]. In Table 3.3, the batch composition for the synthesis of silica microspheres, by means of TEOS in an alcohol (methanol or isopropanol) in the presence of ammonia as a catalyst, with and without double-distilled water (DDW) in the synthesis media, is presented [47].

The source materials for the synthesis are TEOS, DDW, methanol (MeOH), isopropanol, and ammonium hydroxide (40 wt % NH_4OH in water). The batch preparation, to carry out the synthesis following the recipes shown in Table 3.1, was as follows [46–49]:

1. Alcohol + catalysts (base) + DDW (if needed) (mixed with strong agitation)
2. TEOS is added to the reaction mixture
3. The mixture is stirred at room temperature for 1.5 h
4. After the synthetic procedure, the product is heated at 70°C for 20 h

In Figure 3.7, the morphology of the sample 70 (see Table 3.3), which shows a microsphere diameter $D = 220$ nm, is shown [47].

Recently, some modifications to the SFB method were introduced to synthesize silica-based materials of remarkably high specific surface area [11,48,49]. To be precise, in some instances isopropanol was used as the solvent and synthesis media. Besides, DDW in some cases was eliminated [11,48,49]. Also, amines dissolved in water or in a strong base were used, instead of NH_4OH as catalysts, in order to get materials of particularly high specific surface area [11,48,49]. In Table 3.4, the recipes used to synthesize these materials as well the above explained procedure are reported [11,48].

In Figure 3.8 [48,49], the SEM micrographs of two of the synthesized materials (see Tables 3.3 [47] and 3.4 [48]) that is, the silica materials labeled 68C (Figure 3.8a) and 70bs2 (Figure 3.8b) are shown.

It is obvious that sample 68C, which was synthesized using NH_4OH as a catalyst [47], is composed of microspheres with an average particle diameter, $D = 275$ nm.

In contrast, sample 70bs2, which was synthesized using the amine catalysts [49] (see Table 3.4), is composed of particles with an average particle diameter, $D = 100$ nm. This fact indicates that the sphere-packing microstructure was broken [49]. The rest of the materials synthesized using the amine as catalysts have microstructures similar to those reported for sample 70bs2 [49].

TABLE 3.3
Batch Composition for the Synthesis of the Silica Microspheres by Means of the Hydrolysis of Tetraethylorthosilicate

Sample	TEOS [mL]	DDW [mL]	MeOH [mL]	Isopropanol [mL]	NH_4OH [mL]
68F	0.75	0	30	0	3.0
80	1.5	0	30	0	6.0
81C	1.5	8.4	30	0	6.0
68C	1.5	4.5	30	0	6.0
69B	1.5	0.6	0	30	6.0
68E	2.4	0	30	0	6.0
70	0.75	0	30	0	4.5

FIGURE 3.7   SEM micrograph of the sample 70 (bar = 1 µm).

TABLE 3.4
Batch Composition for the Synthesis of Silica in the Presence of an Amine as a Catalyst With and Without DDW in the Synthesis Media

Sample	TEOS [mL]	DDW [mL]	NH₄OH [mL]	Amine [mL]	MeOH [mL]	EtOH [mL]	T [K]
70bs2	0.25	0	0	2	0	10	300
68bs1E	0.25	0	0	1	10	0	300
75bs1	0.35	0	0	2.5	10	0	300
79BS2	0.45	0	0	2.5	0	10	300
74bs5	0.35	2	0	2.5	9	1	300
68C	0.50	1.5	2	0	10	0	300

     (a)   (b)

FIGURE 3.8   SEM micrographs of two silica materials: (a) sample 68C (bar = 1000 nm) and (b) 70bs2 (bar = 500 nm).

3.3.4 Acetate Precipitation

This is a very simple methodology for the synthesis of different materials [41]. The used precursors for the procedure are acetates. This methodology gives an excellent molecular mixing of the reaction components, and provides a reactive milieu throughout the successive heating and decomposition processes required to obtain the end product [41]. This methodology is especially adaptable for the synthesis of thin, dense films for the manufacture of membranes [50].

One possible example of the use of this methodology is the synthesis of perovskites [50]. In this case, the acetates of the different A, B, A′, and B′ perovskite components are separately dissolved in acetic acid under agitation, where the acetates are mixed following rigorous stoichiometric proportions to get the desired product. Then, the mixture is dried and, subsequently, thermally treated at high temperatures, for example, the powder is heated from room temperature up to 1123 to 1573 K at a rate of 2 K/min and kept at this temperature by 5 to 12 h, and then cooled at the same rate in order to get the perovskite.

In Figure 3.9, a SEM micrograph of a powder sample of a $BaCe_{0.95}Yb_{0.05}O_{3-\delta}$ perovskite, which was synthesized following the previously described acetate precipitation procedure, is shown [51].

In Chapter 10, the use of membranes for different applications are described. One of the possible membranes for hydrogen cleaning is an asymmetric membrane comprised of the dense end of a proton conduction perovskite such as $BaCe_{0.95}Yb_{0.05}O_{3-\delta}$ and a porous end to bring mechanical stability to the membrane. In this case, it is possible to take from the slurry, obtained by the acetate procedure, several drops to be released over a porous ceramic membrane, located in the spinning bar of a spin-coating machine. Thereafter, the assembly powder, thin film porous membrane is heated from room temperature up to 1573 K at a rate of 2 K/min, kept at this temperature for 12 h, and then cooled at the same rate in order to get the perovskite end film over the porous membrane [50].

FIGURE 3.9 SEM micrograph of a powder of a $BaCe_{0.95}Yb_{0.05}O_{3-\delta}$ perovskite synthesized by the acetate sol–gel procedure at a magnification of 2000×.

3.4 SYNTHESIS OF MICROPOROUS CRYSTALLINE MATERIALS

3.4.1 ALUMINOSILICATE SYNTHESIS

Aluminosilicate zeolites are in general synthesized in hydrothermal conditions from solutions, including sodium hydroxide, sodium silicate, or sodium aluminate [52–55]. The particular zeolite synthesized is evidently determined by the reactants, and mostly by the synthesis specifications used, such as temperature, pH, and time [52–55]. These materials are synthesized in three steps: induction, nucleation, and crystallization. Nucleation is a step where germ nuclei are obtained from very small aggregates of precursors, becoming larger with time. Crystallization begins involving the germ nuclei from the nucleation step, and other components of the reaction mixture.

The process is affected by several factors that can be modified during the synthesis procedure, that is, presence of cations in the reaction mixture, OH^-, concentration, SiO_2/Al_2O_3 ratio, H_2O, content, temperature, pH, time, aging, stirring of the reaction mixture, order of mixing, and other factors [52,53].

The first hypothesis, proposed by Breck and Flanigen [52,55], to account for the crystallization of aluminosilicate zeolites affirms that it proceeds through the formation of the aluminosilicate gel or reaction mixture, and the nucleation and growth of zeolite crystals from the reaction mixture. This initial model has been almost abandoned, and replaced by the hypothesis of Barrer and others [53,55]. In the framework of this hypothesis, it is assumed that the formation of zeolite crystals occurs in solution. Accordingly, in this model, the nucleation and growth of crystalline nuclei are a consequence of condensation reactions between soluble species, where the gel plays a limited role as a reservoir of matter.

3.4.2 HIGH-SILICA, ALL-SILICA, AND NON-ALUMINOSILICATE ZEOLITES SYNTHESIS

The syntheses methods in the cases of high-silica, all-silica, and other non-aluminosilicate zeolites are similar to the methods used for aluminosilicate zeolites, although the initial gel composition is different [54–64]. The primary purpose of zeolite synthesis is the creation of new zeolitic structures by changing the different aspects involved in the process. To do this, the use of organic substances known as structure-directing agents (SDA) has significantly increased the number of new structures synthesized and accepted as novel materials [54–56,61,63]. These organic compounds can stabilize structures with cavities, and have shapes of comparable dimensions as that of the organic compound [54,55,61]. Amines and related compounds (quaternary ammonium cations), linear or cyclic ethers, and coordination compounds (organometallic complexes) have been the most frequently applied organic templates [54,55,61,63].

The first high-silica zeolites, that is, eta, EU, NU, and the ZSM series, were patented in the late 1960s, or early 1980s [54,55,64,65]. The original all-silica zeolite, silicalite, was, obtained in 1978 [62]. Microporous aluminophosphate molecular sieves AlPO, SAPO, and MeAPO, that is, non-aluminosilicate zeolites, were developed by researchers working at Union Carbide Corporation, New York, USA [56,66]. SAPO molecular sieves were obtained by incorporation of Si [66] in the AlPO framework. MeAPO molecular sieves were obtained by the inclusion of Me (Me = Co, Fe, Mg, Mn, Zn) in the AlPO framework. The family of crystalline microporous phosphates grew considerably; some examples include gallophosphates [67–69], zincophosphates [70,71], beryllophosphates [72], vanadophosphates [73], and ferrophosphates. [74].

Pure silica zeolites have been synthesized essentially in fluoride media using diverse SDA. The following framework types: MFI [65], MEL[75], MTW [76], AFI [77] BEA [78], IFR [79], ITE [80], AST [81], CFI [82], CHA [83], MWW [84], STF [85], STT [86], ISV [87], CON [88], MTT [89], RUT [90], ITW [91], and LTA [92] are represented as members of this series of molecular sieves.

High silica, all silica and non-aluminosilicate zeolites are generally obtained by hydrothermal crystallization of a heterogeneous gel, which consists of a liquid and a solid phase [61]. The reaction media contain the following sources of cations that form the framework (T: Si, Al, P,...), mineralizing agents (HO^-, F^-), mineral cations and/or organic species (cations or neutral molecules), and solvent (generally water) [55,61,93]. Nonaqueous routes [94,95] or dry synthesis methods [96] have also been explored.

The SDAs are often occluded in the microporous cavities and channels of the synthesized material, playing a significant role in the stabilization of the obtained zeolite. The guest-framework stabilizing interactions can be of Coulombic, H-bonding, or van der Waals nature. Nonetheless, guest–guest interactions can also be a factor in the total energy [54,55,61]. Different factors concerning SDA must also be considered, for instance [61], the size of the SDA is directly related to the cavity or pore size of the zeolite, although this effect obviously depends on the temperature.

Besides SDAs and temperature, other components and/or factors in the synthesis are of relevant importance [55,61]. The hydroxide ion acts by controlling the degree of polymerization by increasing the crystal growth. The OH^-/SiO_2 ratio, as well, has been correlated with the pore size, that is, higher ratios imply larger pores [55,61].

On the other hand, many syntheses procedures are carried out using fluoride anions as a substitute for hydroxide ions, resulting in zeolites or related material with higher crystal sizes, and lower structural defects with respect to standard procedures via OH^- [97–115]. Temperature is also a significant factor. Generally, the temperatures used are below 350°C. Besides, high values of temperatures yield more condensed phase species. The pH, reaction time and stirring of the reaction mixture are also important factors.

For the synthesis of pure silica phases, it is possible to use different SDAs in a fluoride media. The SDAs are selected primarily consistent with criteria defined as important in determining a high structure-directing ability (rigidity, size, shape, C/N^+, ratio) [97,116], and the availability of the cation or of the parent amine. Therefore, SDAs with polycyclic moieties (giving rise to SDAs with relatively large and rigid portions) predominate.

After the synthesis procedure, the zeolite should be calcined to remove the organic compounds that are blocking the pores. Calcination entails heating the as-synthesized zeolite in an air flow to a temperature normally ranging from 350°C to 400°C.

In general, zeolite synthesis requires considering many factors, and unfortunately, the relation between these experimental conditions and the nature of the synthesized product is not trivial. To overcome this problem, different strategies have been proposed, and among them the combinatorial approach is a very promising methodology consisting of the use of miniaturized multiautoclave systems to explore a very high number of possibilities.

3.4.3 HYDROTHERMAL TRANSFORMATION OF CLINOPTILOLITE TO PRODUCE ZEOLITES NA-X AND NA-Y

In this section, a hydrothermal treatment that produces phase transformations in clinoptilolite and allows us to synthesize zeolites X and Y is described. Zeolites X and Y have the FAU-type framework; however, the zeolite X has a silicon/aluminum ratio in the range $1 < Si/Al < 1.5$ and zeolite Y has a silicon/aluminum ratio in the range $1.5 < Si/Al < \approx 7.0$.

As previously described, the hydrothermal synthesis of aluminosilicate zeolites is carried out with highly reactive aluminosilicate gels in autogenous conditions. This kind of synthesis may involve the structure-directing role of alkaline cations in solution-mediated crystallization of the amorphous gel [55].

The hydrothermal treatment of natural zeolites [117,118] with highly basic sodium or potassium hydroxide solutions causes their amorphization and change in chemical composition (see Figure 3.10)

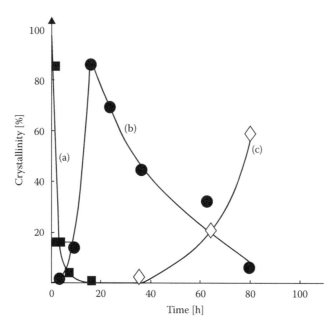

FIGURE 3.10 Crystallization curves: (a) clinoptilolite amorphization curve, (b) faujasite crystallization, and (c) zeolite P_1 crystallization.

[119]. Additionally, the liquid phase produced by the hydrothermal treatment can be employed as a silica source for the preparation of aluminosilicate gels for the posterior synthesis of zeolites [24,120–122].

The hydrothermal transformation process, described with the crystallization curves shown in Figure 3.10, was carried out using zeolite Na-HC (the chemical and phase composition of this sample was given in Tables 4.1 and 4.2; besides, in Figure 3.11, the SEM micrograph corresponding to sample HC [119] is shown) with the following procedure: the Na-HC was hydrothermally treated with a 7.5 M NaOH and 0.6 M NaCl (liquid to solid ratio = 1) solution and statically heated at 373 K in methyl polypropylene bottles [119].

During the recrystallization process (see Figure 3.10), first, a clinoptilolite amorphization curve was observed, then, the faujasite crystallization (see Figure 3.12) and, finally, the zeolite P_1 crystallization curves [119].

To apply this method, a slightly modified procedure, in order to produce adsorbents and acid catalysts, was carried out. The hydrothermal transformation will be carried out using the zeolite Na-CSW as raw material. The sample CSW was mined in Sweetwater, Wyoming. The sample was provided by ZeoponiX Inc., Louisville, Colorado [123]. The mineralogical phase composition of sample CSW was calculated using a previously developed methodology, by means of standards, that is, two well-characterized samples, the clinoptilolite samples [123]. The measured composition was 90 ± 5 wt % clinoptilolite and 10 ± 5 wt % of montmorillonite (2–10 wt %), quartz (1–5 wt %), calcite (1–6 wt %), feldspars (0–1 wt %), magnetite (0–1 wt %), and volcanic glass (3–6 wt %). The chemical composition of the sample Na-CSW is shown in Table 3.5 [123].

The hydrothermal transformation of the sample Na-CSW into Na-faujasite was obtained applying the procedure previously described, but with one change, that is, the sample Na-CSW was mechanically cleaned off with the help of a successive procedure of milling and sieving, which allows the reduction of the concentration of montmorillonite, albite, and calcite. Then the cleaned sample was additionally pulverized by stirring by the homoionization process, previously described, to produce the sample Na-CSW with a grain size of around 0.1 mm. The Na-CSW with a grain size of around

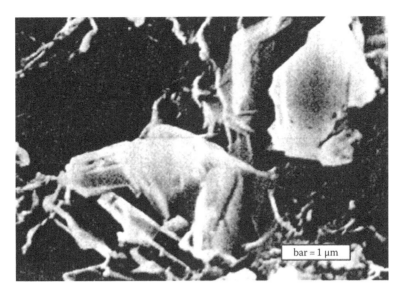

FIGURE 3.11   SEM micrograph of the HC sample (bar = 1 μm).

FIGURE 3.12   SEM micrograph of the Na-Y Zeolite formed after 16 h of hydrothermal transformation (bar = 1 μm).

0.6 mm was hydrothermally treated with a 7.5 M NaOH and 0.6 M NaCl (liquid to solid ratio = 1) solution and statically heated at 373 K in methyl polypentane bottles [119] for 4, 8, 9, 12, and 14 h, as was previously established [119], in order to get sample Na-FCSW(4 h), Na-FCSW(8 h), Na-FHC(9 h), Na-FHC(12 h), and Na-FHC(14 h).

TABLE 3.5
Chemical Composition (in wt %) of the Na-CSW
Sample Determined by EDAX

Sample	O	Si	Al	Fe	Ca	Mg	Na	K
Na-CSW	45.25	37.48	9.25	1.06	0.05	0.58	5.46	0.90

In Figure 3.13, the XRD powder patterns of the transformation products, Na-FCSW(6 h) (Figure 3.13a), Na-FCSW(12 h) (Figure 3.13b), and the pattern of the Na-Y commercial sample CBV100 (Figure 3.13c), with $SiO_2/Al_2O_3 = 5.2$, provided by the PQ Corporation Malvern, PA, USA, which was used as a standard for comparison, are shown [126]. The XRD powder patterns show that the transformation products are Na-faujasite zeolites, with 80% crystallinity compared with the standard (CBV-100). The analysis by x-ray diffraction of all the samples lets us confirm that the maximum yield of the hydrothermal transformation procedure was achieved at approximately 9 h.

The explanation of the procedure for the measurement of the lattice parameters by x-ray diffraction is given in Chapter 4. In Table 3.6, the Si/Al relation, determined with the assistance of the interdependence between the unit cell parameter and the aluminum contents in the zeolite framework for faujasite, that is, using the so-called Breck–Flanigen relationship, is reported [52]

$$N_{Al} = 115.2(a_0 - 24.191)$$

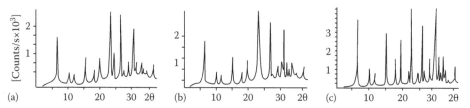

FIGURE 3.13 X-ray diffraction profiles of the transformation products: (a) Na-FCSW(6 h), (b) Na-FCSW(12 h), and (c) the commercial sample CBV100.

TABLE 3.6
Si/Al Relation of the Na-Faujasite
Zeolites Obtained from the
Hydrothermal Transformation of
Clinoptilolite

Sample	(Å)[a]	Si/Al
Na-FCSW(4 h)	24.72	2.1
Na-FCSW(8 h)	24.90	1.4
Na-FHC(9 h)	24.84	1.5
Na-FHC(12 h)	24.89	1.4
Na-FHC(14 h)	24.90	1.4

[a] Unit cell parameter of the cubic lattice of Na-X

where

N_{Al} is the number of framework aluminum atoms per unit cell, that is, per 192 T-atoms per unit cell in the FAU framework of the X and Y zeolites

a_0 is the cell parameter of the cubic lattice of the FAU framework measured for the sample under test 24.191 in angstroms is the intercept

Table 3.6 shows that zeolites Na-X and Na-Y have been produced depending on the reaction time. These materials are similar in phase composition to commercial Na-X and Na-Y zeolites, because, are composed of about 80 wt % of the zeolitic phase. In commercial zeolites because part of the phases are binders [124], the amount of zeolite is about 80 wt %. Applications of these type of materials are described elsewhere [11,25,52,125,126].

3.4.4 SYNTHESIS OF MeAPO MOLECULAR SIEVES

In Section 3.4.2, the aluminophosphate zeolites were described, now a more detailed description of the synthesis methods of these materials ensues. The development of crystalline, microporous phosphate-based oxides was initiated in 1982 with the synthesis of materials with aluminophosphate composition [127]. A new development in this field was the synthesis of silico-aluminophosphates in 1984 [66]. The number of structures and types of phosphate-based crystalline microporous oxides increased with the synthesis of metal aluminophosphates [128–131], obtained by the substitution of Al by Ga [72], and Be [132], Zn [133], and P by As and V [134,135]. Currently, there are about 18 different elements, that is, Li(I), Co(II), Fe(II), Mg(II), Mn(II), Zn(II), Be(II), Ni(II), Sn(II), B(III), Al(III), Ga(III), Fe(III), Cr(III), Si(IV), Ge(IV), Ti(IV), As(V), and V(V) that can be combined with P(V) in different framework types [136].

Crystalline microporous aluminophosphates containing framework metals, that is, MeAPO, constitute an interesting group of molecular sieves [29,137]. These materials are synthesized hydrothermally between 100°C and 250°C using organic templates, and crystallize into different structure types.

MeAPO-5 molecular sieves crystallize into the AFI-type structure [138], and is perhaps one of the most widely studied [136,137]. The isomorphic substitution of Al by Fe in the crystalline framework of $AlPO_4$-5 is currently achieved during crystallization [29], but in some cases generates framework charge and the capacity for the formation of Brönsted acid sites [29,139]. The creation of the MeAPO-5 anionic framework takes place by the substitution of Al(III) by a metallic (Me) species in the $AlPO_4$-5 framework. If the substitution is produced by Me(II) [29,136], the framework needs a charge-compensation mechanism, which is provided by the protonation of the amines used as the template agent [29]. The protonated amine generates Brönsted sites during calcination of the as-synthesized samples. On the other hand, when Al(III) is substituted by Fe(III), the strict alternation of (FeO_2) and (PO_2) tetrahedra [136] takes place, and then the framework is electrically neutral. Therefore, cations are not required for charge balance and the material has no acidity [139].

Here, we describe the synthesis of some MeAPO-5 (Me: Co, Mn, Fe, and Zn) molecular sieves [29]. The synthesis gels for MeAPO-5 are prepared using procedures previously described [29,129–131,139], and alumina (Merck), orthophosphoric acid (85%, BDH), $CoSO_4 \cdot 9H_2O$, $ZnSO_4$, $MnSO_4 \cdot 9H_2O$, and $Fe_2(SO_4)_3 \cdot 5H_2O$ (BDH, Analar) are the source materials used in the synthesis. The template was, in all cases, triethylamine (TEA) (BDH). The gels were crystallized using Teflon-lined autoclaves at 473 K. A reference AlPO4-5 was synthesized similarly but in the absence of the iron salt [29].

A typical synthesis procedure is summarized by the following steps [29]:

1. Alumina is slurried in half quantity of water, which will be used for the synthesis.
2. The metal sulfate is added to this slurry.
3. Phosphoric acid is diluted in another half quantity of water.
4. The phosphoric acid solution is added to the alumina plus Me salt slurry.
5. The obtained precursor is stirred for 20 min.

6. TEA is added to the precursor mixture under rapid agitation.
7. The mixture is charged into Teflon-lined autoclaves and statically heated at 200°C for 24h.

The chemical analysis of the synthesized samples expresses in the framework composition, that is, $(Me_xAl_yP_z)O_2$ provides the as-synthesized sample framework composition, which is shown in Table 3.7, indicating the presence of about 1% of Me in the synthesized aluminophosphate [29].

From the x-ray powder diffractograms (Figure 3.14), it is noted that the crystallized products exhibit all the characteristic reflections of the MeAPO-5 molecular sieves [140] and a high crystallinity and degree of purity [29].

3.4.5 SYNTHESIS OF PILLARED, LAYERED CRYSTALLINE MICROPOROUS MATERIALS

Due to the problems involved in the synthesis of zeolites having extra large channel and cavity sizes, other methodologies for the preparation of microporous three-dimensional crystalline materials, for instance, pillaring of layered systems, such as clays, have been developed [141–146]. In this regard, a variety of ultra-large pore materials consisting of layered structures with pillars in the interlamellar region, specifically, the pillared-layered structures, have been synthesized.

The principal class of swelling clays are the smectites; this type of minerals comprises beidellite, hectorite, fluorhectorite, saponite, sauconite, montmorillonite, and nontronite [147]. In order to create thermally stable pillared materials with relatively large pores with the help of clays, oxyhydroxyaluminum cations as pillaring agents have been applied [148,149]. The common method for the synthesis of these clays consists of exchanging the charge-compensating cations located in the interlamellar position, that is, Na^+, K^+, and Ca^{2+}, with larger inorganic hydroxyl cations [144]. These hydroxy species are polymeric or oligomeric hydroxyl metal cations produced by the hydrolysis of

TABLE 3.7
$(Me_xAl_yP_z)O_2$ Framework Composition of the Synthesis Products

MeAPO Species	x	y	z
CoAPO-5	0.012	0.488	0.500
ZnAPO-5	0.014	0.486	0.500
MnAPO-5	0.010	0.490	0.500
FeAPO-5	0.010	0.490	0.500

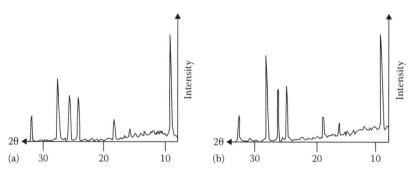

FIGURE 3.14 X-ray powder diffraction profile of (a) $AlPO_4$-5 and (b) a powder pattern representative of the MeAPO-synthesized samples.

salts of metals, such as Al, Zr, Ga, Cr, Si, Ti, Fe, and mixtures of them [141–145]. As soon as the exchanged samples are subject to meticulous thermal treatment, dehydration and dehydroxylation take place, producing stable metal oxide clusters, which operate as pillars, dividing the layers and forming a two-dimensional gallery with an aperture, which, if appropriately made, can be greater than 1.0 nm [142,144].

In the case of pillaring with aluminum, it is possible to use two types of solutions to carry out the pillaring process, that is, an Al_3Cl solution to which 2.33 mol of NaOH per mole of Al is added, or a commercially available solution in which $Al(OH)_3$ or Al metal is dissolved in an Al_3Cl solution until $OH^-/Al = 2.5$ [141]. In both solutions, it has been shown that a major polymeric species know as the Keggin ion, that is, $[Al_{13}O_4(OH)_{24}(H_2O)_{12}]^{7+}$ is formed [151]. During this process, the first step is the increase in the smectite interlayer spacing from about 9.4 Å to about 19 Å, in harmony with the incorporation of the Keggin ion, which has the form of a prolate spheroid. Subsequently, on heating, the Keggin ion loses the protons to the layers to as charge-compensating ions as the pillar becomes an Al_2O_3 particle [149]. In Figure 3.15, the pillaring procedure is schematically shown [141,144,145].

When carrying out the pillaring procedure, it is necessary to prevent overloading the interlayer or hosting an excessively small amount of pillars, otherwise a nonporous intercalate or unstable structure is produced [142]. The pillar altitude should be similar as the lateral separation between pillars, resulting in a product with an almost regular pore size distribution [150].

To improve the stability of the produced materials, researchers have worked on both the nature of the pillars and the nature of the layered silicate. In regard to the pillars, the originally synthesized materials based on alumina were thermally stable, but exhibited inadequate hydrothermal stability [142]. Since the chemistry of Ga is similar to the chemistry of Al, it has been shown that in a pure Ga^{3+} solution, the equivalent of Al_{13} polymer is synthesized upon limited hydrolysis [152,153]. Subsequently, it has been feasible to prepare a mixed $GaAl_{12}$ polymer where the tetrahedral Al^{3+} of the Al_{13} cation is substituted with a Ga^{3+} ion [153]. Then, the resulting structure can be described as the Baker–Figgs, ε-isomer of the Keggin structure [154]. In this instance, the somewhat larger ionic radius of the Ga^{3+} ion turns out the $GaAl_{12}$ structure more symmetrically than the related Al_{13} oligomer; therefore, the $GaAl_{12}$ structure ought to be thermodynamically privileged over the Al_{13} and Ga_{13} and more thermally stable as well [142]. As a result, it has been shown [155,156] that Ga or Ga/Al solutions prepared, as described previously, for Al polymeric solutions, could be used for pillaring montmorillonite. This material has been pillared with GaAl pillars that were synthesized with different Ga/Al ratios [155,156]. The obtained interlayer distance fluctuated from 1.8 to 2.0 nm at room temperature, and between 1.7 and 1.8 nm after calcination at 500°C. It was also estimated that the thermal and hydrothermal stabilities of the GaAl-pillared clays were higher than those of the clay pillared with Al alone [142].

Zirconia hydropolymers obtained through the hydrolysis of $ZrOCl_2 \cdot H_2O$ were between the first to be used for pillaring clays [144]. The starting material for producing the polymer is the zirconyl chloride, which is hydrolyzed by NaOH [145]. The generally accepted structure of the pillaring agent is a $[(Zr(OH)_2 \cdot 4H_2O)_4]^{8+}$ tetramer [157].

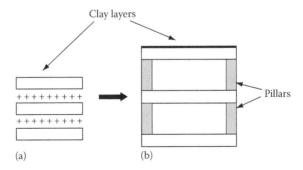

FIGURE 3.15 Schematic representation of the transition from the (a) original clay to the (b) pillared clay.

3.5 SYNTHESIS OF ORDERED SILICA MESOPOROUS MATERIALS

In the past decades, significant efforts have been undertaken on obtaining molecular sieves showing larger pore sizes [58]. But, due to the difficulties in making large-channel and large-cavity-sized zeolites, other materials have been developed. In this regard, supramolecular assemblies, that is, micellar aggregates, rather than molecular species as the SDA's have been introduced to synthesize a new family of mesoporous silica and aluminosilicate compounds [158,159]. The new collection of materials was named mesoporous molecular sieves (MMS). The M41S family was discovered by expanding the concept of zeolite synthesis with small organic molecules, that is, SDAs, to longer-chain surfactant molecules [158,159]. Then, instead of individual molecular-directing means, participating in the ordering of the reagents to shape the porous material, assemblies of molecules are responsible for the formation of these pore systems [45].

The characteristic line of attack for the development of mesostructured materials is the use of the following standard synthesis conditions: low temperatures, coexistence of inorganic and organic moieties, and extensive choice of precursors [61].

The formation mechanism of this family of materials is determined by two features [45]. The first is the dynamics of surfactant molecules to shape molecular assemblies, which leads to micelle, and, ultimately, liquid crystal formation. The second is the capability of the inorganic oxide to undergo condensation reactions to form extended, thermally stable structures.

In a simple binary system of water–surfactant, surfactant molecules show themselves as very active constituents with changeable structures in agreement with concentrations. At low concentrations, they energetically exist as monomolecules; with growing concentration, surfactant molecules combine together to shape micelles in order to decrease the system entropy. The concentration limit at which molecules accumulate to form isotropic micelles is called critical micellization concentration (cmc) [160]. As the concentration process continues, hexagonal, closely packed arrays emerge, producing the hexagonal phases. The following step is the coalescence of the adjacent, mutually parallel cylinders to produce the lamellar phase. In some cases, the cubic phase also appears prior to the lamellar phase [160].

The precise phase present in a surfactant aqueous solution at a particular concentration depends not only on the concentrations but also on its characteristics, that is, the length of the hydrophobic carbon chain, hydrophilic head group, and counterion, and the following parameters: pH, temperature, the ionic strength, and other additives [161]. This is reflected by the effect of the abovementioned matters on cmc. Typically, the cmc reduces with the growth of the chain length of a surfactant, the valency of the counterions, and the ion strength in a solution [160]. In contrast, it increases with growing counterion radius, pH, and temperature.

For example, in an aqueous solution at 25°C, the cmc is about 0.83 mM for the surfactant, $C_{16}H_{33}(CH_3)_3N^+Br^-$; and 11 wt %, small spherical micelles are present. In the concentration range of 11–20.5 wt %, elongated flexible rodlike micelles are formed [162]. Hexagonal liquid crystal phases appear in the concentration region between 26 and 65 wt %, followed by the formation of cubic, lamellar, and reverse phases with increasing concentrations [162]. At 90°C, the hexagonal phase is observed at a surfactant concentration of more than 65 % [163]. This new "organized matter soft chemistry synthesis" is gaining increasing importance [45,142,164,165]. Since, these materials are possible candidates for a diversity of applications, in the fields of catalysis [142,165], optics, photonics, sensors, separation, drug delivery, sorption, acoustic and electrical insulation, and ultra-light structural materials [142,166,167].

The M41S family was initially obtained by hydrothermal synthesis, in basic media, from inorganic gels containing silicate (or aluminosilicate) in the presence of quaternary trimethylammonium cations [158–160]. Pseudomorphic synthesis based on the dissolution–reprecipitation of silica microspheres in alkaline media, in the presence of C_{16}TMABr, permits to produce morphologically controlled MCM-41 [168]. Hexagonal mesoporous silica (HMS) compounds are obtained at neutral pH, according to a new synthesis method using primary amines $C_nH_{2n+1}NH_2$ ($n = 8–18$) as amphiphilic molecules [169,170].

A great amount of work has been dedicated to pore size control. Beck et al. tailored the pore size from 15 to 45 Å by varying the chain length of C_nTMA^+ cations between 8 and 18 carbon atoms [61,159]. The addition of organic molecules such as 1,3,5-trimethylbenzene [159] or alkanes [171] allowed increased pore sizes, up to 100 Å. These swelling agents are soluble in the hydrophobic part of the micelle, increasing the growth of the volume of the template [61]. This technique, although simple in form, is difficult to put into practice, since it lacks reproducibility and produces less organized mesophases. As an alternative to the use of the swelling agent, an efficient method relies on extended hydrothermal treatment in TMA^+ solutions. This procedure makes the pore organization better, as well as increases the pore size [172]. Nevertheless, the pore size of MCM-41 and related materials is limited by the size of the micellar templates; the pore size can be increased by making use of bigger molecules such as polymers or more complex texturing agents.

Amphiphilic block copolymers (ABCs) can be utilized for ordered mesoporous materials synthesis. ABCs represent a new class of functional polymers, with potential for application, mostly because of the high energetic and structural control that can be exerted on the material interfaces [61,173]. The chemical structure of ABCs can be programmed to tailor interfaces between materials of completely different chemical natures, polarities, and cohesion energies [173]. The "traditional" surfactant polymer organized systems (POS) formed by ABCs are excellent templates for the structuring of inorganic networks [61,174]. They have also been used for the growth control of discrete mineral particles [174], diblock (AB) or triblock (ABA). Block copolymers are normally used, in which A represents a hydrophilic block (polyethylene oxide [PEO] or polyacrylic acid [PAA]) and B, a hydrophobic block (polystyrene [PS], polypropylene oxide [PPO], polyisoprene [PI], or polyvinylpyridine [PVP]) [173].

3.6 ACTIVE CARBON AND CARBON NANOTUBE PREPARATION METHODS

Generally, the starting materials used in the commercial production of activated carbons are those with high carbon content, such as wood, lignite, peat, and coal of different ranks [175,176]. But, over the last years, growing interest has now shifted to the use of other low-cost, and abundantly accessible, agricultural byproducts, such as coconut shells, rockrose, eucalyptus kraft, lignin, apricot stone, cherry stone, and olive stone to be converted into activated carbons [176].

In general, two main methods for the preparation of activated carbons are used: physical and chemical activation methods. Physical activation consists of a two-step process carried out at high temperature (800°C–1000°C), that is, carbonization under inert gas, normally nitrogen, and activation under oxidizing agents, as a rule, carbon dioxide or water vapor [176]. On the other hand, chemical activation requires the treatment of the initial material with a dehydrating means, for example, sulfuric acid, phosphoric acid, zinc chloride, potassium hydroxide, or others, at temperatures varying from 400°C to 1000°C, followed by the elimination of the dehydrating agent by meticulous washing [175].

An example of an actual physical activation procedure is as follows [176]: the dried raw material is crushed, and sieved; then the furnace temperature is increased at a rate of 10 [°C/min] up to 600°C for about 3 h, under inert purge gas flow. The resulting chars are then activated at 500°C–900°C for 10–60 min under purified CO_2 flush. In Figure 3.16 [175], a flow chart of the physical activation method is shown.

The standard chemical activation procedure is similar to the physical method of activation. That is, the dried raw material is crushed and sieved to the desired size fraction. Afterward, the obtained powdered material is mixed with a concentrated solution of a dehydrating compound; subsequently, this blend is dried and heated under inert

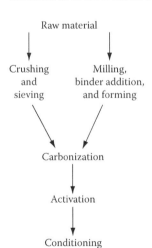

FIGURE 3.16 Flow chart of the physical activation method.

atmosphere in a furnace up to 400°C–700°C, and held at this temperature for several hours; finally, the resulting material is thoroughly washed, dried, and conditioned [175].

In Figure 3.17 [175], a flow chart of the chemical activation method is shown.

In the case of carbon nanotubes (CNTs), numerous syntheses methods have been developed during the last years, for example, the discharge between two graphite electrodes, laser ablation, hydrocarbon decomposition, and catalytic chemical vapor decomposition (CCVD); however, the most applied methods are arc discharge, laser ablation, and CVD [177–179].

In the arc-discharge apparatus, two graphite rods are applied as the cathode and the anode [179]. Between these electrodes, a discharge takes place when a direct current high voltage is supplied. During this process, the electrons from the arc discharge move from the cathode to the anode, colliding with the anodic rod and producing a vapor between the carbon electrodes, with or without a catalyst [177–179]. Thereafter, nanotubes are produced from the carbon vapor, that is, carbon clusters from the anodic graphite rod containing CNTs are deposited on the cathode [177]. In the laser ablation methodology, a high-power laser beam is focused on a carbon-containing feedstock gas, generally, CH_4 or CO [178,179]. These two methods usually generate large amounts of impure material; consequently, the material prepared by these methods must be purified, applying chemical and separation methods. None of these methodologies can be economically industrialized in order to produce the huge amounts required for many applications [178].

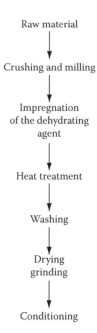

Raw material

↓

Crushing and milling

↓

Impregnation of the dehydrating agent

↓

Heat treatment

↓

Washing

↓

Drying grinding

↓

Conditioning

FIGURE 3.17 Flow chart of the chemical activation method.

In the CCVD process, a hydrocarbon vapor is passed through a tube kept in the furnace in which a catalyst is present at sufficiently a high temperature in order to decompose the hydrocarbon; thereafter the CNTs grow over the catalyst in the furnace, and are then collected after cooling the system to room temperature [179]. The CCVD methodologies, applying catalyst grains and hydrocarbon precursors to grow nanotubes, allow the scale up of the synthesis of CNTs in large quantities [177]. Nevertheless, these methodologies have the disadvantage of producing structures of inferior quality, that is, containing defects, mainly because these CNTs are produced at lower temperatures, specifically, 600°C–1000°C, compared with the arc or laser processes which produce CNTs at temperatures of about 2000°C [178].

3.7 MEMBRANE PREPARATION METHODS

Numerous types of materials, for instance, polymers, glasses, ceramics, and metals can be applied for membrane synthesis [180–183]. The major step in the preparation of a membrane is to adapt the material through an appropriate methodology to get a membrane structure with a morphology suitable for a particular type of separation process [183]. For example, in the special case of membrane reactors (see Section 10.6.2), one of the most important separations is the selective subtraction of hydrogen from the reaction zone; therefore, such membranes must be hydrogen selective [184–186].

3.7.1 CERAMIC METHOD

One of the simplest methods for membrane synthesis is the ceramic technique. This methodology involves the compression of an inorganic solid powder and the ensuing formation of a ceramic material by particle sintering at high temperatures. The process is carried out as follows: the material to be used is appropriately ground with a mechanic grinding machine. Then, the obtained powder is sieved to get particles of average grain diameter, d_p, of several micrometers or less. This fine, grain powder is poured sometimes with an included binder, into a mold, for example, a dye, in order to prepare cylindrical wafers, and afterward pressed at around 50–350 [MPa] with a hydraulic press.

Finally, the obtained wafers are thermally treated in a furnace at around 500°C–2000°C for several hours depending on the raw material and the application of the membrane. This methodology is principally appropriate for materials with high chemical, thermal, and mechanical stabilities [183].

A typical application of the ceramic method is as follows. The natural zeolite sample labeled the CSW sample (see Section 3.4.3 for the detailed description of this natural zeolite rock) is ground with a Bel-Art Micro-Mill mechanic grinding machine [123]. The obtained powders were sieved, using the following sieves, greater than 100 mesh, 100–50 mesh, 50–40 mesh and 40–30 mesh, to obtain particles of average grain diameter, d_p, of 50–100, 220, 340, and 500 μm, respectively [123]. Cylindrical wafers (12.7 mm diameter and 3 mm height) were prepared by pressing 0.55 g of the zeolite powder at 250 MPa using a Carver Manual Hydraulic Press Model 3912 [123]. The wafers were thermally treated in a Fisher Scientific, Isotemp Muffle Furnace 650 Series at 500°C, 600°C, 700°C, 900°C, 1000°C, and 1100°C between 1 and 2 h with a heating rate of 40 [°C/min] and a cooling rate of 1–2°C/min (see Figure 3.18 [123]).

A particular case in the preparation of ceramic membranes is the application of the solgel procedure for membrane preparation [181]. This methodology was the first to permit the creation of ceramic membranes with pore sizes in the nanometer range [183]. In this process, as was previously explained (see Section 3.3.2), an alkoxide precursor is produced into a gel by means of either the colloidal suspension route or the polymeric gel route [181,183]. In the colloidal suspension route, the precursor is hydrolyzed to create a sol, which is a colloidal dispersion of particles in a liquid. Then, by altering the surface charge of the particles in the sol or by increasing the concentration, the particles have a propensity to agglomerate and, consequently, a three-dimensional network structure, that is, a gel, is attained [183]. Thereafter, the gel is carefully dried to avoid crack formation and, finally, the obtained powder is sintered [181,183]. Because of the small particle size obtained with the help of this methodology, porous membranes with very small pore sizes can be obtained.

3.7.2 TEMPLATE LEACHING

Template leaching is another methodology applied for the production of membranes; it can be applied to produce porous glass membranes [180]. The method consists in the formation of a structure with the help of a homogeneous melt of a three-component system, for example, $Na_2O — B_2O_3 — SiO_2$; when

FIGURE 3.18 SEM micrograph of a ceramic membrane prepared with a powder of particle size, $d_p = 500$ μm, and treated at 800°C, During 2 h. (bar = 1 μm).

the melt is cooled, a two-phase system is formed along with a nonsoluble phase, SiO_2, and a soluble phase [180]. The soluble phase or phases are leached out using an acid or base. Then, a broad range of pore diameters can be attained with a minimum of 5 nm [183].

3.7.3 COMPOSITE MEMBRANES

So far we have been describing the preparation of symmetric membranes, that is, those consisting of a single solid phase; alternatively, asymmetric or composite membranes are produced with two ends composed of different solid phases or more slides composed of different solid phases [180]. Specifically, symmetric membranes have a uniform structure, and composite membranes are composed of a number of layers with different structures; as a result, membranes consisting of different layers of different materials are called composite or asymmetric membranes. These kind of membranes are intended to ensure the mechanical stability of the whole membrane, that is, the morphology is arranged in a composite form comprising a highly permeable, usually porous support, and a narrow coating of the nonporous separating material.

In this sense, the supports are covered with films of microporous (pores from 0.3–2 nm) or dense materials, where the support gives mechanical strength while the coating is intended to carry out selective separations [180].

Different methods have been used to deposit thin films, for example, the modification of porous ceramic membranes by pore plugging with the help of a particular CVD methodology, which is named chemical vapor infiltration (CVI) [183,187]. CVI is another form of CVD [187]. CVD involves the deposition onto a surface, while CVI implies deposition within a porous material [188]. Both methods use almost the same equipment [187] and very similar precursors (see Figures 3.3 and 3.19); however, each one functions using different operation parameters, that is, flow rates, pressures, furnace temperature, and other parameters. In the CVI technique, the precursors are carried to the interior of a porous substrate and the products of the reaction out of the porous material by isothermal diffusion. Normally, a forced flow is applied that uses a pressure gradient to compel the gases through the porous substrate; this process is designed to result in a product with a homogeneously distributed coating on the interior [180].

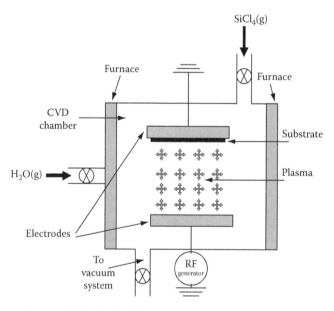

FIGURE 3.19 Chemical vapor infiltration facility.

Generally in CVI, the pores of porous materials are blocked by the reaction product of a precursor, which is normally a gaseous or volatile silica compound, such as silanes, TEOS, and silicon tetrachloride ($SiCl_4$), and the oxidizing agent is usually pure oxygen, air, or steam (see Figure 3.19) [185,189–191]. The porous materials used for CVI are commonly α-alumina supports, γ-alumina membranes [185,189], or porous Vycor glass [189,192].

For example, silica membranes are supported on the all-silica zeolite, silicalite, produced on an α-Al_2O_3 porous membrane of mean pore size 1 μm and wall thickness 0.6 mm [189]. The dense silica membrane was prepared by CVD, that is, CVI using silicon tetrachloride ($SiCl_4$) as the precursor and water, that is, through the process of $SiCl_4$ hydrolysis at temperatures 400°C–600°C [189]. In general, the CVD, that is CVI, methodology is extensively applied for ceramic, dense composite fabrication [185].

Metals can be also be deposited on a substrate layer by means of vapor deposition in this process; the solid material to be deposited is first evaporated in a vacuum system using a thermal source of evaporation or by sputtering methods. Another deposition technique is pyrolysis, which requires the degradation of polymers which are then coated onto porous supports [183].

3.8   POLYMER SYNTHESIS

Polymer synthesis is a complicated process that can be carried out applying different methodologies starting from appropriate monomers [193–196]. The most important methods of polymer synthesis are step-growth polymerization and chain reaction polymerization [194,196].

3.8.1   STEP-GROWTH POLYMERIZATION

This polymerization type, which normally comprises noncatalyzed condensation reactions, is related to chemical reactions involving multifunctional monomers with functional groups, such as –OH, –COOH, and others [193–196]. Where a multifunctional monomer is a molecule which has at least two or more reactive locations, to create intermolecular chemical bonds. The process normally comprises a sequence of condensation reactions that lead to the elimination of low mass products, for example, water [194].

A typical example of a step growth polymerization process is the synthesis of polyesters. In this case, the reaction to produce the ester begins with the interaction of an acid and an alcohol, and water is eliminated [193]. To obtain polyesters, the two reactants must have bifunctionality to form a linear polymer. For example, if an organic acid molecule has two —COOH groups, that is, HOOC — R — COOH, it could react two times with an alcohol with two —OH groups, HO — R′ — OH, which in turn could react two times. Then, if the reacted acids and alcohols have enough length, the polyester will be produced (see Figure 3.20).

Another example of the step-growth polymerization is the synthesis of polyurethanes. Here, linear polyurethanes are produced by the reaction of bifunctional alcohols, HO — R′ — OH, with bifunctional isocyanates, OCN — R — NCO, to produce a polyurethane (see Figure 3.21).

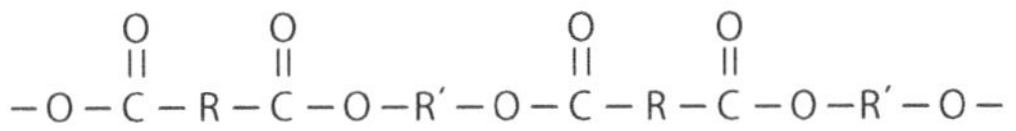

FIGURE 3.20   Polyester.

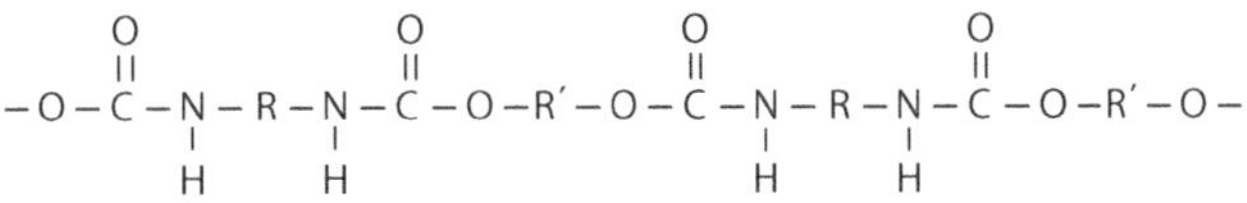

FIGURE 3.21   Polyurethane.

3.8.2 Chain Reaction or Addition Polymerization

The most common chain reaction polymerization is free-radical polymerization. A free radical is merely a molecule with an unpaired electron, which has a tendency to add a supplementary electron in order to form an electron pair which makes it extremely reactive. These molecular complexes could be produced by heat or irradiation, or formed by the addition of a compound, named the initiator (I), for example, dialkyl peroxides (R — O — O — R) or azo compounds (R — N = N — R), which are not strictly, catalysts, since they are chemically altered during the reaction [196].

$$\begin{array}{c} H \\ | \\ I-CH_2=C\bullet \\ | \\ R \end{array}$$

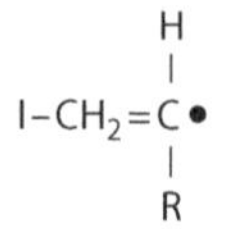

FIGURE 3.22 Initiated monomer.

During addition polymerization, in aqueous medium, free radicals are regularly formed by the division of the initiator into two parts along a single bond [196]. The formed molecular complex can split bonds on another molecule by taking an electron, and then leaving that molecule with an unpaired electron, that is, this molecule becomes another free radical. Therefore, the first step in producing polymers by free-radical polymerization is named initiation. That is, this step starts when the initiator decomposes into free radicals in the presence of monomers. Subsequently, the unsteadiness of carbon–carbon double bonds in the monomer produces a reaction with the unpaired electrons present in the free radical. In this reaction, the active center of the radical captures one of the electrons found in the double bond of the monomer, resulting in an unpaired electron emerging as a fresh active center at the end of the chain [194].

Consequently, the initiation process produces an initiated monomer (see Figure 3.22), which starts the propagation [196].

During this phase, the initiated monomer rapidly adds another monomer to a chain by the addition of a free radical to the double bond of a monomer with the generation of another free radical [196]. That is, in the propagation phase, due to the development of the electron transfer process, the resulting movement of the active center down the chain results in the production of the polymer [195].

In principle, the propagation reaction might persist as far the source of monomers is present. Nevertheless, the enlargement of a polymer chain is stopped by the termination reaction. This reaction takes place by two methods, that is, combination and disproportionation [196]. In the first case, combination takes place when the polymer's enlargement is blocked by free electrons from two developing chains that link and create a single chain. On the other hand, disproportionation stops the propagation reaction when a free radical removes a hydrogen atom from an active chain, and a carbon–carbon double bond acquires the position of the absent hydrogen.

The most common type of addition polymerization reaction is the synthesis of polyethylene from the ethylene monomer or the polymerization of a vinyl monomer (see Figure 2.34).

REFERENCES

1. C.N.R. Rao and J. Gopalakrishnan, *New Directions in Solid State Chemistry* (2nd edition), Cambridge University Press, Cambridge, UK, 1997.
2. W.A. Johnson and R.F. Mehl, *Trans. Am. Inst. Min. Metall. Pet. Eng.*, 135, 416 (1939).
3. M. Avrami, *J. Chem. Phys.* 7, 1103 (1939); 8, 212 (1940); 9, 177 (1941).
4. A.T.W. Kempen, PhD thesis, Department of Chemistry, University of Stuttgart, Stuttgart, Germany, 2001.
5. F.E. Fujita and R.W. Cahn, *Physics of New Materials*, Springer-Verlag, New York, 1998.
6. A.P. Wight and M.E. Davis, *Chem. Rev.*, 102, 3589 (2002).
7. J. Haber, J.H. Block, L. Berlnek, R. Burch, J.B. Butt, B. Delmon, G.F. Froment et al., *Pure and Appl. Chem.*, 63, 1227 (1991).
8. M. Che, O. Clause, and Ch. Marcilly, in *Preparation of Solid Catalysts*, G. Ertl, H. Knozinger, and J. Weitkamp, (editors), Wiley-VCH, Weinheim, Germany, 1999, p. 315.
9. B.C. Gates, in *Preparation of Solid Catalysts*, G. Ertl, H. Knozinger, and J. Weitkamp, (editors), Wiley-VCH, Weinheim, Germany, 1999, p. 371.
10. W.M.H. Sachtler, in *Preparation of Solid Catalysts*, G. Ertl, H. Knozinger, and J. Weitkamp, (editors), Wiley-VCH, New York, 1997, p. 388.

11. R. Roque-Malherbe, *Adsorption and Diffusion of Gases in Nanoporous Materials*, CRC Press, Boca Raton, FL, 2007.

12. C. Louis and M. Che, in *Preparation of Solid Catalysts*, G. Ertl, H. Knozinger, and J. Weitkamp, (editors), Wiley-VCH, Weinheim, Germany, 1999, p. 341.

13. C. Burda, X. Chen, R. Narayanan, and M.A. El-Sayed, *Chem. Rev.*, 105, 1025 (2005).

14. J.W. Geus and A.J. van Dillen, in *Preparation of Solid Catalysts*, G. Ertl, H. Knozinger, and J. Weitkamp, (editors), Wiley-VCH, Weinheim, Germany, 1999, p. 460.

15. Y. Iwasawa, in *Preparation of Solid Catalysts*, G. Ertl, H. Knozinger, and J. Weitkamp, (editors), Wiley-VCH, New York, 1997, p. 427.

16. H.O. Pierson, *Handbook of Chemical Vapor Deposition—Principles, Technology and Applications*, Noyes Publications, New Jersey, 1992.

17. N.D. Parkyns, in *Proceedings of the 3rd International Congress on Catalysis*, Amsterdam, the Netherlands, 1965, p. 914.

18. C. de las Pozas, R. Lopez-Cordero, C. Diaz-Aguila, M. Cora, and R. Roque-Malherbe, *J. Solid State Chem.*, 114, 108 (1995).

19. P.A. Jacobs, in *Metal Clusters in Catalysis*, B.C. Gates, L. Guezi, and H. Knozinger, (editors), Elsevier, Amsterdam, the Netherlands, 1986, p. 358.

20. P.I. Jansen, and R.A. van Santen, *Stud. Surf. Sci. Catal.*, 67, 221 (1991).

21. A. Arcoya, X.L. Seoane, and J. Soria, *Stud. Surf. Sci. Catal.*, 75, 2341 (1993).

22. R.H. Heck and N.Y. Chen, *App. Catal. A Gen.*, 86, 83 (1992).

23. H. Heck and N.Y. Chen, *Ind. Eng. Chem. Res.*, 32, 1003 (1993).

24. C. de las Pozas, PhD thesis, National Center for Scientific Research, Havana, Cuba, 1995.

25. R. Roque-Malherbe, in, *Handbook of Surfaces and Interfaces of Materials*, Vol. 5, H.S. Nalwa, (editor), Academic Press, New York, Chapter 12, 2001, p. 495.

26. G. Onyestyak and D. Kallo, in *Natural Zeolites '93 Conference Volume*, D.W. Ming and F.A. Mumpton, (editors), International Committee on Natural Zolites, Brockport, NY, 1995, p. 437.

27. M. Turco, G. Bagnasco, L. Lisi, G. Russo, D. Sanning, and P. Ciambelli, in *Natural Zeolites 93 Conference Volume*, D.W. Ming and F.A. Mumpton, (editors), International Committee on Natural Zeolites, Brockport, NY, 1995, p. 429.

28. C. de las Pozas, W. Kolodziejskii, and R. Roque-Malherbe, *Microporous Mater.*, 5, 325 (1996).

29. R. Roque-Malherbe, R. Lopez-Cordero, J.A. Gonzales-Morales, J. Oñate-Martinez, and M. Carreras-Gracial, *Zeolites* 13, 481 (1993).

30. P.G. Menon and B. Delmon, in *Preparation of Solid Catalysts*, G. Ertl, H. Knozinger, and J. Weitkamp, (editors), Wiley-VCH, New York, 1997, p. 388.

31. S. Nieto, R. Polanco, and R. Roque-Malherbe, *J. Phys. Chem. C*, 111, 2809 (2007).

32. J. Wu, PhD thesis, California Institute of Technology, Pasadena, CA, 2005.

33. J. Wu, L.P. Li, W.T.P. Espinosa, and S.M. Haile, *J. Mater. Res.*, 19, 2366 (2004).

34. C.-S. Chen, Z.-P. Zhang, G.-S. Jiang, Ch.-G. Fan, W. Liu, and H.J.M. Bouwmeester, *Chem. Mater.*, 13, 2797 (2001).

35. R.A. De Souza and J.A. Kilner, *Solid State Ionics*, 106, 175 (1998).

36. D.W. Susnitzky and C.B. Carter, *J. Mat. Res.*, 9, 1958 (1991).

37. R.E. Carter, *J. Amer. Ceram. Soc.*, 44, 116 (1961).

38. C.J. Brinker, and G.W. Scherer, *Sol-Gel Science*, Academic Press, New York, 1990.

39. B.L. Cushing, V.L. Kolesnichenko, and Ch.J. O'Connor, *Chem. Rev.*, 104, 3893 (2004).

40. J. Lin, M. Yu, C. Lin, and X. Liu, *J. Phys. Chem. C*, 111, 5835 (2007).

41. J. Twu and P.K. Gallagher, in *Properties and Applications of Perovskite Type Oxides*, L.G. Tejuca and J.L.G. Fierro, (editors), Marcel Dekker, New York, 1993.

42. M.P. Pechini, US Patent 3,330,697, July 11, 1967.

43. V. Agarwal and M. Liu, *J. Mater. Sci.*, 32, 619 (1997).

44. A.C. Pierre and G.M. Pajonk, *Chem. Rev.*, 102, 4243 (2002).

45. T.J. Barton, L.M. Bull, G. Klemperer, D.A. Loy, B. McEnaney, M. Misono, P.A. Monson et al., *Chem. Mater.*, 11, 2633 (1999).

46. W. Stobe, A. Fink, and E. Bohn, *J. Colloids Interf. Sci.*, 26, 62 (1968).

47. R. Roque-Malherbe and F. Marquez-Linares, *Mat. Sci. Semicond. Proc.*, 7, 467 (2004); *Surf. Interf. Anal.*, 37, 393 (2005).

48. F. Marquez-Linares and R. Roque-Malherbe, *J. Nanosci. Nanotech.*, 6, 1114 (2006).

49. R. Roque-Malherbe, F. Marquez-Linares, W. del Valle, and M. Thommes, *J. Nanosc. Nanotech.*, 8, 5993 (2008).

50. S. Hamakawa, L. Li, A. Li, and E. Iglesia, *Solid State Ionics* 48, 71 (2002).

51. S. Nieto, R. Polanco, and R. Roque-Malherbe, (in progress).

52. D.W. Breck, *Zeolite Molecular Sieves*, Wiley, New York, 1974.

53. R.M. Barrer, *Hydrothermal Chemistry of Zeolites*, Academic Press, London, 1982.

54. R. Szostak, *Handbook of Molecular Sieves*, Van Nostrand Reinhold, New York, 1992; *Molecular Sieves. Principles of Synthesis and Identification* (2nd edition), Blackie, London, 1998.

55. C.S. Cundy and P.A. Cox, *Chem. Rev.*, 103, 663 (2003).

56. E.M. Flanigen, R.L. Patton, and S.T. Wilson, *Stud. Surf. Sci. Catal.*, 37, 13 (1988).

57. H. Kessler, in *Comprehensive Supramolecular Chemistry, Vol. 7, Solid-State Supramolecular Chemistry: Two- and Three-Dimensional Inorganic Networks*, J.L. Atwood, J.E. Davis, D.D. MacNicol, and F. Vogtle, (editors), Pergamon, London, 1996, p. 425.

58. M.E. Davies, *Nature*, 417, 813 (2002).

59. J.M. Garces, A. Kuperman, D.M. Millar, M. Olken, A. Pyzik, and W. Rafaniello, *Adv. Mat.*, 12, 1725 (2000).

60. M.L. Occelli and Kessler, (editors), *Synthesis of Porous Materials*, Marcel Dekker, New York, 1997.

61. G.J.A.A. Soler-Illia, C. Sanchez, B. Lebeau, and J. Patarin, *Chem. Rev.*, 102, 4093 (2002).

62. E.M. Flanigen, J.M. Bennett, R.W. Grose, J.P. Cohen, R.L. Patton, R.L. Kirchner, J.V. Smith, *Nature*, 271, 512 (1978).

63. H. Robson, (editor), *Verified Synthesis of Zeolitic Materials* (2nd edition), Elsevier, Amsterdam, the Netherlands, 2001.

64. R.L. Wadlinger, G.T. Kerr, and E.J. Rosinski, US Patent 3,308,069 (1967).

65. R.J. Argauer and G.R. Landolt, US Patent 3,702,886 (1972).

66. B.M. Lok, C.A. Messina, R.L. Patton, R.T. Gajek, T.R. Cannan, and E.M. Flanigen, *J. Am. Chem. Soc.*, 106, 6092 (1984).

67. J.B. Parise, *J. Chem. Soc. Chem. Commun.*, 606 (1985).

68. A. Merrouche, J. Patarin, H. Kessler, M. Soulard, L. Delmotte, J.L. Guth, and J.F. Joly, *Zeolites*, 12, 22 (1992).

69. G. Ferey, *J. Fluorine Chem.*, 72, 187 (1995).

70. W.T.A. Harrison, T.E. Martin, T.E. Gier, and J.D. Stucky, *J. Mater. Chem.*, 2, 2175 (1992).

71. M. Wallau, J. Patarin, I. Widmer, P. Caullet, J.L. Guth, and L. Huve, *Zeolites*, 14, 402 (1994).

72. G. Harvey and W.M. Meier, *Stud. Surf. Sci. Catal.*, 49A, 411 (1989).

73. V. Soghomoniam, Q. Chen, R.C. Haushalter, and J. Zubieta, *Angew. Chem., Int. Ed. Engl.*, 32, 610 (1993).

74. J.R.D. Debord, W.M. Reiff, C.J. Warren, R.C. Haushalter, and J. Zubieta, *Chem. Mater.*, 9, 1994 (1997).

75. D.M. Bibby, N.B. Milestone, and L.P. Aldridge, *Nature*, 280, 664 (1979).

76. C.A. Fyfe, H. Gies, G.T. Kokotailo, B. Marler, and D.E. Cox, *J. Phys. Chem.*, 94, 3718 (1990).

77. R. Bialek, W.M. Meier, M. Davis, and M.J. Annen, *Zeolites*, 11, 438 (1991).

78. M.A. Camblor, A. Corma, and S. Valencia, *J. Chem. Soc. Chem. Commun.*, 2365 (1996).

79. P.A. Barrett, M.A. Camblor, A. Corma, R.H. Jones, and L.A. Villaescusa, *Chem. Mater.*, 9, 1713 (1997).

80. M.A. Camblor, A. Corma, P. Lightfoot, L.A. Villaescusa, and P.A. Wright, *Angew. Chem., Int. Ed. Engl.*, 36, 2659 (1997).

81. L.A. Villaescusa, P.A. Barrett, and M.A. Camblor, *Chem. Mater.*, 10, 3966 (1998).

82. P.A. Barrett, M.J. Diaz-Cabanas, M.A. Camblor, and R.H. Jones, *J. Chem. Soc., Faraday Trans.*, 94, 2475 (1998).

83. M.J. Diaz-Cabanas, P.A. Barrett, and M.A. Camblor, *Chem. Commun.*, 1881 (1998).

84. M.A. Camblor, A. Corma, M.J. Diaz-Cabanas, and C. Baerlocher, *J. Phys. Chem. B*, 102, 44 (1998).

85. L.A. Villaescusa, P.A. Barrett, and M.A. Camblor, *Chem. Commun.*, 2329 (1998).

86. M.A. Camblor, M.J. Diaz-Cabanas, J. Perez-Pariente, S.J. Teat, W. Clegg, I.J. Shannon, P. Lightfoot, P.A. Wright, and R.E. Morris, *Angew. Chem., Int. Ed.*, 37, 2122 (1998).

87. L.A. Villaescusa, P.A. Barrett, and M.A. Camblor, *Angew. Chem., Int. Ed.*, 38, 1997 (1999).

88. C. Jones, S. Hwang, T. Okubo, and M. Davis, *Chem. Mater.*, 13, 1041 (2001).

89. P.M. Piccione, B.F. Woodfield, J. Boerio-Goates, A. Navrotsky, and M. Davis, *J. Phys. Chem. B*, 105, 6025 (2001).

90. B. Marler, U. Werthmann, and H. Gies, *Mic. Mes. Mat.*, 43, 29 (2001).

91. P.A. Barrett, T. Boix, M. Puche, D.H. Olson, E. Jordan, H. Koller, and M.A. Camblor, *Chem. Commun.*, 2114 (2003).

92. A. Corma, F. Rey, J. Rius, M.J. Sabater, and S. Valencia, *Nature*, 431, 287 (2001).

93. H. Kessler, J. Patarin, and C. Schott-Darie, *Stud. Surf. Sci. Catal.*, 85, 75 (1994).

94. D.M. Bibby and M.P. Dale, *Nature*, 317, 157 (1985).
95. Q. Huo, R. Xu, S. Li, Z. Ma, Z, J.M. Thomas, R.H. Jones, and A.M. Chippindale, *Chem. Commun.*, 875 (1992).
96. R.F. Lobo, S.I. Zones, and M.E. Davis, *J. Inclusion Phenom. Mol. Recogn. Chem.*, 21, 47 (1995).
97. M.A. Camblor, L.A. Villaescusa, and M.J. Diaz-Cabañas, *Top. Catal.*, 9, 59 (1999).
98. E.M. Flanigen and R.L. Patton, US Patent 4,073,865 (1978).
99. J.L. Guth, H. Kessler, and R. Wey, in *New Developments in Zeolite Science and Technology*, Y. Murakami, A. Iijima, and J.W. Ward, (editors), Elsevier, Amsterdam, the Netherlands, 1986, p. 121.
100. J.P. Gilson, in *Zeolite Microporous Solids: Synthesis, Structure and Reactivity*, E.G. Derouane, F. Lemos, C. Naccache, and F.R. Ribeiro, (editors), NATO ASI Series, No. C352, Kluwer, Dordrecht, 1992, p. 19.
101. H. Kessler, in *Verified Synthesis of Zeolitic Materials* (2nd edition), H. Robson, (editor), Elsevier, Amsterdam, the Netherlands, 2001.
102. P.A. Barrett, M.A. Camblor, A. Corma, R.H. Jones, and L.A. Villaescusa, *J. Phys. Chem. B*, 102, 4147 (1998).
103. P.A. Barrett, M.J. Díaz-Cabañas, and M.A. Camblor, *Chem. Mater.*, 11, 2919 (1999).
104. D.H. Olson, X. Yang, and M.A. Camblor, *J. Phys. Chem. B*, 108, 11044 (2004).
105. C.M. Zicovich-Wilson, M.L. San-Román, M.A. Camblor, F. Pascale, and J.S. Durand-Niconoff, *J. Am. Chem. Soc.*, 129, 11512 (2007).
106. L.A. Villaescusa, F.M. Marquez, C.M. Zicovich-Wilson, and M.A. Camblor, *J. Phys. Chem. B*, 106, 2796 (2002).
107. X. Yang, M.A. Camblor, Y. Lee, H. Liu, and D.H. Olson, *J. Am. Chem. Soc.*, 126, 10403 (2004).
108. Z. Liu, T. Ohsuna, O. Terasaki, M.A. Camblor, M.J. Diaz-Cabañas, and K. Hiraga, *J. Am. Chem. Soc.*, 123, 5370 (2001).
109. H. Koller, A. Wölker, L.A. Villaescusa, M.J. Díaz-Cabañas, S. Valencia, and M. A. Camblor, *J. Am. Chem. Soc.*, 121, 3368 (1999).
110. A. Villaescusa, I. Díaz, P.A. Barrett, S. Nair, J.M. LLoris-Cormano, R. Martínez-Mañez, M. Tsapatsis, Z. Liu, O. Terasaki, and M.A. Camblor, *Chem. Mater.*, 19, 1601 (2007).
111. R. Szostak, *Molecular Sieves: Principles of Synthesis, and Identification* (2nd edition), Blackie, London, 1998.
112. M.A. Camblor, P.A. Barrett, M.J. Diaz-Cabañas, L.A. Villaescusa, M. Puche, T. Boix, E. Perez and H. Koller, *Mic. Mes. Mat.*, 48, 11 (2001).
113. P.A. Barrett, E.T. Boix, M.A. Camblor, A. Corma, M.J. Diaz-Cabañas, S. Valencia, and L.A. Villaescusa, in *Proceedings of the 12th International Zeolite Conference*, Baltimore, MA, 1998; M.M.J. Treacy, B.K. Marcus, M.E. Bisher, and J.B. Higgins, (editors), Materials Research Society, Warrendale, PA, 1999, p. 1495.
114. L.A. Villaescusa and M.A. Camblor, *Recent. Res. Devel. Chem.*, 1, 93 (2001).
115. S.I. Zones, R.J. Darton, R. Morris, and S.-J. Hwang, *J. Phys. Chem. B*, 109, 652 (2005).
116. Y. Kubota, M.M. Helmkamp, S.I. Zones, and M.E. Davis, *Mic. Mater.*, 6, 213 (1996).
117. H.W. Robson, US Patent 3,733,390 (1971).
118. H.W. Robson, K.L. Riley, and D.D. Maness, *ACS Symp. Ser.*, 40, 233 (1977).
119. C. de las Pozas, D. Diaz-Quintanilla, J, Perez-Pariente, R. Roque-Malherbe, and M. Magi, *Zeolites*, 9, 33 (1989).
120. C. de las Pozas, E. Reguera-Ruiz, C. Diaz-Aguila, and R. Roque-Malherbe, *J. Solid State Chem.*, 93, 215 (1991).
121. C. de las Pozas, C. Becker, M. Carreras, and R. Roque-Malherbe, in *Zeolites '91. Memoirs of the 3rd International Conference on the Occurrence, Properties and Utilization of Natural Zeolites. Part 1*, G. Rodriguez and J.A. Gonzales, (editors), International Conference Center Press, Havana, Cuba, 1993, p. 281.
122. C. de las Pozas, C. Becker, M. Carreras, N. Ginarte, L. Lopez-Colado, T. Marquez, and R. Roque-Malherbe, in *Zeolites '91. Memoirs of the 3rd International Conference on the Occurrence, Properties and Utilization of Natural Zeolites. Part 1*, G. Rodriguez and J.A. Gonzales, (editors), International Conference Center Press, Havana, Cuba, 1993, p. 278.
123. R. Roque-Malherbe, W. del Valle, F. Marquez, J. Duconge, and M.F.A. Goosen, *Sep. Sci. Technol.*, 41, 73 (2006).
124. D.M. Ruthven, *Principles of Adsorption and Adsorption Processes*, John Wiley & Sons, New York, 1984.
125. R Roque-Malherbe, L. Lemes, C. de las Pozas, L. Lopez, and A. Montes, in *Natural Zeolites '93 Conference Volume*, D.W. Ming and F.A. Mumpton, (editors), International Committee on Natural Zeolites, Brockport, NY, 1995, p. 299.
126. R. Roque-Malherbe, *Mic. Mes. Mat.*, 41, 227 (2000).

127. S.T. Wilson, B.M. Lok, C.A. Messina, T.R. Cannan, and E.M. Flanigen, *J. Amer. Chem. Soc.*, 104, 1146 (1982).
128. E.M. Flanigen, B.M. Lok, R.L. Patton, and S.T. Wilson, *Pure Appl. Chem.*, 58, 1351 (1986).
129. C.A. Messina, B.M. Lok, and E.M. Flanigen. US Patent 4,544,143 (1985).
130. S.T. Wilson and E.M. Flanigen, US Patent 4,567,029 (1986).
131. E.M. Flanigen, B.M. Lok, R.L. Patton, and S.T. Wilson, *Stud. Surf. Sci. Catal.*, 20, 103 (1986).
132. A. Merrouche, J. Patarin, H. Kessler, and D. Anglerot, French Patent Application No. 91-01106, January, (1991).
133. T.E. Gier and G.D. Stucky, *Nature*, 349, 508 (1991).
134. C. Montes, M.E. Davis, B. Murray, and M. Narayana, *J. Phys. Chem.*, 94, 6431 (1990).
135. M.S. Rigutto and H.J. van Bekkum, *Mol. Catal.*, 81, 77 (1993).
136. J.A. Martens and P.A. Jacobs, *Stud. Surf. Sci. Catal.*, 85, 653 (1994).
137. J.A. Martens and P.A. Jacobs, in *Preparation of Solid Catalysts*, G. Ertl, H. Knozinger, and J. Weitkamp, (editors), Wiley-VCH, Weinheim, Germany, 1999, p. 53.
138. Ch. Baerlocher, W.M. Meier, and D.M. Olson, *Atlas of Zeolite Framework Types* (5th edition), Elsevier, Amsterdam, the Netherlands, 2001.
139. C. de las Pozas, R. Lopez-Cordero, J.A. Gonzales-Morales, N. Travieso, and R. Roque-Malherbe, *J. Mol. Catal.*, 83, 145 (1993).
140. M.M.J. Treacy and J.B. Higgins, (editors), *Collection of Simulated XRD Powder Patterns for Zeolites* (4th edition), Elsevier, Amsterdam, the Netherlands, 2001.
141. A. Clearfield and M. Kuchenmeister, in *Supramolecular Architecture*, T. Bein, (editor), American Chemical Society Symposium Series, Washington, DC, 1992.
142. A. Corma, *Chem. Rev.*, 97, 2373 (1997).
143. S.L. Suib, *Chem. Rev.*, 93, 803 (1993).
144. D.E. Vaughan, *Catal. Today: Pillared Clays*, 2, 187 (1988).
145. J.J. Fripiat, in *Preparation of Solid Catalysts*, G. Ertl, H. Knozinger, and J. Weitkamp, (editors), Wiley-VCH, New York, 1997, p. 284.
146. M.L. Occelli, in *Preparation of Catalysts*, V. Poncelot, G. Jacobs, P.A. Delmon, (editors), Elsevier, Amsterdam, the Netherlands, 1991, p. 287.
147. H.H. Murray, *Applied Clay Mineralogy*, Elsevier Science & Technology Books, the Netherlands, 2007.
148. G.W. Brindley and R.E. Sempels, *Clay Miner.*, 12, 299 (1977).
149. D.E. Vaughan, R.J. Lussier, and J.S. Magee, US Patent 4,176,090 (1979); 4,248,739 (1981); 4,271,043 (1981).
150. J.L. Casci, *Stud. Surf. Sci. Catal.*, 85, 329 (1994).
151. G. Johansson, *Acta Chem. Scand.*, 14, 769 (1960).
152. S.M. Bradley, R.A. Kydd, and R. Yamdagni, *J. Chem. Soc. Dalton Trans.*, 413, 2653 (1990).
153. S.M. Bradley, R.A. Kydd, and R. Yamdagni, *Mag. Res. Chem.*, 28, 741 (1990).
154. L.C.W. Baker and J. S. Figgs, *J. Am. Chem. Soc.*, 92, 3794 (1970).
155. S.M. Bradley and R.A. Kydd, *Catal. Lett.*, 8, 185 (1991).
156. A. Vierra-Coelho and G. Poncelet, *Appl. Catal.*, 77, 303 (1991).
157. S.M. Bradley, *Catal. Today: Pillared Clays*, 2, 233 (1988).
158. C.T. Kresge, M.E. Leonowicz, W.J. Roth, J.C. Vartuli, and J.S. Beck, *Nature*, 359, 719 (1992).
159. J.S. Beck, J.C. Vartuli, W.J. Roth, M.E. Leonowicz, C.T. Kresge, K.D. Schmitt, C.T.-W. Chu et al., *J. Am. Chem. Soc.*, 114, 10834 (1992).
160. X.S. Zhao, G.Q. Lu, and G. J. Millar, *Ind. Eng. Chem. Res.*, 35, 2075 (1996).
161. D. Myers, *Surfactant Science and Technology* (3rd edition), John Wiley & Sons, New York, 2005.
162. C.Y. Chen, H.Y. Li, and M.E. Davis, *Mic. Mat.*, 2, 27 (1993).
163. A. Steel, S.W. Carr, and M.W. Anderson, *J. Chem. Soc. Chem. Com.*, 1571 (1994).
164. S. Mann, S.L. Burkett, S.A. Davis, C.E. Fowler, N.H. Mendelson, S.D. Sims, D. Walsh, and N.T. Whilton, *Chem. Mater.*, 9, 2300 (1997).
165. A. Corma and H. Garcia, *Chem. Rev.*, 103, 4307 (2003).
166. A. Imhof and D.J. Pine, *Nature*, 389, 948 (1997).
167. J.E.G. Wijnhoven and W.L. Vos, *Science*, 281, 802 (1998).
168. T. Martin, A. Galarneau, F. Di Renzo, F. Fajula, and D. Plee, *Angew. Chem., Int. Ed.*, 41, 2590 (2002).
169. P.T. Tanev, M. Chibwe, and T. Pinnavaia, *Nature*, 368, 321 (1994).
170. P.T. Tanev and T. Pinnavaia, *Science*, 267, 865 (1995).
171. N. Ulagappan and C.N.R. Rao, *Chem. Commun.*, 2759 (1996).

172. D. Khushalani, A. Kuperman, G.A. Ozin, K. Tanaka, J. Garces, M.M. Olken, and N. Coombs, *Adv. Mater.*, 7, 842 (1995).
173. S. Förster and T. Plantenberg, *Angew. Chem., Int. Ed.*, 41, 688 (2002).
174. S. Förster and M. Antonioni, *Adv. Mat.*, 10, 195, (1998).
175. F. Rodriguez-Reinoso and A. Sepulveda-Escribano, in *Handbook of Surfaces and Interfaces of Materials*, Vol. 5, H.S. Nalwa, (editor), Academic Press, New York, 2001, p. 309.
176. A.C. Lua, and J. Guo, *Langmuir*, 17, 7112 (2001).
177. P.M. Ajayan, *Chem. Rev.*, 99, 1787 (1999).
178. K. Mukhopadhyay, K. Ram, and K.U.B. Rao, *Def. Sci. J.*, 58, 437 (2008).
179. M. Sharon and M. Sharon, *Def. Sci. J.*, 58, 460 (2008).
180. S. Morooka and K. Kusakabe, *MRS Bull.*, March, 25 (1999).
181. R.W. Baker, *Membrane Technology and Applications*, John Wiley & Sons, New York, 2004.
182. G. Saracco, H.W.J.P. Neomagus, G.F. Versteeg, and W.P.M. van Swaaij, *Chem. Eng. Sci.*, 54, 1997 (1999).
183. S.C.A. Kluiters, Status review on membrane systems for hydrogen separation, Intermediate Report, 5th Research Framework, European Union Project MIGREYD, Contract No. NNE5-2001-670, ECN-C–04-102, 2004.
184. V.M. Gryaznov, *Platinum Metals Rev.*, 30, 68 (1986).
185. A. Nijmeijer, Hydrogen-selective silica membranes for use in membrane steam reforming, PhD dissertation, University of Twente, the Netherlands, 1999.
186. H.P. Hsieh, *Inorganic Membranes for Separation, and Reaction, Membrane Science and Technology Series 3*, Elsevier, Amsterdam, the Netherlands, 1996.
187. G. Cicala, G. Bruno, and P. Capezzutto, in *Handbook of Surfaces and Interfaces of Materials*, Vol. 1, H.S. Nalwa, (editor), Academic Press, New York, 2001, p. 509.
188. T.M. Besmann, *Processing Science for Chemical Vapor Infiltration, Laboratory Industry/Government Briefing*, Oak Ridge National Laboratory, Oak Ridge, TN, 1990.
189. G.R. Gavalas, Ceramic membranes for hydrogen production from coal, Annual Technical Report, DE-FG26-00NT40817, California Institute of Technology, Pasadena, CA, March 18, 2003.
190. G.R. Gavalas, C.E. Megiris, and S.W. Nam, *Chem. Eng. Sci.*, 44, 1829 (1989).
191. B.-K. Sea, K. Kusakabe, and S. Marooka, *J. Memb. Sci.*, 130, 41 (1997).
192. S.W. Nam and G.R. Gavalas, *AIChE Symp. Ser.*, 85, 68 (1989).
193. L.H. Sperling, *Introduction to Physical Polymer Science*, John Wiley & Sons, New York. 2006.
194. D. Braun, H. Cherdron, H. Ritter, B. Voit, and M. Rehahn, *Polymer Synthesis: Theory and Practice* (4th edition), Springer-Verlag, New York, 2005.
195. D.I. Bower, *Introduction to Polymer Physics*, Cambridge University Press, Cambridge, UK, 2002.
196. R.O. Ebewele, *Polymer Science and Technology*, CRC Press, Boca Raton, FL, 2000.

4 Material Characterization Methods

4.1 INTRODUCTION

After the process of synthesis, a thorough characterization of the obtained materials is necessary. It is almost impossible to understand the properties of a material and, consequently, its proper application, if we do not understand its composition, structure, and morphology. A task of fundamental importance in the physical chemistry of materials, as in all material sciences, is "characterization."

The golden rule of material characterization is to apply numerous methods, since only one methodology normally does not bring about a complete understanding of the material. In order to properly characterize a material, various procedures are applied [1,2]. Among them include x-ray diffraction (XRD) [3–5], Mössbauer spectrometry [6], transmission electron microscopy (TEM) [7], scanning electron microscopy (SEM) [8], impedance spectroscopy [9], infrared (IR) and Raman spectroscopy [10,11], nuclear magnetic resonance (NMR), [12,13], x-ray fluorescence (XRF) [14], and energy-dispersive x-ray analysis (EDAX) [8]. To supplement these characterization methodologies, thermal methods are also usually applied [15–18], for example, differential thermal analysis (DTA) [16], thermal gravimetric analysis (TGA) [16], differential scanning calorimetry (DSC) [17], temperature-programmed reduction (TPR) [18], temperature-programmed desorption (TPD) [19], dielectric analysis methods, such as thermodielectric analysis (TDA) and dielectric spectrometry (DS) (as explained in Sections 4.8.3 and 8.6.2, DS is an alternative impedance spectroscopy used in the study of solid electrolytes like anionic and cationic conductors) [15], and adsorption methods [20,21]. Chapter 4 briefly describes the use of XRD, SEM, IR, NMR, XRF, EDAX, DTA, TGA, DSC, TPR, TPD, TDA, DS, and Mössbauer spectrometry. Impedance spectroscopy, as a method to test electrochemical systems, is described in Section 8.6.

4.2 APPLICATION OF XRD IN MATERIAL CHARACTERIZATION

In Chapter 1, the physical principles of XRD were studied; here, those concepts are applied to material characterization.

4.2.1 BRAGG–BRENTANO GEOMETRY POWDER DIFFRACTOMETER

XRD is the most important methodology used in the characterization of materials. It is applied for the determination of the crystalline structure of new materials, the evaluation of the changes in the structural parameters of modified materials, the phase analysis of materials, and particle size and strain determinations [3,4,22–46]. As was described in Chapter 1, the greater part of the applications of the XRD in material characterizations is performed with Bragg–Brentano geometry diffractometers. The principal characteristics of the Bragg–Brentano geometry were explained in Chapter 1. An x-ray diffractometer was schematically shown in Figure 1.24.

4.2.2 Intensity of a Diffraction Peak of a Powdered Sample

Any powdered material consists of a set of randomly oriented crystallites of the material under test. The line intensity of a powder XRD pattern obtained in a Bragg–Brentano geometry diffractometer for a pure sample, comprised of three-dimensional crystallites with a parallelepiped form (see Equation 1.64), is given by [3,4,22,24,28]

$$I_d(\theta) = I_0 K_e \left(\frac{1 + \cos^2(2\theta)}{\sin^2\theta\cos\theta} \right) F_{hkl}^2 m_{hkl} \left(\frac{1}{V_c} \right)^2 B_F(\theta)D(\theta) \left(\frac{\upsilon_\alpha}{\mu} \right) \tag{4.1}$$

where

$I_d(\theta)$ is the diffracted intensity
I_0 is the intensity of the incident beam
K_e is a constant depending on the experimental setup, in which are included: λ the wavelength of the used radiation; r the distance from sample to the detector; m and e the mass and the charge of the electron; c the velocity of light; and ε_0 the permittivity of free space (see Equation 1.51)
F_{hkl}^2 is the structure factor of the reflection (hkl)
m_{hkl} is the multiplicity factor corresponding to the reflection (hkl) of the phase under study
V_c is the volume of the unit cell of the crystal
$B_F(\theta)$ is the peak profile
$D(\theta)$ is the Debye–Waller factor
υ_α is the volume fraction of the phase under test
μ is the linear absorption coefficient of the sample under test

4.2.3 Qualitative Identification of Phases

The simplest application of x-ray powder diffraction is the qualitative identification of phases. This application is very important, because it is the first step to understand more complex uses of the method. However, in some cases, it is enough, because of the nature of the intended application of the material under test, and the previous knowledge about the structure of the concrete material. Figure 4.1 shows an example of the use of XRD for the characterization of a material [21,47]. An MCM-41 mesoporous molecular sieve (MMS), which is the hexagonal phase of the M41S family of materials, is explained as an example of XRD use. The MCM-41 MMS material shows an XRD pattern including three or more low-angle (below 10° in 2θ) peaks that can be indexed to a hexagonal lattice [21].

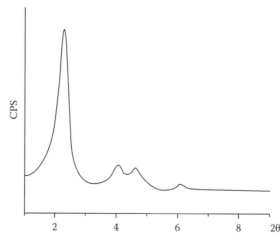

FIGURE 4.1 XRD pattern, counts per second (CPS) vs. 2θ of the mesoporous material MCM-41.

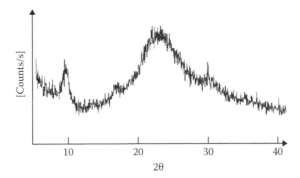

FIGURE 4.2 XRD pattern of the opal sample 81C with an AlPO$_4$-5 microporous molecular sieve synthesized on the surface.

Another example of the use of XRD in material characterization is shown in Figure 4.2. The XRD pattern of a sintered opal membrane covered with a molecular sieve is shown. The membrane was prepared with the opal sample 81C [47]; afterward, the produced membrane was covered with an AlPO$_4$-5 microporous molecular sieve synthesized on the surface [41].

The membrane was produced with the opal powders by applying the following procedure: cylindrical wafers (with a diameter of 12.7 mm and a length (l) of 3 ± 0.1 mm) were prepared by pressing 0.55 ± 0.01 g of the opal powder at 250 MPa with a hydraulic press. The wafers were thermally treated in a furnace at 1100°C for 3 h. The more intense peaks in the XRD profile, shown in Figure 4.2, correspond to the AlPO$_4$-5 crystalline phase [5]. In contrast, the broad peak, shown in Figure 4.2, is related with the amorphous silica constituting the support membrane.

4.2.4 RIETVELD METHOD

A complete understanding of the structure of the material under study or application is a sine qua non condition for the successful research or use of the material. In the case of powders, the best way to decipher the structure of new materials is the Rietveld method. This methodology was initially developed by Hugo M. Rietveld in 1969 [23] as a procedure for refining crystal structures using neutron powder diffraction data. To implement the method in practice, certain information about the estimated crystal structure of the phase or phases of interest in the diffraction profile under test is necessary.

The aim of the method is to find the best agreement between measured y_i^e and calculated y_i^t diffraction patterns. In the refinement procedure, the model is assumed optimum when the sum of the squares of the differences between the experimental and the calculated patterns, R, is a minimum according to a nonlinear optimization process [29]. The sum of the squares of the differences between the experimental and the calculated patterns is given by

$$R = \sum_{1}^{n} w_i \left(y_i^e - y_i^t \right)^2$$

where w_i are the statistical weights assigned to every observation, which is usually equal to the reciprocal of the variance. The calculated intensity profile is given by the previously explained expression for the intensity of a diffraction peak function in a digital form, that is, experimental point by point (see Equation 4.1)

$$y_i^t = s \sum_j L_j F_{hkl}^j \varphi(2\theta_i - 2\theta_j) P_j A + y_i^b$$

where

  j is equivalent to, (hkl) corresponding to a Bragg reflection
  s is a scale factor
  L_j includes the $LP(\theta)$ and multiplicity, m_{hkl}, factors
  F_{hkl}^j is the structure factor, which includes the Debye–Waller factor
  $\varphi(2\theta_i - 2\theta_j)$ is the peak profile function describing the particle size broadening and other sources of peak broadening
  P_j is a preferred orientation correction, which is not included in the previous intensity of a diffraction peak function
  A is an absorption factor
  y_i^b is the pattern background function

The $LP(\theta)$ and multiplicity, m_{hkl}, factors are given by the expressions previously described (see Chapter 1). The structure factor is computed with the help of the following expression [4,22,24,25]

$$F_j = \sum_k f_k N_k e^{2\pi i (hx_k + ky_k + lz_k)} T_k$$

where

  f_k is the scattering factor for an atom k
  N_k is the site occupancy factor
  $hx_k + ky_k + lz_k$ are the positional coordinates for atom k within the unit cell
  T_k is the temperature factor [22,26]

$$T_k = e^{-M_k}, \quad \text{where } M_k = \frac{B_k \sin^2\theta}{\lambda^2}, \quad \text{and} \quad B_k = 8\pi^2 \langle u_k^2 \rangle$$

in which $\langle u_k^2 \rangle$ is the mean square displacement of the vibrating atom or ion, k, in the direction normal to the diffracting planes.

The profile functions applied are Gaussians, Lorentzians, Pearson VII, and other functions. Here, the Gaussian (G) and the Lorentzian (L) functions are described

$$G_{jk} = \frac{C_0^{\frac{1}{2}}}{H_j \pi^{\frac{1}{2}}} e^{-\frac{C_0(2\theta - 2\theta_j)}{H_j^2}} \quad \text{and} \quad L_{jk} = \frac{C_1^{\frac{1}{2}}}{H_j \pi} \left(\frac{1}{1 + C_1 \frac{(2\theta_i - 2\theta_j)}{H_j^2}} \right)$$

where

  C_j is the peak amplitude
  H_j is the full width at half maximum (FWHM)

$H_j = U \tan^2\theta + V \tan\theta + W$, where H, U, and W are parameters that can be refined, and the preferred orientation is given by different expressions, such as [29]

$$P_j = e^{-G_1\alpha_j^2} \quad \text{or} \quad P_j = G_2 + (1 - G_2)e^{-G_1\alpha_j^2}$$

where

  G_1 and G_2 are parameters which can be refined
  α_j is the angle between the (hkl) plane and the preferred orientation vector

Finally, the pattern background can be shaped by different expressions normally consisting of polynomials. One of the applied expressions is given by

$$y_{ib} = \sum_{1}^{n} B_m T_m(2\theta)$$

where

B_m is a refinable parameter

$T_m(2\theta)$ are nonshifted type 1 Chebyshev functions

4.2.5 QUANTITATIVE PHASE ANALYSIS

The powder XRD analysis of the components present in a mixture was one of the first applications carried out with this methodology. In fact, in 1919, Hull was the first to apply the XRD methodology in chemical analysis [43]. However, Klug and Alexander [36] gave a big impetus to the development of the XRD phase analysis method.

The absorption factor for a sample in the form of a plate located in the sample holder of a Bragg–Brentano geometry powder diffractometer is given by [4]

$$A = \frac{1}{2\mu}$$

where μ is the linear absorption coefficient. In this case, a simplified form of Equation 4.1 was proposed by Klug and Alexander to express the intensity of the ith peak of the jth phase included in the mixture under analysis [36]

$$I_{ij} = \frac{K_{ij}x_j}{\rho_j \mu^*} \tag{4.2}$$

where

K_{ij} is a constant

x_j is the weight fraction of the diffracting phase

ρ_j is the density of the diffracting phase

μ^* is the mass absorption coefficient of the mixture under analysis

Then, from Equation 4.2, the integrated intensity of the ith peak in the XRD profile of the mixture is calculated by the addition of all the contributions to this peak from all the phases present in the mixture [30,31,38–40]

$$I_i = \sum_{j=1}^{N} I_{ij} = \sum_{j=1}^{N} \frac{K_{ij}x_j}{\rho_j \mu^*} \quad \text{where } j = 1,\ldots,N \tag{4.3}$$

where N is the number of crystalline phases in the mixture. Equation 4.3 forms a linear system of equations with the complementary normalization condition [30]:

$$\sum_{j=1}^{N} x_j = 1 \tag{4.4}$$

In the simplest case, this system can be solved using the method of multilinear regression [30,31,40]. In some cases, it is necessary to experimentally determine μ^*, the mass absorption coefficient of the mixture [39,40]. To carry out this procedure, a monochromatic CuK_α radiation was used, and with the help of the following equation [39,40]

$$\mu^* = \frac{E}{M} \ln\left[\frac{I_0}{I}\right]$$

where

    E is a constant, characteristic of the experimental arrangement used
    M is the sample mass
    I_0 is the intensity measured without an absorbent
    I is the intensity measured with an absorbent

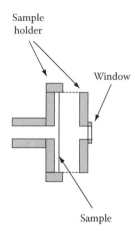

FIGURE 4.3 Schematic representation of the sample holder for the measurement of the mass absorption coefficient of the mixture.

it was determined that $E \approx 1.3$, with the help of a sample holder constructed to embrace a wafer of the powdered sample, and the sample holder was attached to the goniometer arm (see Figure 4.3) [40].

An analysis of Equations 4.2 through 4.4 reveals that the previous method could fail because of the assumption of the absence of microabsorption, when deriving Equation 4.2 [30]. The microabsorption effect for x-rays diffracted from planar granular powder specimens is caused by the presence of large particles with different absorption magnitudes in the mixture, which create bulk porosity, surface roughness, and the difference in absorption coefficient between the different phases present in the mixture. To account for the microabsorption effect, Leroux, Lennox, and Kay proposed a semiempirical correction to Equation 4.1, as follows [37]:

$$I_{ij} = \frac{K_{ij}x_j}{\rho_j(\mu^*)^{\alpha_j}} \tag{4.5}$$

where α_j are empirical parameters. Gonzales et al. [30,31,40], applying Equation 4.5, obtained

$$I_i = \sum_{j=1}^{N} K_{ij}\left(\frac{\mu_j^*}{\mu^*}\right)^{\alpha_j} I_{ij}^0 x_j \tag{4.6}$$

with the complementary condition

$$\sum_{j=1}^{N} x_j = 1$$

Equation 4.6 can now be interpreted as a multilinear regression equation [48]

$$I_i = \sum_{j=1}^{N} K_{ij}\left(\frac{\mu_j^*}{\mu^*}\right)^{\alpha_j} I_{ij}^0 x_j = \sum_{j=1}^{N} \beta_j I_{ij}^0 \tag{4.6a}$$

with independent variables, I_{ij}^0, which are the intensities of the peak, i, of the pure phase, j, and dependent variables, I_i, which is the intensity of the peak, i, of the mixture. The regression coefficients, β_j, can be determined as follows [30,39,48]:

$$\beta_i = \sum_{j=1}^{N} A_{ij}^{-1} B_j$$

where A_{ij}^{-1}; is the inverse matrix of the following matrix [30,31,39,48]

$$A_{ij} = \sum_{k=1}^{P} I_{ik}^0 I_{kj}^0 W_k$$

and

$$B_j = \sum_{k=1}^{P} I_{ik}^0 I_k W_k$$

where [30,31,39,48]

$W_k = 1/\Delta I_k$ are statistical weights

ΔI_k are the errors in the determination of the intensity of peak k

To make a complete calculation of the phase composition of the test sample, this method can take the experimentally measured value of μ^* or the mass absorption coefficient can be calculated with the help of the method of successive approximations [30,31,39]. However, the previously explained method is complex; therefore, it is easier to avoid the microabsorption effect by milling the test sample to get particles of about $1\,\mu m$ [39].

A simple example of the use of XRD for quantitative phase analysis is as follows: the phase composition of a perovskite was determined using the obtained XRD diffraction pattern. The x-ray diffractograms were obtained in a Siemens D5000 x-ray diffractometer, in a vertical setup θ–2θ geometry in the range $15° < 2\theta < 75°$, with a Cu K_α radiation source, Ni filter, and graphite monochromator [32].

Figure 4.4 [32] shows the XRD diffraction pattern of the $BaCe_{0.95}Yb_{0.05}O_{3-\delta}$ perovskite. It is evident, by comparing the obtained pattern with the powder diffraction data contained in the

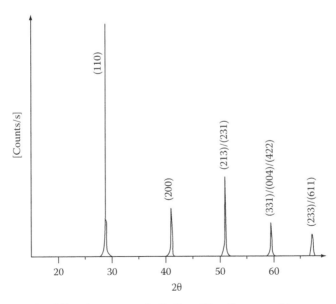

FIGURE 4.4 XRD powder diffraction pattern of a $BaCe_{0.95}Yb_{0.05}O_{3-\delta}$ perovskite.

International Center for Diffraction Data, Powder Diffraction File database, that the analyzed sample is composed of almost 100% $BaCe_{0.95}Yb_{0.05}O_{3-\delta}$ perovskite.

In the case of mixtures, Gonzales et al. [30,31,39,40] have applied the methodology to minerals, for example, laterites [40] and zeolites [44]. In Table 4.1, [44] the results of the phase analysis of several natural zeolites are reported. In Table 4.2, the elemental compositions, in oxide wt %, of the same natural zeolite rocks are reported [44]. To calculate the absolute quantity of the zeolite phase in the samples used as standards, it was necessary to use the adsorption method (see Chapter 6).

4.2.6 LATTICE PARAMETER DETERMINATION

For different studies in materials science, it is necessary to determine the lattice parameters of the obtained or modified materials. In Table 4.3, the equations relating the interplanar distance d, the Miller indexes (hkl), and the lattice parameters for the seven crystalline systems are given [4,22].

As evident in the cubic case, the parameter a of the cubic cell is directly related to the spacing, d_{hkl}, between planes in the (hkl) family. Then, knowing θ_{hkl}, the Bragg angle, and λ, the x-ray wavelength, it is possible to calculate the lattice parameter a with the help of the Bragg law. Consequently, it is the precision in the measurement of $\sin \theta$, not the precision in the determination of θ, that determines the precision in the determination of the cell parameter.

If we made an error in the expansion of the Bragg law,

$$\frac{\Delta d}{d} = \frac{\Delta \lambda}{\lambda} - (\cot \theta)\Delta\theta \approx -(\cot \theta)\Delta\theta \qquad (4.7)$$

TABLE 4.1
Mineralogical Composition (in wt %) of Several Natural Zeolite Rocks Employed to Illustrate Some Properties and Applications of These Materials

Sample	Clinoptilolite	Mordenite	Erionite	Others
HC	85	0	0	15
CMT	42	39	0	19
CMT-C	80	10	0	10
C2	42	31	0	27
C4	0	75	0	25
C5	11	64	0	25
C6	4	71	0	25
MP	5	80	0	15
AD	0	0	85	15
CZ	70	0	0	30
GR	85	15	0	15

Notes: Sample identification (label: deposit name, location). HC: Castillas, Havana, Cuba; CMT: Tasajeras, Villa Clara, Cuba; C1-C6: Camaguey, Cuba; MP: Palmarito, Santiago de Cuba, Cuba; AD: Aguas Prietas, Sonora, Mexico; CZ: Nizni Harabovec, Slovakia; GR: Dzegvi, Georgia.

Montmorillonite (2–10 wt %), quartz (1–5 wt %), calcite (1–6 wt %), feldspars (0–1 wt %), magnetite (0–1 wt %), and volcanic glass (3–6 wt %).

TABLE 4.2
Elemental Composition (in Oxide wt %) of Several Natural Zeolite Rocks Employed to Illustrate Some Properties and Applications of These Materials

Sample	SiO_2	Al_2O_3	Fe_2O_3	CaO	MgO	Na_2O	K_2O	H_2O
HC	66.8	13.1	1.3	3.2	1.2	0.6	1.9	12.1
CMT	66.6	12.5	2.7	2.7	0.7	1.7	0.8	12.9
CMT-C	65.2	11.9	1.6	3.0	0.5	2.2	0.8	14.0
C1	62.3	13.1	1.7	2.8	1.2	0.9	1.4	15.1
C2	64.3	12.9	3.6	5.0	1.3	1.5	1.3	10.4
C4	65.3	10.2	2.1	2.7	0.5	1.7	1.2	14.0
C5	66.2	11.4	3.8	4.2	0.7	1.3	1.0	12.8
C6	67.5	11.0	1.9	4.9	0.5	1.9	0.7	11.9
MP	66.9	11.6	2.7	4.4	0.8	1.8	0.8	12.1
AD	59.6	14.2	2.3	2.2	1.5	2.4	3.3	13.8
CZ	69.8	14.2	1.2	3.0	1.5	2.4	3.3	13.9
GR	62.4	12.0	2.9	4.1	1.8	2.0	1.2	14.1

Notes: Sample identification (label: deposit name, location). HC: Castillas, Havana, Cuba; CMT: Tasajeras, Villa Clara, Cuba; C1-C6: Camaguey, Cuba; MP: Palmarito, Santiago de Cuba, Cuba; AD: Aguas Prietas, Sonora, Mexico; CZ: Nizni Harabovec, Slovakia; GR: Dzegvi, Georgia.

TABLE 4.3
Relations between Interplanar Spacing, Miller Indexes, and Lattice Parameters

Crystalline System	Relation between, d, (hkl), and the Lattice Parameters
Cubic	$\dfrac{1}{d_{hkl}^2} = \dfrac{h^2 + k^2 + l^2}{a^2}$
Tetragonal	$\dfrac{1}{d_{hkl}^2} = \dfrac{h^2 + k^2}{a^2} + \dfrac{l^2}{c^2}$
Hexagonal	$\dfrac{1}{d_{hkl}^2} = \dfrac{4}{3}\left(\dfrac{h^2 + hk + k^2}{a^2}\right) + \dfrac{l^2}{c^2}$
Rhombohedral	$\dfrac{1}{d_{hkl}^2} = \dfrac{(h^2 + k^2 + l^2)\sin^2\alpha + 2[(hk + kl + hl)\cos^2\alpha - \cos\alpha]}{a^2(a - 3\cos^2\alpha + 2\cos^3\alpha)}$
Orthorhombic	$\dfrac{1}{d_{hkl}^2} = \dfrac{h^2}{a^2} + \dfrac{k^2}{b^2} + \dfrac{l^2}{c^2}$
Monoclinic	$\dfrac{1}{d_{hkl}^2} = \dfrac{1}{\sin^2\beta}\left(\dfrac{h^2}{a^2} + \dfrac{k^2\sin^2\beta}{b^2} + \dfrac{l^2}{c^2} - \dfrac{2hl\cos\beta}{ac}\right)$
Triclinic	$\dfrac{1}{d_{hkl}^2} = \dfrac{1}{V^2}\left(S_{11}h^2 + S_{22}k^2 + S_{33}l^2 + 2S_{12}hk + 2S_{23}kl + 2S_{13}hl\right)$

Since the error in the measurement of λ is insignificant. A simple analysis of Equation 4.7 indicates that cot θ comes close to 0, when θ approaches 90° or when 2θ approaches 180°. Consequently, in order to increase the precision in the measurement of lattice parameters, which depends on the measurement of d_{hkl}, we are supposed to use peaks at angles near 180°. However, there are no diffracted beams at these angles. To solve this problem, some extrapolation functions are used. To get these functions, the general approach is to consider the different effects which could result in errors in the measured value of θ. In the case of a diffractometer, the major sources of error are [4,5] misalignment of the diffractometer, the use of a flat sample, absorption in the specimen, displacement of the sample from the instrument axis, and vertical divergence of the incident beam. No single extrapolation function curve is completely satisfactory to take into account the different source of errors, because the source of error has a different dependence on θ; however, an extrapolation of the values calculated for the parameters against $\cos^2\theta$, which is the so-called Bradley–Jay method, is applicable for $\theta > 60$ [4].

Consequently, to determine the cell parameter, the following procedure should be applied [4]: make a correct alignment of the diffractometer, calibrate with an aluminum foil, extrapolate the values calculated for the parameters against $\cos^2\theta$ (see Figure 4.5), and calculate the more precise parameter for $\theta = 90°$ (see Figure 4.5).

In order to calculate the lattice parameter for a noncubic substance, a system of linear equations with the help of the relations reported in Table 4.3 is formed, and then this system is solved. It is easier to understand the procedure with an example; hence, we use the $BaCe_{0.95}Yb_{0.05}O_{3-\delta}$ perovskite that crystallizes in an orthorhombic unit cell [34] to illustrate the methodology [32].

With the help of the measured interplanar distances, corresponding to the following Miller indexes, (110), (200), and (213) (see Figure 4.4), and applying the equation relating the interplanar distance between adjacent planes (d_{hkl}) in the set of planes (hkl) with the a, b, c parameters of the orthorhombic cell (see Table 4.3), a system of three linear equations with three unknowns, that is, the cell parameters, is created. Then, upon solving this system, the cell parameters of the room temperature orthorhombic phase were calculated. The calculated cell parameters were $a = (8.796 \pm 0.001)$ Å, $b = (6.231 \pm 0.001)$ Å, and $c = (6.215 \pm 0.001)$ Å, in good agreement with literature data [34].

In order to make an extrapolation in the case of noncubic crystals, for example, hexagonal or tetragonal crystals, the (hkl) reflections are separated in two groups [4], that is, the ($hk0$) group and the ($00l$) group. With the first group, the extrapolation is carried out to get a_0, and with the second group the extrapolation is carried out to get c_0, following the same procedure previously discussed.

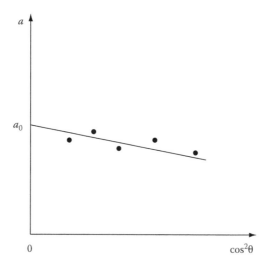

FIGURE 4.5   Schematic extrapolation curve.

TABLE 4.4
Cell Parameters of the MeAPO Species

MeAPO Species	a [nm]	c [nm]
AlPO$_4$-5	13.473	8.351
CoAPO-5	13.624	8.52
ZnAPO-5	13.625	8.51
MnAPO-5	13.624	8.51
FeAPO-5	13.626	8.520

TABLE 4.5
Cell Parameters of Leached Products

Sample	a [Å]	b [Å]	c [Å]
MP(natural)	17.65	20.24	7.48
MP (0.5 M)	17.65	20.24	7.48
MP (1.0 M)	17.65	20.20	7.50
MP (5.0 M)	17.66	20.15	7.51

4.2.6.1 Examples of the Use of Lattice Parameter Determination in the Study of Materials

With regard to the synthesis of the MeAPO-5 (Me: Co, Mn, Fe, and Zn) molecular sieve (see Chapter 3), the chemical analysis of the synthesized samples indicated the presence of about 1% of Me in the synthesized aluminophosphate. Since the incorporation of metals into the AFI structure of AlPO$_4$-5 implies a lattice expansion because, Me(II) has larger ionic radius than Al(III) and P(V) [46]. The measurement of the lattice parameters of the hexagonal lattice corresponding to the AFI structure shows the incorporation of the metal in the molecular sieve framework [46]. In Table 4.4, the cell parameters showing an evident increase in both parameters of the hexagonal lattices from the AlPO$_4$-5 and the MeAPO-5 molecular sieves are reported [46]. Here, a Guinier camera was used to carefully measure the cell parameters [46].

Another example is the acid leaching of a natural mordenite labeled MP, composed of 80 wt % mordenite, 5 wt % clinoptilolite, and 15 montmorinollite (2–10 wt %), quartz (1–5 wt %), calcite (1–6 wt %), feldspars (0–1 wt %), and volcanic glass (see Tables 4.1 and 4.5.) [44].

This sample was acid leached in 0.5, 1.0, and 5.0 M HCl solutions (liquid/solid = 20) for 2 h, and the XRD powder profile was registered. In Table 4.5, the cell parameters showing the changes in the orthorhombic lattice of the MOR framework of the zeolite MP are reported [44].

The results indicate that the mordenite maintains its structure; however, the dealumination process which occurs in the zeolite provokes changes in the cell parameters.

4.2.7 Scherrer–Williamson–Hall Methodology for Crystallite Size Determination

In Equation 4.1, the factor $B_F(\theta)$ is included, which is the peak profile function, that describes particle size broadening and other sources of peak broadening. The XRD method can be used as well for the measurement of the crystallite size of powders by applying the Scherrer–Williamson–Hall methodology [4,35]. In this methodology, the FWHM of a diffraction peak, β, is affected by two types of defects, that is, the dislocations, which are related to the stress of the sample, and the grain size. It is possible to write [35]

$$\beta = \beta_t + \beta_s \tag{4.8}$$

where
β_s is the part of the FWHM contributed by the sample stress
β_t is the portion of the FWHM given by the grain size

Now,

$$\beta_s = 4s\frac{\sin\theta}{\cos\theta} \tag{4.9}$$

and

$$\beta_t = \frac{0.9}{t}\frac{\lambda}{\cos\theta} \tag{4.10}$$

where s measures the sample strain and t is the crystallite size. Therefore,

$$\beta = 4s\frac{\sin\theta}{\cos\theta} + \frac{0.9}{t}\frac{\lambda}{\cos\theta}$$

If we multiply the previous equation by $\dfrac{\cos\theta}{\lambda}$ on both sides, we get

$$\beta\frac{\cos\theta}{\lambda} = 4s\frac{\sin\theta}{\lambda} + \frac{0.9}{t} \tag{4.11}$$

Now, we can set the previous equation in the form of a linear equation, by defining the y-axis as

$$y = \beta\frac{\cos\theta}{\lambda}$$

and the x-axis as

$$x = \frac{\sin\theta}{\lambda}$$

Then,

$$y = ax + b = \beta\frac{\cos\theta}{\lambda} = 4s\frac{\sin\theta}{\lambda} + \frac{0.9}{t}$$

where $a = 4s$ and $b = 0.9/t$.

The procedure to calculate the grain size of a sample is as follows: $\beta(\cos\theta/\lambda)$ is plotted versus $\sin\theta/\lambda$, and then the intercept, $b = 0.9/t$, is computed. Finally, the crystallite size, t, is estimated.

An example of the application of the Scherrer–Williamson–Hall methodology [35] for the calculation of the crystallite size of the $BaCe_{0.95}Yb_{0.05}O_{3-\delta}$ perovskite is given [32]. The selected diffraction peaks ($\theta_1 = 14.167°$, $\theta_2 = 20.527°$, and $\theta_3 = 25.470°$) were scanned at a slow scanning speed of $0.6°/min$. From the recorded XRD pattern, the accurate peak position, integrated intensities, as well as the FWHM, β, of each slow scanned diffraction peak was estimated by fitting them with the Pearson VII amplitude function. The fitting process was carried out with the peak separation and analysis software PeakFit (Seasolve Software Inc., Framingham, Massachusetts) based on the least square procedure [48]. These values were used to calculate the crystallite size and lattice strain of the synthesized powder using the Williamson–Hall equation [35].

The crystallite size of the $BaCe_{0.95}Yb_{0.05}O_{3-\delta}$ perovskite powder was calculated following the method previously explained, that is, the Scherrer–Williamson–Hall methodology [4,35,36]. The calculated radius of the perovskite crystallite, considered as spherical particles, was $a = \frac{t}{2} = 71 \pm 1\,nm$.

4.3 ELECTRON MICROSCOPY

4.3.1 Introduction

The simplest microscope is the light microscope; it consists of an objective lens and an eyepiece. Microscope objectives and eyepieces usually consist of complex lens systems of two or more lenses to correct for lens aberrations (Figure 4.6). The objective lens forms a real intermediate image, which is then magnified by the eyepiece. The objective lens and eyepiece are maintained at a fixed

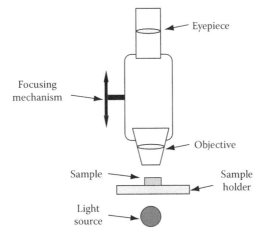

FIGURE 4.6 Diagram of the main parts of a light microscope.

distance and focusing is achieved by moving the whole assembly, up and down, in relation to the sample (see Figure 4.6). High magnification requires very bright illumination of the sample, and a condenser lens is usually placed between the light source and the sample stage to focus light onto the sample.

Resolution is the smallest separation of two points that are visible as distinct entities. The resolving limit of the human eye is 0.1 mm; on the other hand, the resolving limit of the light microscope is 0.2 μm.

4.3.2 Transmission Electron Microscope

The discovery of the wave nature of the electron by de Broglie allowed Ernst Ruska and Max Knoll to invent, in 1931, the transmission electron microscope (TEM) whose maximum resolving power is 0.2 nm. Ruska and Knoll designed and constructed, at the Technical College of Berlin, the first prototype of a TEM. A TEM is analogous in design to a light microscope with one crucial dissimilarity; that is, as a substitute of light, the TEM uses electrons [7,49]. The light source is replaced by a cathode filament that acts as a source of electrons. The electron source must be placed in a vacuum and its position is reversed with respect to the position of the light source, that is, the electron source is on the upper side of the TEM. Electrons are accelerated toward a given sample by a potential difference of several thousand volts, normally 100,000 V and in some cases 1,000,000 V. A series of cylindrical magnets and metal apertures are used to focus the electron beam into a monochromatic beam. This beam collimated by one or two condenser lenses collides with the sample (see Figure 4.7) and interacts with it depending on the density of the material. These interactions are greatly affected by how the specimen is prepared. Thereafter, the electron beam is focused with an objective lens, and is magnified by one intermediate magnetic lens and a projector magnetic lens. Besides, in the equipment, different apertures to help in the alignment of the electron beam are located.

An electron has corpuscular and wave properties, where the wavelength of the electron is related to its velocity, v, as follows:

$$\lambda = \frac{h}{m_e v}$$

where
m_e is the mass of the electron
h is the Planck constant

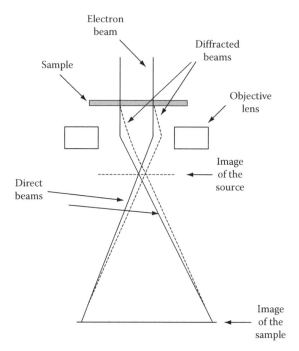

FIGURE 4.7   Schematic representation of a TEM.

Consequently, if the electrons are accelerated toward a given sample by a potential difference of 100,000 V, this will give a velocity of 2×10^8 m/s; then $\lambda = 4 \times 10^{-3}$ nm.

In a TEM image, the detail in the image is formed by the diffraction of electrons from the crystallographic planes of the object being tested (see Figure 4.7). Since the diffraction condition is the Bragg law for the above calculated wavelength, the angles of reflection of the electrons with the test material crystallographic planes are about $\theta_c \approx 0.01$ rad [49]. When a beam of electrons is passed through a sample, it is diffracted by the crystalline planes (Figure 4.7). Some electrons are diffracted and the others pass through the sample without being diffracted. Thereafter, both electron types form an image of the sample (see Figure 4.7).

Electron microscopes are constructed in such a way that it is possible to project either the image of the specimen or the diffraction pattern on to a fluorescent observation screen, and then photograph it on a plate or a film.

This is possible because the projection lens system, which for clarity was not shown in Figure 4.7, is normally included behind the objective lens and below the source image plane. This lens system allows the projection of both the diffraction pattern and the specimen image on the observation screen. In Figure 4.8, [50] the electron diffraction pattern of a Fe thin film is shown. In Figure 4.9, the transmission electron micrograph of the mordenite included in the sample CMT-C (see Table 4.1), where fiber-like crystals of mordenite are seen, is shown [51].

4.3.3   Scanning Electron Microscope

The first scanning electron microscope (SEM) was constructed in 1938 by Manfred von Ardenne, by rastering the electron beam of a TEM over the surface of a sample [8,52]. The initial SEM configuration contained many of the basic principles of modern SEMs. In 1965, Cambridge Scientific Instruments produced the first commercial instrument. Since the original von Ardenne equipment, several design advances have been made, resulting in an improvement of the resolution from 50 nm, in 1942, to ~0.7 nm, today. In modern equipment, in addition to the morphological information given by the secondary electrons (SEs), the SEM can be used to sense X ray fluorescence (XRF) signals

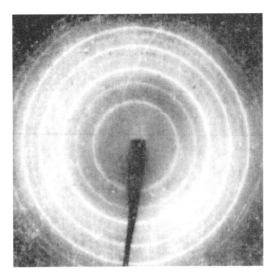

FIGURE 4.8   Electron diffraction pattern of a Fe thin film.

FIGURE 4.9   Transmission electron micrograph of the mordenite included in sample CMT-C (magnification = 10,000×).

from the analyzed sample that provide characteristic x-rays signals, which are used to determine the elemental composition of the sample [8].

XRF and the Auger electrons are used to get analytical information about the sample [52]. XRF is produced by the same mechanism as analyzed in Section 1.6. On the other hand, Auger electrons are emitted in the case when the atom is returning to its minimum energy state, that is, its ground state by filling in the lost electron with another electron from the outer shells and instead of an x-ray photon emission event, the emission of an electron takes place, that is, an Auger electron. Both the XRF and the Auger electrons contain information about the sample composition [8].

In a SEM equipment, the electrons which result from the emission from a filament located in the electron gun are accelerated, with the help of a voltage ranging from 1 to 30 keV (see Figure 4.10) [8,52]. The electron emission event takes place in a vacuum milieu ranging from 10^{-4} to 10^{-10} Torr. Then, the accelerated electrons are directed to the specimen by a series of electromagnetic lenses in the electron column (see Figure 4.10) [8,52].

The resolution and depth of field of the image are established by the electron beam intensity, energy, interaction volume, and the final spot size, which are attuned with one or more condenser lenses and

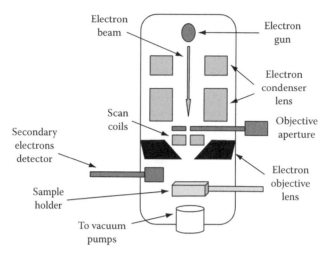

FIGURE 4.10   Schematic representation of the principal components of a SEM.

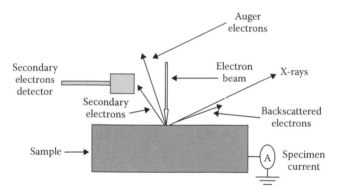

FIGURE 4.11   Graphic description of the principal interactions between the electron beam and the sample.

the objective lens [49]. The lenses are, as well, utilized to shape the electron beam in order to reduce the consequences of spherical aberration, chromatic aberration, diffraction, and astigmatism [8].

There are only two possible interactions between high velocity electrons and materials: the electrons can be elastically scattered, that is, energy and linear momentum are conserved, in which case, if the direction change is more than 90°, these electrons are elastically backscattered, or the electron could suffer an inelastic scattering process, in which case, there is a loss of energy during the process and the electron not only changes direction but also changes its energy [8,49,52].

The mechanisms of loss of energy during an inelastic scattering process are diverse: (1) several electrons lose energy creating phonons, which heat the specimen; (2) other electrons are produced, which create oscillations in the metal's electron gas; (3) another possibility is by breaking the radiation process where the electron emits continuous x-ray radiation spectrum (see Chapter 1). Other interactions provoke the emission of SEs. Finally, other electrons can trigger the emission of inner electrons of the atoms composing the test sample, and then producing the emission of a characteristic x-ray photon or an Auger electron [8,49,52].

In Figure 4.11, the interaction of a high-energy electron beam with materials is represented. The electrons hitting the material surface provoke the emission of electrons from the specimen mainly as backscattered electrons (BSEs) and SEs. SEs are ordinary signals used for the study of the surface morphology of materials. SEs have low kinetic energy, that is, lower than 50 eV.

Therefore, those arising from the first few nanometers of the material's surface have sufficient energy to escape from the material's surface and can thus be detected.

In addition, the portion of the SEs that do not escape from the material's surface, run to ground, and can be detected with an ammeter connected between the specimen and ground (see Figure 4.11) [49]. This signal is usually named the specimen current. In addition, the test sample is grounded to avoid the accumulation of spatial charge, which spoils the SEM image.

SEs and BSEs are typically detected by an Everhart–Thornley (ET) scintillator–photomultiplier secondary electron detector. The SEM image is shaped on a cathode ray tube screen, whose electron beam is scanned synchronously with the high-energy electron beam, so that an image of the surface of the specimen is formed [52]. The quality of this SEM image is directly related to the intensity of the secondary and/or BSE emission detected at each x- and y-point throughout the scanning of the electron beam across the surface of the material [8].

SEM is used to study material morphology by bombarding the specimen with a scanning beam of electrons, and then collecting the slow moving SEs that the specimen generates. These electrons are collected, amplified, and displayed on the picture tube of a television screen. The electron beam and the cathode ray tube scan synchronously so that an image of the surface of the specimen is formed. The specimen preparation includes drying the sample and making it conductive to electricity, if not already. Photographs are taken at a very slow rate of scan in order to capture greater resolution. SEM is typically used to examine the external structure of objects that are as varied as biological specimens, rocks, metals, ceramics, and almost anything that can be observed under a light microscope.

4.3.4 SEM Applications

As an illustration of the use of SEM, micrographs obtained from the characterization of silica materials are described here [47]. This SEM study was carried out with a JEOL 5800 LV model SEM. The acceleration of the electron beam was 20 kV. The sample grains were glued with silver colloid to the sample holder and were coated at vacuum by cathode sputtering with a $30–40\,\mu m$ gold film [47], in order to make the nonconducting silica powder sample conductive.

SEM micrographs of the opal powder 80 (Figure 4.12a) [47] and the membrane obtained with the opal sample 81C (Figure 4.12b) are shown. It is evident that the reported opal powder 80 is composed of microspheres with particle diameters of 200 nm determined by SEM [47].

Figure 4.13 shows another example of the use of SEM in the study of materials, where SEM micrographs of the $BaCe_{0.95}Yb_{0.05}O_{3-\delta}$ perovskite [32], which was previously used as an example for

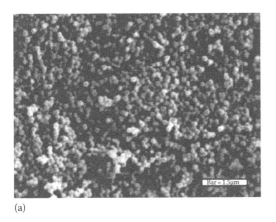

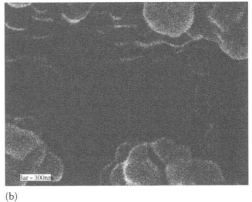

(a)   (b)

FIGURE 4.12   (a) SEM micrograph of the opal sample 80 (bar = 1.5 μm) and (b) SEM micrograph of the opal membrane obtained with the opal sample 80 after the sinterization process (bar = 300 nm).

Bar = 10μm

FIGURE 4.13   SEM micrograph of the $BaCe_{0.95}Yb_{0.05}O_{3-d}$ perovskite (bar = 10 μm).

the application of XRD, are shown. The SEM study was carried out with a JEOL CF35 microscope in the secondary electron (SE) mode at an accelerating voltage of 25 kV to image the surface of the perovskite powders. The sample grains were adhered to the sample holder with an adhesive tape and then coated at vacuum by cathode sputtering with a 30–40 nm gold film prior to observation. The surface morphology was revealed from SEM images and the average grain size was estimated qualitatively [32]. The average grain size estimated with the help of the SEM micrographs (Figure 4.13) corroborates the results obtained with the help of the Scherrer–Williamson–Hall methodology reported in Section 4.2.7.

4.4   ENERGY-DISPERSIVE ANALYSIS OF X-RAYS

SEM is also used to determine the chemical composition of the tested samples. SEMs are normally equipped with EDAX systems for quantitative chemical analysis, which allows direct information of the chemical composition of the selected crystal to be obtained. The analysis is carried out using the x-rays emitted by the sample (Figure 4.11) tested in the SEM sample holder (Figure 4.10).

4.4.1   X-Ray Emission

In order to generate an x-ray beam, a vacuum tube is needed, where an electron beam produced by a heated filament is collimated and accelerated by an electric potential of several kilovolts, that is, from 20 to 45 kV. In a SEM (see Figures 4.10 and 4.11), this beam is directed to a sample, which acts as the anode in the x-ray tube (Figure 4.11). The electrons hitting the sample will convey a fraction of their energy to electrons of the target material, a process resulting in electronic excitation of the atoms composing the sample. The sample then produces two kinds of radiations: the continuous spectrum and the characteristic spectrum (see Figures 1.20 and 1.21). The second type of spectrum (Figure 1.21), called the characteristic spectra, is produced as a result of specific electronic transitions that take place within individual atoms of the anode material. If the energy of the impinging electrons is high enough, some of them will hit a K-shell electron in the anode, and, thereafter, generate an electron vacancy. When such an electron vacancy is generated, it can be quickly filled by an electron from the L-shell or the M-shell of the same atom (see Figure 4.14).

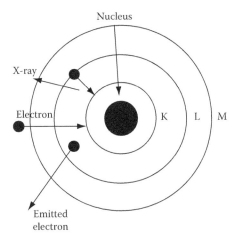

FIGURE 4.14 Electronic transitions during characteristic x-ray emission.

For the hydrogen atom, the wavelength of the emitted radiation linked with a specific electron transition is given by the Rydberg–Bohr equation

$$\frac{1}{\lambda} = R\left(\frac{1}{n_f^2} - \frac{1}{n_i^2}\right)$$

where
R is the Rydberg constant (1.097×10^7 [m^{-1}])
n_i is the principal quantum number of the initial state (higher energy state)
n_f is the principal quantum number of the final state (lower energy state)

XRF spectra were broadly studied by H.G.J. Moseley who confirmed, in 1913–1914, the relationship between the wavelength of characteristic radiation and the atomic number, Z, of the radiation of the emitting anode material (see Table 4.6). Moseley found experimentally that the K_α lines for various anode materials exhibit the empirical relationship:

$$\lambda_{K_\alpha} \propto \left(\frac{1}{Z^2}\right) \quad \text{or} \quad \nu_{K_\alpha} \propto Z^2 \quad \text{where } \lambda_{K_\alpha} = \frac{c}{\upsilon_{K_\alpha}}$$

Moseley's empirical relationship reveals a behavior that is in agreement with the Rydberg–Bohr equation, because the energy levels linked with the outer electron transitions are significantly

TABLE 4.6
K_α Radiation Wavelengths for Some Metallic Anodes

Element Symbol	K_α Wavelength [nm]
Mo	0.0707
Cu	0.1542
Co	0.1790
Fe	0.1937
Cr	0.2291

affected by the screening effect of the inner electrons. The screening effect of the deepest electrons on the nuclear charge is explained by an effective nuclear charge $(Z - 1)$ in the case of the K_α transition, that is, the K–L transition, and, consequently, the Rydberg–Bohr equation assumes the form

$$v_{K_\alpha} = \frac{c}{\lambda_{K_\alpha}} = cR(Z-1)^2 \left(\frac{1}{n_f^2} - \frac{1}{n_i^2} \right) = cR(Z-1)^2 \left(\frac{1}{1^2} - \frac{1}{2^2} \right) = \frac{3cR(Z-1)^2}{4} \tag{4.12}$$

4.4.2 Applications of Energy-Dispersive Analysis of X-Rays

When a sample is bombarded with a particle beam in a SEM, the specimen liberates some of the absorbed energy as x-rays. Then, since each element has its individual exclusive collection of energy levels, the emitted photons are indicative of the element that produced them. Analyzer detectors are then used to characterize the x-ray photons for their energy and abundance to determine the elemental composition of the tested sample. The detectors used in the EDAX analysis are, in general, semiconductor detectors. These detectors include single crystals of either Si or Ge conveniently doped. If the structure of these single crystals were perfect, there would be no confined profusion or lack of electrons. However, there are imperfections, that is, lattice defects and impurities, that produce electron-deficient areas, called holes, and extra free electrons within the crystal lattice. These holes and free electrons work as charge carriers when an electrical field is applied across the crystal. Consequently, a clean crystal, with fewer holes and free electrons, allows less current to pass than an impure crystal does. Normally, Si(Li) semiconductor detectors are then applied as EDAX detectors. It consists of a 2–5 mm thick Si crystal doped with Li, with gold contacts at its endings. A bias is applied across the crystal and a current flow. The crystal is maintained at low temperatures to prevent diffusion of Li from the intrinsic region to the Li-free region.

In order to illustrate the application of the emitted x-rays in chemical analysis, we use a LECA zeolite sample, which was studied by SEM with a JEOL 5800LV model electron microscope equipped with an energy dispersive x-ray analysis accessory, EDAX-DX-4, to obtain the elemental chemical analysis (see Table 4.7) [53]. The acceleration of the electron beam was 20 kV. The LECA zeolite sample grains were glued with silver colloid to the sample holder and was coated under vacuum by cathode sputtering with a 30–40 nm gold film before examination [53].

In order to make an additional illustration of the application of the emitted x-rays in chemical analysis, we will use a homoionic sodium clinoptilolte labeled Na-CSW [41], which was also studied by SEM. The original natural zeolite (labeled CSW), which was homoionized by an ion-exchange procedure [41], was mined in Sweetwater, Wyoming. The sample was provided by ZeoponiX Inc., Louisville, CO. The mineralogical phase composition of sample CSW was calculated using the XRD phase analysis methodology previously described [30,31,38–40], by means of standards, specifically, two very well-characterized samples, the clinoptilolite samples, labeled HC and GR (see Table 4.1) [44]. The results were reported in Section 3.4.3.

The chemical composition was analyzed in the same equipment and with the same conditions previously reported for the LECA zeolite. Table 4.8 reports the chemical composition (in wt %) of the Na-CSW sample, determined by EDAX analysis in a JEOL 5800LV model SEM equipped with an energy dispersive x-ray analysis accessory, EDAX-DX-4 [41].

TABLE 4.7
Chemical Composition (in wt %) of the LECA Zeolite

Sample	C	O	Si	Al	Fe	Ti	Ca	Mg	Na	K
LECA	6.9	36.7	26.8	10.3	6.0	0.5	3.8	1.8	2.3	4.8

TABLE 4.8
Chemical Composition (in wt %) of the Na-CSW Sample Determined by EDAX

Sample	O	Si	Al	Fe	Ca	Mg	Na	K
Na-CSW	45.25	37.48	9.25	1.06	0.05	0.58	5.46	0.90

4.5 INFRARED AND RAMAN SPECTROMETRIES

The IR and Raman spectroscopic methodologies are considered in the present section simultaneously, because of the fact that these are complementary experimental techniques. That is, the best possible analysis of the lattice vibrations of materials and the vibrations of molecules is by applying both methods concurrently [54–58].

Generally, for a vibration to be active in the IR spectra, the IR process must produce a change in the molecular dipole associated with the IR transition. On the other hand, to produce Raman activity, the variation has to be in the polarizability of the molecule during the transition.

4.5.1 INTRODUCTION

The region of the electromagnetic spectrum where IR and Raman spectra are obtained is those between the visible and microwave regions. The electromagnetic spectrum includes an apparently dissimilar set of radiant energy, which comprises cosmic rays, x-rays, ultraviolet, visible, IR, and microwaves. These particular radiations differ in frequency, v, where $v = c/\lambda$, c is the velocity of light, and λ is the wavelength. However, all these type of radiations are composed of photons, which travel at the speed of light, with an energy, E, where $E = hv = hc/\lambda$ and h is the Planck constant.

The IR and Raman spectroscopic methodologies analyze the interaction of electromagnetic radiation in the range $0.78\,\mu m < \lambda < 1000\,\mu m$ in a material [10]. In this range are included the near IR region ($0.78\,\mu m < \lambda < 2.5\,\mu m$), the middle IR region ($2.5\,\mu m < \lambda < 50\,\mu m$), and the far IR region ($50\,\mu m < \lambda < 1000\,\mu m$). The mainly used range is those from 2.5 to $15\,\mu m$, that is, $4000\,cm^{-1} < \bar{v} < 670\,cm^{-1}$, where $\bar{v} = 1/\lambda$ is the wavenumber, which is expressed in per centimeter i.e. cm^{-1} [55].

These are analytical tools since the character of the interaction is related to the structure and composition of the materials under test. When IR radiation goes across a sample, some photons are absorbed or suffer an inelastic scattering process caused by the active vibrations of the atoms, molecules, and ions, which compose the test material. The frequencies of the absorbed, or scattered, radiation are exclusively related to a particular vibration mode. Consequently, the process reveals attributes of the test material. Subsequently, IR (absorption) and Raman (scattering) are vibration-based spectroscopic methods widely used for characterizing materials, because they allow qualitative structural information to be obtained.

An IR spectrum is generally reported as band intensities versus wavenumber. The band intensities are expressed as transmittance, T, or as absorbance A, where the transmittance is

$$T = \frac{I}{I_0}$$

in which
  I_0 is the intensity of the incident radiation
  I is the intensity of the transmitted radiation

Transmittance is frequently expressed in percent:

$$\%T = \frac{I}{I_0} \times 100$$

On the other hand, absorbance is defined as the logarithm (base 10) of the inverse of the transmittance:

$$A = \log_{10}\left(\frac{1}{T}\right) = \log_{10}\left(\frac{I_0}{I}\right)$$

For a monochromatic radiation, the absorbance can be expressed by Beer's law

$$A = \mu L C \tag{4.13}$$

where
   μ is the extinction coefficient
   C is the concentration
   L is the length of the sample crossed by the radiation

The units of μ depends on the units of C.

4.5.2 Differences and Similarities between IR and Raman Phenomena

The initial effect in IR absorption is the transition of a molecule from a ground state to a vibration excited state by an absorption process of an IR photon with energy identical to the difference between the energies of the vibration ground and excited states. The opposite process, that is, IR emission, takes place when a molecule in the excited state emits a photon during the transition to a ground state.

Identical transitions between molecular vibration states are capable of resulting in a Raman scattering process. But a crucial difference between the Raman and the IR effects is that, during the Raman process, the photons concerned are not absorbed or emitted but somewhat shifted in frequency by an amount corresponding to the energy of the particular vibration transition. In the Stokes process, which corresponds to absorption in the IR effect, the scattered photons are shifted to inferior frequencies. The molecules remove energy from the exciting photons. On the other hand, in the anti-Stokes process, which corresponds to emission, the scattered photons are shifted to upper frequencies, because they gather the energy liberated by the molecules during the transition to the ground state [57]. Besides, a considerable number of the scattered photons are not shifted in frequency, because of the Rayleigh elastic scattering, which occurs because of density variations and optical heterogeneities, and is various orders of magnitude more intense than Raman scattering.

4.5.3 Molecular Vibrations

In Section 1.4, we analyzed the vibration modes of solids. These vibrations are in the majority of cases active in the IR region and their study provides information about the structure of the material under investigation. However, the bands related with the solid framework of a material in the middle IR region, which is the region where normally the majority of commercial equipment works, are broad bands with not much information. Nevertheless, always some information can be obtained. In addition, occasionally included in solids are molecules that can be studied with IR spectroscopy, such as occluded molecules, adsorbed molecules, OH groups, and other molecular features. These molecular features are normally of polyatomic character and can be studied with the help of IR spectroscopy.

It is well known that a molecule composed of N atoms has a total of $3N$ degrees of freedom, corresponding to the coordinates of each atom in the molecule. In a linear molecule, two degrees are rotational and three are translational. For a nonlinear molecule, three of these degrees of freedom are rotational and three are translational, and the rest are related to vibrations. Consequently, for a linear molecule, the number of vibration modes are $3N - 5$, and for a nonlinear molecule, $3N - 6$.

If we apply the harmonic model to describe these molecules, following the same methodology used in Section 1.4, it is possible to show that the energy of the system is

$$E = \sum_{\alpha=1}^{N-6}\left(N_\alpha + \frac{1}{2}\right)\hbar\omega_\alpha \tag{4.14}$$

where

N_α is the vibration quantum number
ω_α is the harmonic vibration frequency of a normal mode

and

$$E_n = \left(n + \frac{1}{2}\right)\hbar\omega \tag{4.15}$$

are the energy levels of the quantum harmonic oscillator.

The normal modes of polyatomic molecules in the harmonic approximation can be calculated with the help of computational methods.

4.5.4 Dipole Moment and Polarization

The molecular properties which change during the vibration of molecules are those related to the charge distribution. This happens owing to the fact that the equilibrium geometry of the electronic state will change with the variations in the internuclear separation, due to vibrations of the polyatomic molecule. The changing property related to IR absorption and emission is the dipole moment, μ. On the other hand, the varying property related with the Raman scattering is molecular polarizability, α.

The dipole moment is a vector [56]:

$$\bar{\mu} = \mu_x \bar{i} + \mu_y \bar{j} + \mu_z \bar{k}$$

The three components of the dipole moment vector can be expressed as a series expansion about the equilibrium geometry

$$\mu_{x_i} = \mu_{x_i}^0 + \left(\frac{\partial\mu_{x_i}}{\partial q}\right)_0 q + \frac{1}{2}\left(\frac{\partial^2\mu_{x_i}}{\partial q^2}\right)_0 q^2 + \cdots \tag{4.16}$$

where

$x_i \equiv x, y,$ or z
$\mu_{x_i}^0$ is the equilibrium value of one of the dipole moment components
q is the displacement of the equilibrium geometry ($q = q_0 \cos \omega_0 t$)

In this case, the first derivative is responsible for determining the observation of vibration fundamentals. There are three components for every vibration; then, every one of the possible molecular vibration frequencies has up to three possibilities to be observed in the IR spectrum. Consequently, for a transition of a vibration character to be permitted in the IR spectrum, it is required that as a minimum, one of these three components be different from zero [56].

On the other hand, the polarizability is a tensor, a response function that characterizes the volume and shape of the molecular electronic cloud [11]:

$$\overline{\alpha} = \begin{pmatrix} \alpha_{xx} & \alpha_{xy} & \alpha_{xz} \\ \alpha_{yx} & \alpha_{yy} & \alpha_{yz} \\ \alpha_{zx} & \alpha_{zy} & \alpha_{zz} \end{pmatrix}$$

The matrix representing $\overline{\alpha}$ is symmetric, and the tensor has only six components, and each component can be expanded in series as follows:

$$\alpha_{x_i x_j} = \alpha^0_{x_i x_j} + \left(\frac{\partial \alpha_{x_i x_j}}{\partial q} \right)_0 q + \frac{1}{2} \left(\frac{\partial^2 \alpha_{x_i x_j}}{\partial q^2} \right)_0 q^2 + \cdots \tag{4.17}$$

where
$\alpha^0_{x_i x_j}$ is the equilibrium value of the polarization
q is the displacement of the equilibrium geometry

In this case, as well, the first derivative, $(\partial \alpha / \partial q)_0$, is responsible for determining the observation of vibration fundamentals in the Raman spectrum [11]. Given that the polarizability is a response function of the molecule to an external electric field, then the polarizability and the polarizability derivatives are both symmetric tensors of the second rank. Then, each vibration has six chances to be observed in the Raman spectrum. Therefore, for a vibration transition to be permitted in the Raman spectrum, it is required that at least one of the six components of the derivative tensor is different from zero.

4.5.5 Types of Transitions between States

Consider two states, i and j (see Figure 4.15); for these states, Einstein, in 1917, identified three types of transitions between both states [12]:

- Stimulated absorption
- Stimulated emission
- Spontaneous emission

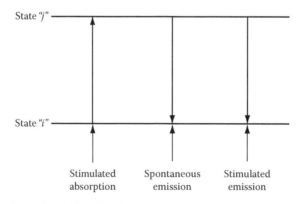

FIGURE 4.15 Absorption and emission of radiation.

Following this model description of the absorption process, the rate at which stimulated absorption transitions are induced is given by

$$N'_{i \to j} = B_{ij}\rho$$

where
B_{ij} is the coefficient of stimulated absorption
ρ is the energy density of photons in the frequency range ν to $\nu + d\nu$, where ν is the frequency of the transition

Now, it is possible to calculate the number of absorption transitions per time interval as follows:

$$N_{i \to j} = N_i B_{ij}\rho$$

where N_i is the number of molecules in the lower state, that is, i.
The rate at which stimulated emission transitions from state j to state i are induced is given by

$$N'_{j \to i} = B_{ji}\rho$$

where B_{ji} is the coefficient of stimulated emission. Einstein, as well, proposed a spontaneous emission process. Consequently, the number of emission transitions per time interval is given by [12,42]

$$N_{j \to i} = N_j(A_{ji} + B_{ji}\rho)$$

where A_{ji} is the coefficient of spontaneous emission.
In the state of thermodynamic equilibrium $N_{j \to i} = N_{i \to j}$; consequently

$$B_{ij}N_j\rho = N_j(A_{ji} + B_{ji}\rho)$$

Now $N_i = Ce^{-\frac{E_i}{kT}}$ and $N_j = Ce^{-\frac{E_j}{kT}}$, then

$$B_{ij}\rho e^{-\frac{E_i}{kT}} = (A_{ji} + B_{ji}\rho)e^{-\frac{E_j}{kT}}$$

Consequently,

$$\rho(\nu) = \frac{A_{ji}/B_{ji}}{(B_{ij}/B_{ji})e^{\frac{E_j - E_i}{kT}} - 1}$$

where $E_j - E_i = h\nu$. This equation is consistent with Planck's radiation law [42], which can be expressed as follows:

$$\rho(\nu) = \frac{8\pi\nu^2}{c^3} \frac{h\nu}{e^{\frac{h\nu}{kT}} - 1}$$

if

$$B_{ji} = B_{ij}$$

and

$$A_{ji} = \frac{8\pi v^3}{c^3} B_{ji}$$

Thereafter, the ratio between the probabilities for spontaneous and stimulated emission coefficients, A_{ji}/B_{ji}, is proportional to v^3; the relative possibility of spontaneous emission augments with the energy of the emitted photon.

4.5.6 IR AND RAMAN TRANSITION PROBABILITIES

In order to calculate the IR and Raman transition probabilities per unit time between the vibration state, i, to the vibration state, j, it is necessary to express in quantum mechanical terms the operator describing the interaction between the molecule and the electromagnetic radiation, which is given by

$$\hat{H}_d = -\overline{p} \cdot \overline{E}$$

where
$\overline{p}$ is the dipole moment
$\overline{E} = \overline{E}_0 e^{i(\overline{k} \cdot \overline{r} - \omega t)}$ is the electric field vector

Then, the probability for the absorption or emission of electromagnetic radiation per unit time for the dipole transition is given by [59]

$$P_{\omega,\omega'} \propto \left| \left\langle \psi_\omega \left| \overline{p} \cdot \overline{E} \right| \psi_{\omega'} \right\rangle \right|^2 \tag{4.18}$$

In the case of the IR process, the dipole moment, $\overline{p} = \overline{\mu}$, is given by Equation 4.16; in the case of the Raman process, the dipole moment is given by $\overline{p} = \alpha \overline{E}$. For molecules in the gas phase with a haphazard orientation and where the average of the square of the angular part is one, the probability for the absorption or emission of electromagnetic radiation per unit time for the dipole transition can be written as follows:

$$P_{\omega,\omega'} \propto \left| \left\langle \psi_\omega \left| \overline{p} \right| \psi_{\omega'} \right\rangle \right|^2 \tag{4.19}$$

Then, in the case of IR absorption neglecting the anharmonic terms, the transition probability can be expressed in the following form [56]

$$P_{\omega,\omega'} \propto \left| \left\langle \psi_\omega \left| \left(\frac{\partial \overline{\mu}}{\partial q} \right)_0 q \right| \psi_{\omega'} \right\rangle \right|^2 \tag{4.19a}$$

or

$$P_{\omega,\omega'} \propto \left| \left\langle \psi_\omega \left| q \right| \psi_{\omega'} \right\rangle \right|^2 \left(\frac{\partial \mu}{\partial q} \right)_0^2 \tag{4.19b}$$

In the Raman scattering case, from Equation 4.17, we get

$$\overline{p} = \alpha_0 \overline{E} + \left(\frac{\partial \alpha}{\partial q}\right)_0 q\overline{E} + \frac{1}{2}\left(\frac{\partial^2 \alpha}{\partial q^2}\right)_0 q^2 \overline{E} + \cdots \tag{4.20}$$

Expressing the electric field as

$$\overline{E} = \overline{E}_0 \cos \omega t \tag{4.21a}$$

and the displacement as

$$q = q_0 \cos \omega_0 t \tag{4.21b}$$

and introducing Equations 4.21 and 4.22 in Equation 4.20 and neglecting the derivative higher than one, we get [56]

$$\overline{p} = \alpha_0 (\overline{E}_0 \cos \omega t) + \left(\frac{\partial \alpha}{\partial q}\right)_0 q_0 \cos \omega_0 t (\overline{E}_0 \cos \omega t) \tag{4.22}$$

Now, with the help of the following trigonometric identity

$$\cos \alpha \cos \beta = \frac{\cos[\alpha + \beta] + \cos[\alpha - \beta]}{2}$$

Equation 4.22 can be reduced to

$$\overline{p} = \alpha_0 (\overline{E}_0 \cos \omega t) + \frac{1}{2}\left\{\left(\frac{\partial \alpha}{\partial q}\right)_0 q_0 \overline{E}_0 \cos\{\omega_0 - \omega\}t + \left(\frac{\partial \alpha}{\partial q}\right)_0 q_0 \overline{E}_0 \cos[\{\omega_0 + \omega\}t]\right\}$$

where
  the first term describes the Rayleigh scattering
  the second term is related to the Stokes process
  the third term is the anti-Stokes term

With respect to the transition probability in the Raman case, it is determined by the following integrals [10]:

$$P_{\omega,\omega'}^{x_i x_j} \propto \left|\left\langle \psi_\omega | \alpha_{x_i x_j} | \psi_{\omega'}\right\rangle\right|^2 \tag{4.23}$$

4.5.7 Selection Rules

The selection rules are restrictions imposed on the quantum transitions, because of the laws of conservation of angular momentum and parity [59]. In the case of IR spectroscopy, within the frame of the harmonic approximation, the applicable rules are the electric dipole selection rules. That is, when the expression in Equation 4.19 has a finite value, the transition is allowed, and when this expression is zero the transition is forbidden. In the Raman case, when one of the integrals given by Equation 4.23 is different from zero, the normal vibration associated is Raman-active.

Since the wave functions included in Equations 4.19a and 4.23 are invariant under symmetry operations, to calculate the selection rules, it is necessary to take into account the symmetry of the absorption system [56].

Symmetry operations are studied with the help of group theory, which discusses the set of operations that satisfies the following four conditions [60]:

- One of the operations is the identity operation.
- Every operation in the group has an inverse.
- The members of the group satisfy the associative law.
- The product of two members of the group is as well a member of the group.

Consequently, in the case of molecules, the symmetry operations that form the point groups convert the molecule into self-coincidence [10].

The symmetry operations in molecules are actions that leave the molecule in a configuration, which is indistinguishable from the initial arrangement. The symmetry operations are [60]

- Rotations about an n-fold axis of symmetry
- Reflection through a plane of symmetry
- Inversion centers
- Roto-reflection, that is, a rotation followed by a reflection through a plane perpendicular to the axis of rotation

Consequently, there are four symmetry elements: the n-fold axis of rotation, labeled C_n; the plane of symmetry, labeled σ; the center of inversion, i, and the n-fold rotation-reflection axis, labeled S_n. Because of mathematical reasons, it is necessary to include the identity symmetry element, I.

A viable arrangement of symmetry operations, whose axes intercept at a point, is a point group. Consequently, the number and nature of the symmetry elements of a given molecule is represented by its point group [60]. Since practically the maximum rotation axis of a molecule is C_6, then, for molecules there are 32 point groups.

The integrals 4.19a and 4.23 do not vanish if the symmetry of the normal vibration and the dipole moment, and the polarizability are of the same species, respectively. But, if the symmetry properties of the vibration and the dipole moment and the polarizability differ in only one element of the group, the integral turns out to be zero; thereafter, the vibration is neither IR nor Raman-active [10]. This means, in very general terms, that for the existence of IR activity, the motion related to a normal mode of vibration ought to be associated with a change in the dipole moment. On the other hand, also in very general terms, the normal modes which are Raman-actives are those accompanied by a change in polarizability.

Let us consider the case of the harmonic approximation (see Equation 4.15). Because of its mathematical properties, the quantum mechanical solution yields that the selection rule for the IR and Raman transitions is [12]

$$\Delta n = \pm 1 \qquad \qquad (4.24a)$$

However, in practice,

$$\Delta n = +1 \qquad \qquad (4.24b)$$

The most populated energy level at normal temperatures is $n = 0$. Then, merely transitions to the next energy level are allowed; consequently, molecules will absorb an amount of energy equal to $E = h\nu$ for a particular transition.

4.5.8 Simplification of the Molecular Vibration Analysis

Through the examination of several IR spectra of a great number of compounds containing the same group of atoms, it was found that despite the remaining part of the molecule, this specific group of atoms absorbs in a specific limited range of frequencies, named the group frequency [10]. The fact that the vibrations of a specific group inside the molecule are more or less independent of the vibrations of the rest of the groups forming the molecule allows the simplification of the molecular vibration analysis, in order to interpret IR and Raman spectra [12,54]. In this regard, it is possible to apply the rule which states that combinations of the atomic displacements that give only two fundamental types of molecular vibrations, that is, stretching (Figure 4.16) and bending (Figure 4.17) [55] can be chosen. Then, for complex molecules, it is reasonable to designate a particular absorption range to bending and stretching vibrations of specific functional groups, which compose the molecule [12,54]. Examples of these groups are those containing moderately light elements such as −OH, −NH$_2$, −CH, CH$_2$, or relatively heavy elements like −CCl, CI, and also multiple bonded groups, for instance C=O, C=C, C≡C, C≡N [10].

In Figures 4.16 and 4.17, all the possible stretching and bending vibrations are shown [12,55]. Stretching implies a variation in the interatomic distance along the axis of the bond amidst two atoms (Figure 4.16), and bending is distinguished by a change in the angle between two bonds (Figure 4.17) [54,55].

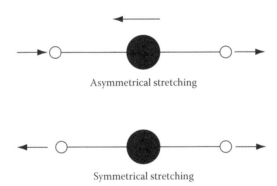

Asymmetrical stretching

Symmetrical stretching

FIGURE 4.16   Stretching vibrations.

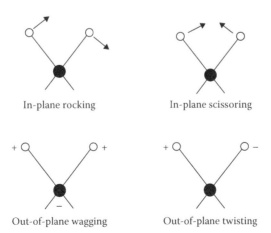

In-plane rocking   In-plane scissoring

Out-of-plane wagging   Out-of-plane twisting

FIGURE 4.17   Bending vibrations.

As a result, specific kinds of vibration modes always appear at similar frequencies for different molecules; then, it is feasible to construct a table of characteristic absorption frequencies. Typical absorption band tables for significant bonds included in polyatomic molecules are reported in ref. [58]. Subsequently, using the typical absorption frequencies, it is relatively simple to recognize the occurrence of functional groups in unidentified compounds, and use this information to get structural information about these substances in order to identify them.

4.5.9   INSTRUMENTATION

4.5.9.1   Fourier Transform Infrared Spectrometer

In Fourier transform infrared (FTIR) spectroscopy, simply one beam is used during the IR analysis of a sample, that is, in this case, all frequencies go through the instrument at one time [61] (see Figure 4.18). The method is called FTIR spectroscopy, because a Fourier transformation is carried out by a computer in order to work out the obtained data and yield a spectrum. FTIR is fast, that is, a spectrum can be obtained in less than a second. Besides, it is a very sensitive technique, because it is possible to make as many scans as necessary to obtain a good spectrum.

The most important component in the FTIR spectrometer is an interferometer (Figure 4.18); this apparatus divides with a beam splitter, that is, a semitransparent mirror, and recombines with a fixed and a moving mirror an IR radiation beam, so that the recombined beam generates a wavelength-dependent interference pattern, that is, an interferogram.

A Michelson interferometer is generally the device used in an FTIR spectrometer. The Michelson interferometer is composed of two mirrors and a beam splitter positioned at an angle of 45° to the fixed and moving mirrors.

In the FTIR, the function of the Michelson interferometer with the moving mirror is to scatter the IR radiation provided by the source into its constituent frequencies. The IR source emits polychromatic radiation. Subsequently, this radiation after passing through the interferometer, with the moving mirror in movement, is transformed. That is, the exit beam after passing through the specimen is an interferogram. This interferogram encompasses the necessary information of a spectrum,

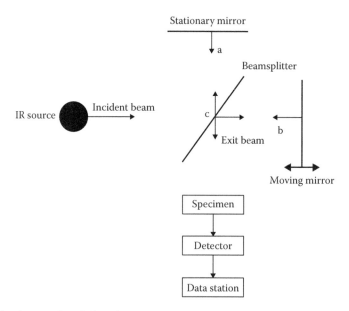

FIGURE 4.18   Fourier transform infrared spectrometer.

but cannot be interpreted without further processing. The interferogram is the cosine Fourier transform of the spectrum [61]. Consequently, with the help of a data station, the obtained information is translated to a recognizable form, that is, a spectrum with the help of Fourier transform methods.

4.5.9.2 Conventional Raman Spectrometer

To get a Raman spectrum, it is necessary to expose a specimen to a monochromatic source of exciting photons, and then measure the intensity of the scattered beam of light. The intensity of the Raman scattered constituent is, by far, inferior than the Rayleigh scattered part. Then, filters and diffraction gratings are utilized to suppress the Rayleigh component. Besides, an extremely sensitive detector is necessary to sense the almost imperceptibly scattered Raman photons [57].

In a conventional Raman spectrometer (Figure 4.19), a visible laser radiation is used as the source of exciting photons; these photons are typically of much higher energies than those of the fundamental vibrations of most chemical bonds or systems of bonds.

In order to get a Raman spectrum, a sample is located in the sample cell. Then, a laser light is focused on the sample using a lens. Usually, the sample cell is a capillary tube, normally made of Pyrex glass, where liquids and solids are sampled in [57]. Or another appropriate sample holder system; where is given the possibility that the light scattered during its interaction with the sample is accumulated using an additional lens and is then focused at the entry slit of the monochromator [32,62].

The fundamental role of the monochromator is to eliminate the Raleigh scattering, the stray light, and operates as a dispersing factor for the incoming radiation. Occasionally, more than one monochromator is utilized to obtain a high resolution and for better suppression of the Rayleigh line. Once the beam leaves the outlet slit of the monochromator, it is accumulated and focused on the surface of a detector, and this optical signal is transformed to an electrical signal inside the detector and further worked out by means of the detector electronics [57,62]. Argon and krypton ion lasers are normally used for sample excitation, but in some cases UV lasers have also been used. Similar detector types are applied in conventional Raman spectroscopy; however, in recent times multichannel detectors mounted on the plane of the exit slit are being increasingly used.

4.5.9.3 Fourier Transform Raman Spectrometer

The Fourier transform Raman spectrometer is constructed around an interferometer (see Figure 4.20) [57]. Normally, a continuous wave Nd:YAG laser (1064 nm) is used for the sample excitation. In relation to the sample arrangement inside the spectrometer, there are two fundamental geometries in which a sample is tested in Raman spectroscopy, that is, the 90° geometry, where the laser beam

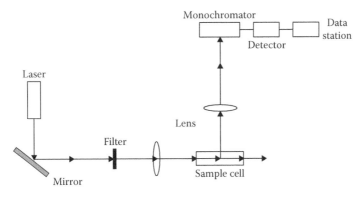

FIGURE 4.19 Schematic representation of a conventional Raman spectrometer.

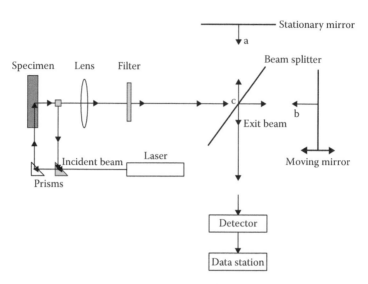

FIGURE 4.20 Schematic representation of a FT Raman spectrometer.

direction and the axis of the collection lens are at 90° to each other (see Figure 4.20), and the 180° scattering geometry.

4.5.10 APPLICATIONS OF FOURIER TRANSFORM INFRARED SPECTROSCOPY AND RAMAN SPECTROSCOPY IN MATERIALS SCIENCE

In the case of silica materials, zeolites and related materials, and mesoporous molecular sieves (MMS), the spectrum in the middle IR region, that is, 250 to 4000 cm^{-1} can be classified in three regions:

1. 250 to 1300 cm^{-1}: This zone can be useful to get information about the structure and to identify changes in composition.
2. 1300 to 2500 cm^{-1}: In this spectral region, there appear bands corresponding to vibrations of organic compounds adsorbed or included within the silica material. The acidic properties of the silica materials are studied in this region making self-supporting wafers that are exposed in vacuum to a test molecule (i.e., pyridine and ammonia) and, subsequently, the excess of this compound (physisorbed molecule) is removed by heating in a vacuum. From the recorded spectra obtained after treatment in a vacuum at different temperatures, Lewis acid sites (at ca. 1450 cm^{-1}) and Brönsted acid sites (at ca. 1545 cm^{-1}) can be characterized.
3. 3000 to 4000 cm^{-1}: This region is associated with OH groups and can be useful to characterize the presence of silanol groups and Al extra-framework associated with OH groups.

In Figure 4.21, the FTIR spectra of the 70bs2 silica sample is reported [64]. The FTIR study was carried out with a Bruker Tensor 27 spectrometer. During the analysis, 32 scans per spectrum were generated; the resolution was 4 cm^{-1}, and the range between 500 and 2000 cm^{-1} was analyzed. The silica as-synthesized powders were pressed with a Carver manual hydraulic press model 3912 in a mixture with KBr (4 mg of silica in 20 mg of KBr) under a pressure of 300 MPa for 15–20 s, to form wafers in order to make the analysis.

The IR spectra of silica displays in the studied range peak around 1100 and 800 cm^{-1}, which are associated to the stretching and bending modes of silica bridges ($\equiv$Si–O–Si$\equiv$), respectively [63]. The water content in the silica sample enclosing physically adsorbed water can be estimated with

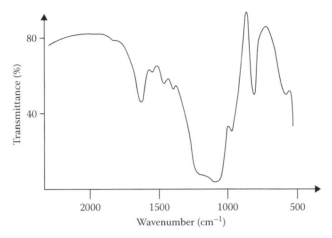

FIGURE 4.21 FTIR spectra of a wafer of silica powders of sample 70bs2 in a mixture with KBr (4 mg of silica in 20 mg of KBr).

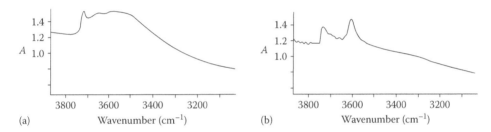

FIGURE 4.22 FTIR spectra of the acid zeolites (a) H-SSZ-24 and (b) H-ZSM-11 dehydrated at 673 K.

a peak corresponding only to water molecules vibration, such as the δH_2O bending mode, at about 1630 cm^{-1}. The analyzed sample contained physically adsorbed water, since it was handled at ambient conditions. The spectra reported in Figure 4.21 show the silica bridges peak around 1100 and 800 cm^{-1}; besides both spectra also display a band at about 1630 cm^{-1} [64].

In Figure 4.22, the FTIR spectra of dehydrated samples of the acid zeolites, H-SSZ-24 and H-ZSM-11, in the range from 3000 to 4000 cm^{-1} are shown [65]. The measurement of the FTIR spectra was carried out in a Bio-Rad FTS 40A FTIR spectrometer with a resolution of 8 cm^{-1} controlled by the Bio-Rad WIN IR software in an Advanced Scan Menu-Kinetics mode [65,66]. To obtain the spectra, 30 scans were made.

The sample dehydration was carried out in a commercial water-cooled AABSPEC cell, which could sustain high temperatures (up to 700°C) and pressures (up to 13 MPa) [65,66]. The windows are made of CaF, and the cell is made of stainless steel. Demountable parts are normally fitted with Viton O-rings to minimize leaking. The complete experimental laboratory assembled setup is shown in Figure 4.23 [21,66]. It is composed of the IR cell connected through stainless steel pipes to a manifold containing the gas inlets for the N_2 carrier, gas grade (99.9999%).

Self-supported wafers obtained by pressing 7–9 mg/cm^2 of the zeolite sample powder at 400 MPa were placed in the cell and then degassed at 673 K.

The FTIR spectra of the acid zeolites H-SSZ-24 and H-ZSM-11 are in the range from 3000 to 4000 cm^{-1}. This region is associated with acid bridging acidic hydroxyl groups. The sample H-SSZ-24 shows a broad band that might be due to the the bridging hydroxyl groups, and an intense band at 3736 cm^{-1}, corresponding to the Si–OH (silanol) groups [65]. Besides, the H-ZSM-11 dehydrated zeolite shows an intense peak at 3609 cm^{-1}, because of acid bridging acidic hydroxyl groups [65].

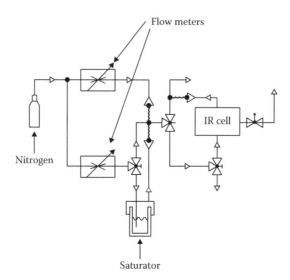

FIGURE 4.23 Representation of the experimental setup used to dehydrate the acid zeolite samples.

Figures 4.24 and 4.25 show the FTIR spectra and FT Raman spectra of two samples, that is, a mesoporous molecular sieve (MMS) and a Ni–Y zeolite where aniline was incorporated and polymerized [67].

In order to incorporate aniline in the zeolite and the MMSs, the host materials were degassed by heating at 573 K for 5 h and cooled in flowing dry air to room temperature [67]. Aniline with a purity of 99.98%, provided by Aldrich, Saint Louis, MO, USA was twice distilled in vacuum before use [67]. Then, the aniline was incorporated in the hosts by saturating a controlled flow of nitrogen of 100 mL/min with aniline, and then exposing this mixture to the degassed host materials for 2 h at room temperature [67]. Subsequently, the zeolites and the MCM-41 molecular sieve were purged with nitrogen for 30 min. Samples were then sealed within quartz cuvettes in a nitrogen atmosphere for further study by FTIR and other methods [67].

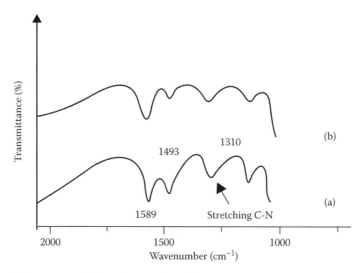

FIGURE 4.24 FTIR spectra of the adducts polyaniline-hosts in KBr pellets: (a) MCM-41; MMS and (b) Ni–Y zeolite.

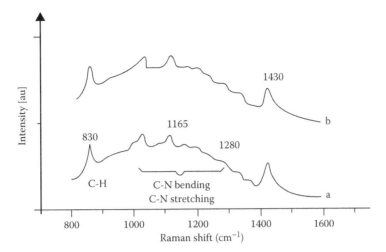

FIGURE 4.25 FT Raman spectra of the adducts polyaniline-hosts in KBr pellets: (a) MCM-41; MMS and (b) Ni–Y zeolite.

The FTIR spectra in the region of framework vibrations, that is, the range from 300 to 1900 cm⁻¹, were recorded on a Nicolet Magna IR-750 FTIR spectrometer using the KBr pellet method, obtaining a spectral resolution of about 4 cm⁻¹ [67]. The FT-Raman spectra were recorded on a Bruker spectrometer (model RFS 100/s). The 1064 nm line of a diode-pumped Nd:YAG laser was used for excitation along with a high-sensitivity germanium diode detector cooled to liquid-nitrogen temperatures. The laser Raman spectra were examined in the 180° scattering configuration using a sample cup specially designed for this study. About 10 mg of each sample was pressed into the sample cup and then mounted on the sample holder. Various laser powers were tried so that the optimum power (130 mW) was selected [67]. The spectral resolution and reproducibility was experimentally determined to be better than 4 cm⁻¹, and the number of scans varied from 700 to 3000 with recording times of 30 min to 2 h. The Raman spectra were corrected for instrumental response using a white light reference spectrum [67].

The FTIR spectra of polyaniline polymerized within zeolites and mesoporous materials are shown in Figure 4.24 [67]. As can be seen, significant changes were observed for different hosts. The spectra presented characteristics of emeraldine salt oxidation state in MCM-41 and faujasite (Y-Ni zeolite) and a mixture of emeraldine (deprotonated form) (see Figure 4.26) and a protonated structure, as deduced from the ratio between the intensities corresponding to the peaks at ca. 1589 cm⁻¹ (quinoid form) and ca. 1493 cm⁻¹ (benzenoid form). These assignments are in agreement with those reported in the literature [67]. The band observed at ca. 1310 cm⁻¹ is assigned to the C–N

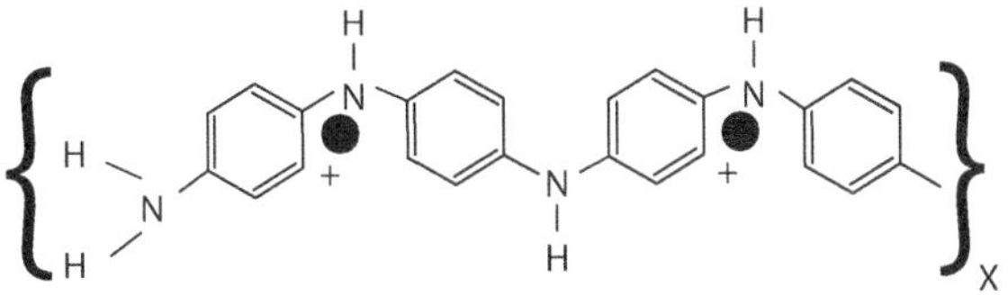

FIGURE 4.26 Emeraldine salt polymer.

stretch of the secondary aromatic amine. On the other hand, the intensity of the peaks corresponding to the polymer incorporated within the hosts can be also correlated with the material type. These bands were normalized with respect to the bands of the hosts.

Figure 4.25 shows the FT Raman spectra of polyaniline included within the two reported hosts [67]. The presence of peaks at ca. 1165 and 1280 cm^{-1} are related with the C–H bending (of the quinoid ring) and the C–N stretching and ring deformations (of the benzoic ring), respectively [67]. The band situated at ca. 1430 cm^{-1} has been assigned to the quinoid structure and that located at ca. 830 cm^{-1} to an aromatic C–H out-of-plane bending [67]. The differences observed in the intensity of the Raman spectra can be correlated with the different levels of polymerization depending on the hosts. Thus, as concluded from IR spectroscopy, the most intense peaks obtained in the same conditions were observed for MCM-41 and faujasite and the low intensity values for the ZSM-5 zeolite. As in the case of IR spectroscopy, the different peaks unambiguously indicate the presence of the polymer. Nevertheless, the high fluorescence background does not allow us to perform the assignment of the different forms of the polymer. The results obtained can be correlated with the different sizes of the hosts. Thus, in the case of MCM-41 and the zeolite Y, the pore sizes are bigger than in the case of the other hosts.

Both materials used as hosts are distinguished by different geometries and pore dimensions, that is, the zeolite Y is constituted of spherical supercages of 1.3 nm of diameter tetrahedrally interconnected through 0.74 nm windows. On the other hand, the structural characteristics of MCM-41 have been expressed in terms of a honeycomb-like structure with a pore diameter of 3.5 nm and a wall thickness of ca. 1.1 nm [67]. The Ni–Y zeolite was prepared by $Ni(NO_3)_2$ exchange of the NaY sample, and the CBV100 was provided by the PQ Corporation Malvern, PA, USA. [67].

Figure 4.27 shows the Raman spectrum of the $BaCe_{0.95}Yb_{0.05}O_{3-x}$ powder perovskite [32]. The Raman spectral measurement was performed using a Jobin-Yvon T64000 conventional Raman spectrophotometer consisting of a double pre-monochromator coupled to a third monochromator/spectrograph with 1800 grooves/mm grating [32]. The 514.5 nm radiation of an Ar laser was focused in a less than 2 μm diameter circular area by using a Raman microprobe with an 80× objective. The same microscope was used to collect the signal in a backscattering geometry and is focussed at the entrance of the pre-monochromator [32] The scattered light dispersed by the spectrophotometer was detected by a charge-coupled device detection system.

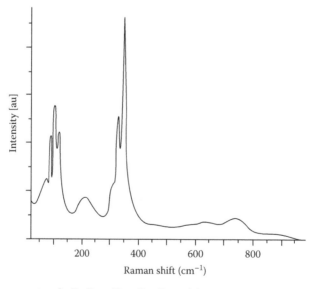

FIGURE 4.27   Raman spectra of a $BaCe_{0.95}Y_{0.05}O_{3-\delta}$ Perovskite.

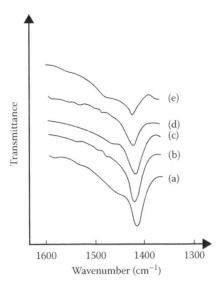

FIGURE 4.28 IR spectra of samples: (a) H-Y, (b) H-CMT, (c) H-HC, (d) H-MP, and (e) H-X after NH_3 adsorption.

The spectrum shows the most intense bands in the frequency range of 300–400 cm^{-1}. The low-frequency region (35–140 cm^{-1}) of the spectrum is dominated by a relatively intense envelop of a peak at 35–140 cm^{-1} [32]. These peaks can be assigned in accordance with the orthorhombic factor group analysis by assigning these bands to stretching vibrational and bending vibrational modes of the CeO_6 octahedra, corresponding to the $BaCe_{0.95}Yb_{0.05}O_{3-x}$ powder perovskite Raman spectrum [32].

Figure 4.28 shows the IR spectra [68] of acid zeolite samples (see Section 2.5.3) obtained in a conventional IR spectrometer after NH_3 was absorbed on the acid zeolites up to an equilibrium pressure of 5×10^4 Pa.

Thin wafers of the zeolite were prepared using KBr, and the relative population of NH_4^+ was measured with IR spectrometry following the increase in the NH_4^+ bending band [68].

4.6 NUCLEAR MAGNETIC RESONANCE SPECTROMETRY

4.6.1 Introduction

In Chapter 1, we have studied the effect of nuclear magnetic resonance (NMR) [69–78]. In this section, we study NMR spectrometry and consider some applications of NMR in materials science.

4.6.2 NMR Spectra

The solution of the Bloch equations (see Section 1.8) allows to calculate the power absorbed, that is, $P(\omega)$, from a rotating magnetic field in an NMR experiment arranged, as illustrated in Figure 1.39. Explicitly, the power absorbed [26,72] is given by the following equation (see Figure 4.29)

$$P(\omega) = \frac{\omega \gamma M_z T_2}{1 + (\omega_0 - \omega)^2 T_2^2} (B_1)^2$$

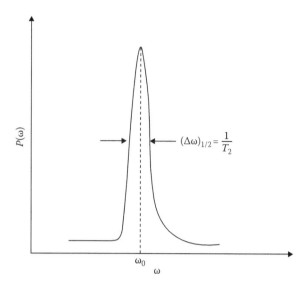

FIGURE 4.29 Power absorbed from a rotating magnetic field in an NMR experiment.

where the half width of the resonance peak is given by [73]

$$(\Delta\omega)_{1/2} = \frac{1}{T_2}$$

Figure 4.29 shows the output of an NMR experiment where the frequency of a spin system placed in a homogeneous static external magnetic field (see Figure 1.35) is scanned and the power absorption measured [26].

However, in modern NMR spectrometers, the spectra measurement is not carried out by the previously explained methodology. Instead the sample is irradiated with a pulse that rotates the magnetization angle of 90°, as explained in Section 1.8.4. A flip angle of 90° is chosen since in this case the magnetization is fundamentally positioned in the xy-plane, that is, the place where the signal is detected [76]. Then, after the termination of the pulse, the magnetization will be restored to its equilibrium state [72,73,76]. That is, the longitudinal magnetization will be restored in the z-direction as a consequence of the spin-lattice interaction process, with a relaxation time T_1. Besides, the transversal magnetization will be randomized owing to the spin–spin interaction process with a relaxation time T_2 [26,72]. The signal from the decaying transversal magnetization process is detected by a detection coil in the form of a free induction decay (FID) time-depending signal, $f(t)$ (see Figure 4.30).

In order to obtain the NMR spectrum in an angular frequency, $F(\omega)$, (Figure 4.30), it is necessary to obtain the Fourier transform (FT) of the FID. The FT is defined by [72,79,80]

$$F(\omega) = \frac{1}{\sqrt{2\pi}} \int\limits_{-\infty}^{\infty} f(t) \exp[i\omega t] d\omega$$

and the inverse by

$$f(t) = \frac{1}{\sqrt{2\pi}} \int\limits_{-\infty}^{\infty} F(\omega) \exp[-i\omega t] d\omega$$

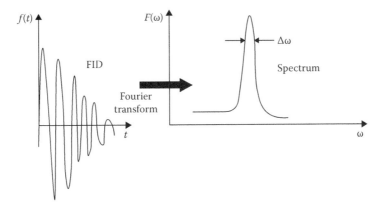

FIGURE 4.30 The FID after a FT is changed to an NMR spectrum.

since the NMR spectra are normally given in frequency, then the spectrum in frequency, $G(\nu)$, is given by the following equation

$$G(\nu) = F(2\pi\nu)$$

where $\omega = 2\pi\nu$.

4.6.3 CHEMICAL SHIFT

If an atom is located in a magnetic field, the electrons of the atom will circulate in the direction of the applied magnetic field. This movement produces a tiny magnetic field at the nucleus that is in opposition to the applied magnetic field [77]. Consequently, the magnetic field at the nucleus is, for this reason normally, less than the applied field by a dimensionless fraction, σ, called the shielding constant. That is, this additional field is directly proportional to the externally applied field:

$$\delta B = -\sigma B_0$$

Then, the electrons that surround each nucleus act to slightly perturb the magnetic field at the spin site. This causes the Larmor precession frequency to be modified by the chemical environment of the spin. This effect, called the chemical shift, is described by the equation

$$B_{local} = B_0 + \delta B_0 = (1 - \sigma)B_0$$

where
  σ is the shielding constant
  B_{local} is the local field

It is evident that this effect modifies the Larmor frequency such that

$$\nu_L = \frac{\omega_L}{2\pi} = \gamma(1 - \sigma)\frac{B_0}{2\pi}$$

In practice, it is common to express the chemical shift of a peak in the spectrum in terms of the relative difference in frequency from some reference peak. The chemical shift in parts per million (ppm) is, therefore, defined as

$$\delta = \frac{\nu - \nu_{ref}}{\nu_{ref}} \times 10^6$$

In NMR spectroscopy, the standards normally used are tetramethylsilane, $Si(CH_3)_4$, shortened TMS for ^1H, ^{13}C, and ^{29}Si, and 85% H_3PO_4 for ^{31}P [71,76,77].

4.6.4 Spin–Spin Coupling

As previously stated, the line width of an NMR peak is given by $(\Delta\omega)_{1/2} = \frac{1}{T_2}$; this is the so-called homogeneous broadening. However, there exist other broadening sources, such as an isotropic distribution of chemical shifts, which is called inhomogeneous broadening.

Nuclei that are identical with the chemical environment or chemical shift are named equivalent nuclei. On the other hand, those nuclei dissimilar with their environments or possessing different chemical shifts are termed nonequivalent. In fact, nuclei which are near to one another exercise an effect on each other's effective magnetic field; this effect is called spin–spin coupling [77].

For liquid substances, where the molecules move fast, this effect is averaged and the NMR spectrum only contains narrow peaks, which is revealed as a fine structure. This fine structure arises because each nucleus contributes to the local field which is felt by the other nuclei, modifying their resonance frequency. In this regard, the power of the spin–spin coupling is expressed by the scalar coupling constant, J [73].

Conversely, for solid samples, the nuclei are placed in fixed sites of the crystal lattice and only oscillate in these fixed sites; then, each nucleus is subjected to different interactions which cannot be averaged by a chaotic motion. Consequently, the NMR spectrum is composed of broad peaks which are difficult to understand.

4.6.5 Magic Angle Spinning-Nuclear Magnetic Resonance

In solids, the averaging due to fast movement will not occur, or even if it does, occurs partially; then, the effects of direct dipolar couplings can be observed [72,73]. The direct magnetic dipolar interaction between nuclear spins causes a nucleus with spin projection, m_I, to produce a magnetic field in the z-axis, B_{nuc}, at a distance, R, given by the following expression [12]

$$B_{nuc} = -\frac{\gamma h \mu_0 m_I}{8\pi^2 R^3}(1 - 3\cos^2\theta)$$

where
  μ_0 is the permeability of vacuum
  θ is the angle between the vector connecting the dipoles and the external field or the angle between the external field and the principal axis of the molecule

Additionally, anisotropic chemical shift effects will also be present in the spectrum due to the nonexistence of molecular tumbling. This means that, since in solids the atoms and molecules are not moving, they do not change the orientation. Since the ability of the applied field to generate electron currents depends on the orientation of the nuclei relative to the applied field, the chemical shift depends on orientation, and since in solids this effect is not averaged out, it will contribute to the peak broadening. The chemical shift anisotropy also varies with the angle, θ,

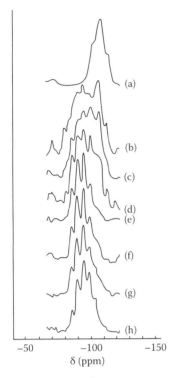

as $(1 - 3 \cos^2 \theta)$ [72]. For this reason, nuclei in solid materials regularly find themselves in a somewhat broad range of magnetic surroundings, which leads to absorption peaks broader than those obtained in the liquid state with a resultant deficit in the spectral resolution [76].

In order to reduce the effect of line broadening in solids, the magic angle spinning-nuclear magnetic resonance (MAS-NMR) method was developed [81]. This methodology is employed fundamentally for abundant magnetic nuclei, and it relies on the fact that dipolar interaction and anisotropic chemical shift effects are, to the first order, proportional to $(1 - 3 \cos^2 \theta)$. Therefore, the dipolar interaction and the anisotropic chemical shift effects disappear for the magic angle $\theta = 54.736°$, which makes [81]

$$(1 - 3\cos^2 \theta) = 0$$

Consequently, fast rotation of a powder sample about an axis, which makes an angle of $\theta = 54.736°$ to the magnetic field direction, should remove both sources of broadening from the spectrum, provided the rotation rate exceeds the magnitude of the broadening in frequency units.

FIGURE 4.31 ^{31}Si MAS-NMR spectra of solid intermediates phases corresponding to the treatment of Na-clinoptilolite as a function of crystallization time in hours. (a) 0 h, (b) 1 h, (c) 4 h, (d) 8 h, (e) 16 h, (f) 24 h, (g) 36 h, and (h) 64 h.

4.6.6 APPLICATIONS OF MAS-NMR

Solid-state NMR is a powerful tool for characterizing material structures. This spectroscopy has been successfully applied to different nuclei allowing the characterization of material frameworks [81–85].

In zeolites, particularly, ^{29}Si, ^{31}P, and ^{27}Al, MAS-NMR has been applied for studying the dealumination processes, determining the Si/Al atomic ratio, characterizing the isomorphic substitutions, and also for identifying reaction intermediates by in situ catalytic reactions [2,81,83–85].

Figure 4.31 [83] shows the ^{31}Si MAS-NMR spectra of intermediary phases resulting from the treatment of Na–clinoptilolite in a 7.5 M NaOH solution at 100°C, as a function of time in hours, that is, 0, 1, 4, 8, 16, 24, 36, and 64 h [83].

At the beginning of the hydrothermal transformation, the spectrum consists of a single, broad low-field-shifted resonance line centered at 100 ppm (chemical shifts were measured from natrolite). As the hydrothermal treatment proceeds, the spectra show five well-resolved lines at −85.2, −89.3, −94,0, −98.9, and −103.5 ppm, corresponding to Si surrounded by 4, 3, 2, 1, and 0 Al, respectively [81].

Figure 4.32 shows the ^{31}Si MAS-NMR spectra of the sample HC (Figure 4.32a) and PHEU(20) (Figure 4.32b); the elemental and phase compositions of the sample HC are given in Tables 4.1 and 4.2.

The sample PHEU(20) is the product of refluxing the sample HC with a 4 M solution of H_3PO_4 at a liquid to solid ratio of 2 mL/g at 373 K for 20 min, and then carefully washing with distilled water [85]. The peaks located at −101.3 and −107.6 ppm were both assigned to Si(1Al) and the peak at −113.0 ppm was assigned to Si(0Al) [85].

Figure 4.33 shows the ^{27}Al MAS-NMR spectra of the sample HC (Figure 4.33a) and PHEU(20) (Figure 4.33b). It is evident that the peaks at about 56 and −14 ppm are from tetrahedral and octahedral Al, respectively [85].

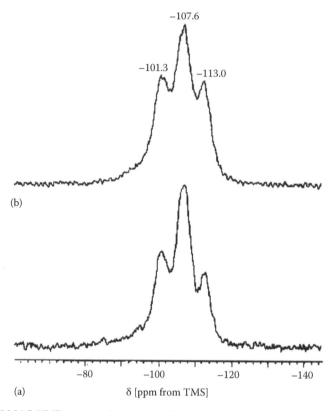

FIGURE 4.32 ^{31}Si MAS-NMR spectra of (a) sample HC and (b) PHEU(20).

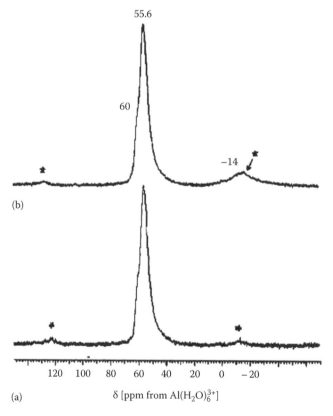

FIGURE 4.33 ^{27}Al MAS-NMR spectra of (a) sample HC and (b) PHEU(20).

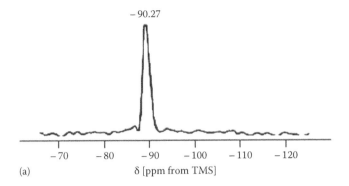

(a) δ [ppm from TMS]

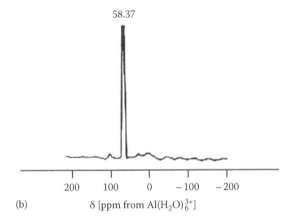

(b) δ [ppm from Al(H$_2$O)$_6^{3+}$]

FIGURE 4.34   (a) ^{31}Si MAS-NMR spectrum of Na-LTA-AlPO$_4$ · 3H$_2$O (20 h) and (b) ^{27}Al MAS-NMR spectrum of Na-LTA-AlPO$_4$ · 3H$_2$O (20 h). The Si spectrum is referred to TMS and the Al spectrum is referred to Al(H$_2$O)$_6^{3+}$.

Figure 4.34a shows the ^{31}Si MAS-NMR spectrum of Na-LTA-AlPO$_4$·3H$_2$O (20 h), and in Figure 4.34b, the ^{27}Al MAS-NMR spectrum of Na-LTA-AlPO$_4$·3H$_2$O (20 h) is shown [84], where the ^{31}Si spectrum is referred to TMS and the ^{27}Al spectrum is referred to Al(H$_2$O)$_6^{3+}$. The sample Na-LTA-AlPO$_4$·3H$_2$O (20 h) is the result of a tribochemical reaction between a Lynde type A zeolite, Na-LTA, synthesized in the author's laboratory, [84] and AlPO$_4$·3H$_2$O, provided by Merck and Co., Inc. White House Station, NJ, USA. [84].

To carry out the tribochemical reaction, 80 wt % of Na-LTA and 20 wt % of AlPO$_4$·3H$_2$O were mechanically ground in an agate ball mill for 20 h. From Figure 4.34a, it is evident that Si is in a tetrahedral coordination surrounded by four Al; the spectrum shown in Figure 4.34b shows that Al is in a tetrahedral coordination surrounded by four Si, similar as in the original Na-LTA sample [84]. However, the signal of ^{29}Si in Figure 4.34a is shifted from −89.2 in the pure Na-LTA to −90.3, which is explained by the introduction of P during the tribochemical reaction [84].

4.7   THERMAL METHODS OF ANALYSIS

The development of thermal analysis methods in materials research has led to a plethora of new methodologies since the elaboration of the first thermal method by by Le Chatelier and Robert-Austen [16,86]. Thermal analysis consists of a group of techniques in which a physical property of a material is measured as a function of temperature at the same time when the substance is subjected to a controlled increase, or in some cases, decrease of temperature. Temperature-programmed techniques, such as DTA [87–89], TGA [87], DSC [53,90], TPR [91,92], and TPD [93–96], contribute to perform a more complete characterization of materials.

4.7.1 Differential Thermal Analysis

In differential thermal analysis (DTA), the temperature difference that develops between a sample and an inert reference material is measured, when both materials are subjected to an identical heat treatment [87]. The related technique of DSA relies on differences in the energy required to maintain the sample and reference at an identical temperature.

DTA is a technique for recording the difference in temperature between a substance and a reference material. The specimens are subjected to identical temperature regimes in an environment heated or cooled at a controlled rate. Therefore, DTA involves heating or cooling a test sample and an inert reference in the same conditions, and at the same time recording any temperature change between the sample and reference. This differential temperature is plotted against temperature and, then, changes in the sample which lead to the absorption or evolution of heat can be detected relative to the inert reference [87,88].

The main characteristics of a DTA equipment are the following:

1. Sample holder comprising thermocouples, sample containers, and a ceramic or metallic block
2. Furnace
3. Temperature programmer
4. Recording system

The important requisites of the furnace are that it should have a stable and sufficiently large hot zone, and must be able to respond rapidly to commands from the temperature programmer. A temperature programmer is necessary in order to get stable heating rates. The recording system must have a low inertia to accurately reproduce variations. The sample holder assemblage consists of a thermocouple, each for the sample and reference, surrounded by a block to guarantee smooth heat dissemination. The sample is contained in a small crucible designed with a cut in the base to be accomodated in the thermocouple. The thermocouples should not be placed in direct contact with the sample to avoid contamination and degradation. The crucible may be made of materials such as silica, alumina, zirconia, nickel, or platinum, depending on the temperature and nature of the tests involved.

Figure 4.35 shows a set of DTA profiles [84]. In Figure 4.35a, the DTA profile of the unmixed $AlPO_3 \cdot 3H_2O$, and in Figure 4.24b the DTA profile of a simple mixture of $AlPO_3 \cdot 3H_2O$ (20 wt %)

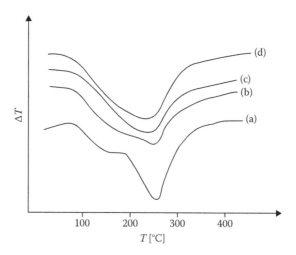

FIGURE 4.35 (a) DTA profiles of $AlPO_3 \cdot 3H_2O$, (b) a mixture of $AlPO_3 \cdot 3H_2O$ (20 wt %) and Na-LTA zeolite (20 wt %), (c) the mixture of $AlPO_3 \cdot 3H_2O$ (20 wt %) and Na-LTA zeolite (20 wt %) after 20 h of tribochemical reaction by milling, and (d) the Na-LTA zeolite.

and the Na-LTA zeolite (20 wt %), are shown. In Figure 4.35c, the DTA profile of the same mixture of $AlPO_3 \cdot 3H_2O$ (20 wt %) and the Na-LTA Zeolite (20 wt %) after 20 h of a tribochemical milling treatment, which caused a change of the phase composition of the mixture to $AlPO_3 \cdot 3H_2O$ (11 wt %) and Na-LTA zeolite (89 wt %), is shown. Figure 4.35d shows the DTA profile of the original Na-LTA zeolite [84].

4.7.2 Thermal Gravimetric Analysis

Thermal gravimetric analysis (TGA) is a simple analytical technique that measures the weight loss or gain of a material as a function of temperature during controlled heating. As the materials are heated, they lose weight due to different processes such as water desorption, or from chemical reactions that release gases. In contrast, some materials can gain weight by reacting with the surrounding atmosphere in the test environment (see Figure 5.26 [32]).

During the TGA testing process, a sample of the analyzed material is placed into, for example, an alumina cup, which is supported on, or suspended from, an analytical balance located outside the furnace chamber. The sample cup is heated according to a predetermined thermal cycle and the balance sends the weight signal to the computer for storage, along with the sample temperature and the elapsed time. The TGA curve plots the TGA signal, converted to percent weight change on the y-axis against the reference material temperature on the x-axis. Therefore, the results of the test is a graph of the TGA signal, that is, weight loss or gain converted to percent weight loss on the y-axis plotted versus the sample temperature in degree Celsius on the x-axis.

Examples of weight loss or weight gain processes are water desorption, structural water release, structural decomposition, carbonate decomposition, gas evolution, sulfur oxidation, fluoride oxidation, rehydration, and other transformations.

The application of thermogravimetric analyzers is explained with the thermogravimetric study of a single-walled carbon nanotube (SWCNT) [89]. TGA is between the standard methods applied for the characterization of carbon nanotubes together with SEM, HRTEM, XRD and Raman spectroscopy. The synthesized SWCNT [89] was studied with the help of a TQ500 thermogravimetric analyzer (TGA) manufactured by TA Instruments, New Castle, DE, USA. The sample was set at room temperature, that is, 23°C, in a flow (100 mL/min) of the purge gas, that is, pure N_2, and after this the temperature was linearly scanned at a heating rate of 10°C/min up to 800°C (Figure 4.36). The data collection, the temperature control, the programmed heating rate, and the gas switching were automatically controlled by the software installed in the intrument. The TGA data were collected as a M_t (wt %) versus T (°C) profile, where, M_t is the sample mass of loss. The gas used in the TGA study was pure N_2 (99.99% purity), provided by Praxair, Inc., Dandury, CT, USA. Figure 4.36 [89] shows the thermogravimetric profile of the synthesized SWCNT.

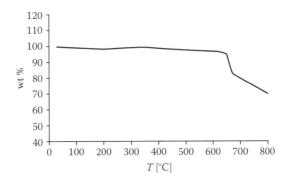

FIGURE 4.36 TG profile of the synthesized SWCNT.

4.7.3 Differential Scanning Calorimetry

Differential scanning calorimetry (DSC) is a method for measuring the energy required to establish a nearly zero temperature difference between a substance and an inert reference material, as both samples are subjected to the same temperature regimes in an environment heated or cooled at a controlled rate [17]. In one of the possible DSC arrangements, the sample and the reference are enclosed in the same furnace, and a highly sensitive sensor is used to measure the difference between the heat flows to the sample and reference crucibles based on the Boersma or heat flux principle [90]. The difference in energy required to maintain them at a nearly identical temperature is provided by the heat changes in the sample. Any excess energy is conducted between the sample and reference through the connecting metallic disc, a feature absent in DTA. In a DSC test, the thermocouples are not embedded in either of the specimens, and the small temperature difference that may develop between the sample and the inert reference is proportional to the heat flow between the two. The fact that the temperature difference is small is important to ensure that both containers are exposed to essentially the same temperature program.

As a demonstration of the use of DSC in the characterization of zeolites and related materials, Figure 4.37 illustrates the DSC profile of the LECA zeolite [53]. The LECA zeolite has a phase composition of 13 ± 4 [%], in wt %, of gismondine and 87 ± 4 [%] of quartz, anorthite, and glass. The DSC profile presented two endothermic peaks at 112°C and 195°C, and an exothermic peak at 312°C. As is very well known, zeolites evolve adsorbed water in the course of heating [97]. Consequently, the endothermic peaks in the DSC profile of LECA zeolite sample were clearly related to adsorbed water located in different sites of the LECA zeolite surface and the framework of the gismondine contained in the LECA zeolite [53].

4.7.4 Temperature-Programmed Reduction

Temperature-programmed reduction (TPR) is normally used in the characterization of catalysts [18,91–93]. In general, to carry out a TPR experiment, a reducing gas mixture, typically 5% hydrogen in nitrogen, flows continuously over the sample [92]. The gas flow rate can be varied precisely using either built-in controls or an optional mass flow controller accessory.

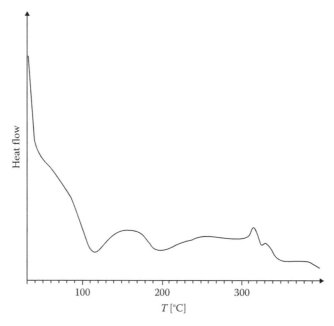

FIGURE 4.37   DSA profile of LECA zeolite.

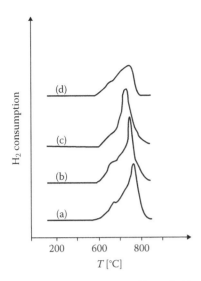

FIGURE 4.38 TPR profiles of (a) MnAPO-5, (b) ZnAPO-5, (c) CoAPO-5, and (d) AlPO₄-5.

A specially designed high-temperature furnace and sample cell with in situ sample temperature sensing is used to heat the tested samples up to more than 1100°C. Linear heating rates are controlled by a temperature controller. Heating rates are operator programmable for maximum flexibility. The reaction between the sample and reducing gas is monitored by a highly stable thermal conductivity detector (TCD) [18,91,92].

At present, this reaction rate signal is presented in real time on the computer display while the PC automatically records signal, temperature, and time. The resulting peak thus formed is a unique, characteristic fingerprint of the sample and the peak maximum represents the temperature of the maximum reaction rate. The reducing gas, or other reactive gas, thus absorbed during a TPR experiment, can be desorbed (TPD) in a separate experiment immediately following this procedure. Temperature-programmed oxidation (TPO) is performed in an analogous manner.

Figure 4.38 [92] shows an example of the use of TPR in the characterization of zeolites and related materials. The TPR profiles reported in Figure 4.38 were recorded in a TPR system, utilizing approximately 0.25 mg of sample in the quartz U-tube flow reactor [92]. The samples were heated at a rate of 10°C/min from a temperature of 23°C up to 550°C and held at this temperature for 4 h in a flow of 20 cm³/min (STP) of dry air. After cooling the sample, the TPR measurements were performed at a heating rate of 10°C/min from a temperature of 23°C up to 1050°C under a reductive flow of 10% H_2/Ar at 25 cm³/min (STP).

The consumption of hydrogen was monitored with a TCD held constantly at 100°C and recorded at a signal rate of 1 point/s. The hydrogen consumption was quantified by means of calibration with pure CuO (Merck >99.9%). The temperature of the sample bed was measured by means of a thermocouple inside the U-tube (with tip within the catalyst bed), and followed an exact linear ramp throughout the TPR run. A cold trap was used to prevent water passing through the TCD. The TPR experimental conditions were selected in agreement with criteria reported elsewhere [89].

4.7.5 TEMPERATURE-PROGRAMMED DESORPTION

Temperature-programmed desorption (TPD) is a very useful methodology in porous materials characterization. A basic feature of the experiment is that, it is necessary to use an appropriate adsorbate, such as CO, NH_3, or H_2O [94–96]; however, recently larger molecules have been applied as adsorbates [97].

The basic experiment is very simple, comprising [96] adsorption onto the sample at a relatively low temperature, normally 300 K. Subsequently, the sample is heated in a controlled fashion, that is, linearly in time, at rates between 0.5 and 20 K/s, and at the same time the evolution of species desorbed from the material into the gas phase are monitored.

In TPD practice, desorption rates are reported, that is, dN/dt, where N is the amount of molecules adsorbed in the material [96]. As the temperature increases and a particular species is capable to desorb from the material, thereafter, the TPD signal will rise; but, as the temperature continues to increase, the amount of adsorbed species on the material will decrease, causing the TPD signal to decrease. This result is expressed as a peak in the TPD signal versus time, or a temperature plot. The temperature of the peak maximum and the shape of the desorption peak provide information about the binding character of the adsorbate/substrate system [96].

Pure carrier gas (typically helium) flows over the sample as the temperature is raised in order to desorb the previously adsorbed gas. Normally, thermoconducting detectors monitor this rate of

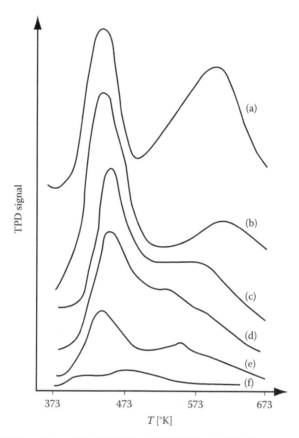

FIGURE 4.39 NH$_3$-TPD profiles of (a) H-MOR, (b) H-HEU, (c) H-LTL, (d) H-FAU, (e) H-MFI, and (f) CoAPO-5 (calcined).

desorption, producing a TPD profile, where the intensity of the desorption signal is proportional to the rate at which the surface concentration of the adsorbed species is changing. Consequently, the area under a peak is proportional to the quantity originally adsorbed. Besides, the kinetics of desorption provides information on the state of aggregation of the adsorbed species. Finally, the position of the peak temperature is related to the enthalpy of adsorption, that is, to the strength of the binding to the surface.

Figure 4.39 [95] shows the NH$_3$-TPD thermograms corresponding to H-MOR, H-HEU, H-LTL, H-FAU, H-MFI, and CoAPO-5 (calcined). The TPD profiles of the samples show two different regions that can be assigned to weak and strong acid sites.

To conclude this section, it is necessary to state that in modern implementations of the technique the detector of choice is a small, quadrupole mass spectrometer and the whole process is carried out under computer control. The data obtained from such an experiment consist of the intensity variation of each recorded mass fragment as a function of time and/or temperature.

4.7.6 FOURIER TRANSFORM INFRARED-TEMPERATURE PROGRAMMED DESORPTION

The author and Wendelbo have developed a new TPR methodology called the Fourier transform infrared-temperature programmed desorption (FTIR-TPD) [97]. These temperature-programmed experiments were performed in an IR high-temperature cell using AABSPEC (model 2000), made of 316 stainless steel, in which the sample under test can be treated in situ up to 973 K, at pressures up to 13 MPa (see Figure 4.23) [97,98]. The temperature of the sample holder was electronically controlled within ±1 K. The optical path length of the cell is 42 mm and the dead volume is 70 cm^3.

The cell is water cooled and equipped with CaF_2 windows. This cell was mounted in the sample section of a Bio-Rad FTS 40A FTIR spectrophotometer.

Adsorption and desorption measurements were carried out as follows: the organic compound to be adsorbed was filled in a stainless steel saturator, which was held thermostatically at 25°C. A flow of N_2 was divided into two and each of these streams passed through a flow controller. One stream (see Figure 4.23) (1–22 mL/min) went through a tube with a sinterplate in the base into the saturator where the N_2 bubbles in the liquid adsorbate, and it is anticipated that the stream through the saturator became saturated with the organic substance. This gas flow was mixed with a bypass stream of pure N_2 (340 mL/min) at the outlet of the saturator, and the unified stream, with a partial pressure of around 15–30 Pa of adsorbate, was then passed through the IR cell.

H-ZSM-5 and H-Beta zeolite powders were pressed to self-sustaining wafers (6–9 mg/cm^2, dry weight) under a pressure of 400 MPa for 3–4 s. The wafers were loaded into the IR cell and dehydrated at 673 K for an interval of 2–3 h in a flow of pure nitrogen (340–400 mL/min) (see Figure 4.23). After this activation, the samples were cooled to the required temperature and kept at this temperature with the help of a thermostat and a background spectrum of the pure degassed zeolite was always collected as a reference before the start of any experiment.

Adsorption of benzene, toluene, and ethylbenzene on the three zeolites at 298 K for partial pressures ranging from about $P/P_0 = 0.01$ to 0.8 were first performed [98], and it appears that all the systems approach saturation at $P/P_0 = 0.1–0.2$. Therefore, we decided to carry out the adsorption of the adsorbate molecules at partial pressures of 15–30 Pa.

The uptake process was followed by purging for 7 min, and subsequently the TPD was carried out. The results were obtained by monitoring the decrease, during desorption, of the absorbance A measured in arbitrary units (au), of a typical IR band of the adsorbate molecules, where A is proportional to N ($A = K \times N$), and where N is the amount of adsorbate in the material and K, a proportionality constant [98,99]. For benzene, the region between 1450 and 1550 cm^{-1} was integrated to obtain a measure of the intensity of the band around 1482 cm^{-1}. For toluene and ethylbenzene, the segment between 1477 and 1517 cm^{-1} was integrated to obtain an intensity measure for the band around 1497 cm^{-1} [97,98].

For the generation of the FTIR-TPD profiles (A vs. T), the decrease in the intensity (absorbance) of one of the selected bands was monitored, during heating at a constant rate [97]

$$\beta = \frac{dT}{dt} = 8.5 \, \text{K/min}$$

from 298 to 500 K with the BIORAD FTS 40A FTIR spectrometer with a resolution of 8 cm^{-1} [97]. That is, one spectrum was measured every minute during the temperature scan to generate the FTIR-TPD profile.

The FTIR-TPD profiles (A vs. T) for the desorption of benzene, toluene, and ethylbenzene from high-silica H-ZSM-5 is reported in Figure 4.40 [97]. These results were fitted with the complementary error function, that is,

$$\text{erfc}(z) = 1 - \text{erf}(z)$$

where $\text{erf}(z)$ is the error function [80]. The equation used for the fitting process was [97]

$$A(T) = \frac{a}{2}(1 - \text{erf}(z)) \qquad (4.25)$$

where $a = A$ (298 K) is the initial value of $A(T)$, $z = T - T_0 / \sqrt{2}\Delta T$, and T in Kelvin is the absolute temperature, and the temperature, T_0, and the temperature range, ΔT, both in Kelvin, are the parameters which characterize the FTIR-TPD profile [97].

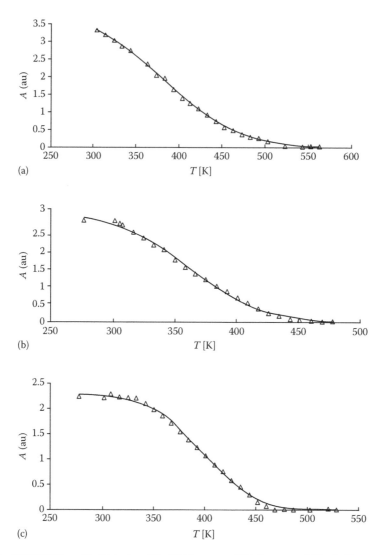

FIGURE 4.40   FTIR-TPD profile (A (au) vs. T) of (a) benzene, (b) toluene, and (c) ethylbenzene in H-ZSM-5 zeolite.

The fitting process was carried out with a program based on a least square procedure [48] that allows us to calculate the best-fitting parameters of the equation defining the relation $A(T)$ versus z, that is, Equation 4.25, specifically, a, T_0, and ΔT. The regression coefficient and the standard errors were also calculated with the least square methodology. The calculated regression coefficients fluctuated between 0.98 and 0.99. The values calculated for the parameters T_0 and ΔT, and the standard errors of the parameters, are reported in Table 4.9.

The values obtained for the parameter $a = A$ (298 K) are not reported in Table 4.9, since absolute adsorption magnitudes are not directly measured by the FTIR methodology [97].

In FTIR-TPD, we are not reporting the desorption rates, dN/dt, as is customary in TPD practice, we are reporting A, which is proportional to N, the amount adsorbed. In this sense, it is evident that [97]

$$\frac{dN}{dt} = K \frac{dA}{dt} \qquad (4.26)$$

where $N = KA$ as stated above. Now, it is easy to show that the desorption rate

TABLE 4.9
FTIR-TPD Parameters

Zeolite	Hydrocarbon	T_0 [K]	ΔT_0 [K]
H-ZSM-5	Benzene	382 ± 2	68 ± 1
H-ZSM-5	Toluene	361 ± 2	48 ± 2
H-ZSM-5	Ethylbenzene	397 ± 1	40 ± 2
H-Beta	Benzene	349 ± 4	47 ± 4
H-Beta	Toluene	356 ± 1	38 ± 1
H-Beta	Ethylbenzene	378 ± 2	44 ± 2

$$\frac{dN}{dt} = K\frac{dA}{dT} \quad \frac{dT}{dt} = K\beta\frac{dA}{dT} \tag{4.27}$$

is proportional to the temperature derivative of the absorbance.

In addition, with the help of Equation 4.1, the temperature derivative of the absorbance becomes

$$\frac{dA}{dT} = F(T) = -b\exp\left[-(z)^2\right]$$

a negative Gaussian function, where b is a constant in arbitrary units. Therefore,

$$\frac{dN}{dt} = K\beta F(T) \tag{4.28}$$

Subsequently, the parameters, T_0 and ΔT, describe the desorption rate since, these parameters also represent the Gaussian function. Therefore, with the obtained results (Figure 4.40 and Table 4.9), it was shown that the parameter T_0 is associated with the adsorption energy of the adsorbate in the zeolite. In addition, the parameter ΔT was linked with the transport of molecules inside the zeolites channels during the nonisothermal desorption process as well with the heterogeneous character of adsorption in zeolites.

4.8 DIELECTRIC ANALYSIS METHODS

4.8.1 Introduction

In Section 1.7, dielectric phenomena in materials were described. Here, we apply dielectric effects in the characterization of materials [88,100–134].

The time dependence of any periodic function can always be expressed in terms of sine and cosine functions; in this sense, it is very convenient to apply the complex number representation using the following equation [135]:

$$V = V_0 e^{i\omega t} = V_0\left(\cos\omega t + i\sin\omega t\right)$$

in which ω is the angular frequency, and a formula introduced in 1748 by Euler is applied in order to represent complex numbers. Evidently, the voltage measured is the real part of the above-written equation.

In order to generalize the concept of resistance in alternate current (AC) circuits, the impedance is defined as follows [9]

$$\tilde{Z} = Z_r(\omega) + iZ_i(\omega) \tag{4.29a}$$

where

Z_r is the real part of the complex number which represents the impedance
$i = \sqrt{-1}$ is the imaginary unit
Z_i is the imaginary part

This parameter is, normally, measured by the calculation of the ratio of the voltage response to the current perturbation which gives the impedance

$$\tilde{Z}(\omega) = \frac{\tilde{V}(\omega, t)}{\tilde{I}(\omega, t)} \tag{4.29b}$$

Usually, the AC impedance experiments are performed over a broad range of frequencies, that is, from some millihertz to some megahertz.

Many materials have the properties of low conductance (high impedance) and low loss. These materials are often referred to as dielectrics. In addition, many materials not normally considered as dielectrics exhibit these properties. It is well known that dielectric methodologies are a good test for the study of molecular and cationic mobilities in materials [88,101–134].

Impedance spectroscopy, applied to electrochemical systems, is described in Section 8.6; here a variant of impedance spectroscopy applied only to electrolytes and not to electrochemical systems as a battery or a fuel cell is described.

The measurements in an impedance spectroscopy test of a simple electrolyte are normally obtained in the hertz to some megahertz frequency range with an impedance analyzer; for this purpose impedance spectroscopy as a methodology is similar with DS (see Section 8.6.2).

The study of materials by dielectric methods is carried out in alternating electric fields and the dielectric behavior is described, in a formal way, by the complex relative permittivity (or complex dielectric constant) (see Section 1.7.2) [15,103,104]

$$\tilde{\varepsilon}(\omega) = (\varepsilon_r'(\omega) - i\varepsilon_r''(\omega))\varepsilon_0 \tag{4.30}$$

where

ε_0 is the permittivity of vacuum
$\varepsilon_r'(\omega)$ is the real part of the complex relative permittivity (or real component of the complex dielectric constant)
$\varepsilon_r''(\omega)$ is the imaginary part of the complex relative permittivity (or imaginary component of the complex dielectric constant)

It is possible to include the study of dielectrics in a single parameter, generalizing the complex permittivity as follows

$$\tilde{\varepsilon}_\sigma(\omega) = \varepsilon_r'(\omega) - i\left(\varepsilon_r''(\omega) + \frac{\sigma_0}{\omega\varepsilon_0} \right) \tag{4.31}$$

in which $\varepsilon_r'(\omega)$ and $\varepsilon_r''(\omega)$ are dielectric processes related by the Kramers–Kronig relations [103,104], if σ_0, the direct-current (DC) conductivity is negligible (see Section 1.7.5). The difference between $\varepsilon_r'(\omega)$ and $\varepsilon_r''(\omega)$, and σ_0 has an important physical meaning, inasmuch as the first two parameters are the result of finite displacements of charge (polarization) and the third is the product of long-range charge transport (conduction) [103].

4.8.2 THERMODIELECTRIC ANALYZER

A thermodielectric profile can be obtained with a thermodielectric analyzer [15,108–110]. This device registers the relationship between the linearly scanned temperature (T) $(30°C < T < 1000°C$, at a rate of 16°C/min) and the rectified output voltage $(0\,V < V_0 < 10\,V)$ of a circuit (dielectric sensor), which differentially compares the impedance of the test sample, in the form of a powder (particle size 0.1–0.2 mm) and a reference (calcined Al_2O_3 powder) (see Figures 4.41 and 4.42 for further information).

The relationship between the output (V_{out}) and input voltages (V_{in}) in the applied circuit (Figure 4.41) is [88]

$$V_{out} = -\frac{|Z_r|}{|Z_x|}V_{in} \tag{4.32}$$

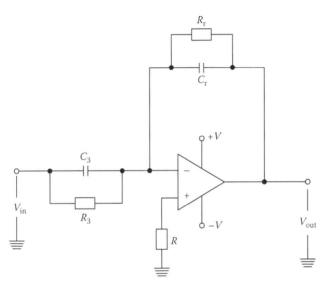

FIGURE 4.41 Dielectric sensor circuit.

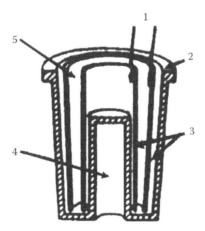

FIGURE 4.42 Inside view of the specimen holder for TDA. 1, electric conductors to the dielectric sensor; 2, ceramic specimen holder for a TDA; 3, nickel made concentric cylindrical electrodes; 4, void where is located a platinum–rhodium thermocouple; 5, hollow place where is located the samples under test in the form of powders (about 1 g of sample).

where

$|Z_r|$ is the complex impedance module of the equivalent circuit formed by the reference capacitor filled with Al_2O_3

$|Z_x|$ is the complex impedance module of the equivalent circuit formed by the capacitor filled with the test sample (see Figures 4.41 and 4.42)

It is necessary now emphasize that was taken as the equivalent circuit for the sample and standard capacitors the model of the resistor and an ideal loss less capacitor in parallel (see Figures 4.41 and 4.48).

The dielectric sensor circuit, comprising an operational amplifier (plugged in for impedance comparison), capacitors formed by the sample and reference specimen holders (represented as real capacitors with an internal resistance, see Figures 4.41 and 4.48), and finally the resistance R, which is added for circuit balancing, is used. This device is fed with an alternating signal (V_{in} 400 Hz of frequency and 25 V of amplitude) [15,88,108–110]. The alternating output voltage (V_{out}), in the case of the thermodielectric analyzer, was rectified up to a direct voltage (V_0) to be registered by the xy-plotter.

The plotter, records the signal of a chromel–alumel thermocouple located inside the heating furnace near to the sample holder in the x-axis; the temperature of the furnace is scanned from 27°C to 1000°C at a rate of 16°C/min. In the y-axis, the rectified output voltage, V_0 (0 V $< V_0 <$ 10 V) is plotted, in order to obtain the thermodielectric profile [15,88].

In the case of DS, the same type of circuit shown in Figure 4.41 is used, fed with a varying frequency alternating signal (V_{in} 0.1 V of amplitude and the frequency in the range 30 Hz $< f <$ 10^6 Hz). Besides, the reference capacitor (C_r) is substituted by a standard capacitor (C_s), and the V_{out} was measured with an oscilloscope [15,109].

It is evident that the explicit form for Equation 4.32 is

$$V_{out} = - \frac{\left| \left[\left(\frac{1}{R_r} \right) + i(\omega C_r) \right]^{-1} \right|}{\left| \left[\left(\frac{1}{R_x} \right) + i(\omega C_x) \right]^{-1} \right|} = - \left[\frac{(\omega C_x)^2 + \left(\frac{1}{R_x} \right)^2}{(\omega C_r)^2 + \left(\frac{1}{R_r} \right)^2} \right] V_{in} \tag{4.33}$$

obtained with the help of the following expression:

$$\tilde{Z} = \frac{1}{\frac{1}{R} + i\omega C} \tag{4.33a}$$

in which, the impedance of the resistance is $Z_R = R$ and the impedance of the ideal capacitor is $Z_C = 1/i\omega C$. Equation 4.33a describes the impedance of a resistance and a capacitor in parallel, where the reference and test sample capacitors are connected in the circuit as real capacitor equivalent circuits (see Figures 4.41 and 4.48), where C_x, C_r and R_r, R_x are the respective capacitance and resistance of the capacitors filled with the powdered sample x and reference compound r.

From Equation 4.33, it is possible to show [15,88]

$$V_{out} \approx \left[\frac{\varepsilon'_x}{\varepsilon'_r} \right] V_{in} \tag{4.34}$$

assuming $1/R_r \approx 0$ and $1/R_x \approx 0$, and using the following relation $C = \varepsilon' C_0$, where C_0 is the capacity of the empty capacitor and ε'_r and ε'_x are the real parts of the complex relative permittivities of the reference and the powdered test sample, respectively.

Equation 4.34 satisfactorily describes the low-temperature effect ($T < 250°C$) in the thermodielectric profile, since the resistance of the sample (at $T < 250°C$) is very high and, therefore, the second term of the impedance module, $[1/R]^2$, can be neglected [15,88].

It is necessary to recognize that ε'_r and ε'_x, are effective values, inasmuch as we are studying powdered zeolites and not monocrystals here.

At a high temperature ($T > 400°C$), the dielectric sensor behaves as a pure resistive circuit because of the decrease of the internal resistance of the capacitors, that is, R_x and R_r (see Figure 4.41). In this instance, the following approximation obtained from Equation 4.33 is valid [15]:

$$V_{\text{out}} \approx \left[\frac{\sigma_x^0}{\sigma_r^0} \right] V_{\text{in}} \tag{4.35}$$

as long as $1/R = \lambda\sigma^0$, where λ is the form factor for the capacitor and σ_r^0 and σ_x^0 are the direct current conductivities of the sample and reference powders, respectively [15,89].

For the real part of the dielectric constant, it is likewise necessary to admit that σ_x^0 and σ_r^0 are effective values, on account of the fact that we are studying powdered zeolites and not monocrystals.

Figure 4.43a shows the block diagram of the dielectric differential thermal analyzer, where the dielectric sensor circuit and the capacitors formed by the sample and reference (C_r) and specimen holders (C_x) are schematically represented, and a photo of the home-made equipment is shown in Figure 4.43b [89,109,110].

4.8.3 THERMODIELECTRIC ANALYSIS

4.8.3.1 First Effect in TDA

In general, in thermodielectric analysis (TDA), we have a first effect in the thermodielectric profile in the temperature range 30°C–250°C, which is related to water polarization [15,113]. The thermodielectric thermographs for Na-AD and Na-MP samples (see Tables 4.1 and 4.2) with different water content are reported in Figures 4.44 and 4.45 [113]. These thermodielectric profiles testify that

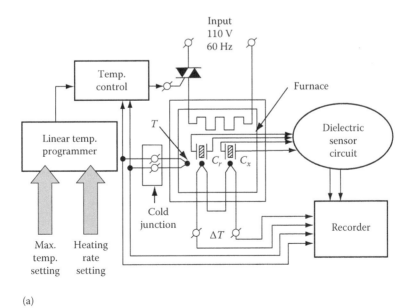

(a)

FIGURE 4.43   (a) Block diagram of the dielectric differential thermal analyzer.

(continued)

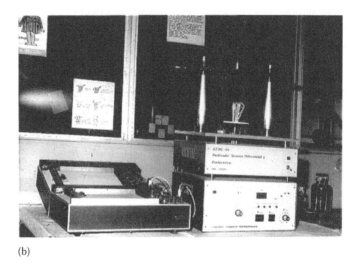

(b)

FIGURE 4.43 (continued)   (b) Photography of the dielectric differential thermal analyzer.

the first effect in TDA is associated with the polarization of the highly polar water molecules contained in the hydrated zeolite. In the illustrations showing the dielectric profiles related to water polarization (Figures 4.44 and 4.45), it can be readily perceived that the intensity of the peak (the effect is in the form of a peak) is associated with the increase in the water content in the zeolite [113]. Thereafter, the above-described phenomenon is caused primarily by water molecule polarization.

The effect has the form of a peak because at the origin of the peak, the increase in temperature is linked with the increasing mobility of water molecules adsorbed in the cavities and the channels of the zeolite. This event is followed by an increment of the permittivity of the zeolite sample, and this increase of the sample permittivity is detected by the thermodielectric analyzer as an increase in V_0 (see Equation 4.34).

The further increase in temperature and the time elapsed during temperature scanning is the reason for a decrease in the water content of the zeolite and, consequently, a decrease in the permittivity. Both the growth of the permittivity with temperature and the following permittivity decrement with the subsequent increase in temperature justify the form of peak exhibited by the first thermodielectric effect.

Thermodielectric experiments were performed using the Na-X zeolite (Si/Al = 1.25 provided by Laporte, London, UK). Within the framework of the cavities of the Na-X zeolite, the following were adsorbed: water (with a dielectric constant $\varepsilon = 78.5$ at 25°C), acetone ($\varepsilon = 24.3$ at 25°C), ethanol ($\varepsilon = 20.7$ at 25°C), and chloroform ($\varepsilon = 4.8$ at 20°C) to produce Na–X(H_2O), Na–X(CH_3COCH_3), Na–X(CH_3CH_2OH), and Na–X($CHCl_3$) samples [15].

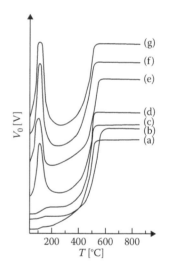

FIGURE 4.44   Thermodielectric profiles (rectified output voltage, V_0 [0–10 V] vs. temperature, T [°C]) of Na-AD with (a) $n_a =$ 0 mmol/g, (b) $n_a = 0.75$ mmol/g, (c) $n_a = 2.19$ mmol/g, (d) $n_a =$ 2.18 mmol/g, (e) $n_a = 3.56$ mmol/g, (f) $n_a = 4.9$ mmol/g, and (g) $n_a =$ 5.8 mmol/g of water adsorbed.

With the help of the dielectric thermographs of the produced Na–X samples, it can be shown that the area, A [V × °C], of the first peak in the thermodielectric profiles of the produced Na–X samples and the dielectric constant of the adsorbed water ($\varepsilon = 78.5$ at 25°C), acetone ($\varepsilon = 24.3$ at 25°C),

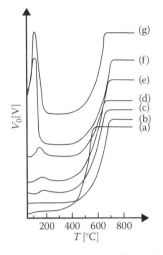

ethanol (ε = 20.7 at 25°C), and chloroform (ε = 4.8 at 20°C) satisfy the following empiric equation [15,119]:

$$A = k\varepsilon \qquad (4.36)$$

where $k \cong 12.8$ [V × °C]. The empiric Equation 4.36 is clearly related to Equation 4.34 and clearly links the first peak in the thermodielectric profile of the zeolites with the orientation polarization of the adsorbed molecules [15,113].

The polarization of water or any other adsorbed molecule is a fundamental contribution to the first effect. However, as previously explained, charge-compensating cations that exist in the zeolite channels and cavities are able to modify the intensity of the peak encountered in the first thermodielectric effect [119]. Figure 4.46 shows that the first effect is more noticeable for monovalent cations (K^+ and Na^+) than for the divalent cation (Ca^{2+}) [120]. This phenomenon may be explained with the help of the hypothesis of cation-hopping polarization. Cationic polarization in zeolites is generated by the cation-hopping polarization mechanism [15,120]. Therefore, it is evident that the amount of monovalent cations existent in the zeolites is twice the content of divalent cations, on account of the charge

FIGURE 4.45 Thermodielectric profiles of Na-MP with (a) $n_a = 0$ mmol/g, (b) $n_a = 1.5$ mmol/g, (c) $n_a = 2.0$ mmol/g, (d) $n_a = 3.8$ mmol/g, (e) $n_a = 4.10$ mmol/g, and (f) $n_a = 4.7$ mmol/g of water adsorbed.

balance principle; hence, samples exchanged with monovalent cations exhibit a much noticeable first effect than samples exchanged with divalent cations.

Besides, the interaction of divalent cations with the zeolite framework and water is stronger on account of the higher charge and lower cationic radius exhibited by Ca^{2+} (Ca^{2+}: 0.99 Å) in contrast with Na^+ and K^+ (Na^+: 0.95 Å and K^+: 1.33 Å). This effect induces a lower mobility of divalent cations, on account of the fact that divalent cations are more intimately linked with the zeolite framework. Hence, the lower cationic mobility is an additional reason for the decrement in the permittivity of tested samples, and this effect is also detected by the thermodielectric analyzer as a decrease in V_0 [110,119].

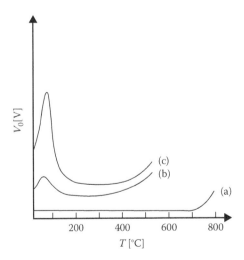

FIGURE 4.46 Dielectric thermograph of (a) Ca-HC, (b) K-HC, and (c) Na-HC with $n_a = 6.2$ mmol/g of water adsorbed.

4.8.3.2　Second Effect in TDA

The existence of a second effect in TDA has ben shown in the following temperature range 150°C–500°C [46,125]. This effect is caused by occluded molecules. As described in Section 3.4.2, the MeAPO-5 and AlPO$_4$-5 molecular sieves and the Na-ZSM-5 zeolite are synthesized in the presence of a template agent, triethylamine (TEA), for the aluminophosphate molecular sieves and tetrabutylammonium hydroxide (TBAOH) for Na-ZSM-5. These amines are occluded in the cavities and channels of MeAPO-5, AlPO$_4$-5, and Na-ZSM-5 during the crystallization process. In this regard, during a thermodielectric analysis test of these type of zeolites, the temperature increase induces the amine molecules, occluded in the framework of the aluminophosphate molecular sieves and the ZSM-5 zeolite, to become mobile and, consequently, increase the permittivity of the zeolite powder sample. Therefore, as already explained for water, the thermodielectric analyzer detects this effect as an increase in V_0. The additional increase of temperature and the time elapsed during the temperature scanning give rise to amine decomposition and desorption of the disintegration products. As a result, the dielectric sensor detects a decrease in permittivity and accordingly a second peak.

In Figure 4.47, the thermodielectric profiles of as-synthesized Na-ZSM-5 [125] and CoAPO-5, ZnAPO-5, MnAPO-5, and AlPO$_4$-5 [46], and a noncrystallized amorphous aluminosilicate gel obtained during the synthesis of Na-ZSM-5 are shown.

In the thermographs shown in Figure 4.47, the presence of water can be observed in the MeAPO-5 and AlPO$_4$-5 molecular sieves, on account of the appearance of the first effect in the thermodielectric profile [46]. Following the first effect, a second peak can be perceived at 150°C–300°C for the aluminophosphate molecular sieves (Figure 4.47) and around 470°C for the as-synthesized Na-ZSM-5 (Figure 4.47); this peak can be assigned to amine desorption [125]. In the case of the noncrystallized amorphous aluminosilicate, since the gel did not crystallize, we obtained an amorphous solid which contained water but did not retain the occluded amine [125].

4.8.3.3　Third Effect in TDA

The existence of a third effect in TDA has also been oberved in the temperature range 400°C–1000°C [15,84,111,114,115,117,121,123,125]. It has been observed that at relatively high temperatures ($T > 400$°C), long-range charge carrier transport [103] is the prevailing cause of the effect found in the thermodielectric profile of materials [15,84,111,114,115,117,121,123,125]. This high-temperature effect, which is evident in the thermodielectric profiles of zeolites (see Figures 4.44 through 4.47), is produced by cationic conduction [15,114].

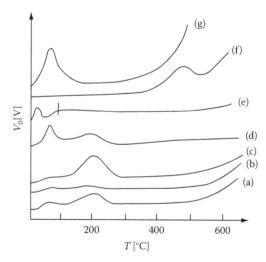

FIGURE 4.47　Thermodielectric profile for (a) CoAPO-5, (b) FAPO-5, (c) ZnAPO-5, (d) MnAPO-5, (e) AlPO$_4$-5, (f) Na-ZSM-5, and (g) a noncrystallized product.

From the thermodielectric profiles shown in Figures 4.44 through 4.47, it is eveident that the present effect did not give rise to a peak. This can be clarified with the help of the relationship between the direct current conductivity, in the case of ionic transport, and temperature [15,114]:

$$\sigma_0 = \sigma_0^* e^{-\frac{E_a}{RT}} \tag{4.37}$$

Substituting Equation 4.37 in Equation 4.35 [114]

$$V_{out} = \frac{\sigma_{0x}^*}{\sigma_{0r}^*} e^{-\frac{E_a^x - E_a^r}{RT}} = \sigma_e e^{-\frac{\Delta E_a}{RT}} \tag{4.38}$$

since V_0 is proportional to V_{out}, because as previously stated, the alternating output voltage (V_{out}), in the case of the thermodielectric analyzer was rectified up to a direct voltage (V_0) to be registered by the xy-plotter [89]. Now, from Equation 4.38, it is possible to obtain the relation [114]

$$\ln V_0 = a - \frac{\Delta E_a}{RT} \tag{4.39}$$

where a is an empiric parameter related with σ_e. Accordingly, if the linear relation between $\ln V_0$ and $1/RT$ fits our experimental data, this experimental match is a support for the proposed mechanism. In Table 4.10, some examples of the satisfactory fit of the obtained experimental data with the linear relation between $\ln V_0$ and $1/RT$ are reported [114].

The agreement between the experimental data and the equation achieved using the assumption of cation-hopping conduction [114] testify that this hypothesis is in concordance with the experimental data. It is interesting to recognize the difference between the measured apparent activation energies for sodium and calcium zeolites (see Table 4.10 [114]). As was hitherto observed for the polarization phenomena, Na^+ cations are more mobile than Ca^{2+} cations; this effect also explains the observed differences in the measured apparent activation energies for cationic conduction for Na^+ and Ca^{2+}.

With regard to ammonium zeolites, a seperate discussion has be carried out. The ammonium ion (NH_4^+) is decomposed during heating to ammonia (NH_3) plus a proton (H^+), and as a result of the heating process an acid zeolite (H–Z) is obtained. These protons are strongly bonded to the oxygen atoms of the zeolite framework forming OH bridge groups [15,114], and therefore have very low mobility in comparison with other charge-compensating cations such as Na^+. The experimental consequence is that the apparent activation energy for NH_4 zeolites is higher than the values reported for other cations.

TABLE 4.10
Numerical Values Obtained for the Parameter, ΔE, of the Equation, $\ln V_0 = a - \Delta E_a/RT$, Obtained with the Help of a Linear Regression Procedure

Sample	ΔE_a[kJ/mol]	r, [Regression Coefficient]
Na-HC	36	0.998
Ca-HC	66	0.989
Na-MP	47	0.998
Ca-MP	60	0.986
NH_4-HC	110	0.989
NH_4-MP	110	0.994

4.8.4 Dielectric Spectroscopy

If a variable (AC) voltage is applied to a dielectric material, the ratio of voltage to current is known as the impedance (see Equation 4.29b). The measured impedance varies with the frequency of the applied voltage in a way that it is related to the properties of the liquid or solid introduced between the capacitor plates. This is due to the physical structure of the material, or the chemical processes within it, or a combination of both [15,46,120,124–130]. The advantages of dielectric measurements over other techniques include rapid acquisition of data, accurate and reproducible measurements, nondestructive analysis, and the ability to differentiate effects due to electrodes, diffusion, mass/charge transfer by analyses over different frequency ranges. There are several models for an electrolyte under an applied voltage. The simplest one, previously applied in the thermodielectric analyzer, is a resistor and an ideal loss-less capacitor in parallel [9] (see Figure 4.48). As previously stated, the impedance of the ideal capacitor is

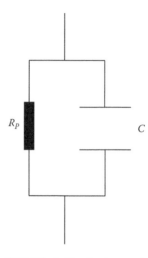

FIGURE 4.48 Real capacitor equivalent circuit.

$$Z_C = \frac{1}{i\omega C} \tag{4.40}$$

where

$$C = \frac{\varepsilon_0 \varepsilon_r'(\omega) A}{d} \tag{4.41}$$

and the impedance of the resistance (see Equation 4.31) of a real capacitor (Figure 4.48) is $Z_R = R_p$, where

$$\frac{1}{R_p} = \frac{\omega A \varepsilon_0 \left(\varepsilon_r'' + \dfrac{\sigma_0}{\omega \varepsilon_0} \right)}{d} \tag{4.42}$$

Now, since the impedance of an equivalent circuit of a parallel connection between a resistance and a capacitor is [132,133]

$$\frac{1}{\tilde{Z}} = \frac{1}{Z_C} + \frac{1}{Z_R} = i\omega C + \frac{1}{R_p} \tag{4.43}$$

given the admittance, $\tilde{Y}$, is the reciprocal of the impedance of the equivalent circuit

$$\tilde{Y} = i\omega C + G_p \tag{4.44}$$

where:

$$G_P = \frac{1}{R_p} \tag{4.45}$$

Consequently, (see Equations 4.30 and 4.31) when measuring the impedance, we are as well determining $\tilde{\varepsilon}_\sigma(\omega)$, since

$$\tilde{Y} = \frac{1}{\tilde{Z}} = \frac{j\omega \varepsilon_0 \tilde{\varepsilon}_\sigma(\omega)}{d} \tag{4.46}$$

The determination of the dielectric dispersion spectra of dehydrated and hydrated zeolites at different temperatures can be attained in a home-made dielectric spectrometer already described [15,120,125,126]. The equipment embodies three parts: a Pyrex glass vacuum system for the exhaustion of the cylindrical capacitor (Figure 4.49a), the cylindrical capacitor (Figure 4.49b), and a dielectric sensor (Figure 4.49c) [120,125]. Figure 4.49a shows a photo of the equipment, and Figure 4.49b shows some details of the capacitor where (1) denotes the electrodes and (2) denotes the connection to the vacuum system. Figure 4.49c shows the diagram of the dielectric sensor circuit, which is included in the metallic box, indicated by the arrow [109,120,125–127].

The powdered zeolite (grain size between 0.2 and 0.5 mm) to be tested is placed between the electrodes of the capacitor, which is also the sample holder. The dielectric sensor circuit, which is fed with an alternating voltage (amplitude 0.1 V: 30 Hz < f < 100 kHz), is similar to that previously described, except having a standard commercial capacitor of known capacitance ($C_s = 110$ pF) in place of the reference capacitor (C_r) (see Figure 4.49c) [15,16,111–113,120]. For the calculation of the real part of the relative permittivity, at different frequencies and temperatures, the following relation was used [15,120]

$$\varepsilon_r'(\omega) = \frac{C_x}{C_0} \qquad (4.47)$$

where C_x is the capacitance of the cylindrical capacitor full of powdered zeolite, and $C_0 = 1.5$ pF is the capacitance of the same capacitor, but empty [16,111]. Finally, C_x, is calculated with the help of the equation [110,125]

$$V_{out} = \frac{C_x}{C_s} V_{in} \qquad (4.48a)$$

That is:

$$C_x = \frac{V_{out}}{V_{in}} C_s \qquad (4.48b)$$

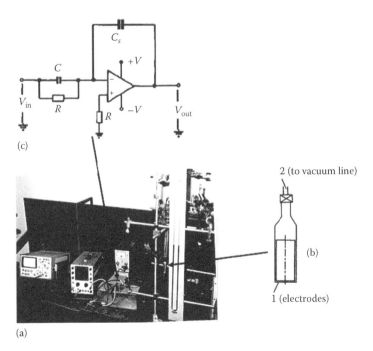

FIGURE 4.49 (a) Vacuum system and the equipment used for the measurements, (b) cylindrical capacitor sample holder, and (c) dielectric sensor circuit.

obtained by applying Equation 4.34, since the resistance of the sample is very high and the second term of the impedance module, $[1/R]^2$, can be neglected.

Equation 4.48 was thoroughly tested using different capacitors in the range of $10\,pF < C_x < 200\,pF$ [120]. These capacitors are located in the sensor circuit in place of the cylindrical sample capacitor [120]. With the help of this procedure, it was shown that Equation 4.48 is really adequate and can be used for the calculation of the capacity of the cylindrical sample capacitor at different frequencies in experiments carried out at a sample temperature $T < 200°C$, where $[1/R]^2$ is negligible [120].

One of the uses of DS in the characterization of high silica zeolites is illustrated in Figure 4.50, where the dielectric spectra of calcined Na-ZSM-5 at different temperatures, that is, (a) 27°C, (b) 100°C, (c) 150°C, and (d) 200°C, are reported [125].

These spectra show low values for the real part of the complex relative permittivity in the whole range of frequencies, indicating low water filling of the Na-ZSM-5 sample and low cationic conduction [125].

Using the dielectric dispersion spectra of Na-ZSM-5 at 27°C, 100°C, 150°C, and 200°C (Figure 4.50) and the Kramers–Kronig relations (see Section 1.7.5), it is possible to calculate [103] the absorption spectra of calcined dehydrated Na-ZSM-5 at 100°C, 150°C, and 200°C (Figure 4.51, assuming that σ_0, the direct current conductivity, is negligible (see Section 1.7.5) [125]. The obtained

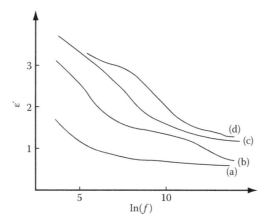

FIGURE 4.50 Dielectric dispersion spectrum of calcined dehydrated ZSM-5 at (a) 27°C, (b) 100°C, (c) 150°C, and (d) 200°C. In the *x*-axis, the natural logarithm of frequency in hertz vs. the real part of the relative permittivity is plotted.

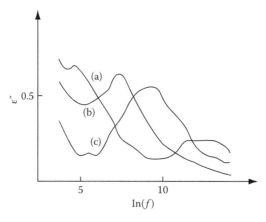

FIGURE 4.51 Dielectric absorption spectra of calcined dehydrated ZSM-5 at (a) 373 K, (b) 423 K, and (c) 473 K.

spectra shows the presence of a maximum for each spectrum (Figure 4.51) where the angular frequency for each maximum (ω_0) in the frequency spectrum depends on temperature.

In Section 1.7.5, the following equation for the real part of $\tilde{\chi}(\omega)$ was obtained and applied [119]

$$\chi'(\omega) = \frac{\chi(0)}{1 + \left(\dfrac{\omega}{\omega_0}\right)^2} \tag{4.49}$$

where $\omega_0 = QD$ is the value of the angular frequency for the absorption spectrum maximum, and Q, a constant,

$$Q = \frac{(Ze)^2 C_0}{\chi \varepsilon_0 kT} \quad \text{and} \quad D = D_0 e^{-\frac{E_a}{kT}} \tag{4.50}$$

is the diffusion coefficient for the cation-hopping mechanism [119,125]. Then

$$\omega_0 = QD_0 e^{-\frac{E_a}{kT}} \tag{4.51}$$

Now,

$$\tilde{\chi}(\omega) = \tilde{\varepsilon}_r(\omega) - \varepsilon_r(\infty) \tag{4.52}$$

then

$$\chi'(0) = \varepsilon'_r(0) - \varepsilon_r(\infty) \tag{4.53}$$

is the static dielectric susceptibility and

$$\chi'(\omega) = \varepsilon'_r(\omega) - \varepsilon_r(\infty) \tag{4.54}$$

where $\varepsilon'_r(0)$ and $\varepsilon_r(\infty)$ are the low frequency limit of the real part of the complex relative permittivity and the high frequency limit of the real part of the complex relative permittivity, respectively. In Table 4.11 [125], the dependence between temperature (T) and the frequency of the maximum (ω_0) is reported, for the spectra reported in Figure 4.51.

If we make an Arrhenius plot of Equation 4.54 using the data reported in Table 4.11, we get that $E_a = 70$ [kJ/mol] [125]. The value measured for the activation energy (E_a) for the cation-hopping mechanism of dielectric relaxation is similar to the values measured for the activation energies for diffusion of atoms in solids.

TABLE 4.11
Temperature Dependence of the Maximum Angular Frequencies for the Absorption Spectra of the Na-ZSM-5 Sample

Temperature [K]	ω_0 [Hz]
373	91
423	1556
473	11339

Figure 4.52 shows the adsorption isotherm of H_2O at 300 K on the natural mordenite, MP (see Table 4.1), where n_a is the magnitude of adsorption in millimoles per gram and P/P_0 is the relative pressure, where P is the equilibrium adsorption pressure and P_0 is the vapor pressure of H_2O at 300 K. The dispersion spectra [126] of the zeolite sample MP with water adsorbed in the zeolite primary and secondary porosity are as follows: (a) $n_a = 0$ mmol/g, (b) $n_a = 1.0$ mmol/g, (c) $n_a = 1.4$ mmol/g, (d) $n_a = 4.0$ mmol/g, (e) $n_a = 5.5$ mmol/g. It is plotted, in the x-axis, the natural logarithm of the frequency (Hz) versus the natural logarithm of the real part of the relative permittivity in the y-axis.

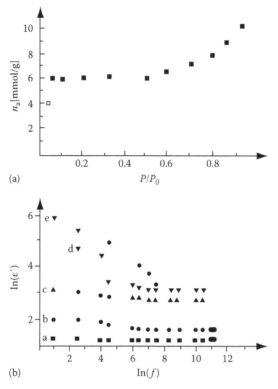

(a)

(b)

FIGURE 4.52 (a) Adsorption isotherm of H_2O at $300\,K$ on the MP sample and (b) dispersion spectra of the MP sample with water adsorbed in the zeolite: (a) $n_a = 0\,mmol/g$, (b) $n_a = 1.0\,mmol/g$, (c) $n_a = 1.4\,mmol/g$, (d) $n_a = 4.0\,mmol/g$, and (e) $n_a = 5.5\,mmol/g$.

The dielectric relaxation mechanism for the processes described by curves (a) to (c) in Figure 4.52b is the same over the whole frequency range, as is evidenced by the constant magnitude of the slope in the log–log plot [103,134]. This mechanism is described by polarization by the dipolar orientation of water molecules and cation hopping [134] of the hydrated cations present in the zeolite channels, with some contribution of the direct current conduction and interfacial polarization [126]. On the other hand, the increment in the magnitude of the real part of the complex relative permittivity with the increase in the water content for the complete frequency range measured is evident (Figure 4.52b). Furthermore, when the quantity of water contained in the zeolite is increased, two relaxation mechanisms are obviously implicated (see the change in the slope of the log–log plot in curves (d) and (e), Figure 4.52b). At a high frequency, the dipolar polarization mechanism is noticed, that is, all the curves (from (a) to (e), Figure 4.52b) exhibit the same slope at a high frequency. At low frequencies and high water content (curves (d) and (e), Figure 4.52b) a strong dispersion, generated by the interfacial mechanism of polarization, is found [103,129]. At a high water content, adsorbed molecules are not only present in the channels and cavities of the zeolitebut are also present in the mesopores created between the zeolite crystallites [126].

Figure 4.53 shows the dispersion spectra of hydrated Ca-HC, K-HC, and Na-HC at $300\,K$ [120]. They are plotted in the x-axis, and the natural logarithm of the frequency (Hz) versus the natural logarithm of the real part of the relative permittivity in the y-axis. This experiment shows once more the higher mobility of monovalent cations in comparison with divalent cations and the higher mobility of Na^+ with respect to K^+. The cause of this effect is due to the inferior cationic radius of Na^+ in comparison with that of K^+.

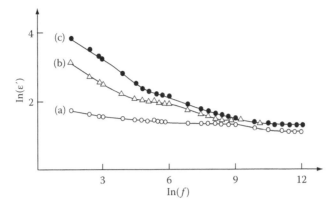

FIGURE 4.53 Dielectric dispersion spectra of (a) Ca-HC, (b) K-HC, and (c) Na-HC with $n_a = 6.2$ mmol/g of water adsorbed.

4.9 MÖSSBAUER SPECTROMETRY

4.9.1 INTRODUCTION

In Section 1.9, we studied the Mössbauer effect, discovered in 1957 by Rudolph Mössbauer [136]. This effect brings about a very high-resolution spectroscopy in the gamma-ray region of the spectrum, the so-called Mössbauer spectroscopy. This methodology has been applied in many areas, such as the measurement of lifetimes of excited nuclear states and nuclear magnetic moments, the investigation of electric and magnetic fields in atoms and crystals, in the analysis of special relativity and the equivalence principle, and in other applications [137–142].

4.9.2 MÖSSBAUER SPECTROMETER

Resonance gamma spectrometry or Mössbauer spectrometry can be used to study the hyperfine interactions between a nucleus and its chemical neighborhood [142]. In order to examine these interactions with the help of a Mössbauer spectrometer, the first-order Doppler effect shift of the wave emitted by a moving source is applied. The arrangement used for a Mössbauer spectrometer consists of a radioactive source containing a Mössbauer isotope in an excited state (see Figure 4.54)

The first-order Doppler frequency shift, $\Delta \nu$, of a wave emitted by a source moving at a velocity, v, is given by the following expression

$$\Delta \nu = \nu \frac{v}{c}$$

where
    ν is the original frequency of the emitted wave
    c is the velocity of light

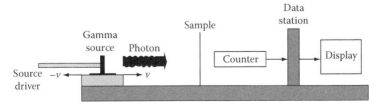

FIGURE 4.54 Schematic representation of a Mössbauer spectrometer.

Since the energy of the photon is given by $E_\gamma = h\nu$, then

$$\Delta E = E'_\gamma - E_\gamma = E\frac{v}{c}$$

in which

E_γ is the energy of the photon emitted by the gamma source at rest

E'_γ is the energy of photon emitted by the gamma source moving at a velocity, v

Figure 4.55 shows a Mössbauer spectrum obtained with a spectrometer similar to that shown in Figure 4.54, that is, in transmission geometry.

The spectrum shown in Figure 4.55 is obtained when there are no hyperfine interactions, that is, the emitter and the absorber have the same energy of emission and absorption. Consequently, the maximum absorption of gamma photons emitted by the source at the thin absorber is $v = 0$, because of resonant absorption. Subsequently, as the velocity of the source is increased in the positive or the negative directions, the resonance is broken and, therefore, the transmission of gamma photons through the absorber increases [135].

In this chapter, we only discuss gamma resonance spectrometry using ^{57}Fe, which is by far the most important Mössbauer isotope [137–142]. Figure 4.56 shows [140,142] the nuclear decay of ^{57}Co to ^{57}Fe, which gives the Mössbauer gamma photons, normally used in gamma resonance spectrometry to study samples containing Fe. The decay consists of a spontaneous electron capture transition experienced by the ^{57}Co isotope which produces ^{57}Fe metastable nuclei, which in sequence decays to their ground states by means of a gamma-ray cascade, comprising the Mössbauer gamma ray [140]. The reasons for the importance of a gamma resonance spectrometry, applying a ^{57}Co source included in a solid sample, are [137–142]

1. High abundance of iron in the earth crust
2. Relatively large half-life, $T_{1/2}$, of the ^{57}Co nuclear transition giving the 14.4 keV Mössbauer gamma ray, specifically $T_{1/2}^{57Co}$ 270 days
3. Small recoil energy generated by the 14.4 keV Mössbauer gamma ray, $E_R = 0.002$ eV
4. Relatively high Debye temperature

The first two reasons are self-evident, the last two give rise to a high probability, f, of a recoil-free process in a solid, approximately given by

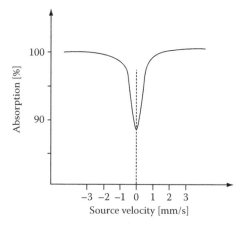

FIGURE 4.55 Mössbauer spectrum in the absence of hyperfine interactions.

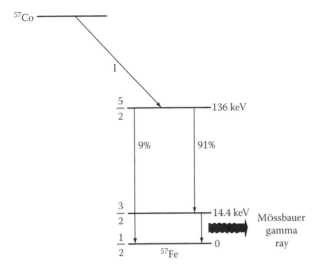

FIGURE 4.56   Decay of ^{57}Co to ^{57}Fe, giving the 14.4 keV Mössbauer gamma ray.

$$f(0) = \exp\left(-\frac{3E_R}{2k\Theta_D}\right) = 0.92$$

a factor which is similar to the Debye–Waller factor previously described.

4.9.3   HYPERFINE INTERACTIONS

A Mössbauer spectrum contains different absorption peaks. The relative intensity, position, and shape of the absorption peaks composing the spectrum are related to the hyperfine interactions affecting the sample nuclei. The ^{57}Co nuclei contained in the source are included in a solid matrix, such as a face-centered cubic (FCC) rhodium, which only affects the position of the emission line but does not split it [140].

4.9.3.1   Chemical or Isomer Shift

All nuclei included in a solid matrix are surrounded by a chemical environment that depends on the composition and structure of the solid. Besides, the ground and excited states of absorbing and emitting nuclei have different volumes. The isomer shift is an effect caused by an interaction which depends on the coulombic electric field exerted by the electric charge in the nucleus. This interaction affects the position of the energy states of the ^{57}Fe nuclei, which gives the Mössbauer gamma ray (see Figure 4.57).

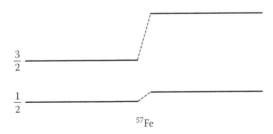

FIGURE 4.57   Effect of the coulombic interaction in the energy levels of ^{57}Fe.

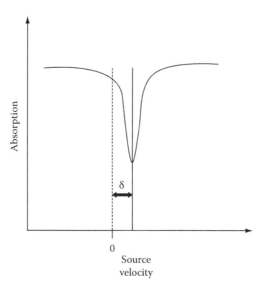

FIGURE 4.58   [57]Fe Mössbauer absorption spectrum resulting from isomer shift.

Consequently, in the absence of other hyperfine interactions, the center of the Mössbauer spectrum (see Figure 4.58) will be displaced with respect to the zero of velocity, $v = 0$, and a quantity, δ, expressed in millimeters per second, called the isomer shift.

The isomer shift is caused because the emitter and the absorber [57]Fe nuclei in the Mössbauer experiment have different electronic charge densities at the position of the nucleus [137–140]. Consequently, we get a small shift in the difference of the energy levels given by [137–140]

$$\Delta E_{i-s} = \Delta E_A - \Delta E_S = \frac{1}{10\varepsilon_0} Ze^2(R_1^2 - R_0^2)\left[\left|\varphi_e^A(0)\right|^2 - \left|\varphi_e^S(0)\right|^2\right]$$

where $\left|\varphi_e^A(0)\right|^2$ and $\left|\varphi_e^S(0)\right|^2$ are the probability densities of the electrons surrounding the nucleus at the origin, R_1 and R_0 are the radius of the first excited and ground states of the nucleus, and Z is the nuclear charge. Consequently, the isomer shift, δ, is given by [140]

$$\delta = \alpha\left[\left|\varphi_e^A(0)\right|^2 - \left|\varphi_e^S(0)\right|^2\right]$$

for [57]Fe, $\alpha < 0$, because the radius of the exited state is lower than the radius of the ground state.

The isomer shift gives information about electron density at the nucleus, the valence of the tested atom, and the structure of the environment of the surroundings of the tested atom.

4.9.3.2 Quadrupole Splitting

In the previous section, it was assumed that the distribution of the nuclear charge is spherical [140]. However, the charge distribution of a nucleus is not always spherically symmetric. In fact, this is not the case for a nucleus with nuclear angular momentum $I > 1/2$; in this case, the nucleus shows nonspherical nuclear charge distributions [142,143]. The electrostatic potential created by a charge distribution localized inside a radius, $|\vec{r}| < R$, can be expressed outside the sphere of radius, R, in rectangular coordinates as follows [144]:

$$\Phi(\overline{r}) = \frac{1}{4\pi\varepsilon_0}\left(\frac{q}{|\overline{r}|} + \frac{\overline{p}\cdot\overline{r}}{|\overline{r}|^3} + \frac{1}{2}\sum_{i,j}Q_{i,j}\frac{x_i x_j}{|\overline{r}|^5}\right)$$

where $\overline{r} = x_1\overline{i} + x_2\overline{j} + x_3\overline{k}$ and the first term is named the monopole term, and is given by q, the total charge of the distribution. The second expressions are the dipole terms, having three components, expressed as the dipole moment vector:

$$\overline{p} = \int \overline{r}'\rho(\overline{r}')d^3r'$$

The third expressions are the quadrupole terms, with nine components, that is, a second rank tensor:

$$Q_{i,j} = \int \rho(\overline{r})(\,3x_i x_j - |r|^2\delta_{i,j})d^3r$$

which are the terms of the quadrupole moment tensor, and which can be reduced to a diagonal matrix characterized by only the diagonal terms $Q_{1,1}$, $Q_{2,2}$, and $Q_{3,3}$ [80].

The quadrupole moment tensor of the nuclear charge distribution is characterized by one term, that is, $Q_{3,3}$, since $Q_{1,1} = Q_{2,2} = -\frac{1}{3}Q_{3,3}$ [144]. Then, the quadrupole moment, Q, of a nuclear state, in units of area, that is, barns (1 barn $= 10^{-28}$ m^2), is expressed as follows [143]

$$Q = \frac{Q_{3,3}}{e} = \frac{1}{e}\left[\int \rho(\overline{r})(3z^2 - |r|^2)d^3r\right] = \frac{1}{e}\left[\int \rho(\overline{r})|r|^2(3\cos^2\theta - 1)d^3r\right]$$

where
  $\rho(\overline{r})$ is the nuclear electric charge distribution
  θ is the polar angle, which is measured with respect to the nuclear spin direction

In quadrupole splitting, the existence of a nonspherical nuclear charge distribution produces an electric quadrupole moment, Q, which indicates that the charge distribution in the nucleus is prolate, when $Q > 0$, or oblate, if $Q < 0$ [137–140].

On the other hand, in some materials which have a symmetry less than cubic, an electric field gradient (EFG) due to the concrete charge distribution in the material can be present [137–140]. If a localized charge distribution is immersed in an external electric field with potential $\Phi(\overline{r})$, the electrostatic energy of the system is [144]

$$E_V = \int \Phi(\overline{r})\rho(\overline{r})d^3r$$

If we make the Taylor expansion of the potential,

$$\Phi(\overline{r}) = \Phi(0) + \overline{r}\bullet\nabla\Phi(0) + \frac{1}{2}\sum_{i,j}x_i x_j\frac{\partial^2\Phi}{\partial x_i\partial x_j}(0) + \cdots$$

Using the definition of electric field, $\bar{E} = -\nabla\Phi$, then

$$\Phi(\bar{r}) = \Phi(0) - \bar{r} \cdot \bar{E}(0) + \frac{1}{2}\sum_{i,j} x_i x_j \frac{\partial E_j}{\partial x_i}(0) + \cdots$$

Now, $\nabla \cdot \bar{E} = 0$ for the external field, it is the possible to subtract $\frac{1}{6}|\bar{r}|^2 \nabla \cdot \bar{E}$ in the previous equation to get

$$\Phi(\bar{r}) = \Phi(0) - \bar{r} \cdot \bar{E}(0) + \frac{1}{6}\sum_{i,j}(3x_i x_j - |\bar{r}|^2)\frac{\partial E_j}{\partial x_i}(0) + \cdots$$

Consequently,

$$E_V = \int \Phi(\bar{r})\rho(\bar{r})d^3r = q\Phi(0) - p \cdot \bar{E}(0) - \frac{1}{6}\sum_i\sum_j Q_{i,j}\frac{\partial E_j}{\partial x_i}(0) + \cdots$$

For nuclear systems with a suitable cylindrical symmetry, the quadrupole interaction energy, E_V^Q, is given by [143]

$$E_V^Q = \frac{1}{4}eQ\left(\frac{3}{2}\cos^2\theta - \frac{1}{2}\right)\left(\frac{\partial^2\Phi}{\partial z^2}\right)_{z=0}$$

where the angle θ is between the symmetry axis of the external electric field and the symmetry axis of the nuclear spin.

Since the transitions involved in gamma absorption resonance are quantum transitions, the quadrupole splitting of the nucleus can be described as follows [143]:

$$E_V^Q = eQ\left(\frac{3m_I^2 - I(I+1)}{4I(2I+1)}\right)\left(\frac{\partial^2\Phi}{\partial z^2}\right)_{z=0}$$

The study of quadrupole interactions in a Mössbauer experiment can be applied as an effective means to describe different atomic sites in a particular material and get information about local valence, coordination type, and symmetry of structural centers in solids, among other electronic, structural, and magnetic properties. Therefore, if the symmetry of the structure of the solid matrix surrounding the nucleus of the test atom in the absorber is less than cubic, it follows that the EFG of the electric charge distribution, which is a second rank tensor, has components different from zero. For this reason a quadrupole interaction between the nonspherical nuclear charge distribution of the $I = 3/2$ state of the ^{57}Fe nucleus and the EFG results [137–140]. This interaction provokes a splitting of the $I = 3/2$ nuclear energy level (see Figure 4.59) [137–140].

In Figure 4.59, it is evident that the quadrupole splitting is added to the isomer shift, that is, the center of the Mössbauer spectrum (see Figure 4.60) will be displaced with respect to the zero of velocity, $v = 0$, and a quantity, δ, expressed in millimeters per second, also called the isomer shift [119–121]. The quadrupole spitting is expressed as two absorption lines (see Figure 4.60) separated by a quantity, Δ, called the quadrupole splitting [137–140].

The quadrupole splitting gives information about the EFG at the nucleus, the valence of the tested atom, and the structure of the environment of the surroundings of the tested atom.

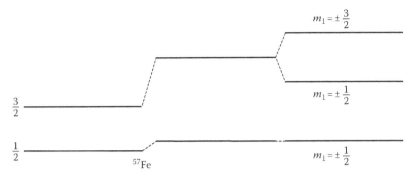

FIGURE 4.59   Effect of the quadrupole interaction in the energy levels of ^{57}Fe.

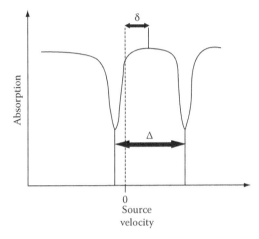

FIGURE 4.60 ^{57}Fe Mössbauer absorption spectrum resulting from quadrupole splitting.

4.9.3.3   Magnetic Splitting

As previously stated, the energy of a nuclear dipole moment in an external magnetic field, $\bar{B}$, is given by

$$E = \bar{\mu} \cdot \bar{B} = -\gamma \bar{I} \cdot \bar{B} = \gamma |\bar{I}| \cdot |\bar{B}| \cos\theta = -\gamma |\bar{B}| m_I \hbar$$

where θ is the angle between $\bar{I}$ and $\bar{B}$, $|\bar{I}| \cos\theta = m_I$, and $\hbar = h/2\pi$. This energy can be considered as a perturbation of the system's energy in the presence of the magnetic field. Similar to some materials, an external magnetic field is present at the ^{57}Fe nucleus. Then, with the help of the above relation, it is possible to calculate the corresponding magnetic energy shift, E_{m_I}. So, the $I = 1/2$ ground state splits into two levels, while the $I = 3/2$ excited state splits into four levels (see Figure 4.61). That is, if the ^{57}Fe nucleus is placed in a magnetic field, then a dipole magnetic interaction is developed [140]. This interaction will remove the degeneracy in the nuclear states with a nuclear angular momentum $I > 0$ and the nuclear energy states split into $2I + 1$ states (see Figure 4.61) [137–140].

Since the selection rule for this transition is $\Delta m_I = 0, \pm 1$, the Mössbauer spectrum will show six absorption lines (see Figure 4.62)[137–140].

Magnetic hyperfine interactions take place in magnetic materials. This effect can be applied to the investigation of the magnetic ordering of materials, the character of the magnetic interactions, the size of the magnetic moment on specific atoms, and some aspects of the electronic structure of the tested atom whose nucleus experiences the magnetic interaction [140].

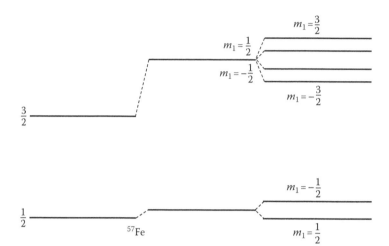

FIGURE 4.61   Effect of the magnetic interaction in the energy levels of ^{57}Fe.

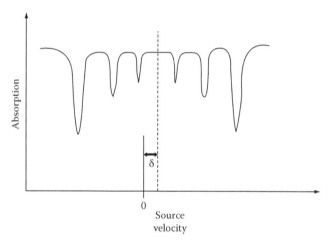

FIGURE 4.62 ^{57}Fe Mössbauer absorption spectrum resulting from magnetic splitting.

4.9.4   APPLICATIONS OF ^{57}FE MÖSSBAUER SPECTROMETRY

Mössbauer spectrometry is a powerful means for the elucidation of the state of iron in materials [44,138–142,145]. Figure 4.63 [44] shows the ^{57}Fe Mössbauer spectra of the natural zeolite rocks, such as MP, C2, C1, and C4 (see Table 4.1). In Table 4.12, the Mössbauer parameters calculated with the help of the numerical resolution of the spectra presented in Figure 4.63 are reported. That is, with the help of the recorded spectra, the accurate peak positions, integrated intensities, as well as the FWHM of each peak were calculated. This calculation was carried out by fitting the spectra with three quadrupole doublets: one for site 1, another for site 2, and a last one for site 3 [44]. The peaks were simulated with Gaussian functions and the fitting process for the numerical resolution of the spectra was carried was carried out with a peak separation and analysis software, developed for this purpose [44,145] based on a least square procedure [48].

From the obtained data, it was possible to conclude that iron found in the natural zeolite is high-spin Fe^{3+} in place of Al^{3+} in the framework tetrahedral sites. Besides, Fe^{3+} is located in extra-framework octahedral sites as $Fe(H_2O)_6^{3+}$. Finally, Fe^{2+} is present in the octahedral coordination in extra-framework sites or in other aluminosilicates present in the zeolite rock [44].

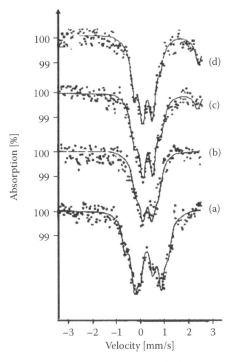

FIGURE 4.63 ^{57}Fe Mössbauer spectra: transmission [%] vs. velocity [mm/s] of natural zeolite rock samples (a) MP, (b) C2, (c) C1, and (d) C4.

TABLE 4.12
Mössbauer Parameters Obtained from the Numerical Resolution of the Spectra Shown in Figure 4.63

Sample	δ_1 [mm/s]	δ_2 [mm/s]	δ_3 [mm/s]	Δ_1 [mm/s]	Δ_2 [mm/s]	Δ_3 [mm/s]	A_1 [%]	A_2 [%]	A_3 [%]
MP	0.66	0.62	0	1.23	0.44	0	20	80	0
C2	0.65	0.62	1.13	1.07	0.46	2.82	32	57	11
C1	0.61	0.62	1.42	1.20	0.42	2.62	25	61	14
C4	0.62	0.60	1.39	1.08	0.40	2.74	16	59	25

Figure 4.64 [145] shows the ^{57}Fe Mössbauer spectra of intermediary phases resulting from the hydrothermal treatment of Na-clinoptilolite in 7.5 M NaOH solution at 100°C as a function of time (in hours), that is, 0, 8, 16, 36, and 64 h [83,145] (see Section 3.4.3). In Table 4.13, the Mössbauer parameters calculated with the help of the numerical resolution of the spectra presented in Figure 4.64 are reported. The spectra were assessed with the help of the results previously described for natural zeolites [44]. That is, two sites were proposed: the octahedral site 1 and the tetrahedral site 2. With the help of the obtained Mössbauer spectra, the precise peak position, integrated intensities, as well as the FWHM of all peaks were calculated by fitting them with two doublets caused by the quadrupole splitting for site 1 and the other for site 2. The peaks were again replicated with Gaussian functions and the fitting process for the numerical resolution of the spectra was carried out with the procedure described elsewhere [44,145], which is based on the least square method [48].

These samples were previously studied with the help of MAS-NMR (see Section 4.6.6). It was shown that at the beginning of the hydrothermal transformation, the spectrum consists of a single,

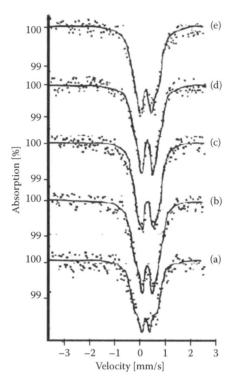

FIGURE 4.64 ^{57}Fe Mössbauer spectra: transmission [%] vs. velocity [mm/s] of samples of Na-clinoptilolite hydrothermally treated in 7.5 M NaOH solution at 100°C, as a function of time in hours, that is, (a) 0 h, (b) 8 h, (c) 16 h, (d) 36 h, and (e) 64 h.

TABLE 4.13
Mössbauer Parameters Obtained from the Numerical Resolution of the Spectra Shown in Figure 4.64

Time [h]	δ_1 [mm/s]	δ_2 [mm/s]	Δ_1 [mm/s]	Δ_2 [mm/s]	A_1 [%]	A_2 [%]
0	0.56	0.55	0.93	0.35	40	60
8	0.58	0.59	1.01	0.43	34	66
16	0.61	0.59	0.90	0.53	42	58
36	0.62	0.60	1.06	0.47	27	73
64	0.59	0.59	0.94	0.42	38	62

broad low-field-shifted characteristic, like that of the natural zeolite used as raw material for the synthesis. In Figure 4.64, the spectrum at $t = 0$ is as well characteristic of the natural zeolite [145]. As the hydrothermal treatment proceeds, the spectra show a dependence on the decrease of the Si/Al ratio in the zeolite [83,145], shown with the help of MAS-NMR.

Figure 4.65 shows the ^{57}Fe Mössbauer transmission spectra of thin iron foils oxidized with nitric oxide (NO) at 500°C and 600°C [146]. In the spectra, the lines corresponding to the position of the Mössbauer transmission spectra of Fe, Fe_2O_3, and Fe_3O_4 are indicated.

The obtained results indicate that the oxidation product was Fe_3O_4 under the same conditions; the oxidation with O_2 gives Fe_2O_3 as the oxidation product, because the oxygen molecule has twice the amount of oxygen in comparison to the NO molecule [146].

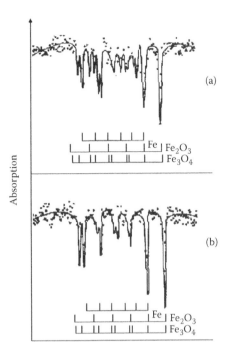

FIGURE 4.65 ^{57}Fe Mössbauer spectra: transmission vs. velocity [mm/s] of Fe thin foils oxidized with NO at (a) 500°C and (b) 600°C.

4.10 MERCURY POROSIMETRY

Mercury porosimetry is the most suitable method for the characterization of the pore size distribution of porous materials in the macropore range that can as well be applied in the mesopore range [147–155]. To obtain the theoretical foundation of mercury porosimetry, Washburn [147] applied the Young–Laplace equation

$$\Delta P = \gamma \left(\frac{1}{r_\mathrm{I}} + \frac{1}{r_\mathrm{II}} \right) \qquad (4.55)$$

where
   ΔP is the pressure difference in a sector, across the interface between two neighboring phases
   γ is the surface tension
   r_I and r_II are the two curvature radii describing the surface [148]

In this regard, Washburn [147,150] supposed that for cylindrical pores of equivalent radius, r, wetted with a liquid of contact angle, θ [149] (see Figure 4.66), a liquid/gas interface enclosed inside a pore will at equilibrium take on the shape of the uniform average curvature. That is, in a uniform cylindrical pore of radius r or in a parallel-sided slit of width r, the mean curvature, $1/\bar{r}$, is equal to [148]

$$\frac{1}{\bar{r}} = -\frac{2\cos\theta}{r}$$

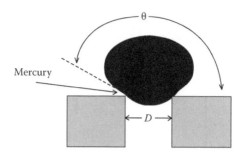

FIGURE 4.66   Mercury making contact with a cylindrical pore.

The pressure difference across the interface between two contiguous phases is given by

$$\Delta P = -\frac{2\gamma\cos\theta}{r} \tag{4.56}$$

an expression named the Washburn equation, which provides an appropriate relationship between the applied pressure and the pore size, whose mathematical simplicity is related to the selection of the cylindrical pore geometry to avoid the complications of dealing with pores of irregular cross sections [150].

Contrary to gas adsorption and pore condensation, in which the pore fluid wets the pore walls, that is, the contact angle is in the range of $0 < \theta < \pi/2$ $\pi/2$; mercury is a nonwetting fluid, that is, the contact angle is in the range of $\pi/2 < \theta < \pi$ [151]. Consequently, for a nonwetting liquid, it is necessary to apply a positive excess of hydrostatic pressure, ΔP, to allow the fluid to penetrate the pores of radius, r.

Mercury is the only fluid appropriate for a porosimetry-type measurement, given it does not wet the majority of materials and cannot penetrate pores by capillary action. The surface tension and the contact angle for pure mercury reported in the literature vary. Values of $\gamma = 484\,mN/m$ and $\theta = 141°$ and $\gamma = 480\,mN/m$ and $\theta = 140°$ [153] are applied; even a value of $485\,mN/m$ is usually used, and a value of $139°$ is also recommended [156]. Afterward, for a pressure span of $0.01-200\,MPa$, the corresponding cylindrical pore radii will be found between $75\,\mu m$ and $3.5\,nm$ [148], that is, a broad range including macro- and mesopores. With the help of mercury porosimetry, the study of particle size distributions, tortuosity factors, permeability, fractal dimension, and compressibility, and pore shapes and network effects in any porous solid is feasible [151].

To perform the mercury porosimetry test, some effects which influence the measurement should be taken into consideration. In this regard, at high pressures, a blank correction (i.e., subtracting an empty cell run from an actual test) is frequently considered as a methodology to correct for the apparent volume intruded due to compression, to counterweigh the compressibility of mercury, and for the elastic distortion of the cell and other component parts [148,149]. In addition, the high hydraulic pressure applied on the material during the intrusion can, in some situations, cause structural damage to the test specimen; subsequently, the mercury intrusion methodology can undervalue the pore volume and overvalue the pore size, if the standard Washburn equation is used [152,155]. A satisfactory regulation of temperature is also needed during the experiment, since the pressure inside a liquid-filled cell is notably temperature dependent [148,150]. An additional important characteristic of mercury porosimetry curves is the appearance of hysteresis between the intrusion and extrusion branches, that is, mercury extrusion curves do not overlap intrusion curves [151,152,154]. Finally, a significant feature is the entrapment of mercury in the porous network after extrusion [151,152,154].

The specific method for the determination of the pore size distribution with a mercury porosimeter is by determining the amount of nonwetting mercury intruded into the pores of the test sample,

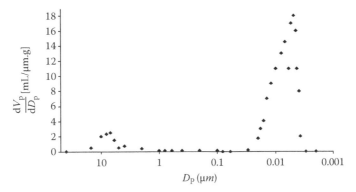

FIGURE 4.67   Differential intrusion of mercury (dV_p/dD_p) vs. log D_p plot for a sample of a highly porous open-cell polymer [poly-(HIPE)].

as a function of increasing applied pressure, and calculating afterward the pore size distribution curves from this information, by applying the Washburn equation [149–151]. To carry out the pore size analysis in a mercury porosimeter, at first, the gas contained in the sample cell is evacuated. Mercury cannot intrude into the sample when pores are filled with another liquid, and thereafter mercury is transferred into the sample cell under vacuum, and, subsequently, pressure is applied to force mercury into the test sample. Throughout the process, the applied pressure, P, normally measured in megapascals or pounds per square inch, and the intruded volume of mercury, V, measured in cubic centimeters, are monitored during the measurement, and an intrusion–extrusion curve is obtained. That is, with the help of this process, a pore size distribution curve is obtained and is represented by plotting $\Delta V_p/\Delta \log D_p$ versus log D_p and/or $\Delta V_p/\Delta D_p$ versus log D_p (see Figure 4.67), where V_p is the accumulated volume of mercury intruded in the pore volume up to the pore of width D_p.

The poly-[HIPE]* sample intrusion mercury porosimetry study reported in Figure 4.67 was carried out in a Micromeritics, Atlanta, GA, USA, AutoPore IV-9500 automatic mercury porosimeter.† The sample holder chamber was evacuated up to 5×10^{-5} Torr; the contact angle and surface tension of mercury applied by the AutoPore software in the Washburn equation to obtain the pore size distribution was 130° and 485 mN/m, respectively. Besides, the equilibration time was 10 s, and the mercury intrusion pressure range was from 0.0037 to 414 MPa, that is, the pores size range was from 335.7 to 0.003 μm. The poly-(HIPE) sample was prepared by polymerizing styrene (90%) and divinylbenzene (10%) [157].

4.11   MAGNETIC FORCE IN NONUNIFORM FIELDS PHASE ANALYSIS METHOD

The measurement of the magnetic force exerted by a magnetic field gradient on a magnetic phase, included in a powder mixture of a material or mineral, is a useful phase analysis method to determine the concentration of this magnetic phase. This methodology is particularly helpful in the case where the concentration of the magnetic phase is below the range of detection of XRD, and even Mössbauer spectrometry, that is, below about 2 wt % [44].

The equipment to carry out this test consists of a pair of Helmholtz coils, which creates the nonuniform magnetic field, and a small, nonmagnetic sample holder, included between the coils and

---

* The author gratefully acknowledges Dr. Michael Silverstein, Israel Institute of Technology, Haifa, Israel, for kindly providing the poly-[HIPE] sample [157].

† The author express his acknowledgment to Luis Roberto Cordero (TTC Analytical Service Corp. Caguas, PR, USA.), Michael L. Strickland (Micromeritics Instrument Corporation), and Micromeritics Analytical Services for the mercury porosimetry intrusion study of the poly-[HIPE] sample.

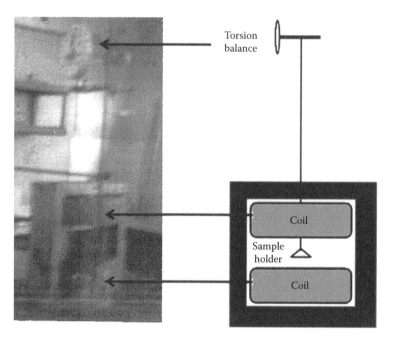

FIGURE 4.68 Magnetic force measurement facility.

suspended by a tiny cable connected to the torsion balance bar (Figure 4.68) [44]. The measurement is based on the well-known fact that the axial magnetic field gradient of a pair of Helmholtz coils, dB/dz, interacts with the magnetic dipole moment of the sample included in the sample holder, generating a force. The torsion balance to measure this force is composed of a wire attached at one end, and set in a form so that a force applied by a bar attached to the wire at the other end twists the wire. In this condition, when a force is applied at right angles to the mobile end, the bar will turn the wire up to equilibrium, where the torque of the wire equilibrates the applied force. Subsequently, the amount of the force is proportional to the twisting angle measured by a goniometer placed in front of the balance. Then, the measurement of the bar twist can be applied to calculate the magnetic force acting on the sample holder.

This method is applied for the determination of the amount of magnetite present in the natural zeolite rocks C2 and C4 (see Table 4.1). The magnetite present in this zeolite was magnetically separated from the grounded samples and identified by XRD and Mössbauer spectrometry [44].

The home-made arrangement for the measurement of the force in a magnetic field gradient consists of two Helmholtz coils (19,000 turns each) with a field strength of 0.15 T, a magnetic field gradient of 2×10^{-4} T/m, and a torsion balance which allows the measurement of forces within $\pm 2 \times 10^{-5}$ N (Figure 4.68) [44]. The magnetic force measurement facility was calibrated with a mixture prepared from a zeolite free from magnetic phases and then the magnetite extracted from the tested zeolites [44].

REFERENCES

1. E.N. Kaufmann, (editor), *Characterization of Materials*, Vols. I and II, John Wiley & Sons, New York, 2003.
2. P.E.J. Flewitt and R.K. Wild, *Physical Methods for Materials Characterization*, Taylor & Francis LTD, London, 1994.
3. A. Guinier, *X-Ray Diffraction in Crystals, Imperfect Crystals and Amorphous Bodies*, Dover Publications Inc., New York, 1994.
4. B.D. Cullity and S.R. Stock, *Elements of X-Ray Diffraction* (3rd edition), Prentice Hall, Upper Saddle River, NJ, 2001.

5. M.M.J. Treacy and J.B. Higgins, *Collection of Simulated XRD Powder Patterns for Zeolites* (4th edition), Elsevier, Amsterdam, the Netherlands, 2001.
6. M. Mashlan, M. Miglierini, and P. Schaaf, (editors), *Material Research in Atomic Scale by Mossbauer Spectroscopy*, Springer-Verlag, New York, 2003.
7. D.B. Williams and C.B. Carter, *Transmission Electron Microscopy*, Plenum Press, New York, 1996.
8. M. Staniforth, J. Goldstein, P. Echlin, E. Lifschin, and D. Newbury, *Scanning Electron Microscopy and X-Ray Microanalysis*, Springer-Verlag, New York, 2003.
9. M.E. Orazem and B. Tribollet, *Electrochemical Impedance Spectroscopy*, John Wiley & Sons, New York, 2008.
10. K. Nakamoto, *Infrared and Raman Spectra of Inorganic and Coordination Compounds: Part A: Theory and Applications in Inorganic Chemistry*, John Wiley & Sons, New York, 1997.
11. D.A. Long, *The Raman Effect*, John Wiley & Sons, New York, 2001.
12. P.W. Atkins, *Physical Chemistry* (6th edition), W.H. Freeman & Co., New York, 1998.
13. P.T. Callaghan, *Principles of Nuclear Magnetic Resonance*, Clarendon Press, Oxford, 1991.
14. R. Jenkins, (editor), *X-Ray Fluorescence Spectrometry*, John Wiley & Sons, New York, 1999.
15. R. Roque-Malherbe, in *Handbook of Surfaces and Interfaces of Materials*, Vol. 2, H.S. Nalwa, (editor), Academic Press, New York, Chapter 13, 2001, p. 509.
16. R.F. Speyer, *Thermal Analysis of Materials*, Marcel Dekker, New York, 1993.
17. P. Gabhot, (editor), *Principles and Applications of Thermal Analysis*, John Wiley & Sons, New York, 2008.
18. A. Jones and B. Mac Nicol, *Temperature-Programmed Reduction for Solid Materials Characterization*, Marcel Dekker, New York, 2001.
19. P. Malet and A. Caballero, *J. Chem. Soc. Faraday Trans.*, 84, 2369 (1988).
20. F. Rouquerol, J. Rouquerol, and K. Sing, *Adsorption by Powder Porous Solids*, Academic Press, New York, 1999.
21. R. Roque-Malherbe, *Adsorption and Diffusion in Nanoporous Materials*, CRC Press/Taylor & Francis, Boca Raton, FL, 2007.
22. M. Birkholz, *Thin Film Analysis by X-Ray Scattering*, John Wiley & Sons, New York, 2006.
23. H.M. Rietveld, *Acta Crystallogr.*, 2, 65 (1969).
24. R. Jenkins and R.L. Snyder, *Introduction to X-Ray Powder Diffractometry*, John Wiley & Sons, New York, 1996.
25. B.E. Warren, *X-Ray Diffraction*, Dover Publications, New York, 1990.
26. Ch. Kittel, *Introduction to Solid State Physics* (8th edition), John Wiley & Sons, New York, 2004.
27. H.P. Myers, *Introduction to Solid State Physics* (2nd edition), CRC Press, Boca Raton, FL, 1997.
28. J.I. Gersten and F.W. Smith, *The Physics and Chemistry of Materials*, John Wiley & Sons, New York, 2001.
29. R.A. Young, in *The Rietveld Method*, R.A. Young, (editor), International Union of Crystallography, Oxford University Press, Oxford, 1993, p. 1.
30. C. Gonzales, R. Roque-Malherbe, and E.D. Shchukin, *J. Mater. Sci. Lett.*, 6, 604 (1987).
31. C. Gonzales and R. Roque-Malherbe, *Acta Crystallogr.*, A43, 622 (1987).
32. S. Nieto, R. Polanco, and R. Roque-Malherbe, *J. Phys. Chem. C*, 111, 2809 (2007).
33. J. Wu, L.P. Li, W.T.P. Espinosa, and S.M. Haile, *J. Mater. Res.*, 19, 2366 (2004).
34. K.S. Knight, *Solid State Ionics*, 74, 109 (1994).
35. K. Williamson and W.H. Hall, *Acta Metall.*, 1, 222 (1953).
36. H.P. Klug and L.E. Alexander, *X-Ray Diffraction Procedures for Crystalline and Amorphous Solids* (2nd edition), John Wiley & Sons, New York, 1974.
37. J. Leroux, D.H. Lennox, and K. Kay, *Anal. Chem.*, 25, 740 (1953).
38. R. Roque-Malherbe, A. Dago, and C. Diaz, *Rev. Cubana Fis.*, 3, 105 (1983); *Chemical Abstracts*, Vol. 101, No. 182704.
39. C.R. Gonzalez and R. Roque-Malherbe, *KINAM*, 5, 67 (1983).
40. C.R. Gonzalez. PhD dissertation, Laboratory of Zeolites, National Center for Scientific Research, Havana, Cuba, 1986.
41. R. Roque-Malherbe, W. del Valle, F. Marquez, J. Duconge, and M.F.A. Goosen, *Separ. Sci. Technol.*, 41, 73 (2006).
42. A. Beiser, *Concepts of Modern Physics* (6th edition), McGraw-Hill Education, New York, 2003.
43. A.W. Hull, *J. Amer. Chem. Soc.*, 41, 1168 (1919).
44. R. Roque-Malherbe, C. Díaz, E. Reguera, J. Fundora, L. López-Colado, and M. Hernández-Vélez, *Zeolites*, 10, 685 (1990).

45. M. Hernández-Vélez, O. Raymond, A. Alvarado, A. Jacas, and R. Roque-Malherbe, *J. Mater. Sci. Lett.*, 14, 1653 (1995).

46. R. Roque-Malherbe, R. Lopez-Cordero, J.A. Gonzales-Morales, J. de Onate-Martinez, and M. Carreras-Gracial, *Zeolites*, 13, 481 (1993).

47. R. Roque-Malherbe and F. Marquez-Linares, *Surf. Interf. Anal.*, 37, 393 (2005).

48. N.R. Draper and H. Smith, *Applied Regression Analysis* (3rd edition), John Wiley & Sons, New York, 1998.

49. R.E. Reed-Hill and R. Abbaschian, *Physical Metallurgy Principles* (3rd edition), International Thomson Publishing, Boston, MA, 1994.

50. R. Roque-Malherbe and J. Buttner, *Revista CNIC*, 6, 1 (1975); *Chemical Abstracts*, Vol. 84, No. 93517w.

51. G. Rodríguez, C. Lariot-Sanchez, J.C. Romero, and R. Roque-Malherbe, *Stud. Surf. Sci. Catal.*, 25, 275 (1985).

52. J.I. Goldstein, D.E. Newbury, P. Echlin, D.C. Joy, C. Fiori, and E. Lifshin, *Scanning Electron Microscopy and X-ray Microanalysis*, Plenum Publishing Co., New York, 1981.

53. R. Roque-Malherbe, W. del Valle, J. Ducongé, and E. Toledo, *Int. J. Environ. Pollut.*, 31, 292 (2007).

54. W.E. Smith and G. Dent, *Modern Raman Spectroscopy. A Practical Approach*, John Wiley & Sons, New York, 2005.

55. D.A. Skoog, F.J. Holler, and T.A. Nieman, *Principles of Instrumental Analysis* (5th edition), Saunders College Publishing, Philadelphia, PA, 1998.

56. R. Aroca, *Surface Enhanced Vibrational Spectroscopy*, John Wiley & Sons, New York, 2006.

57. U.P. Agarwal and R.H. Atalla, in *Surface Analysis of Paper*, T.E. Conners and S. Banerjee, (editors), CRC Press, Boca Raton, FL, 1995, p. 152.

58. L.J. Bellamy, *The Infrared Spectra of Complex Molecules* (3rd edition), Chapman & Hall, London, Vol. I, 1975; Vol. II, 1980.

59. A. Messiah, *Quantum Mechanics*, North-Holland Publishing Co., New York, 1961.

60. R.L. Carter, *Molecular Symmetry and Group Theory*, John Wiley & Sons, New York, 1997.

61. P.R. Griffith and J.A. de Haseth, *Fourier Transform Infrared Spectrometry*, John Wiley & Sons, New York, 1986.

62. A. Dixit, PhD dissertation, University of Puerto Rico, Río Piedras, Puerto Rico, 2003.

63. A. Burneau and J.P. Gallas, in *The Surface Properties of Silicas*, A.P. Legrand, (editor), John Wiley & Sons, New York, 1998, p. 147.

64. R. Roque-Malherbe, F. Marquez, W. del Valle, and M. Thommes, *J. Nanosci. Nanotechnol.*, 8, 5993 (2008).

65. R. Roque-Malherbe and V. Ivanov, *Micropor. Mesopor. Mat.*, 47, 25 (2001).

66. R. Roque-Malherbe, R. Wendelbo, A. Mifsud, and A. Corma, *J. Phys. Chem.*, 99, 14064 (1995).

67. F. Marquez, R. Roque-Malherbe, J. Duconge, and W. del Valle, *Surf. Interface Anal.*, 36, 1060 (2004).

68. M. Fuentes, J. Magraner, C. de las Pozas, R. Roque-Malherbe J. Pérez-Pariente, and A. Corma, *App. Catal.*, 47, 367 (1989).

69. P.J. Hore, *Nuclear Magnetic Resonance*, Oxford University, Oxford, UK, 1995.

70. J.H.H. Nelson, *Nuclear Magnetic Resonance Spectroscopy*, Prentice Hall, Upper Saddle River, NJ, 2002.

71. R.K. Harris, E.D. Becker, S.M. Cabral de Menezes, R. Goodfellow, and P. Granger, *Pure Appl. Chem.*, 73, 1795 (2001).

72. H. Günther, *NMR Spectroscopy* (2nd edition), John Wiley & Sons, New York, 2001.

73. I.N. Levine, *Physical Chemistry* (5th edition), McGraw-Hill, Boston, MA, 2002.

74. R. Reif, *Fundamentals of Statistical and Thermal Physics*, McGraw-Hill, Boston, MA, 1965.

75. D.A. McQuarrie, *Statistical Mechanics*, University Science Books, Sausalito, CA, 2000.

76. A.C. Larsson, PhD thesis, Lulea University of Technology, Department of Chemical Engineering and Geosciences, Division of Chemistry, Porson, Lulea, Sweden, 2004.

77. J.P. Hornak, *The Basics of Nuclear Magnetic Resonance*, www.cis.rit.edu/htbooks/ nmr/inside.htm, 2002.

78. B. Cowan, *Nuclear Magnetic Resonance and Relaxation*, Cambridge University Press, Cambridge, UK, 2005.

79. M.L. Boas, *Mathematical Methods in the Physical Sciences*, John Wiley & Sons, New York, 1966.

80. G.B. Arfken and H.J. Weber, *Mathematical Methods for Physicists* (5th edition), Academic Press, New York, 2001.

81. E. Lipma, E. Magi, A. Somoson, M. Tormak, and A.R. Grimer, *J. Amer. Chem. Soc.*, 102, 4889 (1980).

82. E.O. Stejskal and J.D. Memory, *High Resolution NMR in the Solid State: Fundamentals of CP/MAS*, Oxford University Press, New York, 1994.

83. C. de las Pozas, D. Díaz-Quintanilla, J. Pérez-Pariente R. Roque-Malherbe, and M. Magi, *Zeolites*, 9, 33 (1989).

84. R. Roque-Malherbe, J. Oñate, and J. Fernandez-Bertran, *Solid State Ionics*, 34, 193 (1991).

85. C. de las Pozas, W. Kolockiewics, and R. Roque-Malherbe, *Microporous Mater.*, 5, 325 (1996).

86. W.W. Wendlandt, *Amer. Lab.*, 9, 59 (1977).

87. M.E. Brown, *Introduction to Thermal Analysis: Techniques and Applications* (2nd edition), Springer-Verlag, New York, 2001.

88. A. Montes, R. Roque-Malherbe, and E.D. Shchukin, *J. Thermal Anal.*, 31, 41 (1986).

89. V. Lopez, F. Marquez-Linares, C. Morant, C. Domingo, E. Elizalde, F. Zamora, and R. Roque-Malherbe, *Nano Today* (submitted).

90. G. Hohne, W. F. Hemminger, and H.J. Flammersheim, *Differential Scanning Calorimetry* (2nd edition), Springer-Verlag, New York, 2003.

91. S. Besselmann, C. Freitag, O. Hinrichsen, and M. Muhler, *Phys. Chem. Chem. Phys.*, 3, 4633 (2001).

92. C. de las Pozas, R. López-Cordero, C. Díaz-Aguilas, M. Cora, and R. Roque-Malherbe, *J. Sol. State Chem.*, 114, 108 (1995).

93. J.L. Falconer and J.A. Schwarz, *Catal. Rev. Sci. Eng.*, 25, 141 (1983).

94. R.J. Cvetanovic and Y. Amenomiya, *Adv. Catal.*, 17, 103 (1967).

95. C. de las Pozas, R. López-Cordero, J.A. González-Morales, N. Travieso, and R. Roque-Malherbe, *J. Mol. Catal.*, 83, 145 (1993).

96. P. Malet, *Stud. Surf. Sci. Catal.*, 57B, 333 (1990).

97. R. Roque Malherbe and R. Wendelbo, *Thermochim. Acta*, 400, 165 (2003).

98. R. Wendelbo and R. Roque-Malherbe, *Mic. Mat.*, 10, 231 (1997).

99. H.G. Karge, and W. Niessen, *Catal. Today*, 8, 451 (1991).

100. Z. Stoynov and D. Vladikova, *Differential Impedance Analysis*, Akademicno Izdatelstvo, Sofia, Bulgaria, 2005.

101. E. Barsoukov and J.R. Macdonald, (editors), *Impedance Spectroscopy: Theory, Experiment, and Applications* (2nd edition), John Wiley & Sons, New York, 2005.

102. E. Schosler and A. Schönhals, *Colloid Polym. Sci.*, 267, 963 (1989).

103. A.K. Jonscher, *Dielectric Relaxation in Solids*, Chelsea Dielectric Press, London, 1983.

104. K.C. Kao, *Dielectric Phenomena in Solids*, Elsevier, Amsterdam, the Netherlands, 2004.

105. R. Roque-Malherbe, R. Roque and F. Morales, Patent Certificate No. 21059, ONIITEM, Havana, Cuba (1982).

106. R. Roque-Malherbe and F. Morales, *Rev. CNIC: Ciencias Quimicas*, 12, 235 (1981).

107. R. Roque-Malherbe and J.C. Antuña, *Rev. CNIC: Ciencias Quimicas*, 16, 89 (1985).

108. Rolando Roque-Malherbe, Reinaldo Roque-Malherbe, and F. Morales, Patent Certificate No. 21059, ONIITEM, Havana, Cuba (1982).

109. M. Hernandez, A. Rodriguez, A. Montes, and R. Roque-Malherbe, Patent Certificate No. 21746, ONIITEM, Havana, Cuba (1987).

110. A. Montes and R. Roque-Malherbe, Patent Certificate No. 21922, ONIITEM, Havana, Cuba (1990).

111. R. Roque-Malherbe and A. Montes, *J. Thermal Anal.*, 31, 517 (1986).

112. R. Roque-Malherbe, C. de las Pozas, and J.J. Castillo, *Thermal Anal.*, 32, 321 (1987).

113. M. Carreras, R. Roque-Malherbe, and C. de las Pozas, *J. Thermal Anal.*, 32, 1271 (1987).

114. R. Roque-Malherbe, C. de las Pozas, and M. Carreras, *J. Thermal Anal.*, 34, 1113 (1988).

115. J.A. Alonso, R. Roque-Malherbe, C. Gonzales, and C. de las Pozas, *J. Thermal Anal.*, 34, 865 (1988).

116. M. Hernandez-Velez and R. Roque-Malherbe, *Rev. Cubana de Fisica*, 8, 83 (1988).

117. N. Vega, R. Roque-Malherbe, and C. Gonzales, *J. Thermal Anal.*, 37, 1358 (1989).

118. C. de las Pozas, D. Diaz, J. Perez-Pariente, R. Roque-Malherbe, and M. Magi, *Zeolites*, 9, 33 (1989).

119. R. Roque-Malherbe, L. Lemes-Fernandez, L., Lopez-Colado, C. de las Pozas, and A. Montes-Caraballal, in *Natural Zeolites '93 Conference Volume*, D.W. Ming, and F.A. Mumpton, (editors), International Committee on Natural Zeolites, Brockport, NY, 1995, p. 299.

120. R. Roque-Malherbe and M. Hernandez-Velez, *J. Thermal Anal.*, 36, 1025 (1990); 36, 2455 (1990).

121. R. Gonzales and R. Roque-Malherbe, *J. Thermal Anal.*, 37, 787 (1991).

122. R. Roque-Malherbe, C. de las Pozas, and M. Carreras, *J. Thermal Anal.*, 37, 2423 (1991).

123. O. Vigil, J. Fundora, H. Villavicencio, M. Hernandez-Velez, and R. Roque-Malherbe, *J. Mater. Sci. Lett.*, 11, 1725 (1992).

124. R. Roque-Malherbe, J. Oñate-Martinez, E. Reguera, and E. Navarro, *J. Mater. Sci.*, 28, 2321 (1992).

125. F. Fernandez, M. Hernandez-Velez, and R. Roque-Malherbe, *Stud. Surf. Sci. Catal.*, 84A, 833 (1994).

126. M. Hernandez-Velez and R. Roque-Malherbe, *Mater. Sci. Lett.*, 14, 1112 (1995).

127. M. Hernandez-Velez, R. Blanco-Montes, R. Roque-Malherbe, H. Villavicencio-Garcia, F., Fernandez-Gutierrez, A., Berazain-Iturralde, and J.M. Abela-Martinez, *Bol. Soc. Espanola de Ceramica y Vidrio*, 34, 409 (1995).

128. A.R. Haidar and A.K. Jonscher, *J. Chem. Soc. Faraday Trans. I.*, 82, 132 (1986).

129. P. Pissis and D. Dauaukaki-Diamanti, *J. Phys. Chem. Solids*, 54, 701 (1993).

130. A. Szas and J. Liszi, *Zeolites*, 11, 517 (1991).

131. S-M. Park and Y-S. Yoo, *Anal. Chem.* 75, 455A (2003).

132. S.O. Kasap, *Principles of Electronic Materials* (2nd edition), McGraw-Hill, New York, 2002.

133. J. Wu, PhD dissertation, California Institute of Technology, Pasadena, CA, 2005.

134. A.K. Jonscher, *The Universal Dielectric Response. A Review of Data and Their New Interpretation*, Chelsea Dielectric Group, Pulton Place, London, UK, 1978.

135. A.C. Melissinos and J. Napolitano, *Experimental Modern Physics* (2nd edition), Academic Press, New York, 2003.

136. R. Mössbauer, *Z. Physik*, 151, 124 (1958).

137. H. Frauenfelder, *The Mössbauer Effect*, Benjamin, New York, 1962.

138. G.K. Wertheim, *Mössbauer Effect Principles and Applications*, Academic Press, New York, 1964.

139. U. Gonser, (editor), *Mössbauer Spectroscopy*, Springer-Verlag, New York, 1975.

140. D.P. Dickson and F.J. Berry, in *Mössbauer Spectroscopy*, D.P. Dickson and F.J. Berry, (editors), Cambridge University Press, Cambridge, UK, 1986, p. 1.

141. A.G. Maddock, *Mössbauer Spectroscopy: Principles and Applications of the Techniques*, Horwood Publishing Limited, Westergate, Chichester, UK, 1997.

142. R.V. Parish, in *Mössbauer Spectroscopy*, D.P. Dickson and FJ. Berry, (editors), Cambridge University Press, Cambridge, UK, 1986, p. 17.

143. K. S. Krane, *Introductory Nuclear Physics*, John Wiley & Sons, New York, 1987.

144. J.D. Jackson, *Classical Electrodynamics* (2nd edition), John Wiley & Sons, New York, 1975.

145. C. de las Pozas, C. Díaz-Aguila, E. Reguera-Ruiz, and R. Roque-Malherbe, *J. Solid State Chem.*, 93, 215 (1991).

146. R. Roque-Malherbe, B.S. Bosktein, and A.A. Zhujovitskii, *Rev. Metal. CENIM*, 15, 287 (1979).

147. E.W. Washburn, *Phys. Rev.*, 17, 273 (1921); *Proc. Nat. Acad. Sci. USA*, 7, 115 (1921).

148. J. Rouquerol, D. Avnir, C.W. Fairbridge, D.H. Everett, J.H. Haynes, N. Pernicone, J.D.F. Ramsay, K.S.W. Sing, and K.K. Unger, *Pure Appl. Chem.*, 66, 1739 (1994).

149. A.W. Adamson, and A.P. Gast, *Physical Chemistry of Surfaces* (6th edition), John Wiley & Sons, New York, 1997.

150. C.A. Leon, *Adv. Coll. Int. Sci.*, 76, 341 (1998).

151. F. Porcheron, M. Thommes, R. Ahmad, and P.A. Monson, *Langmuir*, 20, 6482 (2007).

152. F. Porcheron, P. Monson, and M. Thommes, *Langmuir*, 23, 3372 (2004).

153. A.V. Kiseliov, in *Curso de Fisica Quimica*, Guerasimov, Ya, (editor), Mir, Moscow, 1971. p. 522.

154. C. Felipe, F. Rojas, I. Kornhauser, M. Thommes, and G. Zgrablich, *Adsorpt. Sci. Technol.*, 24, 623 (2006).

155. C.C. Egger, C. du Fresne, V.I. Raman, V. Schädler, T. Frechen, S.V. Roth, and P. Müller-Buschbaum, *Langmuir*, 24, 5877 (2008).

156. J. Shu, Master thesis, Technical University of Delft, Delft, the Netherlands, 2006.

157. A.Y. Sergienko, H. Tai, M. Narkis, and M.S.J. Silverstein, *App. Polym. Sci.*, 94, 2233 (2004).

5 Diffusion in Materials

5.1 INTRODUCTION

Diffusion is the random movement of molecules or small particles taking place due to the motion caused by thermal energy [1–20]. It is a general property of matter linked with the propensity of systems to occupy all accessible states [20]. In a more simple way, diffusion is a spontaneous tendency of all systems to equalize concentration, if any external influence does not impede this process. Specifically, atoms, molecules, or any particle chaotically moves in the direction where less elements of its own type are located.

Diffusion in dense materials is a significant area of materials science. Diffusion plays a crucial role in the kinetics of numerous processes taking place during the processing of dense materials, such as phase transformation, high-temperature oxidation, permeation, precipitation, ion conduction, sintering, and other processes.

Besides, diffusion of gases in porous materials is an important topic as well, because this process is crucial in catalysis, gas chromatography, and gas separation.

In this chapter, diffusion in solid materials, that is, metals, oxides, and nanoporous crystalline, ordered, and amorphous materials is discussed. We first study diffusion in a phenomenological, general form; afterward the diffusion of atoms in crystals by means of knowledge obtained from studies of diffusion in metals is discussed. Thereafter, those phenomena that are exclusive to oxides are separately discussed. Finally, diffusion in nanoporous materials is described.

5.2 FICK'S LAWS

The quantitative study of diffusion started in 1850–1855 with the works of Adolf Fick and Thomas Graham. From the conclusion of his studies, Fick understood that diffusion obeys a law isomorphic to the Fourier law of heat transfer [17]. This fact allowed him to propose his first equation in order to macroscopically describe the diffusion process, that is, Fick's first law:

$$\bar{J} = -D\bar{\nabla}C \tag{5.1}$$

where
$\bar{J}$ is the matter flux
$\bar{\nabla}C$ is the concentration gradient
D is the Fickean diffusion coefficient, or transport diffusion coefficient, which is the proportionality constant

The units of the above described parameters in the International System (SI) are D [m²/s] for the Fickean diffusion coefficient, C [mol/m³] for the concentration, and $\bar{J}$ [mol/m²·s] for the flux.

It is necessary to state that the flux, and, consequently, the diffusion coefficient, as discussed later, have to be chosen relative to a frame of reference, since the diffusion flux, $\bar{J}$, gives the number of species crossing a concrete unit area in the medium per unit of time [13].

Under the influence of external forces, the particles move with an average drift velocity, v_F, which gives rise to a flux:

$$J_F = Cv_F$$

where C is the concentration of diffusing species
and

$$\bar{v}_F = M_F \bar{F}_F$$

where
M_F is the mobility
$\bar{F}_F$ is the generalized force

Then, in the presence of an external force, the total flux is given by the relation

$$J = -D\nabla C + C v_F \qquad (5.2)$$

Fick's second law:

$$\frac{\partial C}{\partial t} = D\nabla^2 C \qquad (5.3)$$

is an expression of the law of conservation of matter, that is,

$$\frac{\partial C}{\partial t} = -\nabla \cdot \bar{J}$$

If the diffusivity is independent of position.
In one dimension, Fick's first equation is a linear relation between the matter flux, J, and the concentration gradient, $\frac{\partial C}{\partial x}$:

$$\bar{J} = -D\frac{\partial C}{\partial x} \qquad (5.4)$$

It is necessary to clarify now that as diffusion is the macroscopic expression of the tendency of a system to move toward equilibrium, the real driving force should be the gradient of the chemical potential, μ, as explained in the next section.

5.3 THERMODYNAMICS OF IRREVERSIBLE PROCESSES

Let us now provide a brief summary of the application of the thermodynamics of irreversible processes to diffusion [6,12,21,22].

In an irreversible process, in conformity with the second law of thermodynamics, the magnitude that determines the time dependence of an isolated thermodynamic system is the entropy, S [23–26]. Consequently, in a closed system, processes that merely lead to an increase in entropy are feasible. The necessary and sufficient condition for a stable state, in an isolated system, is that the entropy has attained its maximum value [26]. Therefore, the most probable state is that in which the entropy is maximum.

Irreversible processes are driven by generalized forces, X, and are characterized by transport (or Onsager) phenomenological coefficients, L [21,22], where these transport coefficients, L_{ij}, are defined by linear relations between the generalized flux densities, J_i, which are the rates of change with time of state variables, and the corresponding generalized forces X_i:

$$J_k = \sum_{i=1}^{N} L_{ki} X_i \qquad (5.5)$$

where the Onsager reciprocity relations

$$L_{ij} = L_{ji}$$

are fulfilled; besides, it is postulated that the rate of entropy production per unit volume, due to internal processes, may be expressed as [6]

$$T\left(\frac{\partial s}{\partial t}\right) = \sum_{i=1}^{N} J_i X_i$$

where s denotes the entropy per unit volume of the system.

Familiar examples of the relation between generalized fluxes and forces are Fick's first law of diffusion, Fourier's law of heat transfer, Ohm's law of electricity conduction, and Newton's law of momentum transfer in a viscous flow.

In the realm of the previously stated principles, the real driving force behind mass transport is the gradient of chemical potential; in this sense, in the absence of an external Newtonian force like that exerted in a charged species by an electric field, it is expressed as

$$X_i = -T\left[\nabla\left(\frac{\mu_i}{T}\right)\right] \tag{5.6}$$

for the transport of particles of the type i.

On the other hand, heat conductivity means the transfer of energy caused by temperature gradients. Fourier's law of heat transfer is expressed as

$$\overline{Q} = -\kappa \nabla T \tag{5.7}$$

where

$\overline{Q}$ is the energy flux

∇T is the temperature gradient

κ is the thermal conductance coefficient, which is the proportionality constant

In terms of the thermodynamics of irreversible processes, the generalized force for the energy flux is

$$X_q = T\left(\nabla\frac{1}{T}\right) = -\frac{\nabla T}{T} \tag{5.8}$$

With the help of Equation 5.6, if we consider that the only flux in the system is that of the particles labeled i, that is, $\overline{J}_i$, then the corresponding thermodynamic force at constant temperature is given by [6]

$$X_i = -(\nabla\mu_i)_T \tag{5.9}$$

Applying Equation 5.5 now, we get

$$J_i = -L_{ii}(\nabla\mu_i)_T \tag{5.10}$$

If the expression for the chemical potential of an ideal system is substituted [27,28],

$$\mu_i = \mu_i^0 + RT \ln C_i$$

we get

$$J_i = -L_{ii}RT(\nabla \ln C_i)$$

If the mobility, M_i, of the diffusing species is defined as the velocity under the action of a unit force, then

$$M_i = L_{ii}RT \tag{5.11}$$

Then

$$J_i = -M_i(\nabla \ln C_i) = -\frac{M_i}{C_i}\nabla C_i = -D_i\nabla C_i \tag{5.12}$$

where the diffusion coefficient in the case of an ideal system is

$$D_i = \frac{M_i}{C_i} = \frac{L_{ii}RT}{C_i} \tag{5.13}$$

5.4   DIFFUSION COEFFICIENTS

5.4.1   Tracer-Diffusion Coefficient and Self-Diffusion Coefficient

Fick's first law, in the form given by Equation 5.12, allows us to define the tracer- and the self-diffusion coefficients. Diffusion of a tracer isotope is the case when a diffusing atom, which is marked by their radioactivity of by their isotopic mass (see Figure 5.1) [7], is introduced in an extremely dilute concentration in an otherwise homogeneous crystal with no driving force [4]. In this case, the tracer gradient of concentration will give rise to a net flow of tracer atoms. Consequently,

$$\overline{J}^* = -D^* \frac{\partial C^*}{\partial x}$$

Thin film of tracer particles

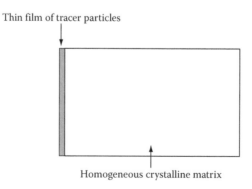

Homogeneous crystalline matrix

FIGURE 5.1   Thin layer of tracer-diffusing particles on the host.

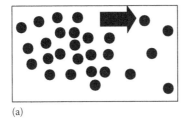 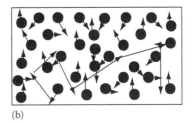

(a)   (b)

FIGURE 5.2 Schematic representation of the differences between: (a) tracer self-diffusion and (b) self-diffusion.

where D^* is the tracer diffusion coefficient.

If the tracer is composed of the same species as that of the solid host, then the diffusion coefficient is named the tracer self-diffusion coefficient, where D_A^* is the tracer self-diffusion coefficient. It is necessary to clarify that self-diffusion is a particle transport process that takes place in the absence of a chemical potential gradient [13]. This process is described, as explained later, by following the molecular trajectories of a large number of molecules, and determining their mean square displacement (MSD).

The difference in the microphysical situations between the tracer self-diffusion coefficient and the self-diffusion coefficient is schematically represented in Figure 5.2 [12].

In Figure 5.2a, the tracer self-diffusion process, and in Figure 5.2b, the self-diffusion process, are represented.

5.4.2 INTRINSIC DIFFUSION COEFFICIENT: THE KIRKENDALL EFFECT

The intrinsic diffusion coefficients, $\bar{D}_A$ and $\bar{D}_B$, of a binary alloy A–B express the diffusion of the components A and B relative to the lattice planes [7]. Therefore, during interdiffusion, a net flux of atoms across any lattice plane is present, where, normally, the diffusion rates of the diffusing particles A and B are different. Subsequently, this interdiffusion process provokes the shift of lattice planes with respect to a fixed axis of the sample, result which is named the Kirkendall effect [9].

The diffusion flux is proportional to the gradient of chemical potential [6], where the chemical potential of species i in a binary alloy is given by [7]

$$\mu_i = \left(\frac{\partial G}{\partial n_i} \right)_{P,T,n_{j\neq i}}$$

where i = A,B. For an ideal binary solution

$$\mu_i = \mu_i^0 + RT \ln \left(\frac{C_i}{C_A + C_B} \right) = \mu_i^0 + RT \ln X_i \tag{5.14}$$

For a nonideal solution, the gradient of the chemical potential is given by [9]

$$\mu_i = \mu_i^0 + RT \ln a_i \tag{5.15}$$

in which

$$a_i = \gamma_i X_i \tag{5.16}$$

in which

a_i is the activity of the component i

γ_i is the activity of the coefficient

Now, neglecting correlation effects, since (see Equations 5.11 through 5.13)

$$J_i = -C_i M_i \nabla \mu_i \qquad (5.17)$$

where M_i is the mobility of the diffusing species, i, in the binary solution. Then, substituting Equation 5.15 in Equation 5.17, and taking into consideration Equation 5.16, we get [9]

$$J_i = -RTC_i M_i \nabla \ln \gamma_i C_i = RTM_i \left(1 + \frac{\partial \ln \gamma_i}{\partial \ln X_i}\right) \nabla C_i = \overline{D}_i \nabla C_i \qquad (5.18)$$

where

$$\overline{D}_i = RTM_i \left(1 + \frac{\partial \ln \gamma_i}{\partial \ln X_i}\right) = D_{AB}^i \Phi \qquad (5.19)$$

are the intrinsic diffusion coefficients and

$$D_{AB}^i = RTM_i \qquad (5.20)$$

are the tracer diffusion coefficients of alloy components determined in a homogeneous alloy (see Figure 5.3) [7]. The so-called thermodynamic factor is given by [13]

$$\Phi = \left(1 + \frac{\partial \ln \gamma_i}{\partial \ln X_i}\right) \qquad (5.21)$$

Now, we describe the Kirkendall effect [9]. The flux, as well as the diffusion coefficient, has to be chosen relative to a frame of reference. In Figure 5.4, the laboratory frame of reference, X, which is the observer frame of reference, and the moving frame of reference, x, which moves with the inert markers, are shown.

Consequently, the interdiffusion in one dimension of two components A and B is given by the following expressions [4,7,9]:

Thin film of the tracer particles of species A or B

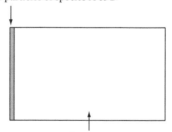

Homogeneous crystalline matrix of the $A_x B_{1-x}$ alloy

FIGURE 5.3 Thin layer of a diffusing particle A* or B* on the alloy $A_x B_{1-x}$.

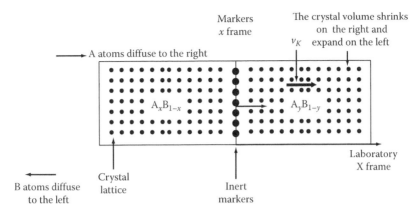

FIGURE 5.4 Motion of markers during single-phase interdiffusion.

$$J_A = -\bar{D}_A \frac{\partial C_A}{\partial x}$$

and

$$J_B = -\bar{D}_B \frac{\partial C_B}{\partial x} \tag{5.22}$$

which describe the particle fluxes of species A and B in relation to the markers' moving frame of reference (see Figure 5.4). Afterward, the interdiffusion of the two species will lead to a net flow relative to the flowing lattice, that is, the moving frame [9],

$$J_{net} = J_A + J_B = -\bar{D}_A \frac{\partial C_A}{\partial x} - \bar{D}_B \frac{\partial C_B}{\partial x} \tag{5.23}$$

Now, it is possible to calculate the marker velocity, v_K. The accumulation rate of mass provoked by the net diffusion, transversely to the marker plane, may be related to the marker speed as follows:

$$v_K = -J_{net} \bar{V} \tag{5.24}$$

where Ω is the molar volume of the alloy. Therefore, from Equations 5.23 and 5.24, it is possible to get

$$v_K = \left(\bar{D}_A \frac{\partial C_A}{\partial x} + \bar{D}_B \frac{\partial C_B}{\partial x} \right) \bar{V} \tag{5.25}$$

At this moment, if both components of the alloy are supposed to have equal partial molar densities in the alloy, subsequently

$$C_A + C_B = \frac{1}{\bar{V}}$$

or

$$\bar{V} C_A + \bar{V} C_B = 1 \tag{5.26}$$

The mole fraction is given by

$$\bar{V}C_A = X_A \tag{5.27a}$$

$$\bar{V}C_B = X_B \tag{5.27b}$$

Then, substituting Equations 5.27 in Equation 5.26, and upon differentiation of the result, we get

$$\frac{\partial X_A}{\partial x} = -\frac{\partial X_B}{\partial x} \tag{5.28}$$

Combining Equations 5.27 with Equations 5.28 and 5.25, it is shown that [4,7,9]

$$v_K = (\bar{D}_A - \bar{D}_B)\left(\frac{\partial X_A}{\partial x}\right) \tag{5.29a}$$

or in other terms

$$v_K = (\bar{D}_B - \bar{D}_A)\left(\frac{\partial X_B}{\partial x}\right) \tag{5.29b}$$

5.4.3 INTERDIFFUSION OR CHEMICAL DIFFUSION COEFFICIENT

This coefficient describes a diffusion process under the influence of a gradient in the chemical composition. When two diffusing species mix together, their rate of mixing depends on the diffusion rates of both species. Consequently, the interdiffusion coefficient is defined to measure this rate of mixing in relation to a laboratory frame of reference [13]. In this sense, the relation defining the interdiffusion coefficient, deduced by Darkens [29], is

$$\tilde{D} = (X_B\bar{D}_A + X_A\bar{D}_B) \tag{5.30a}$$

named the Darkens' second equation. Therefore,

$$\tilde{D} = (X_B D_{AB}^A + X_A D_{AB}^B)\Phi \tag{5.30b}$$

5.5 MICROSCOPIC DESCRIPTION OF DIFFUSION

5.5.1 INTRODUCTION

Diffusion is the random migration of molecules or small particles arising from motion due to thermal energy. A very simple derivation of Fick's first law, based on the random walk problem, can be obtained in one dimension. In this case, $J_x(x,t)$, that is, the number of particles, N, that move across unit area, A, in unit time, τ, can be defined as

$$J_x(x,t) = \frac{N}{A\tau}$$

The number of particles at position x and $(x + \Delta)$ at time t is $N(x)$ and $N(x + \Delta)$ (see Figure 5.5), respectively. Subsequently, half of the particles at x at time t move to the right (i.e., to $x + \Delta$); at the

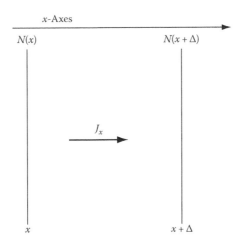

FIGURE 5.5   One-dimensional diffusion.

same time, the other half moves to the left (i.e., to $x - \Delta$) during the time-step, τ. Likewise, half of the particles at $(x + \Delta)$ at time t moves to the right (i.e., to $x + 2\Delta$), while the other half moves to the left (i.e., to x). Subsequently, the net number of particles that move from x to $(x + \Delta)$, when $t = t + \tau$ is

$$N = -\frac{[N(x+\Delta,t)-N(x,t)]}{2}$$

Now

$$J_x(x,t) = -\frac{\Delta^2}{2\tau}\left(\frac{1}{\Delta}\right)\left(\frac{N(x+\Delta,t)}{A\Delta} - \frac{N(x,t)}{A\Delta}\right) = -D\frac{[C(x+\Delta)-C(x,t)]}{\Delta}$$

Since

$$C(x,t) = \frac{N(x,t)}{A\Delta}$$

Then, for, $\Delta \rightarrow 0$

$$J_x(x,t) = \lim_{\Delta \rightarrow 0} -D\frac{[C(x+\Delta)-C(x,t)]}{\Delta} = -D\frac{\partial C(x,t)}{\partial x}$$

where

$$D = \frac{\Delta^2}{2\tau} \tag{5.31}$$

is the self-diffusion coefficient in one dimension.

5.5.2   RANDOM WALKER IN ONE DIMENSION

Consider now a random walker in one dimension, with probability, R, of moving to the right, and, L, for moving to the left. At, $t = 0$, we place the walker at $x = 0$, as indicated in Figure 5.6. The walker can then jump, with the above probabilities, either to the left or to the right for each time-step. Every

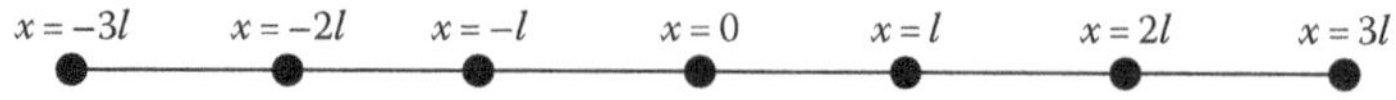

$$x = -3l \qquad x = -2l \qquad x = -l \qquad x = 0 \qquad x = l \qquad x = 2l \qquad x = 3l$$

FIGURE 5.6   One-dimensional random walker that can jump either to the left or to the right.

step has a length, $\Delta x = l$, and we have a jump either to the left or to the right at every time-step. Let us now assume that we have equal probabilities for jumping to the left or to the right, that is, $L = R = 1/2$. Then, the average displacement after N time steps is

$$\langle x(N) \rangle = \left\langle \sum_{i=1}^{N} \Delta x_i \right\rangle = \sum_{i=1}^{N} \langle \Delta x_i \rangle = 0, \quad \text{for } \Delta x_i = \pm l$$

Now, since we have an equal opportunity of jumping either to the left or to the right, thereafter, the value of $\langle x(N)^2 \rangle$ is

$$\langle x(N)^2 \rangle = \left\langle \left(\sum_{i=1}^{N} \Delta x_i \right)^2 \right\rangle = \left\langle \sum_{i=1}^{N} \Delta x_i \sum_{j=1}^{N} \Delta x_j \right\rangle = \sum_{i=1}^{N} \langle \Delta x_i^2 \rangle + \sum_{i \neq j} \langle \Delta x_i \Delta x_j \rangle = l^2 N$$

Since, the steps are not correlated, i.e., $\langle \Delta x_i \Delta x_j \rangle = \langle \Delta x_j \Delta x_i \rangle$, then:

$$\sum_{i \neq j} \langle \Delta x_i \Delta x_j \rangle = 0$$

Now, since the time between jumps, or mean residence time, is τ. Therefore, the jump frequency is $\Gamma = \frac{1}{\tau}$; consequently, if Γ is constant, then for N steps where $N = \frac{t}{\tau}$, the MSD is

$$\langle x^2(N) \rangle = N l^2 = \left(\frac{l^2}{\tau} \right) t \tag{5.32}$$

5.5.3   Fokker–Planck Equation

A Markov process is a stochastic process, where the time dependence of the probability, $P(x,t)dx$, that a particle position at time, t, lies between x and $x + dx$ depends only on the fact that $x = x_0$ at $t = t_0$, and not on the entire history of the particle movement. In this regard, the Fokker–Planck equation [11]

$$\frac{\partial P(x,t)}{\partial t} = -\frac{\partial}{\partial x} \left[f(x) P(x,t) \right] + \kappa \frac{\partial^2}{\partial x^2} \left[g(x) P(x,t) \right]$$

is a partial differential equation for $P(x,t)$, which accounts for the time development of a Markov process.

A very important application of the Markov dynamics is random walk. In the special case of random walk $f(x) = 0$ and $g(x) = 1$, then the diffusion equation for a random walk in one dimension is

$$\frac{\partial P(x,t)}{\partial t} = D \frac{\partial^2 P(x,t)}{\partial x^2} \tag{5.33}$$

in which D is the self-diffusion coefficient and $P(x,t)$, where $\int_{-\infty}^{\infty} P(x,t)dx = 1$ is the probability density to find a diffusing particle at the position x, during the time t, if this particle was at $x = 0$ at $t = 0$.

The solution of this equation with the following initial and boundary conditions

$$\left(\frac{\partial P(x,t)}{\partial x}\right)_{-\infty,t} = 0$$

$$\left(\frac{\partial P(x,t)}{\partial x}\right)_{\infty,t} = 0$$

$$P(x,0) = \delta(x)$$

is [5]

$$P(x,t) = \left(\frac{1}{(4\pi Dt)}\right)^{1/2} \exp\left(-\frac{x^2}{4Dt}\right) \tag{5.34}$$

Therefore, it is very easy to show that the one-dimensional MSD is

$$\langle x^2 \rangle = \int x^2 P(x,t)dx = 2Dt \tag{5.35}$$

Then,

$$D = \frac{\langle x^2 \rangle}{2t} \tag{5.36}$$

Consequently, from Equation 5.32

$$D = \frac{l^2}{2\tau} \tag{5.37}$$

5.5.4 Diffusion Mechanisms in Crystalline Solids

A number of diffusion mechanisms in crystalline solids are possible. Atoms vibrate in their equilibrium sites; after that, periodically, these oscillations turn out to be large enough to give rise to a jump from one site to the other. The order of magnitude of the frequency of these oscillations is about 10^{12}–10^{13} Hz. In this regard, it has been shown that the jump rate at which an atom jumps into an empty neighboring site is given by [30]

$$\omega = \omega_0' e^{-\frac{G_M}{RT}} = \omega_0' e^{\frac{S_M}{R}} e^{-\frac{H_M}{RT}} \tag{5.38}$$

where G_M, S_M, and H_M are the free energy, entropy, and enthalpy of migration, respectively. Regarding the type of elementary jump that takes the atom from one equilibrium site to the other, it is possible to distinguish the following types [4]: exchange mechanism, ring mechanism, relaxation mechanism, interstitial mechanism, interstitialcy mechanism, vacancy mechanism, divacancy mechanism, dislocation pipe mechanism, grain boundary mechanism, and surface mechanism.

We do not discuss all these mechanisms here. However, the most important mechanisms, the vacancy mechanism and the interstitial mechanism, are described.

5.5.4.1 Vacancy Mechanism

As is well known, in thermal equilibrium, every crystal, at a temperature above absolute zero encloses a certain number of vacant lattice sites, where the probability, P, of finding a vacancy in a solid at equilibrium at an absolute temperature, T, is given by

$$P = \frac{N_V}{N} = K_V e^{-\frac{H_V}{RT}} \tag{5.39}$$

where

N_V is the equilibrium number of vacancies

N is the total number of sites

H_V represents the enthalpy of formation of a vacancy

$K_V = e^{\frac{S_F}{R}}$ is a pre-exponential factor, in which S_V represents the entropy of formation of a vacancy

Consequently, whenever a vacancy is present, the atoms that surround this vacant site can jump to it (see Figure 5.7).

For a vacancy mechanism, in a self-diffusion process, the jump frequency, Γ, of an atom to a given adjacent site is given by [4]

$$\Gamma = \omega N_V \tag{5.40}$$

If there are z adjacent sites, then the total jump frequency is given by

$$\nu = z\Gamma \tag{5.41}$$

Then, combining Equations 5.38 through 5.41, we get

$$\nu = z \left(\omega_0' e^{\frac{S_M}{R}} e^{-\frac{H_M}{RT}} \right) \left(N e^{\frac{S_V}{R}} e^{-\frac{H_V}{RT}} \right) = z\omega_0 e^{-\frac{H_M + H_V}{RT}} = \nu_0 e^{-\frac{E_a}{RT}}$$

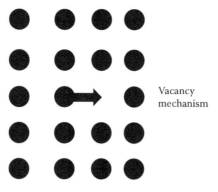

Vacancy mechanism

FIGURE 5.7 Schematic representation of the vacancy mechanism.

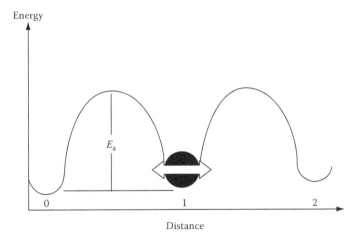

FIGURE 5.8 Activation energy.

where the quantity E_a is called the activation energy (see Figure 5.8), a parameter which is experimentally calculated with the help of the Arrhenius plot.

5.5.4.2 Interstitial Mechanism

Solute atoms, which are smaller than the solvent atoms in binary interstitial alloys, such as C, H, N, and O are usually incorporated as interstitials in the void sites of the lattice, for example, in octahedral and tetrahedral sites in the close-packed cubic and close-packed hexagonal lattices (see Figures 1.6, 1.7, and 2.12), and in the body-centered cubic lattices (Figure 5.9) [7].

The interstitial solute diffuses by jumping from one interstitial site to the other (see Figure 5.10). This mechanism is also called the direct interstitial mechanism in order to differentiate it from the interstitialcy mechanism.

The interstitial mechanism is fundamentally seen in solute atoms which, as was previously stated, are smaller than the solvent atoms in binary interstitial alloys. Then, in the present mechanism

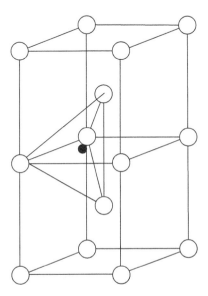

FIGURE 5.9 Tetrahedral interstitial site in the BCC lattice.

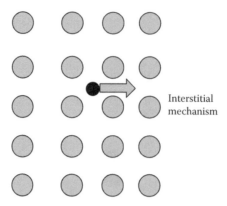

FIGURE 5.10 Schematic representation of the interstitial mechanism.

$$\Gamma = z\omega = z\left(\omega_0' \; e^{\frac{S_M}{T}} e^{-\frac{H_M}{RT}}\right) = v_0^i e^{-\frac{E_a^i}{RT}}$$

5.5.5 RANDOM WALKER IN A CUBIC CRYSTALLINE STRUCTURE

Let us now consider a random walker in a three-dimensional cubic lattice. The atom will jump between sites of the normal lattice for a substitutional diffuser, and from interstitial to interstitial site for an interstitial diffuser. In the present case, the Einstein–Smoluchovskii equation for the diffusion coefficient in three dimensions which is a generalization of Equation 5.36, that is,

$$D = \frac{1}{6t}\left\langle \bar{R}^2 \right\rangle \tag{5.42}$$

where $\left\langle \bar{R}^2 \right\rangle$ is the MSD in three dimensions, is fulfilled.

The displacement after N time steps, that is, in time t is

$$\bar{R}(N) = \sum_{i=1}^{N} \bar{r}_i$$

Hence, the MSD is given by

$$\left\langle \bar{R}(N)^2 \right\rangle = \left\langle \left(\sum_{i=1}^{N} \bar{r}_i\right)^2 \right\rangle = \left\langle \left(\sum_{i=1}^{N} \bar{r}_i\right)\cdot\left(\sum_{j=1}^{N} \bar{r}_j\right) \right\rangle = \sum_{i=1}^{N} \left\langle \bar{r}_i^2 \right\rangle + \sum_{i \neq j} \left\langle (\bar{r}_i)\cdot(\bar{r}_j) \right\rangle$$

In the first approximation, we ignore the correlation effects that are contained in the double summation, that is, since the steps are not correlated, consequently

$$\left\langle (\bar{r}_i)\cdot(\bar{r}_j) \right\rangle = \left\langle (\bar{r}_j)\cdot(\bar{r}_i) \right\rangle$$

Therefore,

$$\sum_{i \neq j} \left\langle (\bar{r}_i)\cdot(\bar{r}_j) \right\rangle = 0$$

Hence, for a true random walker, the MSD is

$$\left\langle \overline{R}_{random}(N)^2 \right\rangle = \sum_{i=1}^{N} \left\langle \overline{r}_i^2 \right\rangle \tag{5.43}$$

In the general case, correlation effects are present during the diffusion process. This fact implies that

$$\sum_{i \neq j} \left\langle (\overline{r}_i) \cdot (\overline{r}_j) \right\rangle \neq 0$$

Consequently, it is possible to define the correlation factor as follows [4,7]:

$$f = \lim_{n \to \infty} \left(\frac{\left\langle \overline{R}^2(N) \right\rangle}{\left\langle \overline{R}_{random}^2(N) \right\rangle} \right) = \lim_{n \to \infty} \frac{\sum_{i=1}^{N-1} \sum_{j=i+1}^{N} \left\langle (\overline{r}_i) \cdot (\overline{r}_j) \right\rangle}{\sum_{n=1}^{N} \left\langle \overline{r}_i^2 \right\rangle} \tag{5.44}$$

In the case of diffusion in a cubic lattice, it is possible to only consider nearest-neighbor jumps of equal lengths, l, into, z, adjacent sites. Besides, the time between jumps is, τ, then $t = N\tau$; in addition, the jump frequency is

$$\Gamma = \frac{\langle n \rangle}{zt} \tag{5.45}$$

where $\langle n \rangle$ is the average number of jumps. Thus, the MSD can be written as follows [4,7,9]:

$$\left\langle \overline{R}_{random}^2(N) \right\rangle = \langle n \rangle l^2 \tag{5.46}$$

Consequently, combining Equations 5.42, 5.45, and 5.46, we get

$$D = \frac{1}{6} l^2 z \Gamma \tag{5.47}$$

The mean residence time can be written as follows:

$$\tau = \frac{1}{z\Gamma}$$

Then,

$$D = \frac{l^2}{6\tau} \tag{5.48}$$

which is the self-diffusion coefficient without correlation. If the jumps are correlated, then the self-diffusion coefficient can be expressed in the following form [4,7]:

$$D = \frac{1}{6} f l^2 z \Gamma \tag{5.49}$$

5.6 SOME DIFFUSION PROCESSES IN METALS

5.6.1 Hydrogen Diffusion in Metals

In Pd and other transition metals, hydrogen has a high solubility and diffuses very fast, possibly because of the high d-electron density in the band structure of these metals. During absorption, the hydrogen molecule is first dissociated in the Pd surface; subsequently, the adsorbed hydrogen atoms are ionized, and are incorporated directly into the material as protons and electrons, e^-, as follows [31,32]

$$\left(\frac{1}{2}\right) H_2(g) \rightarrow H_i^{\bullet} + e' \tag{5.50}$$

in which

$H_i^{\bullet}$ is an interstitial proton

e' is a conduction electron, where the proton is interstitially located in tetrahedral and/or octahedral sites [33]

Therefore, it is possible to consider that a neutral dissociation of hydrogen occurs as follows [32]:

$$\left(\frac{1}{2}\right) H_2(g) \rightarrow H_i^{\bullet} + e' \cong H \tag{5.51}$$

That is [9],

$$\left(\frac{1}{2}\right) H_2(gas) \rightarrow H \text{ (solid solution)}$$

where the equilibrium constant for this reaction is

$$K_H = \frac{C_H}{\sqrt{P_{H_2}}} \tag{5.52}$$

named the Sievert's law

where

C_H is the hydrogen concentration in the solid, which is equivalent to the proton concentration C_{H^+} and P_{H_2} are the pressures of hydrogen in the gas phase

In Figure 5.11, the procedure for hydrogen purification using a palladium thin sheet is illustrated.

In this case, it is well known that the process occurs in steady state. To understand this process, one must consider it as a special case of binary diffusion, where the diffusivity of the Pd atoms is zero. Consequently, the frame of reference is the fixed coordinates of the solid Pd thin film. The interdiffusion or chemical diffusion coefficient is the diffusivity of the mobile species [20], that is, hydrogen. Then, the hydrogen flux in the Pd thin film is given by

$$J_H^{Pd} = -\bar{D}_H^{Pd} \frac{\partial C_H}{\partial x} \tag{5.53}$$

where $\bar{D}_H^{Pd}$ is the intrinsic diffusion coefficient of H in Pd. In Figure 5.12, the linear concentration profile in the case of steady-state diffusion is shown. Consequently,

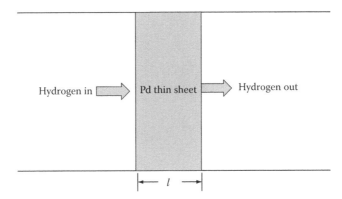

FIGURE 5.11 Diagram of the system for hydrogen purification using a Pd thin sheet.

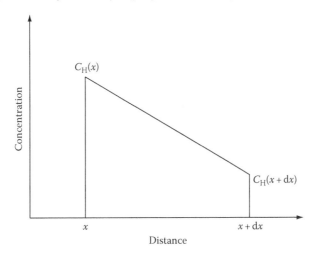

FIGURE 5.12 Linear concentration profile in the case of steady-state diffusion.

$$J_{\mathrm{H}}^{\mathrm{Pd}} = -\bar{D}_{\mathrm{H}}^{\mathrm{Pd}} \frac{\Delta C_{\mathrm{H}}}{\Delta x} = \bar{D}_{\mathrm{H}}^{\mathrm{Pd}} K_{\mathrm{H}} \frac{\sqrt{P_{\mathrm{H}_2}^{\mathrm{in}}} - \sqrt{P_{\mathrm{H}_2}^{\mathrm{out}}}}{l} \qquad (5.54)$$

As dense membrane processes are not described in detail in this section, Chapter 10 discusses them in detail.

5.6.2 Formation of a Surface Fe–Ni Alloy

A Ni thin film of $5\,\mathrm{mg/cm^2}$ is electrodeposited over an iron (99.8 wt % of Fe) sheet of area $5\,\mathrm{cm^2}$ and width 0.5 mm, and the obtained composite system (see Figure 5.13) is thermally treated in high vacuum (10^{-5} Torr) at different temperatures, 500°C, 600°C, 700°C, and 900°C, for different time intervals. Then, a Fe–Ni alloy is obtained on the surface (see Figures 5.13 and 5.14). The high vacuum furnace was a home-made high vacuum system composed of a quartz container included in a furnace. The container was evaluated with the high vacuum system composed of a diffusion pump and a mechanical pump [34,35].

Afterward, the x-ray diffraction profiles of the treated samples were obtained, by irradiating the side where the Ni was electrodeposited [34], and the intensity A_α of the (221) peak of the α-Fe, body-centered cubic phase, and the intensity A_γ of the (220) peak corresponding to the γ-FeNi, face-centered cubic alloy were measured [34,35].

For the XRD peak intensity, the following parameter was defined

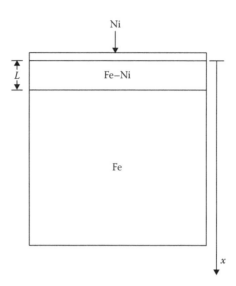

FIGURE 5.13   Graphic representation of the formation of the Fe–Ni surface alloy.

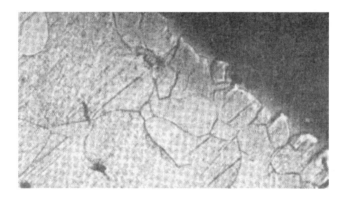

FIGURE 5.14   Optical micrograph of the Fe–Ni surface alloy.

$$\xi = \frac{A_\gamma}{A_\alpha + A_\gamma} \tag{5.55}$$

which characterizes the kinetics of the formation of the Fe–Ni alloy. It is then evident from Figure 5.15 that the kinetics obeys a parabolic relation [34]:

$$\xi^2 = Kt \tag{5.56}$$

Consequently, it is evident that diffusion is controlling the kinetics of the process. Besides, the saturation presented at 700°C and 900°C is due to the limited amount of Ni present in the electrodeposited film.

If we now make an Arrhenius plot, that is, $\ln K$ versus $\frac{1}{T}$ of the kinetic constant, K, of Equation 5.56 (see Figure 5.16), it is possible to calculate the activation energy. The following value, $E_a = 121 \pm 15\,\text{kJ/mol}$, for the activation energy of the diffusion process, results [34].

A similar study was carried out using a sheet of carbon steel (2317 AISI) as support instead of a sheet of iron [35]. In this case, a Ni thin film of 6 mg/cm^2 is electrodeposited over the steel sheet.

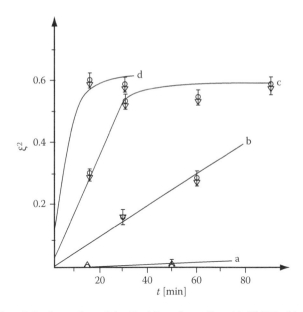

FIGURE 5.15 Kinetics of the formation of the Fe–Ni surface alloy. (a) 500°C, (b) 600°C, (c) 700°C, and (d) 900°C.

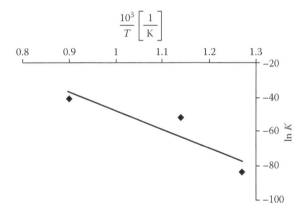

FIGURE 5.16 Arrhenius plot.

Subsequently, the obtained composite system, that is, a Ni-electrodeposited thin film over the steel sheet, is thermally treated in high vacuum (10^{-5} Torr) at different temperatures, 740°C and 1050°C, for different time intervals. Thereafter, a steel–Ni alloy is obtained on the surface (see Figure 5.17).

In order to explain the results reported in Figure 5.15, a simple diffusion model was proposed [34]. In Figure 5.13, the model system is shown, and from this representation of the diffusion process, it is possible to state that [34]

$$\frac{dL}{dt} = \alpha J_{Ni} \tag{5.57}$$

where

α is a parameter that characterizes the increase of the γ phase Fe–Ni alloy

J_{Ni} is the flux of Ni through the Fe–Ni alloy (see Figure 5.13)

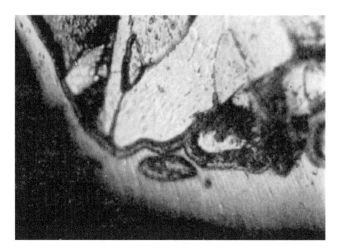

FIGURE 5.17   Optical micrograph of the steel–Ni surface alloy.

It is obvious from a detailed inspection of Figures 5.14 and 5.17 that the increase of the Fe–Ni alloy advances through the grain boundary diffusion and bulk diffusion through the crystallites composing the host metal. Consequently, the diffusion coefficient in the present case is an effective diffusion coefficient involving the grain boundary diffusion and bulk diffusion [34–36]. In order to calculate J_{Ni}, that is, the flux of Ni through the Fe–Ni alloy, we suppose the steady-state diffusion, and consequently [34]

$$J_{Ni} = -D_{Ni}^{e} \frac{K^*}{L} \tag{5.58}$$

where D_{Ni}^{e} is the effective interdiffusion coefficient, since the analyzed process can be considered as a particular case of binary diffusion, where the diffusivity of the Fe atoms is zero and K^*/L is the concentration gradient. Consequently, the frame of reference is the fixed coordinates of the sheet, and then the interdiffusion coefficient is the diffusivity of the mobile species, that is, Ni [20] and K^*, which is a constant depending on the concentration of Ni in the alloy. Now, combining Equations 5.57 and 5.58, we get [34]

$$\frac{dL}{dt} = \frac{\alpha D_{Ni}^{e} K^*}{L} = \frac{K}{L} \tag{5.59}$$

Now, integrating Equation 5.59, we get

$$L^2 = Kt$$

The result coincides with the experimentally obtained Equation 5.56.

Previously, with the help of Figure 5.15, the activation energy of the diffusion process was calculated; in this case, the value obtained was $E_a = 121 \pm 15\,kJ/mol$. This value is about half the activation energy reported for the diffusion of Ni in bulk Fe [37]. The grain diffusion coefficient also obeys the Arrhenius law, and the activation energy for these processes are lower that those occurring in bulk [36]. Consequently, this supports our previous statement that the diffusion coefficient here is an effective diffusion coefficient involving the grain boundary diffusion and bulk diffusion [34,35].

5.6.3 Effect of the Diffusion of Fe in a Fe–Ni Alloy during the Oxidation of the Alloy with Nitric Oxide

A set of Fe–Ni alloys were prepared at 1500 in an Ar atmosphere using extremely pure Fe and Ni [38]. The obtained alloys were pulverized to get powders of about 0.5 mm of grain size. Then, the alloy powders were degassed for 1 h at 100°C in high vacuum (10^{-5} Torr). After this, the oxidation of the different Fe–Ni alloys with 10, 20, 25, and 30 wt % of Ni in the alloy, with NO at a pressure of 20 Torr, at a temperature of 225°C and 275°C, for 20–340 min, was carried out [38,39].

In Figure 5.18, the kinetics of the increase in the width of the Fe_3O_4 oxide formed in the different alloys during the oxidation process is shown [38]. In Figure 5.19, a model of the oxidation process

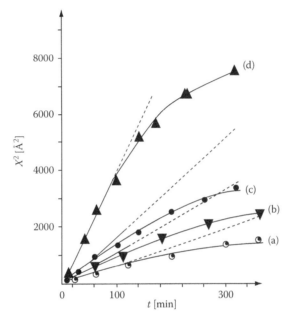

FIGURE 5.18 Kinetics of the formation of Fe_3O_4 during the oxidation of different Fe–Ni alloys with (a) 10, (b) 20, (c) 25, and (d) 30 wt % of Ni in the alloy with NO at a pressure of 20 Torr and at a temperature of 225°C.

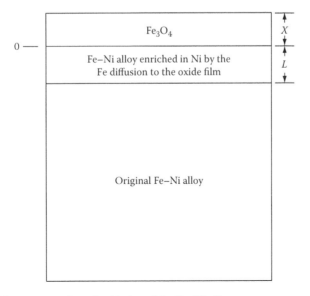

FIGURE 5.19 Graphic representation of oxidation of the Fe–Ni alloy.

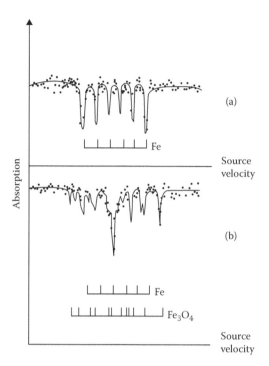

FIGURE 5.20 Mössbauer spectra. (a) Original Fe–Ni alloy (20 wt %) and (b) after the oxidation process.

that explains the decrease in the oxidation rate, observed in Figure 5.18, because of the formation of an intermediate alloy enriched in Ni by the Fe interdiffusion to the oxide film is shown [36].

The formation of the intermediate alloy enriched in Ni was tested with the help of the Mössbauer spectra of the original Fe-Ni (20 wt % of Ni in the alloy) and the same alloy oxidized with NO at a pressure of 20 Torr at 500°C for 5h (Figures 5.20a and 5.20b) [38]. In Figure 5.20a, the Mössbauer spectrum of the original Fe–Ni alloy (20 wt %) is shown; this spectrum only reveals the peaks corresponding to the magnetic splitting of the ^{57}Fe, while in Figure 5.20b, the formation of the paramagnetic Fe–Ni alloy formed after the oxidation process, as a result of the diffusion of iron, due to the enrichment of the Fe–Ni alloy in Ni is shown [38]. In Figure 5.20b, the formation of magnetite (Fe_3O_4) is depicted [38].

The oxidation process was carried out in a home-made, vacuum furnace which included a quartz container included in a furnace. The quartz container was evacuated with a high-vacuum system composed to diffusion pump and a mechanical pump. The system was connected to a sensitive manometric manifold to measure the pressure of the oxidant gas [39].

5.7 DIFFUSION IN OXIDES

5.7.1 Defect Chemistry of Oxides

It is a well-known fact that it is not feasible to manufacture crystals that are perfect in all aspects, because of the presence of crystal defects. This effect occurs because of the fact that over absolute zero, the existence of defects in crystals is thermodynamically necessary.

In Section 2.3, the structure of oxides were studied. For a stoichiometric oxide, such as AO_δ, where $\delta = 0$, we will have thermal disorder, and the concentration of vacancies and interstitials will be determined by the Frenkel, anti-Frenkel [40,41], and Schottky [40,42] mechanisms (see Figures 5.21 through 5.23).

A Frenkel defect is formed when an ion leaves its position in the lattice, leaving behind a vacancy, and moves to a nearby interstitial site (see Figure 5.21). The Frenkel [41] equilibrium is expressed as follows

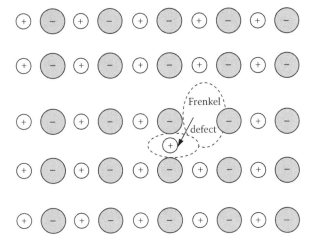

FIGURE 5.21 Frenkel point defect in ionic crystals.

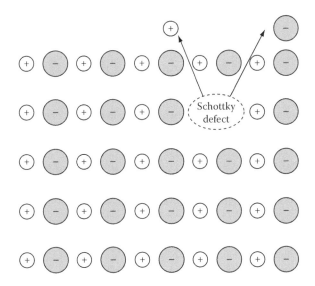

FIGURE 5.22 Schottky point defect in ionic crystals.

$$A_A^x + V_i^x \leftrightarrow V_A'' + A_i^{\bullet\bullet}$$

with the help of the Kroger–Vink notation [43], which describes the formation of a cation, V_A'' vacancy and a cation, $A_i^{\bullet\bullet}$, interstitial from the neutral lattice, A_i^x, cation position and a neutral lattice, V_i^x, interstitial site. The cation vacancies are twice negatively charged, and the cation interstitial vacancy is twice positively charged with respect to the neutral lattice. The equilibrium constant in the case of the Frenkel mechanism (Figure 5.21) is given by

$$K_F = C_{I_C} C_{V_C} \tag{5.60}$$

where
C_{I_C} is the equilibrium concentration of cation interstitials
C_{V_C} is the total concentration of cation vacancies

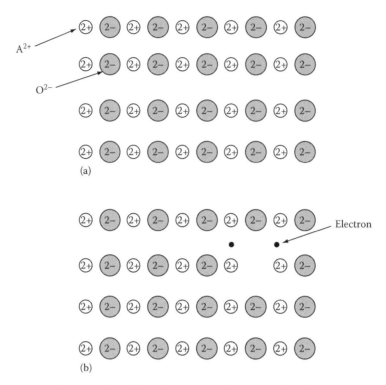

FIGURE 5.23 (a) Stoichiometric AO oxide and (b) nonstoichiometric AO oxide.

On the other hand, Schottky defects occur when ions of different charges leave their sites in the crystalline lattice leaving behind two vacancies. In the case of the Schottky [42] mechanism, the following equilibrium, described with the help of the Kroger–Vink notation, is expressed [43]

$$\text{nil} \leftrightarrow V_A'' + V_O^{\bullet\bullet} \tag{5.61}$$

which describes the formation of a cation, V_A'', vacancy and an anion, $V_O^{\bullet\bullet}$, vacancy, from a neutral lattice, nil. As was previously stated, the anion vacancies are twice negatively charged, and the cation vacancies are twice positively charged with respect to the neutral lattice. The equilibrium constant in the case of the Schottky mechanism (Figure 5.22) is given by

$$K_S = C_{V_A} C_{V_C} \tag{5.62}$$

where
C_{V_A} is the equilibrium concentration of anion vacancies
C_{V_C} is the total concentration of cation vacancies

If oxygen is incorporated into or removed from the oxide lattice, it results in the formation of a nonstoichiometric oxide, where holes or electrons are produced (see Figure 5.23b for the removal of oxygen). The process of incorporation of oxygen is described by [8]

$$\frac{1}{2}O_2(g) + V_O^{\bullet\bullet} \leftrightarrow O_O^x + 2h^{\bullet} \tag{5.63}$$

where this equation describes that gaseous oxygen, $O_2(g)$, occupies an oxygen vacancy, $V_O^{\bullet\bullet}$, producing a neutral lattice, O_O^x, oxygen anion and two holes for charge compensation.

In the case of perovskites, a material of interest in the context of this book, oxygen defects may possibly be the prevailing ones while cations are only minority defects. Notwithstanding the fact that the perovskite formula is ABO_3 and not AO, the equilibrium equations previously expressed are as well satisfied.

5.7.2 Oxygen Transport in Oxides

Generally, oxygen diffusion occurs by means of oxygen vacancies (see Figure 5.23). For example, in the case of perovskites, there is no evidence of oxygen interstitials [44]. Therefore, only oxygen vacancies are considered here. Accordingly, in oxides, the self-diffusion coefficient or tracer-diffusion coefficient is given by [8].

$$D_O^* = f_O D_V C_{V_O^{\bullet\bullet}} \tag{5.64}$$

where

f_O is the geometrical correlation factor for oxygen tracer diffusion in the oxygen sublattice of the oxide structure

D_V is the self-diffusion coefficient of the oxygen vacancies

$C_{V_O^{\bullet\bullet}}$ is the concentration of oxygen vacancies

Following a phenomenological approach, the driving force for chemical diffusion, in oxides, is the gradient of the electrochemical potential, η_i

$$\eta_i = \mu_i + z_i F \Phi \tag{5.65}$$

where

μ_i is the chemical potential

z_i is the charge number of the charge carrier

Φ is the electric potential

F is the Faraday constant

The diffusion flux is given by

$$J_i = -L_{ii} \cdot \nabla(\mu_i + z_i F \Phi) \tag{5.66}$$

where L_{ii} is the Onsager transport coefficient. To preserve local electrical neutrality, the dynamic electroneutrality condition, whose general form is,

$$z_1 J_1 + z_2 J_2 = 0 \tag{5.67}$$

must be fulfilled. Now, during oxygen diffusion, if index 1 signifies O^{2-} and index 2 means e^-, we then have

$$-2J_{O^{2-}} - J_e = 0 \tag{5.68}$$

where applying Equation 5.66

$$J_{O^{2-}} = -L_{00} \cdot \nabla(\mu_{O^{2-}} - 2F\Phi) \tag{5.69}$$

and

$$J_{e^-} = -L_{ee} \cdot \nabla(\mu_{e^-} - F\Phi) \tag{5.70}$$

Assuming that the chemical equilibrium can be described as follows

$$\frac{1}{2}O_2 + 2e^- \leftrightarrow O^{2-} \tag{5.71}$$

which implies that

$$\nabla\mu_{O^{2-}} = \frac{1}{2}\nabla\mu(O_2) + 2\nabla\mu_{e^-} \tag{5.72}$$

Thereafter, combining Equations 5.68 through 5.70 and Equation 5.72, the flux of oxygen can be written with the help of the ambipolar diffusion equation [45]

$$J_{O^{2-}} = -L_{00}\frac{t_e}{2}\nabla\mu(O_2) \tag{5.73}$$

where t_e is the electron transference number given by

$$t_e = \frac{L_{ee}}{4L_{00} + L_{ee}} \tag{5.74}$$

Now,

$$\nabla\mu(O_2) = \left(\frac{\partial\mu(O_2)}{\partial C(O_2)}\right)\nabla C(O_2) \tag{5.75}$$

Combining Equations 5.73 and 5.75, we get

$$J_{O^{2-}} = -L_{00}\frac{t_{el}}{2} \cdot \left(\frac{\partial\mu_O}{\partial C_O}\right)\nabla C_O = -\tilde{D}\nabla C_O \tag{5.76}$$

As we have commented previously for metals, the diffusion in concentration gradients is described with the chemical diffusion coefficient, or interdiffusion coefficient. In this case, it is possible to consider that the interdiffusion and the intrinsic diffusion coefficients are equivalent, since we have only the movement of one species, that is, oxygen, by the vacancy mechanism. Subsequently, applying the Einstein relation

$$L_{00} = \frac{D_0 C(O_2)}{RT} \tag{5.77}$$

and the expression for the chemical potential of oxygen is

$$\mu(O_2) = \mu^O(O_2) + RT \ln a(O_2)$$

The chemical diffusion coefficient is given by [8]

$$\tilde{D} = D_O \frac{t_{el}}{2} \frac{\partial \ln a(O_2)}{\partial \ln C(O_2)} \qquad (5.78)$$

where

D_O is the oxygen self-diffusion coefficient

$a(O_2)$ is the oxygen activity

$C(O_2)$ is the oxygen concentration

t_{el} is the electronic transference number

When membranes prepared with the mixed-conducting perovskites are located in an oxygen partial pressure, $[P_{O_2}]$, gradient at elevated temperatures, a stream of oxygen molecules from the high to the low P_{O_2} side are produced [46]. During this process, oxygen molecules are reduced to oxide ions at the membrane surface in contact with the high P_{O_2} side, and are then transported through the oxide membrane to the low P_{O_2} side where the oxide ions are transformed to oxygen molecules. The electrical current ensuing from the transport of oxide ions is compensated by a current of electrons, so that neither electrode nor external circuitry is necessary to transport oxygen inside the membrane [47]. It is known that for relatively thick membranes, the oxygen permeation is limited by the counterdiffusion of oxide ions and electrons in the bulk of the membrane, and a decrease in the thickness of membrane leads to a proportional increase in the flux.

5.7.3 Defect Chemistry in Proton-Conducting Perovskites

The defects in doped $BaCeO_3$ perovskites, which are good proton conductors, are described here [32,33,48–54]. It has been recognized that temperature and atmosphere significantly affect the transport properties of the majority of protonic conductors [33]. Certainly, under a definite temperature range and specific atmosphere, doped $BaCeO_3$ has significant protonic conductivity [48–54].

The introduction of defects into the perovskite structure and their distribution in the structure are key factors that determine the protonic conductivity [50,51]. The inclusion of trivalent dopants ideally takes place, as described in the Kroger–Vink notation, [43] by

$$2Ce^x_{Ce} + O^x_O + Me_2O_3 \rightarrow 2M^\bullet_{Ce} + V^{\bullet\bullet}_O + 2CeO_2 \qquad (5.79)$$

where

Ce^x_{Ce} is a neutral (with respect to the lattice) Ce^{4+} cation in a lattice site

O^x_O is a neutral (as well with respect to the lattice) O^{2-} anion in a lattice site

$M^\bullet_{Ce}$ is the doping metal M^{3+} included in the lattice which has a unitary positive charge with respect to the lattice

$V^{\bullet\bullet}_O$ is a twice positively charged oxygen vacancy [50,51]

The introduction of protons into the perovskite is typically carried out with the help of gas streams containing H_2O (g) or H_2. Applying the Kroger–Vink notation again, it is possible to describe that the oxygen vacancies, $V^{\bullet\bullet}_O$, react with water to fill lattice positions with oxide ions, O^x_O, and produce interstitial protons, $H^\bullet_i$, according to [53,54]

$$H_2O(g) + V^{\bullet\bullet}_O \rightarrow O^x_O + 2H^\bullet_i \qquad (5.80)$$

Protons are retained in the material by combining with oxide ions at normal lattice sites, according to

$$O_O^x + H_i^\bullet \rightarrow OH_O^\bullet$$

Consequently, the net reaction describing the interaction of oxygen vacancies with water vapor to produce proton charge carriers can be written as follows:

$$H_2O(g) + V_O^{\bullet\bullet} + O_O^x \rightarrow OH_O'$$

If hydrogen is included in the gas stream, it is incorporated directly into the material as protons and electrons, e⁻, through interaction with oxide ions in the absence of moisture, according to [53]

$$\left(\frac{1}{2}\right)H_2 + O_O^x \rightarrow OH_O^\bullet + e'$$

5.7.4 Proton Transport Mechanisms

Normally in oxides, like perovskites, during hydrogen transport, the molecule is first dissociated in the surface of the oxide; then, the adsorbed hydrogen atoms are ionized and incorporated directly into the material as protons and electrons. Thereafter, two fundamental types of proton transport mechanisms in oxides are recognized: free migration and vehicle mechanisms. During the first mechanism, that is, the free migration mechanism or Grotthuss mechanism, the charge carrier, that is, the proton, moves by cation hopping or jumping between immobile host oxygen ions [31,33,53]. On the other hand, the second mechanism, that is, the vehicle mechanism, entails the movement of proton as a passenger on a larger ion like (OH)⁻ or (H₃O)⁺ [31,33,53]. The main proton conduction mechanism in oxides entails proton transfer between adjacent (OH)⁻ and O²⁻ and OH reorientation, that is, the Grotthuss mechanism, rather than OH diffusion [33]. Specifically, proton conduction in a very important group of proton conductors, that is, the perovskites, occurs by proton migration through the free migration mechanism [49,53–56].

Quantum molecular dynamic simulation studies [56–59] of proton conduction in $BaCeO_3$, $BaZrO_3$, $LaAlO_3$, $LaMnO_3$, and $CaZrO_3$ indicated that the proton, locally, relaxes the lattice to allow the transitory arrangement of hydrogen bonds, and, subsequently, proton transfer occurs between adjacent oxygen ions [56,60] (see Figure 5.24).

The calculated energy difference between the ground state and the barrier state is less than 0.2 eV for most of the perovskites studied, which is much less than the experimentally observed

Ground state   Barrier state

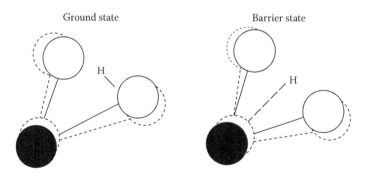

FIGURE 5.24 Arrangements for proton transfer between neighboring oxygen ions. The relaxed lattice is represented by solid lines and the ideal lattice is symbolized by the dashed lines.

activation energy (E_a) in ABO_3 perovskites, in the range of 0.35–1.1 eV for proton conductors [60]. For $BaCeO_3$, the quantum MD calculations [61–63] indicated that the amount of covalence between B site cations and oxygen anions, and the degree of hydrogen bonding within the lattice, are responsible for proton transport. Then, materials with a open crystal structure denotes larger separation between oxygen anions. Consequently, smaller B–O covalence provides softer B–O vibrations, assisting the transfer of protons between oxygen sites. Hence, a compromise between oxygen–oxygen separation and the stiffness of the B–O bonds must be achieved to maximize proton conductivity [61,62].

5.7.5 BAND STRUCTURE OF PROTON-CONDUCTING PEROVSKITES

The electronic structures of proton-conducting perovskites have been investigated using photoemission spectroscopy (PES) and x-ray absorption spectroscopy (XAS) [64–66]. For various proton conductors, such as $CaZrO_3$, $SrTiO_3$, and $SrCeO_3$, the Fermi levels (E_F) of dry-annealed conductors are located at the valence band side. However, the Fermi level is higher in H_2-annealed conductors. In dry-annealed proton conductors, it has been found that the hole state and the acceptor-induced level are observed at the top of the valence band and just above E_F, respectively [66]. Similar phenomena might be expected in $BaCeO_3$-doped perovskites. Thus, in $BaCe_{0.9}Y_{0.1}O_{3-\delta}$ dry-annealed and H_2-annealed conductors, the Fermi level (E_F) is in the band gap region as expected from the rigid band model, previously reported in the cases of $CaZrO_3$, $SrTiO_3$, and $SrCeO_3$. [66].

5.7.6 PROTON TRANSPORT MECHANISM IN OXIDES

The self-diffusion coefficient for protons can be written as follows [67,68]

$$D = D_0 \exp\left(-\frac{\Delta H_m}{RT}\right)$$

where

$$D_0 = \left(\frac{zN\nu l^2}{6}\right)\exp\left(-\frac{\Delta S_m}{R}\right)$$

in which
   z is the number of possible jump directions
   l is the jump distance
   N is the fraction of vacant jump destinations
   ν is the vibration frequency
   ΔS_m is the jump entropy
   $E_a = \Delta H_m$ is the activation energy for proton migration

However, if we follow a phenomenological approach, the diffusion in concentration gradients is described with the chemical diffusion coefficient or interdiffusion coefficient. Here, the chemical diffusion coefficient for the case where only protons and electrons are moving in the proton-conducting oxide is calculated. As was previously stated, the driving force for chemical diffusion in oxides is the gradient of the electrochemical potential, η_i, and the diffusion flux is then given by Equation 5.66. In this case, in order to preserve local electrical neutrality, the dynamic electroneutrality condition, whose general form is,

$$z_1 J_1 + z_2 J_2 = 0$$

must be fulfilled and the index 1 signifies H^+ and the subscript 2 means e^-. Then, we have

$$J_{H^+} + J_e = 0 \tag{5.81}$$

where applying Equation 5.66

$$J_{H^+} = -L_{HH} \cdot \nabla(\mu_{H^+} + F\Phi) \tag{5.82a}$$

and

$$J_{e^-} = -L_{ee} \cdot \nabla(\mu_{e^-} - F\Phi) \tag{5.82b}$$

The chemical equilibrium can be described as follows

$$\frac{1}{2}H_2 \leftrightarrow H^+ + e^-$$

which implies that

$$\nabla\mu_{H^+} = \frac{1}{2}\nabla\mu(H_2) - \nabla\mu_{e^-} \tag{5.83}$$

Thereafter, combining Equations 5.81 through 5.83; the flux of hydrogen can be written with the help of the ambipolar diffusion equation [45,69]

$$J_{H^+} = -L_{HH}\frac{t_e}{2}\nabla\mu(H_2)$$

where t_e is the electron transference number given by

$$t_e = \frac{L_{ee}}{L_{HH} + L_{ee}}$$

Now,

$$\nabla\mu(H_2) = \left(\frac{\partial\mu(H_2)}{\partial C(H_2)}\right)\nabla C(H_2)$$

Combining Equations 5.81 through 5.83 we get

$$J_{H^+} = -L_{HH}\frac{t_e}{2}\cdot\left(\frac{\partial\mu(H_2)}{\partial C(H_2)}\right)\nabla C(H_2) = -\tilde{D}\nabla C(H_2) \tag{5.84}$$

As we have commented previously for metals, the diffusion in concentration gradients is described with the chemical diffusion coefficient or the interdiffusion coefficient. Here, it is possible to consider that the interdiffusion and the intrinsic diffusion coefficients are equivalent, since we have the movement of only one species, that is, hydrogen. Subsequently, applying the Einstein relation

$$L_{HH} = \frac{D_H C(H_2)}{RT}$$

and the expression for the chemical potential of oxygen

$$\mu(H_2) = \mu^O(H_2) + RT \ln a(H_2)$$

can be obtained that the chemical diffusion coefficient is given by

$$\tilde{D} = D_H \frac{t_e}{2} \frac{\partial \ln a(H_2)}{\partial \ln C(H_2)} \tag{5.85}$$

where

D_H is the proton self-diffusion coefficient
$a(H_2)$ is the proton activity
$C(H_2)$ is the hydrogen concentration

5.7.7 ABSORPTION AND DIFFUSION OF HYDROGEN IN NANOCRYSTALS OF THE $BaCe_{0.95}Yb_{0.05}O_{3-\delta}$ PROTON-CONDUCTING PEROVSKITE

Samples of polycrystalline powders of composition $BaCe_{0.95}Yb_{0.05}O_{3-\delta}$ were synthesized using the conventional solid-state reaction method (see Section 3.3.1) [32]. A highly crystalline perovskite that does not include other crystalline phases was obtained (see Sections 4.2 through 4.5). The absorption kinetics, that is, the amount of hydrogen uptake in proton-conducting perovskites, was measured with a thermogravimetric analyzer (TQ500 Thermogravimetric Analyzer (TGA) produced by TA Instruments New Castle, DE, USA.) [32].

Prior to the measurement of the diffusion coefficient and the equilibrium absorption magnitude, the tested samples were carefully degassed at 1273 K for 6 h in a flow of the pure purge gas, that is, N_2. After degassing, the sample was set at the desired experimental temperature in a flow of the pure purge gas, and maintained at this temperature. Subsequently, the flowing gas is changed to a mixture of 4 wt % H_2 + 96 wt % N_2. The data collection, temperature control, programmed heating rate constantly maintained at 5 K/min, and gas switching were automatically controlled by using the TQ500 TGA software. The TGA data were collected as a M_t [mg] versus t [s], plot where M_t is the mass of hydrogen absorbed at time t. With the help of the M_t versus t plot, the chemical diffusion coefficient is evaluated using a solution of Fick's second law for a geometry appropriate for the experimental setup [20]. For a spherical geometry, the crystal geometry that we are considering here [20], and the appropriate boundary and initial conditions, the solution of Fick's second law is [5]

$$\frac{M_t}{M_\infty} = 1 - 3\frac{\tilde{D}}{\beta a^2}\exp(-\beta t)\left\{1 - \left(\frac{\beta a^2}{\tilde{D}}\right)^{1/2}\cot\left(\frac{\beta a^2}{\tilde{D}}\right)^{1/2}\right\}$$

$$+\left(\frac{6\beta a^2}{\tilde{D}\pi^2}\right)\sum_1^\infty\left(\frac{\exp\left(\dfrac{-\tilde{D}n^2\pi^2 t}{a^2}\right)}{n^2\left[n^2\pi^2 - \left(\dfrac{\beta a^2}{\tilde{D}}\right)\right]}\right) \tag{5.86}$$

where

M_t is the mass absorbed at time t
M_∞ is the mass absorbed at equilibrium
$\tilde{D}$ is the chemical diffusion coefficient in the case of the single-component diffusion of hydrogen

In addition, $r = a$ is the radius of the perovskite crystallite and β is a time constant that describes the evolution of the absorptive partial pressure in the dead space of the TGA furnace, that is, $P = P_0[1 - \exp(-\beta t)]$, where P_0 is the steady-state partial pressure and P is the partial pressure at time, t [20]. Consequently, the initial nonstationary partial pressure of the adsorbate in the gas stream is accounted for with the help of the parameter β. Cutting the series in Equation 5.86, and using only the first four terms, an approximation to the solution of Fick's diffusion equation is obtained, which is numerically fitted to experimental data. Then, for each experiment, the numerical values of the chemical diffusion coefficient, $\tilde{D}$, are calculated with the help of a nonlinear regression method [32]. The fitting process was carried out with the peak separation and analysis software program PeakFit [70] based on a least-square procedure, which allows to calculate the best-fitting parameters [71], that is, the numerical values of the relation between the chemical diffusion coefficient and the square of the particle radius, $\tilde{D}/a^2$, the equilibrium absorption mass, M_∞, the parameter, β, the regression coefficient, and the standard errors [32].

The calculated crystal radius is used for the computation of the chemical diffusion coefficient with the help of Equation 5.86, since the parameter determined during the fitting process is $A_1 = \tilde{D}/a^2$; consequently, $\tilde{D} = A_1 a^2$.

As was previously explained for metals, during hydrogen absorption, the molecule is first dissociated on the surface of the oxide. Subsequently, the adsorbed hydrogen atoms are ionized, and are incorporated directly into the material as protons and electrons, e^-, through interaction with the oxide ions, and, as explained later by another mechanism, interstitially located in tetrahedral and octahedral sites. Besides, since the proton will interact with the neighboring electron density, it, consequently takes, in a certain way, the form of an hydrogen atom [33]. Therefore, it is possible to consider that in this case, a neutral dissociation of hydrogen occurs as follows

$$\left(\frac{1}{2}\right) H_2(g) \rightarrow H_i^\bullet + e' \cong H$$

that is

$$\left(\frac{1}{2}\right) H_2(gas) \rightarrow H(solid\ solution)$$

The equilibrium constant for this reaction is given by Equation 5.52, which is equivalent to Sievert's law, where C_H is the hydrogen concentration in the perovskite, equivalent to the proton concentration, C_{H^+}, and P_{H_2} is the pressure of hydrogen in the gas phase. However, in our case, the number of sites for hydrogen absorption in the perovskite is limited. Therefore, the hydrogen concentration in the perovskite is restricted to a maximum value, C_H^0. Consequently, the relation between the hydrogen concentration and the pressure of hydrogen in the gas phase is given by a Langmuir-type absorption isotherm equation [27]:

$$C_H = \frac{C_H^0 K \sqrt{P_{H_2}}}{1 + K \sqrt{P_{H_2}}} \tag{5.87}$$

This is because we are assuming that in our system, the absorption process occurs in a finite number of immobile sites [20]. In the case of a surface, it is possible to show that it fulfills a Langmuir-type absorption isotherm, since, one of the conditions for Langmuir-type absorption or adsorption, is the existence of a fixed number, N, of absorption immobile sites, or adsorption immobile sites [20,72]. In this case, we have volume filling and not surface recovery; however, the final result is a Langmuir-type absorption isotherm [20] as in Equation 5.87, which also considers the dissociation of the hydrogen molecule [27].

In Figure 5.25 [32], the absorption magnitude (A_m, in wt %) at different temperatures, where $A_m = M_\infty/S_w$, M_∞ is the hydrogen mass absorbed at equilibrium, and S_w is the initial perovskite sample weight included in the TGA ceramic sample holder, is reported.

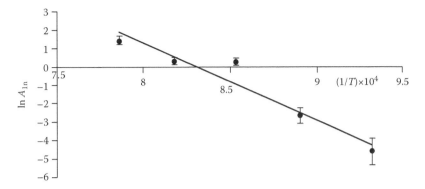

FIGURE 5.25 Plot of $\ln A_m$ vs. $10^4/T$ for the calculation of the enthalpy of absorption.

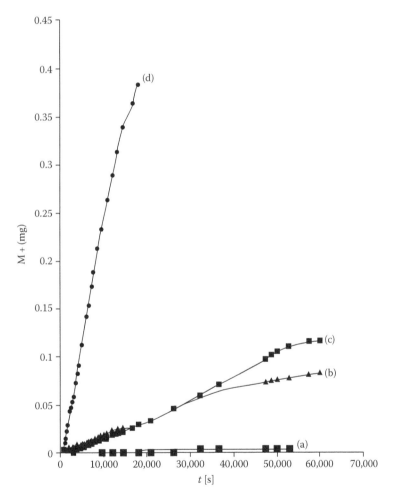

FIGURE 5.26 Absorption kinetics of hydrogen in $BaCe_{0.95}Yb_{0.05}O_{3-\delta}$ at (a) 1123, (b) 1173, (c) 1223, and (d) 1273 K.

The numerical value of M_∞ was calculated with the help of Equation 5.86 using the M_t [mg] versus t [s] data collected with the TGA, where M_t is the mass of hydrogen absorbed at time, t (see Figure 5.26 [32]). The relative error for A_m, calculated with the values of the standard error of M_∞ computed with the nonlinear regression methodology, [70,71] was around 25%. In the case of the test at 1073 K, the absorption was so small that the reported values are only estimations.

Currently, it is accepted that hydrogen absorption into a perovskite is carried out through the proton interaction with oxide ions according to [53,73]

$$\left(\frac{1}{2}\right)H_2 + O_O^x \rightarrow OH_O^\bullet + e'$$

Then, the proton-conducting perovskite $BaCe_{0.95}Yb_{0.05}O_{3-\delta}$, following the previously explained mechanism, can accommodate, on average, about 2.95 protons per cell in the ideal perovskite crystal lattice, since in the perovskite cell 2.95 oxide ions (see Figure 2.19b) are included. Consequently, since the perovskite cell has 1 Ba, 0.95 Ce, 0.05 Yb, and 2.95 O, the maximum amount of hydrogen that can be absorbed in this perovskite is in atomic wt % of H

$$N_A^m = \frac{2.97}{329.27} \times 100 = 0.00903 \times 100 = 0.9\,wt\,\%\,H$$

The atomic weights of Ba, Ce, Yb, O, and H, in atomic mass units (amu) per atom, are 37.34, 140.12, 173.04, 15.999, and 1.008, respectively. Then, the molecular weight of the perovskite in this case is 329.27 amu per cell.

In Table 5.1, experimental results which indicate higher absorption magnitudes are reported.

In metals, the conduction electrons are delocalized in the whole crystal, and consequently, the H^+ neighborhood results in the electron density of the conduction band [31,33]. Then, hydrogen in metals and small band gap semiconductors exist as a screened proton [9]. As a result, it is possible that the proton could have a high coordination number, and can be interstitially located in tetrahedral and/or octahedral sites [32].

Consequently, in order to explain our results, that is, the high absorption magnitudes measured, we advance the following hypothesis: at a high temperature, as is our case here, the proton can have a high coordination number, and can be interstitially located in tetrahedral and octahedral sites. Therefore, the reaction of hydrogen with the perovskite is given by [32]

$$\left(\frac{1}{2}\right)H_2(g) \rightarrow H_i^\bullet + e' \cong H_i$$

For the absorption in interstitial sites, [32]

$$\left(\frac{1}{2}\right)H_2 + O_O^x \rightarrow OH_O^\bullet + e' \cong H_O$$

TABLE 5.1
Experimental Values of Mass Absorbed at Equilibrium (M_∞) and the Absorption Magnitude ($A_m = M_\infty/S_w \times 100$) of H_2 in $BaCe_{0.95}Yb_{0.05}O_{3-\delta}$ at Different Temperatures and Sample Weights of the Tested Samples (S_w)

T [K]	1273	1223	1173	1123	1073
M_∞ [mg]	0.52 ± 0.09	0.14 ± 0.03	0.16 ± 0.04	0.005 ± 0.001	0.001 ± 0.001
A_m [wt %]	4.3 ± 0.7	1.3 ± 0.3	1.2 ± 0.3	0.06 ± 0.01	0.01 ± 0.01
S_w [mg]	12.131	10.674	13.304	8.208	10.960

in the case of interaction with the oxide anions. Notwithstanding the fact that currently the accepted paradigm is that the proton needs an oxygen to exist inside a perovskite [31,33,53], we need to accept the previously stated hypothesis in order to explain our results.

To validate our hypothesis, we can use the existing knowledge in relation to the band structure of the H_2-annealed $BaCe_{0.9}Y_{0.1}O_{3-\delta}$ perovskite [66]. For these materials, the Fermi level (E_F) is in the band gap region [66]. The band gap is approximately $5\,eV$ [66]. Besides, the E_F is located at approximately $1\,eV$ above the top of the valence band, that is, shifted to the conduction band [66]. All these facts indicate the introduction of electrons during the incorporation of H_2 into the perovskite. In addition, in H_2-annealed [66] $BaCe_{0.9}Y_{0.1}O_{3-\delta}$, the hole states are absent, and the intensities of acceptor and Ce 4f defect-induced levels decrease, indicating that the doped hydrogen exchanges with the hole at the top of the valence band. Besides, a new hydrogen-induced level is introduced just below E_F [32].

In this case, we have introduced more hydrogen than those incorporated in the previously referred paper [66]. Therefore, more electrons were introduced during the H_2 incorporation into the perovskite. Consequently, we can conclude that the studied material at 1023–1273 K behaves as a small band gap semiconductor, because of the increase of electrons with temperature in the conduction band, due to the shifting to the conduction band of the Fermi level and the hydrogen-induced level. Then, there will be a sufficient number of electrons in the conduction band to screen the proton and allow it to have a high coordination number, and be interstitially located in the tetrahedral and octahedral sites.

Subsequently, the results reported in Figure 5.25 and Table 5.1 can be explained if we include, besides the three oxygen, all the sites in the ideal perovskite structure as possible absorption sites for hydrogen (see Figure 2.19b).

That is, if we incorporate, besides the three oxygen, the eight tetrahedral sites and three of the four octahedral sites, one octahedral site is occupied by Ce, that is, the atom *B*. We then have a maximum of 13.95 hydrogen sites in the perovskite. Since the molecular weight of the synthesized perovskite is 326.296 and the maximum weight of the hydrogen absorbed is 14.112, the weight of the perovskite plus absorbed hydrogen is 340.408, which in wt % is [32]

$$N_A^m = \frac{14.112}{340.408} \times 100 = 0.0415 \times 100 = 4.15\,\text{wt}\,\%$$

The above stated hypothesis is supported by the existence of a high positive enthalpy of absorption, ΔH_{ab}^0, in the studied material (see Table 5.1 and Figure 5.25), which is consistent with the increase of electrons with temperature in the conduction band [32]. The value measured for the enthalpy of absorption, ΔH_{ab}^0, for this material uses the following expression (see Figure 5.25)

$$A_m = A_m^0 e^{\frac{-\Delta H_{ab}^0}{kT}} \tag{5.88a}$$

which in linear form is

$$\ln A_m = \ln A_m^0 - \frac{\Delta H_{ab}^0}{k}\left(\frac{1}{T}\right) \tag{5.88b}$$

This gives $\Delta H_{ab}^0 = (3.6 \pm 0.5)$ eV, and the linear regression coefficient, $r^2 = 0.93$. It is necessary at this point to affirm that positive values for the enthalpy of absorption are possible in metals [7] and oxides [53].

The justification of Equation 5.88a is simple. It is based on the equilibrium reaction $(1/2)H_2(gas) \rightarrow H(solid\ solution)$, whose equilibrium constant is [9]

TABLE 5.2
Chemical Diffusion Coefficient ($\tilde{D}$) and
Sample Coverage $\left(\theta = \dfrac{A_m}{N_A^m}\right)$ for Hydrogen
Diffusion in the $BaCe_{0.95}Yb_{0.05}O_{3-\delta}$ Perovskite

T [K]	$\tilde{D} \times 10^{18}$ [m²/s]	$\theta = A_m/N_A^m$
1273	26 ± 10	1.000
1223	8 ± 3	0.316
1173	7 ± 3	0.289
1123	2 ± 1	0.015
1072	1 ± 1	0.002

Source: Nieto, S. et al., *J. Phys. Chem. C*, 111, 2809, 2007.

$$K_H^0 = e^{\frac{-\Delta G_{ab}^0}{kT}} \tag{5.89}$$

where ΔG_{ab}^0 is the standard Gibbs energy change during the absorption process. Therefore, since $\Delta G_{ab}^0 = \Delta H_{ab}^0 - T\Delta S_{ab}^0$, Equation 5.88a results from Equations 5.52 and 5.89, where the pre-exponential factor, A_m^0, includes the square root of the hydrogen pressure, which is constant, together with the entropic factor and a constant for dimensional consistency. It is necessary now to acknowledge that Equation 5.88b allows us to get an approximate value of the enthalpy of absorption, ΔH_{ab}^0, because the equilibrium is described by the Langmuir-type absorption isotherm.

In Figure 5.26, the absorption kinetic data of hydrogen in $BaCe_{0.95}Yb_{0.05}O_{3-\delta}$ at 1123 K (Figure 5.26a), 1173 K (Figure 5.26b), 1223 K (Figure 5.26c), and 1273 K (Figure 5.26d) are reported. In Table 5.2, the chemical diffusion coefficient ($\tilde{D}$) for hydrogen diffusion in $BaCe_{0.95}Yb_{0.05}O_{3-\delta}$ at different temperatures and the sample coverage or fractional saturation of the perovskite, $\theta = A_m/N_A^m$, where A_m is the amount absorbed and N_A^m is the maximum amount absorbed, are reported.

The chemical diffusion and absorption parameters were calculated by fitting Equation 5.86 with the obtained experimental data [20] (see Table 5.2). The regression coefficient, r^2, for the nonlinear regression fitting process [71], calculated by the Peakfit software, was $r^2 = 0.97 \pm 0.01$. The relative error for $\tilde{D}$ calculated with the values of the standard error of $A_1 = \tilde{D}/a^2$, computed with the help of the nonlinear regression methodology [20,71], was around 35%, including the error in the determination of the particle size. In the case of the test at 1073 K, the absorption was so small that the reported values are only estimations.

5.8 DIFFUSION IN POROUS MEDIA

5.8.1 TRANSPORT MECHANISMS IN POROUS MEDIA

We have discussed previously diffusion in dense crystalline materials. Now, we study the transport of molecules in porous media. According to the classification scheme proposed by the International Union of Applied Chemistry (IUPAC), pores are divided into three categories on the basis of size: macropores (more than 50 nm), mesopores (from 2 to 50 nm), and micropores (less than 2 nm) [74,75].

There are four well-known types of diffusion in solids [10]: gaseous or molecular diffusion [75], Knudsen diffusion [76–80], liquid diffusion [10], and atomic diffusion. In Figure 5.27, the possible transport mechanisms in porous media are schematically shown [77]. Gaseous flow (Figure 5.27a)

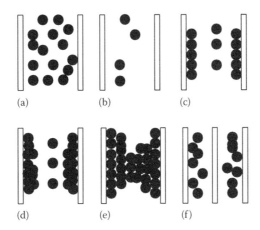

FIGURE 5.27 Transport mechanisms in porous media: molecular or gaseous flow. (a) Knudsen flow, (b) surface diffusion, (c) multilayer diffusion, (d) capillary condensation, and (e) configurational diffusion.

takes place when the pore diameter is larger than the mean free path of the fluid molecule. Then, collisions between molecules are more frequent than those between molecules and the pore surfaces. As the pore dimension decreases (Figure 5.27b), or the mean free path of the molecule increases, owing to the lowering of pressure, the flowing species tend to collide more and more with the pore walls that among themselves. Molecules, then, flow almost independently from one another according to the Knudsen flow [20,79].

In addition, surface flow (Figure 5.27c) is attained when the diffusing molecules can preferentially be adsorbed on the pore surfaces. An extension of this mechanism is multilayer diffusion (Figure 5.27d) [81], which can be considered as a transition flow regime between the capillary and surface flows. Now, if capillary condensation is attained (Figure 5.27e), and the diffusing component is condensed within the pore, the flow fills the pore and then evaporates at the other end of the pore. The last mechanism (Figure 5.27f) is configurational diffusion, which is active when pore diameters are small enough to let only small molecules to diffuse along the pores, while preventing the larger ones to get into the pores [10,12,20].

Configurational diffusion is the term coined to describe diffusion in zeolites and related materials, and is characterized by very small diffusivities (10^{-12} to 10^{-18} m^2/s), which is strongly dependent on the size and the shape of the guest molecules, high activation energies (10–100 kJ/mol), and concentration [82]. Zeolites and related materials are microporous crystalline solids of special interest in the chemical and the petroleum industries as catalysts and sorbents. For these applications, migration or diffusion of adsorbed molecules through the pores and cages within the crystals plays a dominant role.

If the diffusion process takes place at high temperatures, as is normally the case in applications, we have essentially three regimes with different diffusivities according to the pore diameter (see Figure 5.28 [82]). For macropores, that is, pores with diameters of 500 Å or larger, collisions between the molecules occurs much more frequently than collisions with the wall, and molecular diffusion is the dominant mechanism. Typically, the diffusion constants of gases are around 10^{-5} m^2/s [10]. At the same time, as the size of the pores decreases, the number of collisions with the wall increases; at this point, Knudsen diffusion takes over and the mobility starts to depend on the dimensions of the pore. At even smaller pore sizes, in the range of 20 Å or less, that is, when the pore diameter turns out to be similar to the size of the molecules, the molecules will constantly experience interaction with the pore surface. Consequently, diffusion in the micropores of a zeolite or related materials typically takes place in the configurational diffusion regime [20].

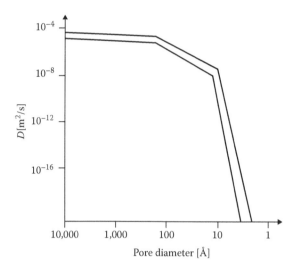

FIGURE 5.28 Relation between diffusivity and pore diameter.

5.8.2 VISCOUS VERSUS KNUDSEN FLOWS

For an ideal gas modeled as rigid spheres, the mean free path of the molecules, λ, can be related to the temperature, T, and pressure, P, via the following equation [2,3]

$$\lambda = \frac{kT}{\sqrt{2}\pi\sigma P} \qquad (5.90)$$

where

$\sigma_c = \pi d_m^2$ is the collision cross section of the molecules
d_m is the molecular diameter
k is the Boltzmann constant

To classify the different types of gas-phase flow, the ratio between the mean free path, λ, and the characteristic length of the flow geometry, L, commonly referred to as the Knudsen number K_n [80,83,84]

$$\frac{\lambda}{L} = K_n \qquad (5.91)$$

is applied. That is, according to the magnitude of K_n, three main flow regimes can be defined: viscous flow when $K_n \ll 1$, Knudsen flow for $K_n \gg 1$, and transition flow for $K_n \approx 1$ [80]. More precisely for $K_n < 10^{-2}$, the continuum hypothesis is correct; on the other hand, for $K_n > 10$, the continuum approach fails completely, and the regime can then be described as being a free molecular flow [83,84]. Under such a situation, the mean free path of the molecules is far greater than the characteristic length scale, and as a result, molecules reflected from a solid surface travel, on average, many length scales before colliding with other molecules [83,84]. In conclusion, in the large K_n region, continuum models, such as the compressible Navier–Stokes equation, do not hold [14,85]. In other words, when λ becomes comparable to L, the linear transport relationship for mass, diffusion, viscosity, and thermal conductivity is no longer valid [85]. For this reason, discrete models are proposed to examine the behavior of the rarefied gas flow [85]. A method to treat the large K_n region is the so-called direct simulation Monte Carlo (DSMC) method, which gives a solution to the Boltzmann equation without any assumptions on the form of the distribution function [86].

5.8.3 Viscous Flow in a Straight Cylindrical Pore

The gaseous self-diffusion coefficient is [2]

$$D^* = \frac{1}{3} <v> \lambda \tag{5.92}$$

Now,

$$<v> = \left(\frac{8kT}{\pi M} \right)^{1/2}$$

is the mean velocity, where M is the mass of the gas molecule and

$$\lambda = \frac{kT}{2^{1/2} \pi d_m^2 P}$$

is the mean free path. As a result [2,10],

$$D^* = \frac{2}{3\pi d_m^2} \left(\frac{kT}{P} \right) \left(\frac{kT}{\pi M} \right)^{1/2} \tag{5.93}$$

If the formerly discussed conditions for viscous diffusion are satisfied for a cylindrical macropore, that is, a pore of diameter larger than 50 nm, as soon as the pore diameter is large relative to the mean free path, collisions between diffusing molecules will take place considerably more often than collisions between molecules and the pore surface [2,20]. Under these circumstances, the pore surface effect is negligible, and, consequently, diffusion will take place by basically the same mechanism as in the bulk gas. Therefore, the pore diffusivity is equal to the molecular gaseous diffusivity (Equation 5.93).

5.8.4 Knudsen Flow in a Straight Cylindrical Pore

In the Knudsen regime, the rate at which momentum is transferred to the pore walls surpasses the transfer of momentum between diffusing molecules. The rate at which molecules collide with a unit area of the pore wall is [2]

$$\omega = \frac{C <v>}{4}$$

Now, if the average velocity in the flow direction is $<v_z>$, subsequently, the momentum flux per unit time to an element of area of the wall in the z-direction is (see Figure 5.29) [12,20]

$$F_z = \frac{C <v>}{4} (m <v_z>) (2\pi r dz)$$

Now, this force must be equalized by

$$F_z = -\pi r^2 dP$$

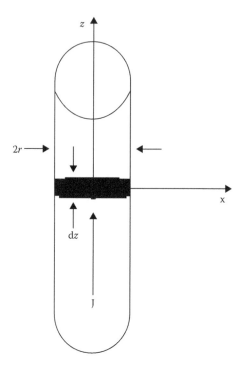

FIGURE 5.29   Knudsen flow in a straight cylindrical pore of diameter $d_p = 2r$.

where P is the gas pressure in the element of volume (see Figure 5.29). It is possible to calculate the flux in the pore with the following expression

$$J = C<v_z> = -D_K \frac{\partial C}{\partial z}$$

in which D_K is the Knudsen diffusivity [39]. It is easy to calculate D_K knowing the value of $<v> = (8kT/\pi M)^{1/2}$, from the kinetic theory of gases, and that $P = kCT$ [2,3] for an ideal gas. Consequently,

$$D_K^* = \frac{d_P}{2} \left(\frac{\pi kT}{2M} \right)^{1/2}$$
(5.94)

where
   d_p is the pore diameter
   M is the mass of the gas molecule

Then, if the formerly discussed conditions for Knudsen diffusion are satisfied for a mesopore, that is, a pore of diameter in the range between 20–50 nm, the diffusion coefficient for the Knudsen flow in a straight cylindrical mesopore is described by Equation 5.94.

5.9   DIFFUSION IN MICROPORES

5.9.1   Mechanism of Diffusion in Zeolites

We have already explained the structure of zeolites and related materials (see Section 2.5.1). Diffusion in zeolites is a very important industrial problem and, consequently, it has been comprehensively studied [12,20,87–103]. In several processes using zeolites, the rate of diffusion of adsorbed molecules inside the zeolite pore system plays an important, and sometimes critical, role in determining the overall observed performance [96]. However, diffusion in zeolites is inadequately understood owing to the sensitivity of zeolite diffusivities on the dimensions of the diffusing molecules, zeolite pores and cavities, and on energetic interactions, such as those between adsorbates and the zeolitic framework, and charge-compensating cations [98]. In particular, multicomponent diffusion has not received the necessary consideration in comparison to single-component diffusion, even though such multicomponent behavior is of importance in the practical applications of zeolites and related materials.

When a molecule diffuses inside a zeolite channel, it becomes attracted to and repelled by different interactions, such as the dispersion energy, repulsion energy, polarization energy, field dipole energy, field gradient quadrupole, adsorbate–adsorbate interactions, and the acid–base interactions with the active site if the zeolite contains hydroxyl bridge groups [20]. This transport can be pictured as an activated molecular hopping between fixed sites [12,20,88]. Therefore, during the transport of gases through zeolites, both diffusion between localized adsorption sites and the gas translation diffusion will contribute to the overall process [88].

Subsequently, it is possible to consider that the adsorbate–adsorbent interaction field inside these structures is characterized by the presence of sites of minimum potential energy for the interaction of adsorbed molecules with the zeolite framework and charge-compensating cations. A simple model of the zeolite–adsorbate system is that of the periodic array of interconnected adsorption sites, where molecular migration at adsorbed molecules through the array is assumed to proceed by thermally activated jumps from one site to an adjacent site, and can be envisaged as a sort of lattice-gas.

In order to describe the adsorption and diffusion in the zeolites in the framework of a modified lattice-gas, which takes into account the crystalline structure of the zeolite, the interaction among adsorbed molecules and the possibility of a transition of adsorbed molecules among different adsorption sites in the same unit cell and different unit cells that follows the model description of molecular diffusion in zeolites were previously proposed [88,104].

In the framework of the present model's description of diffusion, the zeolite is considered to be a three-dimensional array of N identical cells, i, centered at R_i, each containing N_0 identical sites localized at $R_{i\alpha} = R_i + U_\alpha$, where the potential energy is a minimum (see Figure 5.30 [104]). If a molecule is localized at $R_{i\alpha}$, its energy would be $-\varepsilon$ and the interaction energy between molecules

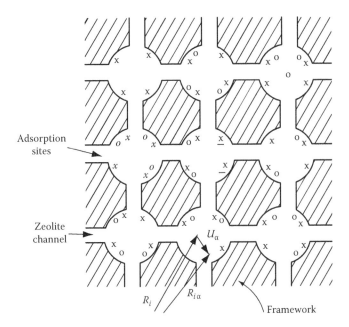

FIGURE 5.30 Schematic representation of the zeolite and a related material.

localized at different sites, that is, $R_{i\alpha}$ and $R_{i\beta}$, is $-U_{\alpha\beta}$ (see Figure 5.30 [104]). In the frame of the model the motion of molecules between sites, that is, the jumping of molecules inside a cell, is considered, and the motion between cells, through zeolite channels and cavities, is also taken into account [20,88,104]. The solution of the motion equation for a molecule in the present system leads to two energy states an $N(N_0 - 1)$-fold degenerate state, E_2, where molecules are adsorbed in a site and, N, delocalized states, E_1, where the molecule moves through the zeolite, jumping from site to site through the zeolite channels and cavities [104]. Between adsorption and delocalized states, there exists an energy gap E_g [88,104]. For low coverage of the adsorption space, that is, at low pressures in the Henry's law region of adsorption, $E_g = E_1 - E_2$, which means that diffusion is an activated process in the frame of the present model [88,104].

The delocalized state can be considered to be a transition state, and transition state theory [105], a well-known methodology for the calculation of the kinetics of events, [12,88,106–108] can be applied. In the present model description of diffusion in a zeolite, the transition state methodology for the calculation of the self-diffusion coefficient of molecules in zeolites with linear channels and different dimensionalities of the channel system is applied [88]. The transition state, defined by the delocalized state of movement of molecules adsorbed in zeolites, is established during the solution of the equation of motion of molecules whose adsorption is described by a model Hamiltonian, which describes the zeolite as a three-dimensional array of N identical cells, each containing N_0 identical sites [104]. This result is very interesting, since adsorption and diffusion states in zeolites have been noticed [88].

Hence, consider an arrangement of adsorption sites in a linear channel in which each site is an energy minimum with respect to the neighboring space [88]. Besides, the number of adsorbed molecules is supposed to be small relative to the number of sites, so that each molecule can be considered to be an independent subsystem with enough free sites for jumping [88]. The canonical partition function for a molecule considered to be an independent subsystem in one of the adsorption sites with energy E_2 is [107,108]

$$Z_a = Z_x^a Z_y^a Z_z^a Z_i^a \exp\left(-\frac{E_2}{RT}\right) \tag{5.95}$$

where

Z_x^a, Z_y^a, and, Z_z^a represent the canonical partition functions for movement in directions x, y, and z of the adsorbed molecule

Z_i^a represents the partition function for the internal degrees of freedom of the molecule

E_2 is the energy of the adsorption state

We now deal with the translation of the molecule in the z-direction through the transition state, that is, the delocalized state of movement. The partition function for a molecule in the transition state becomes [12]

$$Z^* = Z_x^* Z_y^* \left(\frac{2\pi MkT}{h^2}\right)^{1/2} \Delta z Z_i^* \exp\left(-\frac{E_1}{RT}\right) \tag{5.96}$$

where

Z_x^* and Z_y^* are the partition functions for the movement in the directions x and y of the molecule in the transition state

Δz is the distance of movement in the transition state, in which $\Delta z = l$, where l is the jump distance

M is the mass of the molecule

R is the gas constant

k is the Boltzmann Constant

h is Planck's constant

T is the absolute temperature

If we suppose now, as is the case in classical transition state theory, that the molecules in the ground (adsorbed state) and transition states (delocalized movement state) are in equilibrium, it can be easily shown that the equilibrium constant for this equilibrium process is [12,109]

$$K = \frac{N^*}{N_a} = \frac{L^*}{L} \exp\left(-\frac{E_g}{RT}\right)\left(\frac{2\pi MkT}{h^2}\right)^{1/2} \frac{\Delta z}{Z_z^a} \qquad (5.97)$$

If

$$\left(\frac{Z_x^* Z_y^*}{Z_x^a Z_y^a}\right)\left(\frac{Z_i^*}{Z_i^a}\right) \approx 1$$

where N^* is the number of molecules in the transition state, N_a is the number of adsorbed molecules, L^* is the maximum number of transition states, L is the maximum number of adsorption sites, and Z_z^a is the partition function for the motion of the molecule in the z-direction in the adsorption state [88]. The average velocity of translation of the molecule through the transition state is [105]

$$<v_z> = \left(\frac{2kT}{\pi M}\right)^{1/2}$$

Consequently, the time of residence of the molecule in the transition state [88] is

$$T = \left(\frac{<v_z>}{\Delta z}\right) = \frac{<v_z>}{l}$$

The number of molecules passing through the transition state per unit time is [88]

$$\vartheta = \left(\frac{dN^*}{dt}\right) = \frac{N^*}{T} \qquad (5.98)$$

In addition, the jump frequency of the molecules through the transition state is [88]

$$\Gamma = \frac{\vartheta}{N_a} \qquad (5.99)$$

Considering the dynamic equilibrium between the adsorbed state and the diffusion state [12], Γ, the jump frequency of molecules between sites is [12]

$$\Gamma = \frac{L^*}{L}\left(\frac{2kT}{h}\right)\exp\left(-\frac{E_g}{RT}\right)\frac{1}{Z_z^a} \qquad (5.100)$$

where $\dfrac{L^*}{L}$ = 1, 2, or 3 depending on the dimensionality of the channel system [12]. Now, the self-diffusion coefficient for a molecule in a zeolite is

$$D^* = \left(\frac{l^2}{2K\tau} \right) \tag{5.101}$$

in which

l is the jump distance

$K = 1, 2,$ or 3 is the dimensionality

$\tau = \dfrac{1}{\Gamma}$ is the time between jumps

We must now take into account two extreme situations. The first is where the adsorption in the site is very strong, that is, localized adsorption, and the partition function for the movement in the z-direction in the adsorption site is a vibration partition function for a molecule in a potential energy well [110]

$$Z_z^a = \left(\frac{kT}{h\upsilon} \right) \tag{5.102}$$

where

υ is the vibration frequency

k is the Boltzmann constant

The second situation is where the adsorption is delocalized; the molecule can move in the neighborhood of the site, and the partition function for the movement in the z-direction in the adsorption site is the translational partition function [109]:

$$Z_z^a = \left(\frac{2\pi MkT}{h^2} \right)^{1/2} l \tag{5.103}$$

We can obtain a diffusion coefficient for the case of localized adsorption on the sites by introducing in Equation 5.101 the expression for τ, described in Equation 5.100, and using Equation 5.102 for Z_z^a, the vibrational partition function [88]:

$$D_i^* = \upsilon l^2 \exp\left(-\frac{E_g}{RT} \right) \tag{5.104}$$

The diffusion coefficients for mobile adsorption can also be calculated by introducing in Equation 5.56 the expression for τ, described in Equation 5.55, and using Equation 5.103 for Z_z^a, the translational partition function:

$$D_m^* = \left(\frac{kT}{2\pi M} \right)^{1/2} l \, e^{-E_g/RT} \tag{5.105}$$

Equations similar to Equations 5.104 and 5.105 for the self-diffusion coefficients are calculated using different approaches, and they result in [10]

$$D^* = gul \exp\left(-\frac{E}{RT} \right) \tag{5.106}$$

where

$g = \dfrac{1}{z}$, where z is the coordination number

u is the velocity at which the molecule travels, where $u = \upsilon l$ during localized adsorption and

$u = \left(\dfrac{8RT}{\pi M} \right)^{1/2}$ during mobile adsorption

l is the jump distance or diffusional length
M is the molecular weight

5.9.2 SINGLE-COMPONENT DIFFUSION IN ZEOLITES

In this section, single-component diffusion in zeolites [20,87–92], with the help of the case study of the diffusion of p-xylene and o-xylene in H-ZSM-11 and H-SSZ-24 zeolites, is discussed [90]. SSZ-24 is a 12-MR zeolite that was first obtained as the silicon counterpart of $AlPO_4$-5, and, later, was obtained as a borosilicate (B-SSZ-24), which could be exchanged with Al to yield the H-SSZ-24 zeolite [111]. This zeolite exhibits the AFI framework, which embodies a one-dimensional channel network without cavities, consisting of parallel 12-MR channels with a free-channel diameter, $\sigma_w = 7.0\,\text{Å}$ [112]. Additionally, ZSM-11 encloses an intersecting two-directional 10-MR channel system, where the two-dimensional channel system is characterized by the free-channel diameter, $\sigma_w = 5.8\,\text{Å}$ [112].

An experimental facility was described in Section 4.5.9 (see Figure 4.22) that was used to carry out the characterization of the groups present on the surface of a porous material or the channels and/or cavities of a microporous material applying the FTIR methodology. With this methodology, it is also possible to measure different diffusion coefficients in microporous materials with the help of the FTIR method [87–92]. Here, a laboratory-assembled facility similar to that reported in Section 4.5.9 that has two manifolds (Figure 5.31) instead of one, for the introduction of the diffusing molecules, and thus has the capability to deliver two different hydrocarbons to the IR cell, is described [90].

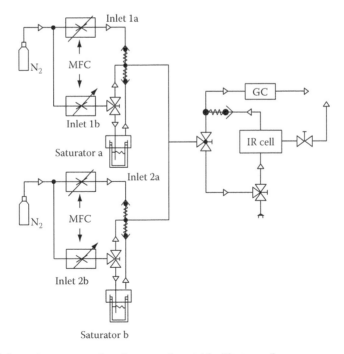

FIGURE 5.31 Schematic representation of an experimental facility to perform measurements of the Fickean diffusion coefficient by the FTIR method.

The equipment is composed of the IR cell connected through stainless steel pipes to two manifolds containing two thermostated saturators (see Figure 5.31). Both manifolds have two gas inlets for the carrier gas (helium [87] or nitrogen [88–92], with purity grade 99.99%). Through inlet 1, the carrier gas bubbles through the saturator in which the adsorbate is located (see Figure 5.31). The carrier gas coming from inlet 1 (a or b), which is saturated with the corresponding hydrocarbon, is mixed with a measured flow of pure carrier gas coming from inlet 2 (a or b). Sensitive mass-flow controllers (MFC) must be used for both the inlets. The simultaneous variations of both flow controls enable us to vary the relative partial pressure of the hydrocarbon in the range $0.01 < P/P_0 < 0.9$.

The measurements were carried out as follows: the compound to be tested was filled in one or both of the stainless steel saturators, which were held thermostatically at 25°C. The flow of the gas carrier was divided in two, and each of these streams were passed through a flow controller [90]. One stream went through a tube with a sinterplate in the base into the saturator where the carrier gas bubbles into the liquid adsorbate. This gas flow, that is, the stream which has passed through the saturator and becomes saturated with the test substance, was then mixed with the bypass stream of pure carrier gas at the outlet of the saturator, and the unified stream was then passed through the IR-cell [90].

The measurement process is normally carried out by monitoring the change in the intensity, that is, the absorbance, of a band of the adsorbate molecule in an FTIR spectrometer obtaining spectra consisting of 1 scan per spectrum at 0.85 s per scan, without delay between scans, selecting the proper range for the tested adsorbate. For p-xylene the range from 1480 to 1550 cm^{-1} and the band around 1517 cm^{-1} were measured and the double peak with bands at 1467 and 1497 cm^{-1} and the range from 1420 to 1520 cm^{-1} for o-xylene, was used [88,90].

The key equipment for testing is the water-cooled IR high-temperature cell (see Figure 5.31 [88,90]). In these cells, demountable parts are normally fitted with Viton O-rings to minimize leaking. The temperature of the sample holder is controlled electronically with very low variation, normally $\Delta T < 1$°C [88–92].

Self-supported wafers obtained by pressing 7–9 mg/cm^2 of the zeolite sample powder at 400 MPa are placed on the sample holder and introduced in the cell. These wafers counteract the limitations due to the absence of macropores for the transport of the diffusing molecules during the adsorption process, enabling measurement of intracrystalline diffusion [87,88]. Prior to the measurement of the diffusion coefficient, the samples must be carefully degassed at 450°C for 2 h in a flow of the pure carrier gas. After degassing, the sample is cooled to the desired temperature and maintained at this temperature with the help of the temperature control. The flow rate is then adjusted to get the desired relative partial pressure. A background spectrum of the pure degassed zeolite is obtained as a reference, and then the flow coming from the saturator after mixing with the flow of pure carrier gas is admitted at a precisely defined pressure to the IR cell. The collection of the set of IR spectra is started at the same moment of the admittance of the diffusing molecule, and the spectra are stored as the difference spectra obtained by subtracting from each measurement a background spectrum of the pure degassed zeolite [87–92].

This equipment can be used for the study of a single-component diffusion, and the measurement of the corresponding Fickean diffusion coefficient made using a solution of Fick's second law for a geometry appropriate for the experimental setup [87–92]. In this case, the flow rates were adjusted to get the desired partial pressure (6.7 Pa, $P/P_0 = 0.006$) [90].

For a spherical geometry, which is the case if the zeolite crystals which conform the wafer are approximately spherical with a variable surface concentration and the initial concentration inside the sphere equals zero, the solution of Fick's second law is given by an equation very similar to Equation 5.86 [5]

$$\frac{M_t}{M_\infty} = 1 - 3\frac{D}{\beta a^2}\exp(-\beta t)\left\{1 - \left(\frac{\beta a^2}{D}\right)^{1/2}\cot\left(\frac{\beta a^2}{D}\right)^{1/2}\right\}$$

$$+ \left(\frac{6\beta a^2}{D\pi^2}\right)\sum_1^\infty\left(\frac{\exp\left(\frac{-Dn^2\pi^2 t}{a^2}\right)}{n^2\left[n^2\pi^2 - \left(\frac{\beta a^2}{D}\right)\right]}\right) \tag{5.107}$$

where M_t is proportional to the absorbance A [au] (where A is reported in arbitrary units, that is, $M_t \sim A_t^T$, is the amount of adsorbate taken at time, t, and M_∞ also proportional to the absorbance, A, that is, $M_\infty \sim A_\infty^T$ is the equilibrium adsorption value, and D is the Fickean diffusion coefficient in the case of single-component diffusion [87,88]. In addition, $r = a$ is the radius of the zeolite crystallite, and β is a time constant that describes the evolution of the adsorptive partial pressure in the dead space of the IR cell, that is [5]:

$$P = P_0[1 - \exp(-\beta t)]$$

where

P_0 is the steady-state partial pressure

P is the partial pressure at time, t

Consequently, the initial nonstationary partial pressure of the adsorbate in the gas stream is accounted for with the help of the parameter β [87]. Cutting the series included in Equation 5.107, and using only the first four terms, an approximation to Equation 5.107 is obtained, which is numerically fitted to the experimental data [88,90]. In Figure 5.32, some examples of the uptake curves fitted with Equation 5.107 are reported (see Ref. [88] for more details).

With the help of Equation 5.107, as was previously done with Equation 5.86, we obtain a transport or chemical diffusion coefficient that is a result of Fick's laws. We now interpret the meaning of this coefficient: if we consider diffusion in a microporous solid, as a special case of binary diffusion, where A is the mobile species and the diffusivity of the microporous framework atoms is zero, then, the frame of reference are the fixed coordinates of the porous solid; consequently, we have a particular case of interdiffusion where the diffusion coefficient is simply the diffusivity of the mobile species [12,20].

The driving force of diffusion is the gradient of the chemical potential. In the case of single-component diffusion in a zeolite, the chemical potential can be related to the concentration by considering the equilibrium vapor phase [12,20]

$$\mu_A = \mu_A^0 + RT\ln P_A \tag{5.108}$$

in which

P_A is the partial pressure of the component, A

μ_A^0 is the chemical potential of the standard state of component, A

Mass transport is described by an atomic or molecular mobility, M_A, which is defined by [13]

$$\bar{v}_A = M_A\bar{F}_A \tag{5.109a}$$

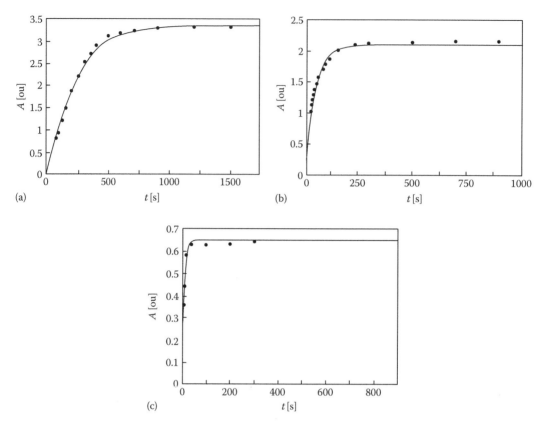

FIGURE 5.32 Examples of uptake curves.

where $\bar{v}_A$ is the average drift velocity, and

$$\overline{F}_A = -\nabla\mu_A \tag{5.109b}$$

is the force on particle A, in the case where the only driving force is a concentration gradient. In addition, it is well known that

$$J_A = v_A C_A \tag{5.110}$$

Now, with the help of the previous relations, we can obtain for an one-dimensional diffusion [12,20]:

$$J_A = -RTM_A \left(\frac{d \ln P_A}{d \ln C_A} \right) \left(\frac{dC_A}{dx} \right) \tag{5.111}$$

Consequently, the chemical diffusion coefficient is given by [12,20]

$$D_A = RTM_A \left(\frac{d \ln P_A}{d \ln C_A} \right) = D_0 \left(\frac{d \ln P_A}{d \ln C_A} \right) = D_0 \Psi \tag{5.112}$$

in which

$D_0 = RTM_A$ is the intrinsic or corrected diffusion coefficient

Ψ is the thermodynamic factor

If we apply the Barrer and Jost model for the calculation of the self-diffusion coefficient [72] that is valid for a Langmuir system where the probability of a successful jump is proportional to the number of vacant sites [12], we get the corrected or intrinsic diffusion coefficient

$$D_0 = D_A(1 - \theta) \tag{5.113}$$

where $\theta = n_a/N_a$ is the fractional saturation of the adsorbent, in which n_a is the amount adsorbed and N_a is the maximum amount adsorbed.

In Table 5.3 (see Figure 5.33), the corrected diffusion coefficients of p-xylene and o-xylene in H-ZSM-11 and H-SSZ-24 zeolites, calculated by the fitting of Equation 5.107 to the uptake data measured with the FTIR spectrometer, and then corrected with Equation 5.113, are reported [90]. In Figure 5.33, two examples of uptake curves, that is, the sorption kinetics for the single-component diffusion of p-xylene and o-xylene in H-SSZ-24 zeolite at 400 K, are shown [90]. The use of Equation 5.107 is completely justified only for a set of uniform spherical particles of radius, a. In the case of the H-ZSM-11 and H-SSZ-24 zeolites, the scanning electron microscopy (SEM) study of the samples was carried out in a JEOL-ISM 6300 Electron Microscope to determine crystal size and morphology [90]. It was shown that the sample of the H-SSZ-24 zeolite exhibits crystals with "starburst" morphology whose average sizes are $a = 1.4\,\mu m$, $b = 0.7\,\mu m$, and $c = 0.7\,\mu m$; likewise, the H-ZSM-11 sample presented regular "coffin-" shaped crystals with the following average sizes: $a = 1.8\,\mu m$, $b = 0.6\,\mu m$, and $c = 0.7\,\mu m$ [90]. In this case, it is possible to define an equivalent spherical radius by means of the equation [103]

$$r = \frac{2}{3}\left(\frac{1}{a} + \frac{1}{b} + \frac{1}{c}\right) \tag{5.114}$$

The probe molecules used in the single diffusion are p-xylene and o-xylene. The kinetic diameter of these probe molecules (σ_m) are $\sigma_m = 5.8\,\text{Å}$ for p-xylene, and $\sigma_m = 7.0\,\text{Å}$ for o-xylene [11].

On the other hand, the Eyring equation:

$$D_0 = D_0^* e^{-Ea/RT} \tag{5.115}$$

TABLE 5.3
Corrected Diffusion Coefficients
($D^* \times 10^9$ cm²/s) of H-ZSM-11 and
H-SSZ-24 Zeolites at Different Temperatures

Sortive	H-ZSM-11	H-SSZ-24	T [°C]
p-Xylene	0.2	1.0	350
	1.4	1.2	375
	1.1	1.0	400
	3.0	1.4	425
o-Xylene	0.02	–	375
	0.04	2.4	400
	0.12	3.5	425

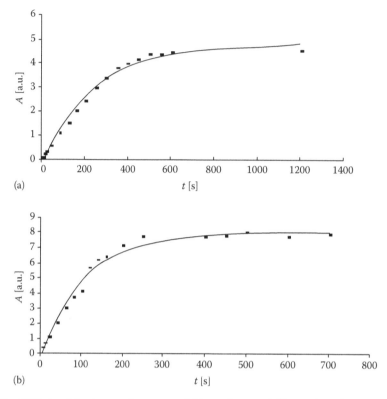

FIGURE 5.33 Diffusion kinetics uptake curve of (a) p-xylene and (b) o-xylene in H-SSZ-24 at 400 K. Experimental points (−) fitted with the continuous theoretical curve.

was used for the calculation of the activation energy, E_a, and the pre-exponential factor, D_0^*. The calculated values for the diffusional activation energy, E_a, and the pre-exponential factor are reported in Table 5.4 for the studied systems [90].

The diffusion of guest molecules through zeolite cavities and channels is strongly influenced by geometrical factors [20]. A parameter that geometrically characterizes diffusion in zeolites is η, which is given by the relation $\eta = \sigma_m/\sigma_w$, where σ_m is the kinetic diameter of the guest molecule [11] and σ_w is the window or channel diameter [11,12]. Neither σ_m nor σ_w can be precisely described by a number because the guest molecules cannot be considered as rigid molecules nor the windows

TABLE 5.4
Diffusional Activation Energies and Pre-Exponential Factors of the Eyring Equation for the Diffusion of H-ZSM-11 and H-SSZ-24 Zeolites at Different Temperatures

Zeolite	Sortive	E_a [kJ/mol]	$D_0^* \times 10^4$ [cm²/s]
H-ZSM-11	p-Xylene	41	3.3
	o-Xylene	48	0.9
H-Beta	p-Xylene	4	0.00007
	o-Xylene	16	0.003

of the channels can be regarded as rigid doors [11,20]. Nevertheless, the accumulated experimental knowledge indicates that generally if $\eta < 1$, the diffusion process is relatively free and if $\eta > 1$ the transport process in zeolites is usually obstructed by geometric factors.

For the systems described in this chapter, the values for σ_m and σ_w are $\sigma_m = 5.8$ and $\sigma_m = 7.0\,\text{Å}$ for p-xylene and o-xylene, respectively [11,12] and $\sigma_w = 7\,\text{Å}$ for the SSZ-24 channel windows and $\sigma_w = 5.8\,\text{Å}$ for the ZSM-11 zeolite [112]. Therefore, p-xylene and o-xylene relatively, freely move in H-SSZ-24 during single-component diffusion, inasmuch as $\eta_{p-x} = 0.83$ and $\eta_{o-x} = 1.00$. In addition, for H-ZSM-11, the single-component diffusion of o-xylene is hindered by steric factors inasmuch as $\eta_{o-x} = 1.21$, but the single-component diffusion for p-xylene is relatively free since $\eta_{p-x} = 1$. These facts are reflected on the reported single-component diffusion coefficients (see Table 5.3). Besides, the results reported in Table 5.3 reasonably agree with data previously reported in the literature for the diffusion of xylenes in zeolites with 10- and 12-MR channels [88,116–120].

Both p- and o-xylene experience interaction with the Brönsted sites of acid zeolites, since the relative basicity for p- and o-xylene are 1.00 and 1.13, respectively [121–123]. In this regard, a possible mechanism for the diffusion process in H-SSZ-24 and H-ZSM-11 is composed of jumps of the xylene molecules between acid sites; this is the so-called hopping between sites or the jump diffusion mechanism [11,12,88]. In the realm of the jump diffusion mechanism, it is possible to calculate the self-diffusion coefficient [88]. Then, the numerical evaluation of the pre-exponential factors in Equations 5.104 and 5.105, was possible because all the terms included in the equations are well defined, that is $R = 8.3\,\text{kJ/(mol K)}$, $T = 300\text{–}400\,\text{K}$, $M = 106\,\text{g/mol}$, $l = 10\,\text{Å}$ [88], and $\nu = 10^{12}\text{–}10^{13}\,\text{s}^{-1}$ [72]. The calculations result in a pre-exponential term for localized adsorption in the range $10^{-1}\,\text{cm}^2/\text{s} < D_0^* < 10^{-2}\,\text{cm}^2/\text{s}$, and a pre-exponential term for mobile adsorption in the range $3 \times 10^{-4}\,\text{cm}^2/\text{s} < D_0^* < 5 \times 10^{-4}\,\text{cm}^2/\text{s}$ [90]. If the approximation $D^* = D_0$, [115] is made, then it is possible to compare the values reported in Table 5.4 and the calculated values. The comparison indicates that the pre-exponential term for localized adsorption does not agree with the experimentally obtained pre-exponential terms, and, consequently, we can conclude the diffusion of aromatic hydrocarbons in highly siliceous acid zeolites is not related to strong adsorption [28]. This conclusion was also reached by others for the diffusion of benzene and toluene in the ZSM-5 zeolite [116,117].

In the case of H-SSZ-24, the values of the pre-exponential factor experimentally obtained (see Table 5.4) do not agree with the values theoretically predicted by the equation for a jump diffusion mechanism of transport in zeolites with linear channels, in the case of mobile adsorption [6,26]. Furthermore, the values obtained for the activation energies are not representative of the jump diffusion mechanism. As a result, the jump diffusion mechanism is not established for H-SSZ-24. This affirmation is related to the fact that in the H-SSZ-24 zeolite Brönsted acid sites were not clearly found (see Figure 4.4.); consequently p- and o-xylene do not experience a strong acid–base interaction with acid sites during the diffusion process in the H-SSZ-24 channels, and, therefore, the hopping between sites is not produced.

In the case of H-ZSM-11, the results reported in Table 5.4 indicate that the mechanism for p- and o-xylene diffusion is by hopping between sites or jump diffusion, since the values reported for D_0 are within the limits $1 \times 10^{-4} < D_0^m < 5 \times 10^{-4}\,\text{cm}^2/\text{s}$. Hence, the values of the pre-exponential factor experimentally obtained agree reasonably well with the values calculated with the help of Equation 5.105. Furthermore, the values obtained for the activation energies are also representative of the jump diffusion mechanism [12]. Consequently, the jump diffusion mechanism is established for H-ZSM-11.

5.9.3 Two-Component Diffusion in Zeolites

This section discusses counterdiffusion in a binary system as well in zeolites, with the help of the case study of the counterdiffusion of p-xylene plus o-xylene in H-ZSM-11 and H-SSZ-24 zeolites [90].

In a counterdiffusion experiment of two molecules, A and B, the sample is initially saturated with a stream of the adsorbate, A, at a fixed partial pressure; then, to this stream of carrier gas plus gas, A, the carrier gas saturated with gas, B, is admitted to finally obtain the same partial pressures for both components. [113,114]

The Onsager irreversible thermodynamics approach in terms of the Fick's law methodology is, between the most frequently applied procedures to describe diffusion in mixtures. For a binary mixture (i.e., two types of diffusing molecules: A and B) it is given by [124,125]:

$$J_A = -D_{AA}\rho\nabla n_A - D_{AB}\rho\nabla n_B$$
$$J_B = -D_{BA}\rho\nabla n_A - D_{BB}\rho\nabla n_B$$

(5.116)

If the cross coefficients are neglected in equation "5.116" [115]; then, for codiffusion, where the species, A and B, are diffusing in parallel, two diffusivities, $D_{AA} \approx D_A$, and, $D_{BB} \approx D_B$, can be defined [114]. Thus, to follow the sorption kinetics, during the codiffusion experiment, it is necessary to track both components in the mixture [125]. However, in the case of counterdiffusion, where molecule, A, moves "in" and molecule, B, moves "out", one effective diffusivity $D_e = D_A \approx D_B$ can be defined to describe the process [90,125]. Thus, to track the sorption kinetics, during the counterdiffusion experiment, it is necessary to follow only of one of the components in the mixture. Consequently, to measure these diffusivities equation "5.107" was applied. In the present cases, D, was substituted in equation "5.107" by: D_e, for counterdiffusion.

That is, the kinetics of the process is governed by the same equation used in the single-component experiment, but now we measure the effective diffusivity, D_e. As a result, we need only to follow the

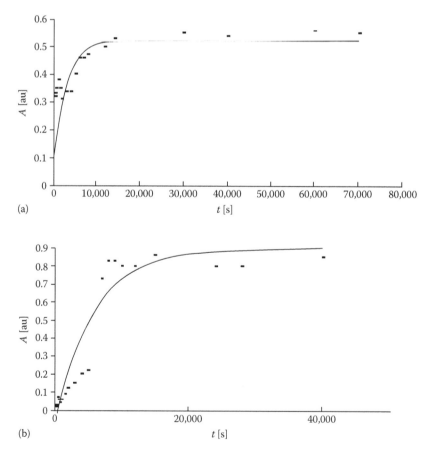

(a)

(b)

FIGURE 5.34 Counterdiffusion kinetics of p- + o-xylene in H-ZSM-11 at (a) 375 K, (b) 400 K. Experimental points (–) fitted with the continuous theoretical curve.

TABLE 5.5
Effective Diffusion Coefficients ($D_e \times 10^9$ [cm²/s])
in p-Xylene Plus o-Xylene Counterdiffusion in
H-ZSM-11 and H-SSZ-24 Zeolites at Different
Temperatures at a Gas Phase Concentration,
c_{p-x} [%] = c_{o-x} [%] = 50 [%]

Sortive	H-ZSM-11	H-SSZ-24	T [°C]
$p + o$-xylene	0.05	0.2	375
$p + o$-xylene	0.02	0.4	400
$p + o$-xylene	0.14	1.5	425

evolution of one component in the counterdiffusion experiment, then fit the results with Equation 5.107, following the procedure previously explained.

In the case of the measurement of the diffusivity in the p-xylene + o-xylene counterdiffusion experiment, the sample was initially saturated with a stream of p-xylene at a partial pressure of 6.7 Pa; then, to this stream of carrier gas plus p-xylene, the carrier gas saturated with o-xylene was admitted, to finally obtain the same partial pressure, 6.7 Pa, for both hydrocarbons. The composition of the final hydrocarbon mixture, that is, the gas phase concentration of p-(c_{p-x}) and o-(c_{o-x}) xylene, obtained was checked with a gas chromatograph (FISONS 8000) coupled to the gas outlet of the IR cell (see Figure 5.34). The gas phase concentration, for p-(c_{p-x}) and o-xylene (c_{o-x}) in the fed mixture of the counterdiffusion experiment was the same: c_{p-x} [%] = c_{o-x} [%] = 50 [%] [90]. If Figure 5.34, the uptake curves corresponding to the counterdiffusion kinetics of *para + ortho* xylene in H-ZSM-11 at 375 K and 400 K are shown [90].

In Figure 5.34, the counterdiffusion kinetic data for p-xylene plus o-xylene in H-ZSM-11 at 375 and 400 K and a concentration relation, c_{p-x} [%] = c_{o-x} [%] = 50 [%], are shown [90].

In Table 5.5, the effective diffusivity, D_e, for p-xylene plus o-xylene counterdiffusion in H-SSZ-24 and H-ZSM-11 zeolites at different temperatures and a concentration relation, c_{p-x} [%] = c_{o-x} [%] = 50 [%], are reported [90]. It is evident that the kinetics is governed by ordinary diffusion. Additionally, the study of the counterdiffusion of p-xylene + o-xylene and the reverse case o-xylene + p-xylene in a zeolite with a 10 member ring plus 12 member ring interconnected channel-like CIT-1 gives experimental evidence for the existence of molecular traffic control [125].

REFERENCES

1. W. Jost, *Diffusion in Solids, Liquids and Gases*, Academic Press, New York, 1960.
2. R. Reif, *Fundamentals of Statistical and Thermal Physics*, McGraw-Hill, Boston, MA, 1965.
3. W. Kauzmann, *Kinetic Theory of Gases*, Addison-Wesley, Reading, MA, 1966.
4. J.R. Manning, *Diffusion Kinetics for Atoms in Crystals*, Van Nostrand, Princeton, NJ, 1968.
5. J. Crank, *The Mathematics of Diffusion* (2nd edition), Oxford University Press, Oxford, UK, 1975.
6. B.S. Bokstein, M.I. Mendelev, and D.J. Srolovitz, *Thermodynamics and Kinetics in Materials Science*, Oxford University Press, Oxford, UK, 2005.
7. H. Mehrer, in *Diffusion in Condensed Matter*, P. Heithans and J. Karger (editors), Springer-Verlag, Berlin, 2005, p. 3.
8. M. Martin, in *Diffusion in Condensed Matter*, P. Heithans and J. Karger (editors), Springer-Verlag, Berlin, 2005, p. 209.
9. M.E. Glicksman, *Diffusion in Solids*, John Wiley & Sons, New York, 2000.
10. J. Xiao and J. Wei, *Chem. Eng. Sci.*, 47, 1123 (1992).
11. Ch. Kittel, *Elementary Statistical Physics*, John Wiley & Sons, New York, 1958.
12. J. Karger and D.M. Ruthven, *Diffusion in Zeolites and Other Microporous Solids*, John Wiley & Sons, New York, 1992.
13. M. Kizilyalli, J. Corish, and R. Metselaar, *Pure Appl. Chem.*, 71, 1307 (1999).

14. R.B. Bird, W.E. Stewart, and E.N. Lightfoot, *Transport Phenomena* (2nd edition), John Wiley & Sons, New York, 2002.
15. P. Brauer, S. Fritzsche, J. Karger, G. Schutz, and S. Vasenkov, Diffusion in channels and channel networks, *Lect. Notes Phys.*, 89, 634 (2004).
16. J. Karger and F. Stallmuch, in *Diffusion in Condensed Matter*, P. Heithans and J. Karger (editors), Springer, Berlin, 2005, p. 417.
17. A. Fick, *Ann. Phys.*, 94, 59 (1855).
18. L. Onsager, *Phys. Rev.*, 37,405 (1931); 38, 2265 (1932).
19. A. Einstein, *Ann. Phys.*, 17, 549 (1905).
20. R. Roque-Malherbe, *Adsorption and Diffusion in Nanoporous Materials*, CRC Press, Boca Raton, FL, 2007.
21. S.R. De Groot and P. Mazur, *Non-Equilibrium Thermodynamics*, Elsevier, Amsterdam, the Netherlands, 1962.
22. I. Prigogine, *Thermodynamics of Irreversible Process*, John Wiley & Sons, New York, 1967.
23. D.A. McQuarrie, *Statistical Mechanics*, University Science Books, Sausalito, CA, 2000; *Problems in Thermodynamics, and Statistical Physics*, P.T. Lansberg (editor), PION, London, 1971, Chapter 25.
24. H. Kreuzer, *Nonequilibrium Thermodynamics, and Its Statistical Foundations*, Clarendon Press, Oxford, 1981.
25. R. Kubo, M. Toda, and N. Hashitsume, *Statistical Physics II. Non-Equilibrium Statistical Mechanics*, Springer-Verlag, Berlin, 1991.
26. L. Landau and E.M. Lifshits, *Statistical Physics*, Addison & Wesley, Reading, MA, 1959.
27. P.W. Atkins, *Physical Chemistry* (6th edition), W.H. Freeman & Co., New York, 1998.
28. I. Levine, *Physical Chemistry* (5th edition), McGraw Hill, New York, 2002.
29. L.S. Darken, *Trans. AIME*, 175, 184 (1948).
30. G.H. Vineyard, *J. Phys. Chem. Sol.*, 3, 121 (1957).
31. R. Oesten and R.A. Higgins, *Ionics*, 1, 427 (1995).
32. S. Nieto, R. Polanco, and R. Roque-Malherbe, *J. Phys. Chem. C*, 111, 2809 (2007).
33. K.-D. Kreuer, *Chem. Mater.*, 8, 610 (1996).
34. R. Roque-Malherbe, F. Ruiz, and A. Perez-Reyes, *Rev. Metal CENIM*, 15, 373 (1979). *Chemical Abstracts*, Vol. 93, No. 77057b.
35. A. Perez-Reyes, C. Diaz-Aguila, and R. Roque-Malherbe, *Revista de Metalurgia CENIM*, 20, 343 (1984). *Chemical Abstracts*, Vol. 102, No. 153010a.
36. R.E. Reed-Hill and R. Abbaschian, *Principles of Physical Metallurgy* (3rd edition), International Thomson Publishing, Boston, MA, 1994.
37. M.B. Bronfin, *Fiz. Met. Metalved.*, 40, 363 (1975).
38. R. Roque-Malherbe, *Rev. CENIC*, 11, 79 (1980). *Chemical Abstracts*, Vol. 96, No. 39015r.
39. R. Roque-Malherbe, B.S. Bokstein, and A.A. Zhujovitskii, *Revista de Metalurgia CENIM*, 15, 28 (1979). *Chemical Abstracts*, Vol. 93, No. 13843x.
40. S.O. Kasap, *Principles of Electronic Materials* (2nd edition), McGraw-Hill Higher Education, New York, 2002.
41. J. Frenkel, *Z. Physik*, 53, 652 (1926).
42. C. Wagner and W. Schottky, *Z. Phys. Chem. B*, 11, 163 (1930).
43. F. Kroger and V. Vink, *Solid State Phys.*, 3, 307 (1956).
44. R.A. De Souza and J.A. Kilner, *Solid State Ionics*, 106, 175 (1998).
45. C. Wagner, *Z. Phys. Chem. B.*, 21, 25 (1933) and *Prog. Solid State Chem.* 10, 3 (1975).
46. H.J.M. Bouwmeester and A.J. Burggraaf, in *Fundamentals of Inorganic Membrane Science and Technology*, A.J. Burggraaf and L. Cot (editors), Elsevier, Amsterdam, the Netherlands, 1996, p. 435.
47. C.-S. Chen, Z.-P. Zhang, G.-S. Jiang, Ch-G. Fan, W. Liu, and H.J.M. Bouwmeester, *Chem. Mater.*, 13, 2797 (2001).
48. H. Iwahara, K. Uchida, and K. Ogaki, *J. Electrochem. Soc.*, 135, 529 (1988).
49. N. Bonanos, *Solid State Ionics*, 53–56, 967 (1992).
50. J. Wu, L.P. Li, W.T.P. Espinosa, and S.M. Haile, *J. Mater. Res.*, 19, 2366 (2004).
51. J. Wu, R.A. Davies, M.S. Islam, and S.M. Haile, *Chem. Mater.*, 17, 846 (2005).
52. L. Li, A. Li, and E. Iglesia, *Stud. Surf. Sci. Catal.*, 136, 357 (2001).
53. T. Norby, M. Wideroe, R. Glöckner, and Y. Larring, *Dalton Trans.*, 19, 3012 (2004).
54. J. Wu, Defect chemistry and proton conductivity in Ba-based perovskites, PhD thesis, California Institute of Technology, Pasadena, CA (2005).
55. K.D. Kreuer, *Solid State Ionics*, 97, 1 (1997).
56. M.S. Islam, *J. Mater. Chem.*, 10, 1027 (2000).
57. R. Glockner, M.S. Islam, and T. Norby, *Solid State Ionics*, 122, 145 (1999).
58. R.A. Davies, M.S. Islam, and J.D. Gale, *Solid Stat Ionics*, 126, 323 (1999).

59. R.A. Davies, M.S. Islam, A.V. Chadwick, and G.E. Rush, *Solid State Ionics*, 130, 115 (2000).
60. M. Cherry, M.S. Islam, J.D. Gale, and C.R.A. Catlow, *J. Phys. Chem.*, 99, 14614 (1995).
61. W. Munch, G. Seifert, K.D. Kreuer, and J. Maier, *J. Solid State Ionics*, 86–88, 647 (1996).
62. W. Munch, G. Seifert, K.D. Kreuer, and J. Maier, *J. Solid State Ionics*, 97, 39 (1997).
63. T. Scherban, Yu.M. Baikov, and E.K. Shalkova, *Solid State Ionics*, 1, 66 (1993).
64. T. Higuchi, T. Tsukamoto, N. Sata, M. Ishigame, Y. Tezuka, and S. Shin. *Phys. Rev. B*, 57, 6978 (1998).
65. S. Yamaguchi, K. Kobayashi, T. Higuchi, S. Shin, and Y. Iguchi. *Solid State Ionics*, 136–137, 305 (2000).
66. T. Higuchi, H. Matsumoto, T. Shimura, K. Yashiro, T. Kawada, J. Mizusaki, S. Shin, and T. Tsukamoto, *Jpn. J. App. Phys.*, 43, 731 (2004).
67. P. Kofstad, *Non-Stoichiometry, Diffusion and Electrical Conductivity of Binary Metal Oxides*, Wiley, New York, 1972.
68. K.Ch. Kao, *Dielectric Phenomena in Solids*, Elsevier, Amsterdam, the Netherlands, 2004.
69. T. Norby and R. Hausgrud, in *Non-Porous Inorganic Membranes*, A.F. Sammells and M.V. Mundschau, (editors), Wiley-VCH Verlag GmbH and Co., Weinheim, Germany, 2006, p.1.
70. PeakFit®, *Program System, Seasolve Software Inc.*, Framingham, MA.
71. N.R. Draper and H. Smith, *Applied Regression Analysis* (3rd edition), John Wiley & Sons, New York, 1998.
72. R.M. Barrer and W. Jost, *Trans. Faraday Soc.*, 45, 928 (1949).
73. L. Li and E. Iglesia, *Chem. Eng. Sci.*, 58, 1977 (2003).
74. K.S.W. Sing, D.H. Everett, R.A.W. Haul, L. Moscou, R.A. Pirotti, J. Rouquerol, and T. Siemieniewska, *Pure Appl. Chem.*, 57, 603 (1985).
75. R.W. Baker, *Membrane Technology and Applications*, John Wiley & Sons, New York, 2004.
76. R. Roque-Malherbe, W. del Valle, F. Marquez, J. Duconge, and M.F.A. Goosen, *Sep. Sci. Technol.*, 41, 73 (2006).
77. G. Saracco and V. Specchia, *Catal Rev.-Sci. Eng.*, 36, 305 (1994).
78. C.N. Satterfield, *Heterogeneous Catalysis in Practice*, McGraw-Hill, New York, 1980.
79. M.R. Wang and Z.X. Li, *Physical Review E*, 68, 046704 (2003).
80. J.-G. Choi, D.D. Do, and H.D. Do, *Ind. Eng. Chem. Res.*, 40, 4005 (2001).
81. R.J.R. Ulhorn, K. Keizer, and A.J. Burggraaf, *J. Membrane Sci.*, 66, 271 (1992).
82. M.F.M. Post, *Stud. Surf. Sci. Catal.*, 58, 391 (1991).
83. R.W. Barber and D.R. Emerson, in *Advances in Fluid Mechanics IV*, M. Rahman, R. Verhoeven, and C.A. Brebbia (editors), WIT Press, Southampton, U.K., 2002, p. 207.
84. S.A. Schaaf and P.L. Chambre, *Flow of Rarefied Gases*, Princeton University Press, Princeton, NJ, 1961.
85. H. Mizuseki, Y. Jin, Y. Kawazoe, and L.T. Wille, *J. App. Phys.*, 87, 6561 (2000).
86. G. Bird, *Annu. Rev. Fluid Mech.*, 10, 11 (1978).
87. W. Niessen and H.G. Harge, *Stud. Surf. Sci. Catal.*, 60, 213 (1991).
88. R. Roque-Malherbe, R. Wendelbo, A. Mifsud, and A. Corma, *J. Phys. Chem.*, 99, 14064 (1995).
89. R. Roque-Malherbe, *Mic. Mes. Mat.*, 56, 321 (2002).
90. R. Roque-Malherbe, and V. Ivanov, *Mic. Mes. Mat.*, 47, 25 (2001).
91. G. Sastre, N. Raj, C. Richard, C. Catlow, R. Roque-Malherbe, and A. Corma., *J. Phys. Chem. B*, 102, 3198 (1998).
92. R. Wendelbo and R. Roque-Malherbe, *Mic. Mat.*, 10, 231 (1997).
93. D.N. Theodorou and J. Wei, *J. Catal.*, 83, 205 (1983).
94. P.H. Nelson, A.B. Kaiser, and D.M. Bibby, *J. Catal.* 127, 101 (1991).
95. R.O. Snurr, A.T. Bell, and D.N. Theodorou, *J. Phys. Chem.*, 97, 13742 (1993).
96. A. Corma, *Chem. Rev.*, 97, 2373 (1997).
97. F. Marquez-Linares and R. Roque-Malherbe, *Facets-IUMRS J.*, 2, 14 (2003) and 3, 8 (2004).
98. R. Snurr and J. Karger, *J. Phys. Chem. B*, 101, 6469 (1997).
99. H. Ramanan, S.M. Auerbach, and M. Tsapatsis, *J. Phys. Chem. B*, 108, 17171 (2004).
100. A.I. Skoulidas and D.S. Sholl, *J. Phys. Chem. B*, 105, 3151 (2001).
101. J. Valyon, G. Onyestyak, and L.V.C. Rees, *Langmuir*, 16, 1331 (2000).
102. D.W. Breck, *Zeolite Molecular Sieves*, Wiley, New York, 1974.
103. R.M. Barrer, *Zeolite and Clay Minerals as Sorbents and Molecular Sieves*, Academic Press, London, UK, 1978.
104. J. de la Cruz, C. Rodriguez, and R. Roque-Malherbe, *Surf. Sci.*, 209, 215 (1989).
105. S. Glasstone, K.J. Laidler, and H. Eyring, *The Theory of Rate Process*, McGraw-Hill, New York, 1964.
106. J. Karger, H. Heifer, and R. Haberlandt, *Chem. Soc., Faraday Trans.*, 76 1569 (1980).
107. D.M. Ruthven and R.I. Derrah, *J. Chem. Soc., Faraday Trans.*, 68, 2322 (1972).
108. R.L. Larry, A.T. Bell, and D.N. Theodorou, *J. Phys. Chem.*, 95, 8866 (1991).

109. T.L. Hill, *An Introduction to Statistical Thermodynamics*, Dover Publications Inc., New York, 1986.
110. W. Rudzinskii and D.H. Everett, *Adsorption of Gases on Heterogeneous Surfaces*, Academic, New York, 1992.
111. R.F. Lobo and M.E. Davis, *Mic. Mat.*, 3, 61 (1994).
112. Ch. Baerlocher, W.M. Meier, and D.M. Olson, *Atlas of Zeolite Framework Types* (5th edition), Elsevier, Amsterdam, the Netherlands, 2001.
113. J. Karger and M. Bulow, *Chem. Eng. Sci.*, 30, 893 (1975).
114. D.M. Ruthven, *Principles of Adsorption and Adsorption Processes*, John Wiley & Sons, New York, 1984.
115. J. Karger, *Surf. Sci.*, 36, 397 (1973).
116. J. Xiao and J. Wei, *Chem. Eng. Sci.*, 47, 1143 (1992).
117. H.G. Karge and W. Niessen, *Mic. Mat.*, 1, 1 (1993).
118. M. Bulow, J. Caro, B. Rohl-Kuhn, and B. Zibrowius, *Stud. Surf. Sci. Catal.*, 46, 505 (1989).
119. D. Sheen and L.V.C. Rees, *Zeolites*, 11, 666 (1991).
120. M. Eic and D.M. Ruthven, *Zeolites*, 8, 258 (1988).
121. D. Barthomeuf and A. Mallmann, *Ind. Eng. Chem. Res.*, 29, 1435 (1990).
122. J.A. Rabo and G.J. Gajda, in *Acidity and Basicity of Solids*, J. Fraissard, J. and L. Petrakis (editors), Kluwer Academic Publishers, the Netherlands, 1994, p. 127.
123. J. Sauer, P. Upliengo, E. Garrone, and V.R. Saunders, *Chem. Rev.*, 94, 2095 (1994).
124. Y. Wang and M.D. Le Van, *Ind. Eng. Chem. Res.* 46, 2141 (2007).
125. R. Roque-Malherbe and V. Ivanov. *J. Mol. Cat. A.* (In Press) (2009).

6 Adsorption in Nanoporous Materials

6.1 INTRODUCTION

The term "adsorption" was proposed by Kayser in 1881 to describe the increase in concentration of gas molecules on neighboring solid surfaces, a phenomenon earlier noticed by Fontana and Scheele in 1777.

The majority of adsorbents applied in industry has porous sizes in the nanometer region. In this pore-size territory, adsorption is an important method for the characterization of porous materials. To be precise, gas adsorption provides information concerning the microporous volume, the mesopore area, the volume and size of the pores, and the energetics of adsorption. Also, gas adsorption is an important unitary operation for the industrial and sustainable energy and pollution abatement applications of nanoporous materials.

Nanoporous materials like zeolites and related materials, mesoporous molecular sieves, clays, pillared clays, the majority of silica, alumina, active carbons, titanium dioxides, magnesium oxides, carbon nanotubes and metal-organic frameworks are the most widely studied and applied adsorbents. In the case of crystalline and ordered nanoporous materials such as zeolites and related materials, and mesoporous molecular sieves, their categorization as nanoporous materials are not debated. However, in the case of amorphous porous materials, they possess bigger pores together with pores sized less than 100 nm. Nevertheless, in the majority of cases, the nanoporous component is the most important part of the porosity.

On the other hand, in addition to adsorption properties, nanoporous materials are a group of advanced materials with other excellent properties and applications in many fields, for example, optics, electronics, ionic conduction, ionic exchange, gas separation, membranes, coatings, catalysts, catalyst supports, sensors, pollution abatement, detergency, and biology [1–42].

6.2 DEFINITIONS AND TERMINOLOGY

6.2.1 SOME DEFINITIONS

6.2.1.1 Adsorption and Desorption

The clean surface of any solid is distinguished by the fact that the atoms which make the surface do not have all their bonds saturated. This produces an adsorption field over this surface. The adsorption field produces an accumulation of molecules near the solid surface [1–10]. This phenomenon, that is, adsorption, is a general tendency of surface systems, since during its occurrence a decrease of the surface tension is experienced by the solid [2]. The term adsorption is used for this process; for the reverse, the term desorption is used [1]. Adsorption, for gas–solid and liquid–solid interfaces, the cases that are considered in this chapter, is on one hand an increase in the concentration of gas molecules in a solid surface, and on the other, an increase in the concentration of a dissolved substance at the interface of a solid and a liquid phase, where both phenomena are caused by the existence of surface forces [2].

6.2.1.2 Pore Size

The adsorbents normally applied are in general porous. The classification of the different pore widths of porous adsorbents was carried out by the International Union of Pure and Applied Chemistry (IUPAC) [1]. IUPAC classified these materials as microporous, with pore diameters between 0.3 and 2 nm, mesoporous with pore diameters between 2 and 50 nm, and macroporous with pore diameters greater than 50 nm [1]. The pore width, D_p, is defined as equal to the diameter in the case of cylindrical-shaped pores, and as the distance between opposite walls in the case of slit-shaped pores.

6.2.1.3 Adsorption Space Filling

The adsorption of vapors in complex porous systems takes place approximately as follows [1–3]: at first, micropore filling occurs, where the adsorption behavior is dominated nearly completely by the interactions of the adsorbate and the pore wall; after this, at higher pressures, external surface coverage occurs, consisting of monolayer and multilayer adsorption on the walls of mesopores and open macropores, and, at last, capillary condensation occurs in the mesopores.

6.2.1.4 Dynamic Adsorption

It is a mass transfer between a mobile, solid, or liquid phase, and the adsorption bed packed in a reactor. To carry out adsorption, a reactor, where a dynamic adsorption process will occur, is packed with an adsorbent [2]. The adsorbents normally used for these applications are active carbons, zeolites and related materials, silica, mesoporous molecular sieves, alumina, titanium dioxide, magnesium oxide, clays, and pillared clays.

6.2.1.5 Adsorption Isotherm

To study adsorption, in practice, the relationship between the amount adsorbed, n_a, and the equilibrium pressure, P, at constant temperature, T, is measured in order to get an adsorption isotherm [1]

$$n_a = F(P)_T \tag{6.1}$$

6.2.1.6 Physical and Chemical Adsorptions

The gas adsorption process is normally considered a physical process, named physical adsorption, since the molecular forces involved in this process are usually of the van der Waals type [2–10]. Physical adsorption of gases in solid surfaces takes place in the case where during the adsorption process a reaction with exchange of electrons between the solid surface and the gas molecules with the formation of chemical bonds is not necessary [1,2]. In a situation where during the adsorption process, a reaction by means of electron exchange between the solid surface and the gas molecules takes place, then the phenomenon is named chemical adsorption [1,2].

6.2.1.7 Mobile and Immobile Adsorptions

The physical adsorption of gases and vapors in solids could be as well be categorized as mobile adsorption, which takes place when the adsorbed molecule acts as a gas molecule in the adsorption space, or immobile adsorption, which takes place when the adsorbed molecule is forced to vibrate around an adsorption site [2].

6.2.1.8 Monolayers and Multilayers

For open surfaces, adsorption consists of a layer-by-layer loading process, where the first layer is filled as in the case when $\theta = n_a/N_m = 1$, where θ is the surface recovery and N_m is the monolayer capacity. As a result, it is understood that we have monolayer adsorption when $\theta = n_a/N_m < 1$, and multiplayer adsorption when $\theta = n_a/N_m > 1$.

6.2.1.9 Parameters Characterizing Porous Adsorbents

The parameters characterizing porous adsorbents are the following. Specific surface area is represented by S and measured in [m²/g] (the surface area is the outer surface, to be precise, the area outside of the micropores; if the adsorbent does not present micropores, the surface area and the outer surface area coincide [2]); the micropore volume is denoted by W^{MP} and measured in [cm^3/g]; the pore volume is designated W, which is the sum of the micropores and mesopore volumes of the adsorbent, and measured in [cm^3/g]; and finally the pore size distribution (PSD) [1,8]. The PSD is a plot of $\dfrac{\Delta V_p}{\Delta D_p}$ versus D_p, where V_p is the pore volume accumulated up to the pore width D_p measured in [cc-STP/g Å] [1,2]. The unit, cc-STP, denotes the quantity adsorbed, measured in cubic centimeters at standard temperature and pressure (STP), that is, 273.15 K and 760 Torr, that is, 1.01325×10 Pa.

6.2.2 Magnitude of Adsorption

The interfacial layer is the inhomogeneous part of a system that is intermediate between two contacting bulk phases. The properties characterizing this region are drastically different from, but related to, the properties of the bulk phases [2,7]. To thermodynamically treat this system is considered that in the ideal reference system, the concentration should stay constant up to the Gibbs dividing surface (GDS) (see Figure 6.1). But, in an actual system, the concentration changes across the interface, or thickness,

$$\gamma = z_\beta - z_\alpha$$

from phase α to phase β (Figure 6.1) [2,7]. For gas–solid adsorption (see Figure 6.1), since $c_s^g = 0$ in the solid phase [1,2,4]; the surface excess amount is:

$$n^\sigma = n - V^{\alpha_0} c^g = n - c_g^g (V^g + V^a) \tag{6.2}$$

in which

$$V^{\alpha_0} = V^g + V^a$$

where
  n is the total amount of gas present in the system
  V^a is the adsorption space
  V^g is the volume of the gas phase
  c_g^g is the concentration in the gas phase and c_s^g is the concentration of the gas in the solid

Now if A is the adsorbent surface area and t the thickness of the adsorbed layer, then the volume of the adsorbed layer, or adsorption space is [4]:

$$V^a = At \tag{6.3}$$

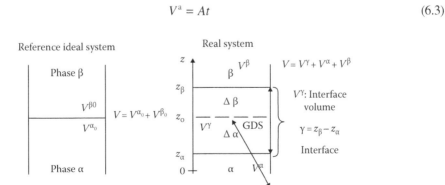

FIGURE 6.1 Gibbs dividing surface.

We may now define the amount adsorbed as:

$$n^{\mathrm{a}} = \int_0^{V^{\mathrm{a}}} c \, dV = A \int_0^t c \, dz \tag{6.4}$$

Then, the total quantity of gas molecules in the system is

$$n = n^{\mathrm{a}} + V^{\mathrm{g}} c_{\mathrm{g}}^{\mathrm{g}} \tag{6.5}$$

so

$$n^{\mathrm{a}} = n - V^{\mathrm{g}} c_{\mathrm{g}}^{\mathrm{g}} \tag{6.6}$$

As a result, combining Equations 6.2 and 6.6, we get

$$n^{\mathrm{a}} = n^{\sigma} + V^{\mathrm{a}} c_{\mathrm{g}}^{\mathrm{g}} \approx n^{\sigma}$$

given that

$$V^{\mathrm{a}} c_{\mathrm{g}}^{\mathrm{g}} \approx 0$$

We now precisely define some surface parameters. If the mass of a degassed adsorbent is defined as m_{s} and measured in grams, subsequently, the specific surface area is

$$S = \frac{A}{m_{\mathrm{s}}}$$

which is measured in [m²/g]. Besides, the specific surface excess amount is defined by

$$n_{\mathrm{a}} = \frac{n^{\sigma}}{m_{\mathrm{s}}} \approx \frac{n^{\mathrm{a}}}{m_{\mathrm{s}}}$$

where n_{a} is the amount adsorbed measured in [mol/g].

The specific surface excess amount is what is typically measured in practical adsorption studies (see Section 6.4). This parameter is almost equal to the amount adsorbed, n_{a}, and is dependent on the equilibrium adsorption pressure, P, at constant adsorbent temperature, T. Accordingly, gas adsorption data is, in practice, expressed by the adsorption isotherm:

$$n_{\mathrm{a}} = \frac{n^{\sigma}}{m_{\mathrm{s}}} \approx \frac{n^{\mathrm{a}}}{m_{\mathrm{s}}} = F(P)_T$$

6.3   ADSORPTION INTERACTION FIELDS

When a molecule contacts the surface of a solid adsorbent, it becomes subjected to diverse interaction fields, such as the dispersion energy, ϕ_{D}, repulsion energy, ϕ_{R}, polarization energy, ϕ_{P}, field dipole energy, $\phi_{\mathrm{E\mu}}$, field gradient quadrupole energy, ϕ_{EQ}, sorbate–sorbate interaction energy, ϕ_{AA}, and some specific interactions as the acid–base interaction with the active site, ϕ_{AB}, if the surface contains hydroxyl bridge groups [2–10].

The dispersion or London forces [43] between adsorbed nonpolar molecules and any adsorbent emerges when the transient dipoles, become correlated. As a result, the instantaneous dipole of the adsorbed nonpolar molecule induces a dipole in the adsorbent atoms, and, subsequently, both interact to lower the energy of the adsorbate–adsorbent system. Due to the correlation, the attraction between the instantaneous dipoles, developed in the entire system, does not vanish, and produces an induced-dipole–induced-dipole interaction energy, described by the following equation:

$$\phi_D = -\frac{C}{r^6}$$

As it is obvious, the intensity of the dispersion interaction is dependent on the polarizability of the adsorbate molecule and the adsorbent surface atom; thereafter, the constant C can be calculated with the help of the London formula

$$C = \frac{2\alpha_S\alpha_A}{3}\left(\frac{I_S I_A}{I_S + I_A}\right)$$

where
α_A and α_S are the polarizabilities
I_A and I_S are the ionization energies of the adsorbate molecules and the adsorbent surface atoms

The constant, C, can also be calculated with the help of the Kirkwood–Muller equation, as later explained, or with the Slater–Kirkwood expression.

When the adsorbed molecules are compressed against a solid surface, the nuclear and electronic repulsions, and the increasing electronic kinetic energy, start to overcome the attractive forces, and, therefore, the repulsion forces sharply augment in a very complicated form [43,44]. This complication is avoided by proposing approximate repulsion potentials. In this sense, the terms $\phi_D + \phi_R$ for the potential energy of interaction of a nonpolar molecule with a surface could be represented by the Lennard–Jones (L–J) (12–6) potential [10,45]

$$\phi_D + \phi_R = 4\varepsilon\left(\left(\frac{\sigma}{z}\right)^{12} - \left(\frac{\sigma}{z}\right)^6\right)$$

which is a special case of the Mie, $n - m$, potential, where $n > m$, ε is the potential energy minimum, and σ is the gas–solid separation at maximum interaction. These potentials, that is, the dispersion and repulsion, are present in the interaction of all molecules with any adsorbent. With the help of these potentials, it is possible to describe the adsorption of nonpolar molecules in nanoporous materials, whenever the electrostatic interaction is negligible [24].

The electrostatic contribution to the potential can be calculated as

$$\phi_E = \phi_P + \phi_{E\mu} + \phi_{QE}$$

It is required to calculate the electrostatic interaction between molecules, which possess a localized charge distribution, mathematically described by $\rho(\bar{r})$ (where $\bar{r} = x_1\bar{i} + x_2\bar{j} + x_3\bar{k}$ in rectangular coordinates, and $x_1 = x$, $x_2 = y$, and $x_3 = z$), and are immersed in an external electric field with a potential, $V(\bar{r})$, given by a solid adsorbent. Subsequently, the electrostatic energy of the adsorbate–adsorbent system can be determined with the help of the following expression [46]:

$$\phi_E = \int V(\bar{r})\rho(\bar{r})d^3r$$

If we now make the Taylor expansion of the potential,

$$V(\bar{r}) = V(0) + \bar{r} \cdot \nabla V(0) + \frac{1}{2}\sum_i \sum_j x_i x_j \frac{\partial^2 V}{\partial x_i \partial x_j}(0)$$

Then, applying the definition of the electric field, $\bar{E} = -\nabla V$, we get

$$V(\bar{r}) = V(0) - \bar{r} \cdot \bar{E}(0) - \frac{1}{2}\sum_i \sum_j x_i x_j \frac{\partial E_j}{\partial x_i} + \cdots$$

Now, since $\nabla \cdot \bar{E} = 0$ for the external field, it is the possible to subtract in the preceding expression $\frac{1}{6}|\bar{r}|^2 \nabla \cdot \bar{E}$ to get [46]

$$V(\bar{r}) = V(0) - \bar{r} \cdot \bar{E}(0) - \frac{1}{6}\sum_i \sum_j (3x_i x_j - |\bar{r}|^2)\frac{\partial E_j}{\partial x_i} + \cdots$$

As a result

$$\phi_E = \int V(\bar{r})\rho(\bar{r})d^3r = qV(0) - \bar{P} \cdot \bar{E}(0) - \frac{1}{3}\sum_i \sum_j Q_{i,j}\frac{\partial E_j}{\partial x_i} + \cdots$$

where the dipole terms, with three components, are [47]

$$\bar{p} = \int \bar{r}\rho(\bar{r})d^3r$$

and the quadrupole terms, with nine components, are given by [47,48]

$$Q_{i,j} = \frac{1}{2}\int \rho(\bar{r})(3x_i x_j - |\bar{r}|^2\delta_{i,j})d^3r$$

that is, a symmetrical

$$Q_{i,j} = Q_{j,i}$$

and traceless

$$Q_{x,x} + Q_{y,y} + Q_{z,z} = 0$$

tensor that can be reduced to a matrix characterized by only the diagonal terms $Q_{x,x}$, $Q_{y,y}$, and $Q_{z,z}$ [47]. Additionally, for molecules with a cylindrical symmetry about the z-axis, the quadrupole moment tensor of the charge distribution is determined by only one term, that is, $Q_{z,z}$ [46]. For a molecule, $q = 0$, and, subsequently,

$$\phi_E = \frac{1}{2}\int V(\bar{r})\rho(\bar{r})\mathrm{d}^3r = -\bar{p}\cdot\bar{E}(0) - \frac{1}{3}\sum_i\sum_j Q_{i,j}\frac{\partial E_j}{\partial x_i}(0) + \cdots$$

In this regard, the electrostatic polarization term arises in the case of nonpolar molecules. These nonpolar molecules when within an electric field are polarized, and afterward produce an induced dipole moment. This induced dipole moment, $\bar{p}_i$, interacts with the adsorbent, and the interaction potential is given by

$$\phi_P = -\bar{p}_i\cdot\bar{E}$$

where $\bar{E}$ is the intensity of the electric field of the adsorbent, and the induced dipole moment is given by the following equation

$$\bar{P}_i = \alpha\bar{E}$$

in which α is the average polarizability of the adsorbed molecule. As a result,

$$\phi_P = -\alpha(\bar{E}\cdot\bar{E}) = \alpha|\bar{E}|^2$$

The field dipole energy appears in the instance of polar molecules interacting with an adsorbent that have a crystalline electric field. Subsequently, for polar molecules, the field dipole term could be written as follows

$$\phi_\mu = -\bar{\mu}\cdot\bar{E} = -\mu E\cos\varphi$$

where
  $\bar{p} = \bar{\mu}$ is the permanent dipole moment of the adsorbed polar molecule
  φ is the angle between the electric field and the dipole moment

The field gradient quadrupole energy is given by the following equation

$$\phi_{EQ} = \frac{Q}{2}\left(\frac{\partial E}{\partial r}\right)$$

where

$$Q = \frac{1}{2}\int\rho(r,\theta)(3\cos^2\theta - 1)r^2\mathrm{d}V$$

is the quadrupole moment of the adsorbed molecule with cylindrical symmetry and $\left(\frac{\partial E}{\partial r}\right)$ is the electric field gradient of the adsorbent.

The electrostatic interactions between the adsorbed molecule and the adsorbent framework depend on the structure and composition of the adsorbed molecule and the adsorbent itself. For example, for H_2O, H_2S, SO_2, and NH_3 (molecules with a high dipole moment), and CO_2 (a molecule with a high quadrupole moment), the electrostatic, attractive interactions are stronger than the dispersion interactions [2]. Alternatively, dispersion is the fundamental attractive force present during adsorption in all adsorbents, in the case of molecules like H_2, Ar, CH_4, N_2, and O_2. Given the dipole moments of these molecules is zero, the quadrupole moment is very low or absent, and the polarization effect will only be noticeable in the case of adsorbents with high electric fields, as explained later [34]. The dispersion and repulsion interactions are present in all adsorption gas–solid systems; therefore, they are nonspecific interactions [28]. However, the interaction between the adsorbent and dipolar molecules, such as H_2O, NH_3, SO_2, and H_2S and quadrupolar ones like CO_2, is responsible for specific interactions. This specificity is used in the application adsorbents in gas drying and purification [28].

6.4 MEASUREMENT OF ADSORPTION ISOTHERMS BY THE VOLUMETRIC METHOD

The volumetric adsorption experiment [6] shown in Figure 6.2 consists of a thermostated sample cell of volume V_g at an experimental temperature, T, a container with a precisely determined volume, named the calibrated volume, V_c, a connection to the gas reservoir, and a transducer for pressure measurement. In addition, the volume at stopcock 3, at ambient temperature, T_r, and the thermostat is denoted V_2, and the volume between stopcocks 1, 2, and 3, also at ambient temperature, T_r, is denoted V_1.

The adsorbate from the gas container is introduced into the manifold, or dose volume, V_1, of the formerly evacuated vacuum system and the resultant pressure, P_1, is measured with the transducer; subsequently, stopcock 2 is opened resulting in a pressure P_2. Then [6]

$$V_1 = \frac{V_c P_2}{P_1 - P_2}$$

Upon opening stopcock 3, the gas contacts the adsorbent at temperature T. The volume at the sample-side is called the dead volume V_d

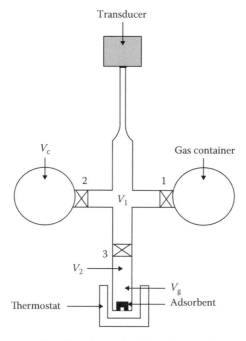

FIGURE 6.2 Schematic representation of a volumetric adsorption experiment.

$$V_d = V_2 + \frac{V_g T_r}{T}$$

These volumes, V_2 and V_g, must be measured with the adsorbent introduced in the sample cell using He, which is a gas not adsorbed normally at the experimental temperature. To measure V_d, the same methodology earlier explained for the measurement of V_1 is applied. In order to determine V_d, volume V_2 is experimentally made approximately zero ($V_2 \approx 0$).

The application of the volumetric method for the determination of adsorption isotherms can be performed as follows. First, the adsorbate gas is introduced into the manifold volume, V_1, and the quantity dosed is measured, usually in cubic centimeters at STP, that is, standard temperature and pressure, specifically 273.15 K and $1.01325 \times 10 \, Pa$

$$n_{dose}^i = \frac{P_1^i V_1}{RT_r} + \frac{P_2^{i-1} V_d}{RT_r}$$

where

P_2^{i-1} is the equilibrium pressure of the previous adsorption step ($i-1$th, step)
T_r is the ambient temperature
V_1 is the dose volume
V_d is the dead volume
P_1^i is the initial pressure (ith, step)
n_{dose}^i is the initial number of moles during the ith adsorption step

Subsequently, when equilibrium is achieved, the amount of gas not adsorbed is calculated by

$$n_{final}^i = \frac{P_2^i (V_1 + V_d)}{RT_r}$$

where P_2^i is the equilibrium pressure of the current measurement (ith step). The amount adsorbed in the ith isotherm point, $\Delta^i n_a$, is then calculated with the help of the following equation

$$\Delta^i n_a = \frac{n_{dose}^i - n_{final}^i}{m_s}$$

in which m_s is the mass of degassed adsorbent. To conclude, the isotherm is calculated with the help of the sum of the different adsorption steps

$$n_a^i = \sum_{j=1}^{i} \Delta^j n_a$$

where n_a^i is the magnitude of adsorption or amount adsorbed up to the ith adsorption step. Then plotting n_a^i versus P_2^i the experimental isotherm is obtained.

6.5 THERMODYNAMICS OF ADSORPTION

6.5.1 ISOSTERIC AND DIFFERENTIAL HEATS OF ADSORPTION

Adsorption is a general tendency of matter and during its occurrence, a decrease in the surface tension is experienced by the solid. For this reason, adsorption is a spontaneous process, where a decrease in the Gibbs free energy is observed, that is, $\Delta G < 0$. In addition, during physical adsorption, molecules from a disordered bulk phase pass to a more ordered adsorbed state, because in this adsorbed state the molecules are restricted to move in a surface or a pore [2]. Subsequently, during adsorption by the whole system, a decrease of entropy is experienced, that is, $\Delta S < 0$. Consequently, it is then possible to assert that since [2]

$$\Delta G = \Delta H - T\Delta S$$

then

$$\Delta H = \Delta G + T\Delta S < 0$$

Accordingly, adsorption is an exothermic process, and thereafter the a process favored by a decrease in temperature.

The fundamental equation of thermodynamics for a bulk mixture is

$$dU = TdS - PdV + \sum_i \mu_i dn_i$$

in which
U is the internal energy of the system
S is the entropy
V is the volume
T is the temperature
μ_i are the chemical potentials
n_i is the number of moles of the components contained in the system

The thermodynamic approach applied here considers the adsorbent plus the adsorbed gas, or vapor, as a solid solution (system aA). Applying this description, it is feasible to get the fundamental thermodynamic equation for the aA system [2,15,16,25]

$$dU_{aA} = TdS_{aA} - PdV_{aA} + \mu_a dn_a + \mu_A dn_A$$

where
U_{aA}, S_{aA}, and V_{aA} are the internal energy, entropy, and volume of the system aA, respectively
μ_a, μ_A are the chemical potentials of the adsorbate, a, and the adsorbent, A.
n_a and n_A are the number of moles of the adsorbate and the adsorbent in the system aA, respectively

If we define $\Gamma = \dfrac{n_a}{n_A}$, then $\mu_a = \mu_a(T, P, \Gamma)$ and $\mu_A = \mu_A(T, P)$. As a result [15]

$$d\mu_a = -\bar{S}_a dT + \bar{V}_a dP + \left(\frac{\partial \mu_a}{\partial \Gamma}\right)_{T,P} d\Gamma$$

in which $\bar{S}_a$ and $\bar{V}_a$ are the partial molar entropy and volume of the adsorbate in the aA system, respectively. During equilibrium, the chemical potential of the adsorbate in the aA phase and the gas phase is equal, then

$$d\mu_a = d\mu_g = -\bar{S}_g dT + \bar{V}_g dP$$

Subsequently, for $\Gamma = $ constant

$$\left[\frac{d \ln P}{dT}\right]_\Gamma = \frac{\overline{H}_g - \overline{H}_a}{RT^2} = \frac{q_{iso}}{RT^2} \tag{6.7a}$$

where $\overline{H}_g$ and $\overline{H}_a$ are the partial molar enthalpies of the adsorbate in the gas phase and in the aA system, respectively. Now, applying Equation 6.7a, it is possible to define the isosteric enthalpy of adsorption [10]

$$\Delta H(n_a) = -(\overline{H}_g - \overline{H}_a) = -q_{iso} \tag{6.7b}$$

where q_{iso} is the enthalpy of desorption, or the isosteric heat of adsorption. The isosteric heat of adsorption is calculated with the help of adsorption isotherms. An additional significant adsorption heat is the differential heat of adsorption, which is defined as follows [6]

$$q_{diff} = \frac{\Delta Q}{\Delta n_a} \tag{6.8}$$

where ΔQ is the evolved heat during the finite increment, Δn_a, in the magnitude of adsorption that gives the evolution of heat. It can be also approximately determined with the help of the following expression:

$$q_{diff} \approx q_{iso} - RT \tag{6.9}$$

But, it is required to acknowledge that Relation 6.9 is only exactly satisfied in the case of inert adsorbents, and porous adsorbent systems are not generally inert.

6.5.2 CALORIMETRY OF ADSORPTION

As was explained in the previous section, when an adsorbate contacts an adsorbent, heat is released. The thermal effect produced can be measured with the help of a thermocouple placed inside the adsorbent and referred at the room temperature (see Figure 6.3) [3,31,34,49]. This is a version of the Tian–Calvet heat-flow calorimeter [50]. This calorimetric technique is distinguished by the fact that the temperature difference between the tested adsorbent and a thermostat is measured. Consequently, in the Tian–Calvet heat-flow calorimeter, the thermal energy released in the adsorption cell is allowed to flow without restraint to the thermostat [3,31,34,49].

In this type of calorimeter, the heat flows through a thermocouple, and then the voltage potential, produced by the thermocouple and which is proportional to the thermal power, is amplified and recorded in an x–y plotter (see Figure 6.3) [3,31,34,49]. The concrete thermal effect produced is the integral heat of adsorption, which is measured with the help of the heat-flow calorimeter using the equation [50]

$$\Delta Q = \kappa \int \Delta T dt \tag{6.10}$$

where
ΔQ is the integral heat of adsorption evolved during the finite increment [6,31]
κ is a calibration constant
ΔT is the difference between thermostat temperature and the sample temperature during adsorption
t is time [50]

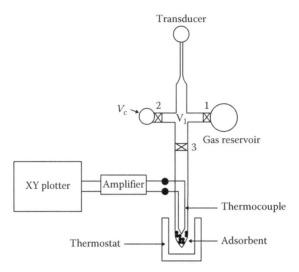

FIGURE 6.3   Schematic representation of a calorimeter and a volumetric system.

The home-made heat-flow calorimeter used consisted of a high vacuum line for adsorption measurements applying the volumetric method. This equipment comprised of a Pyrex glass, vacuum system including a sample holder, a dead volume, a dose volume, a U-tube manometer, and a thermostat (Figure 6.3). In the sample holder, the adsorbent (thermostated with 0.1% of temperature fluctuation) is in contact with a chromel–alumel thermocouple included in an amplifier circuit (amplification factor: 10), and connected with an *x–y* plotter [3,31,34,49]. The calibration of the calorimeter, that is, the determination of the constant, κ, was performed using the data reported in the literature for the adsorption of NH_3 at 300 K in a Na-X zeolite [51].

The differential heat of adsorption was calculated with the help of Equation 6.8 in increment form [2,3,6,31]. The error in q_{diff} is ± 2 kJ/mol and $\theta = \dfrac{n_a}{N_a}$, where n_a is the magnitude of adsorption, and N_a is the maximum magnitude of adsorption in the zeolite [2,3]. In Figure 6.4, the plot of the differential heat of adsorption, q_{diff}, versus the zeolite micropore volume recovery, θ, for the adsorption of CO_2 in

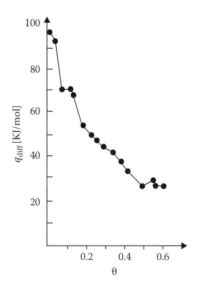

FIGURE 6.4   Plot of q_{diff} vs. θ for the adsorption of CO_2 at 300 K in an MP zeolite.

a natural zeolite labeled MP is illustrated [49]. The MP zeolite is a natural zeolite sample from deposits in Palmarito, Santiago de Cuba, Cuba, which is composed of a mordenite (80 wt %), clinoptilolite (5 wt %), and other phases (15 wt %), where the other phases include montmorillonite (2–10 wt %), quartz (1–5 wt %), calcite (1–6 wt %), feldspars (0–1 wt %), and volcanic glass (see Table 4.1).

The data reported in Figure 6.4 [49] indicate that the adsorption process of CO_2 in this natural zeolite is a heterogeneous process [10]. That is, the heat of adsorption is a decreasing function of the zeolite micropore volume recovery, θ. In the plot of q_{diff} versus θ, the first two steps, one at 90 kJ/mol and the other at 70 kJ/mol, are evident, that is, relatively high values of the adsorption heats, which indicates that the CO_2 molecules strongly interact through their quadrupole moments with the mordenite framework.

Thereafter, a decrease of the adsorption heat is obvious, up to a value equivalent to the bulk heat of condensation of the CO_2 molecules at 40 kJ/mol [49].

The data reported in Figure 6.5 indicate that the adsorption process of NH_3 in the natural mordenite, MP, is a heterogeneous process. In the plot, a first value at very high energy is evident, which should be related to the acid–base interaction of the NH_3 molecule with acidic sites generated in the natural mordenite during the activation [31,49].

After that, it is seen one step at 110 kJ/mol, that is, a high value of adsorption heat; this fact indicates that the NH_3 molecules strongly interact through their dipole moments with the mordenite framework [31,49]. Thereafter, a decrease of the adsorption heat up to a value of the adsorption heat, equivalent to the liquefaction heat of NH_3 at 42 kJ/mol, is seen [49].

In Figure 6.6, the q_{diff} versus θ profiles describing the heat release during the adsorption of NH_3 at 300 K in both a Na-C3 zeolite and a Ca-C3 zeolite are shown (the phase and elemental composition of sample C3 are described in Table 4.1) [31].

In Figure 6.7, the q_{diff} versus θ figures describing the heat liberation during the adsorption of NH_3 at 300 K in a Na-C4 zeolite and a Ca-C4 zeolite are shown (the phase and elemental composition of sample C3 are described in Table 4.1) [31].

In zeolites, the adsorption process is influenced by the charge-compensating cation present in the zeolite framework [2,8,29,31,34]. In Figures 6.6 and 6.7, these effects are clearly demonstrated for the HEU framework of sample C3 and the MOR framework of sample C4, where the interaction with Ca^{2+} is higher that the interaction with Na^+. This effect is caused by the higher charge and lower ionic radius of Ca^{2+} ($R_{Ca^{2+}} = 1.14\,\text{Å}$, in hexagonal coordination [52]) in relation to Na^+ ($R_{Ca^{2+}} = 1.16\,\text{Å}$, in hexagonal coordination [52]).

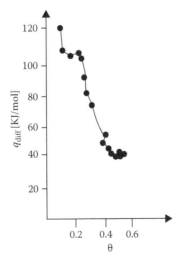

FIGURE 6.5   Plot of q_{diff} vs. θ for the adsorption of NH_3 at 300 K in an MP zeolite.

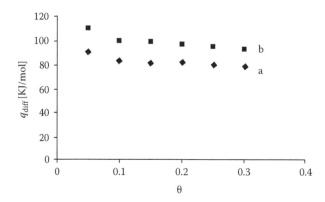

FIGURE 6.6   Plots of q_{diff} vs. θ for the adsorption of NH_3 at $300\,K$ in (a) Na-C3 and (b) Ca-C3 zeolites.

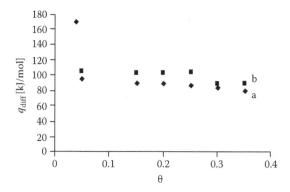

FIGURE 6.7   Plots of q_{diff} vs. θ for the adsorption of CO_2 at $300\,K$ in NH_3 at $300\,K$ in (a) Na-C4 and (b) Ca-C4 zeolites.

On the other hand, during the dehydration process of the zeolite, in order to measure the adsorption heats, the Ca^{2+} cation hydrolyzes the zeolitic water and creates Brönsted acid sites in the zeolite (see Section 2.5.3) [53]. As is well known, the NH_3 molecule suffers a strong interaction with these acid sites [31].

6.5.3   SOME RELATIONS BETWEEN MACROSCOPIC AND MICROSCOPIC ADSORPTION PARAMETERS

The molar integral change of free energy at a temperature T during adsorption is [2,13]

$$\Delta G^{ads} = \Delta H^{ads} - T\Delta S^{ads} \tag{6.11}$$

Besides, it is possible to show that that the relation between the enthalpy of adsorption (ΔH^{ads}) and the differential heat of adsorption (q_{diff}) for porous systems is [2,15,16,25]

$$\Delta H^{ads} \approx -q_{diff} - RT + \frac{T}{\Gamma}\left(\frac{\partial \vartheta}{\partial T}\right)_{\Gamma} \tag{6.12}$$

where

$$\Gamma = \frac{n_a}{n_A}$$

$$q_{\text{diff}} \approx q_{\text{iso}} - RT$$

n_a and n_A are the number of moles of the adsorbate and the adsorbent in the aA system, respectively and

ϑ is defined by [14]

$$\vartheta = RT \int_0^P \Gamma d \ln P \qquad (6.13)$$

Now assuming that the change in entropy on adsorption is negligible in comparison with the rest of terms in Equation 6.11, and taking into account that the change in the free energy of adsorption (ΔG^{ads}) can be expressed by

$$\Delta G^{\text{ads}} = RT \ln \left(\frac{P}{P_0} \right) \qquad (6.14)$$

and [6,13]

$$-q_{\text{diff}} \approx U_0 + P_a \qquad (6.15)$$

where U_0 and P_a are the adsorbate–adsorbent, and adsorbate–adsorbate interaction energies, respectively. Then, from Equations 6.11, 6.12, 6.14 and 6.15, we obtain

$$RT \ln \left(\frac{P}{P_0} \right) + \left(RT - \frac{T}{\Gamma} \left(\frac{\partial \vartheta}{\partial T} \right) \right) = U_0 + P_a \qquad (6.16)$$

If the adsorbed phase is ideal, that is, $\Gamma = KP$, then from the definition of ϑ

$$\vartheta = RT \int_0^P \Gamma d \ln P = RTK \int_0^P \frac{P dP}{P} = RTKP \qquad (6.17)$$

Consequently, from Equations 2.14 and 2.15 [2]

$$RT \ln \left(\frac{P}{P_0} \right) + \left(RT - \frac{T}{\Gamma} (R\Gamma) \right) = U_0 + P_a$$

or

$$RT \ln \left(\frac{P}{P_0} \right) = U_0 + P_a \qquad (6.18)$$

6.6 SYSTEMS FOR THE AUTOMATIC MEASUREMENT OF SURFACE AREA AND POROSITY BY THE VOLUMETRIC METHOD

6.6.1 EQUIPMENT

The adsorbed amount as a function of pressure can be obtained fundamentally by volumetric (see Figure 6.2 [20]) and gravimetric methods [6] The volumetric method used for studying microporosity and surface area, and pore size analysis is based fundamentally on nitrogen and argon adsorption isotherms obtained with liquid nitrogen (77.35 K) and liquid argon (87.27 K), respectively. Since, the shape of the adsorption and desorption isotherms depends on pore size, temperature, as well as on the chemical and geometrical heterogeneities of the tested porous material, the adsorption method is very useful for the characterization of porous materials.

A typical commercial volumetric adsorption apparatus is shown in Figure 6.8 [20]. The volumetric sorption equipment is equipped with pressure transducers in the dosing volume compartment of the apparatus and high-precision pressure transducers dedicated to measure the pressure in the sample cell [21,22]. Hence, the sample cell is isolated throughout equilibration, which assures a very small void volume, and as a result a highly accurate determination of the adsorbed amount [21].

The vapor pressure of the adsorbate at the temperature of the adsorption experiment, P_0, is measured during the entire analysis using a saturation pressure transducer, which permits the vapor pressure to be monitored for each data point [20,21]. This methodology produces great accuracy and precision in the determination of the relative pressure $x = P/P_0$, and, consequently, in the measurement of the PSD [2].

As a final point, it is necessary to state that the vacuum system of a standard commercial volumetric adsorption apparatus uses a diaphragm pump, as a fore-pumping system for the turbo-molecular pump, in order to ensure a completely oil-free environment for the adsorption measurement and the outgassing of the sample prior to the analysis [21].

6.6.2 POROUS MATERIAL CHARACTERIZATION BY ADSORPTION METHODS

Porous materials are of huge practical significance for applications in industry, energy production, and pollution abatement. In this regard, microporous materials, such as zeolites and related materials,

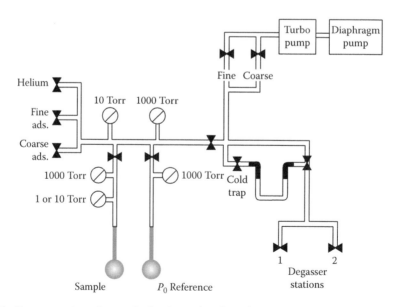

FIGURE 6.8 Representation of a standard volumetric adsorption apparatus. (Taken from Thommes, M., in *Nanoporous Materials: Science and Engineering*, Lu, G.Q. and Zhao, X.S. (eds.), Imperial College Press, London, UK, 2004, 317. With permission.)

are extensively applied in the petrochemical industry as heterogeneous catalysts, in cracking and other applications. Micro/mesoporous materials and mesoporous materials, for example, silica gels, porous glasses, mesoporous titania, and alumina, active carbon, and others materials, are extensively applied in separation processes, catalysis, and in other areas. The successful operation of adsorption processes in industry, energy production, and pollution abatement require a comprehensive characterization of these porous materials with regard to micropore volume, surface area, and PSD [1–10].

Porous materials, such as silica gels, active carbons, alumina, titania, and porous glasses, are amorphous. On the other hand, materials like zeolites are crystalline. Besides, mesoporous molecular sieves, such as MCM-41 and SBA-15, are not crystalline, although they are ordered. Such an order is not present in amorphous materials. Accordingly, a more complete characterization can be performed in the case of crystalline and ordered materials, while for amorphous nanoporous materials, a broad characterization is more difficult. However, some distinctive properties, such as the microporous volume, total pore volume, specific surface area, and PSD, and other properties, can be determined [2,4,5].

Gas adsorption measurements are extensively applied for the characterization of the surface and the porosity of porous materials [2,4,5]. This method is particularly applied for the calculation of the surface area, pore volume, and PSD of porous materials [2,4,5]. During the adsorption of vapors in complex porous systems, the adsorption process takes place generally as follows: initially, micropore filling, where the adsorption performance is dominated virtually totally by the interactions of the adsorbate and the pore wall [2,3]. As a matter of fact, we consider in this book that adsorption in the micropores could be considered as a volume filling of the microporous adsorption space, and not as a layer-by-layer surface coverage [11,14,26]. Later, at higher pressures, external surface coverage, consisting of monolayer and multilayer adsorption on the walls of mesopores and open macropores, and capillary condensation occurs in the mesopores [2,20].

Vapor adsorption in micropores is the most important method for measuring the micropore volume using the Dubinin adsorption isotherm, the t-plot method, and other adsorption isotherms [2,4,5]. Surface coverage is usually described with the help of the BET adsorption isotherm, which provides an algorithm for the calculation of the specific surface area of the porous solid [1,2,4,5]. On the other hand, capillary condensation of vapors is the main methodology of assessment of PSD in the range of the mesopores [1,2,4,5]. Capillary condensation is related with a shift of the vapor–liquid coexistence in pores compared with bulk fluid. This fact indicates that a confined fluid in a pore condenses at a pressure lower than the saturation pressure at a given temperature. This phenomenon, in the greater part of the studied systems, is accompanied by hysteresis [2,4,5].

During the past five decades, the standard method for determining the PSD in the mesoporous range with the help of adsorption isotherms was the Barret–Joyner–Hallenda (BJH) method [2,4]. However, this method does not estimate the PSD correctly. Therefore, a novel method of adsorption isotherm assessment based on the non-local density functional theory has revolutionized the method of PSD calculation in porous materials [2,41].

6.7 ADSORPTION IN ZEOLITES

6.7.1 Introduction

The study of zeolites as adsorbent materials began in 1938 when Professor Barrer published a series of papers on the adsorptive properties of zeolites [28]. In the last 50 years, zeolites, natural and synthetic, have turned out to be one of the most significant materials in modern technology [27–37]. Zeolites have been shown to be good adsorbents for H_2O, NH_3, H_2S, NO, NO_2, SO_2, CO_2, linear and branched hydrocarbons, aromatic hydrocarbons, alcohols, ketones, and other molecules [2,31,34]. Adsorption is not only an industrial application of zeolites but also a powerful means of characterizing these materials [1–11], since the adsorption of a specific molecule gives information about the microporous volume, the mesoporous area and volume, the size of the pores, the energetics of adsorption, and molecular transport.

6.7.2 Some Examples of Adsorption Systems in Zeolites

Highly dealuminated Y zeolite, commercially labeled as DAY (dealuminated Y), with an FAU framework type is an excellent adsorbent for the removal of organic compounds in wastewaters, as a consequence of the hydrophobic surface properties of this zeolite, owing to the low concentration of Al (see Section 2.5.1). As an example of adsorption in a zeolite, in Figure 6.9, the N_2 adsorption isotherm at 77 K of the DAY zeolite, DAY-20 F (Si/Al = 20), provided by Degussa AG, Düsseldorf, Germany, is shown. The N_2 adsorption isotherms were measured with an Accelerated Surface Area and Porosimetry System (Autosorb-1) from Quantachrome, Boynton Beach, FL, USA [21] which is a volumetric automatic equipment similar to that reported in Figure 6.8.

In Section 3.4.3, the hydrothermal treatments that produce phase transformations in natural zeolites, especially in clinoptilolite, which allows the possibility of synthesizing zeolites X and Y, was described. Also, the hydrothermal treatment of natural zeolites with highly basic sodium hydroxide solutions was explained.

The treatments were carried out during 4, 8, 9, 12, and 14 h in order to get samples of Na-FCSW (4 h), Na-FCSW (8 h), Na-FHC (9 h), Na-FHC (12 h), and Na-FHC (14 h), respectively. In Figure 6.10, the N_2 adsorption isotherms at 77 K of Na-FCSW (6 h), Na-FCSW (9 h), and Na-FCSW (12 h) samples are reported. The N_2 adsorption isotherms at 77 K were measured with an Accelerated Surface Area and Porosimetry System (ASAP 2000) from Micromeritics [22], which is a volumetric automatic equipment similar to that reported in Figure 6.8.

6.7.3 Determination of the Micropore Volume

6.7.3.1 Dubinin Adsorption Isotherm Equation

The Dubinin adsorption isotherm equation is a good tool for the measurement of the micropore volume. This isotherm can be deduced with the help of Dubinin's theory of volume filling, and Polanyi's adsorption potential [11,26]. The Dubinin adsorption isotherm equation has the following form [11]

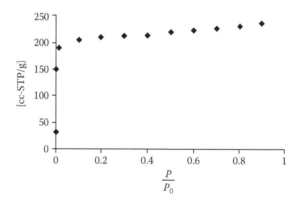

FIGURE 6.9 N_2 adsorption isotherm at 77 K of the DAY zeolite.

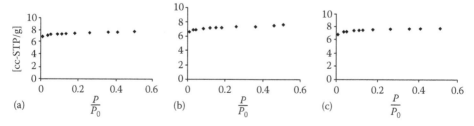

FIGURE 6.10 N_2 adsorption isotherm at 77 K of the synthetic zeolites (a) Na-FCSW (6 h), (b) Na-FCSW (9 h), and (c) Na-FCSW (12 h).

$$n_a = N_a \exp\left(-\frac{RT}{E}\ln\left[\frac{P_0}{P}\right]\right)^n \tag{6.19}$$

where

n_a is the magnitude of adsorption in the micropore volume, which is expressed in mole adsorbed/mass of dehydrated adsorbent

P_0 is the vapor pressure of the adsorptive at the temperature, T, of the adsorption experiment

P is the equilibrium adsorption pressure

E is a parameter called the characteristic energy of adsorption

N_a is the maximum amount adsorbed in the volume of the micropore, and $n(1 < n < 5)$ is an empirical parameter

It is possible, as well, to express the Dubinin adsorption isotherm equation in linear form

$$\ln(n_a) = \ln(N_a) - \left(\frac{RT}{E}\right)^n \left[\ln\left(\frac{P_0}{P}\right)\right]^n \tag{6.20}$$

This equation is a powerful tool for the description of the adsorption data in microporous material. In Figure 6.11, the Dubinin plot of the adsorption isotherm in the range $0.001 < P/P_0 < 0.03$, describing the adsorption of NH_3 at 300 K in the natural clinoptilolite sample HC is shown (see Table 4.1) [25]. The adsorption data reported in Figure 6.11 were determined volumetrically in a home-made Pyrex glass vacuum system, consisting of a sample holder, a dead volume, a dose volume, a U-tube manometer, and a thermostat [25,31]. It is evident that, in the present case, the experimental data is accurately fitted by Equation 6.20.

The Dubinin plot is the following linear plot

$$y = \ln(n_a) = \ln(N_a) - \left(\frac{RT}{E}\right)^n \left[\ln\left(\frac{P_0}{P}\right)\right]^n = b - mx$$

where

$y = \ln(n_a)$

$b = \ln(N_a)$

$m = \left(\frac{RT}{E}\right)^n$

$x = \left[\ln\left(\frac{P_0}{P}\right)\right]^n$

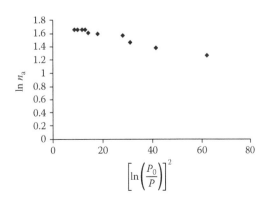

FIGURE 6.11 Dubinin plot for the adsorption of NH_3 at 300 K in the sample HC.

With the Dubinin plot, shown in Figure 6.11, it is possible to calculate the maximum adsorption capacity of this zeolite, which is $N_a = 5.67$ mmol/g and the characteristic energy $E = 28.3$ kJ/mol [25].

The fitting process of the Dubinin equation can be also carried out with the help of a nonlinear regression method, where the fitting can be carried out with a program based on a least-square procedure [54], which allows us to calculate the best-fitting parameters of Equation 6.19. Besides, the program calculates the regression coefficient and the standard errors.

6.7.3.2 Osmotic Adsorption Isotherm Equation

One more isotherm equation that could be helpful for the determination of the micropore volume is the osmotic isotherm of adsorption. Within the framework of the osmotic theory of adsorption, the adsorption process in a microporous adsorbent is regarded as the "osmotic" equilibrium between two solutions (vacancy plus molecules) of different concentrations. One of these solutions is generated in the micropores, and the other in the gas phase, and the function of the solvent is carried out by the vacancies; that is, by vacuum [26]. Subsequently, if we suppose that adsorption in a micropore system could be described as an osmotic process, where vacuum, that is, the vacancies are the solvent, and the adsorbed molecules the solute, it is possible then, by applying the methods of thermodynamics to the above described model, to obtain the so-called osmotic isotherm adsorption equation [55]:

$$n_a = \frac{N_a K_0 P^B}{1 + K_0 P^B} \tag{6.21a}$$

Equation 6.21a reduces, for $B = 1$, to a Langmuir-type isotherm equation describing a volume filling:

$$n_a = \frac{N_a K_0 P}{1 + K_0 P} \tag{6.21b}$$

Equation 5.21a is known in literature [10] as the Sips or Bradleys isotherm equation. This isotherm equation describes fairly well the experimental data of adsorption in zeolites and other microporous materials [26]. The linear form of the osmotic equation can be expressed as follows

$$y = P^B = N_a \left(\frac{P^B}{n_a} \right) + \frac{1}{K} = mx + b \tag{6.22}$$

where
$y = P^B$
$x = P^B / n_a$
$m = N_a$ is the slope
$b = 1/K$ is the intercept

In Figure 6.12 [2,25], the plot of the linear form of the osmotic isotherm equation, with $B = 0.5$, using adsorption data of NH_3 adsorbed at 300 K in an homoionic magnesium natural zeolite sample labeled CMT (see Table 4.1), is shown. The adsorption data reported in Figure 6.12 were determined volumetrically in a Pyrex glass vacuum system, previously described in the case of the Dubinin equation [25,31]. With this plot, it is possible to calculate the maximum adsorption capacity of this zeolite, which is $m = N_a = 5.07$ mmol/g and $b = 1/K = -0.92$ (Torr)$^{0.5}$.

The fitting process of the osmotic isotherm equation can be also carried out with the help of a nonlinear regression method [54].

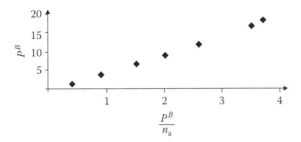

FIGURE 6.12 Linear osmotic plot with $B = 0.5$ of the adsorption data of NH_3 at $300\,K$ in a magnesium homoionic CMT zeolite.

6.7.3.3 Langmuir-Type and Fowler–Guggenheim-Type Adsorption Isotherm Equations

Applying the grand canonical ensemble method, it is possible to deduce a set of isotherm equations to mathematically treat the adsorption process in zeolites and related materials [2,3,25,56–60]. These isotherms are deduced applying the following model: the zeolite is visualized as a grand canonical ensemble (GCE) [2,3,25,56–60]. To be precise, the zeolite cavities or channels are considered as independent open subsystems belonging to the GCE [2,3,25,56–60]. In this situation, it is feasible to apply statistical thermodynamics to describe the adsorption of gases in zeolites. The zeolite is considered, in the frame of the current model, as a GCE, where the cavities or channels are the systems belonging to the ensemble [2,3,25,56–60]. In addition, the adsorption space is considered energetically homogeneous, that is, the adsorption field is identical in any position within the adsorption space [2,3,25,56].

The mathematical deduction of these isotherms is carried out with the help of the laws of statistical thermodynamics as described elsewhere [2,3,25,56]. In this sense, the isotherm equation obtained in the case of immobile adsorption with lateral interactions is [2,3,25,56]

$$\theta = \frac{\overline{N}}{m} = \frac{K_1 P}{1 + K_1 P} \tag{6.23a}$$

$$K_I = \left\{ \frac{Z_a^i}{Z_g^i} \right\} \left[\frac{1}{RT\Lambda} \right] \exp\left(\frac{[(E_0^g - E_0^a) + \Omega\theta)]}{RT} \right) \tag{6.23b}$$

or

$$K_I = K_o^1 \exp\left(\frac{\Omega\theta}{RT} \right) \tag{6.23c}$$

in which Z_g^i and Z_a^i are the canonical partition functions for the internal degrees of freedom in the gas and the adsorbed phases respectively. In addition, E_0^g is the reference energy state for the gas molecule, and $\Omega = cE_i/2$. In addition, $\Lambda = (2\pi MRT/h^2)^{3/2}$, in which $M = N_A m$ is the molar mass of the adsorptive molecule, m is the mass of the adsorptive molecule, N_A is the Avogadro number, and h is the Planck constant. Lastly, it is necessary to affirm that the average adsorption field in the volume filled by the adsorbed molecules, in this case of immobile adsorption, is $\xi(\theta) \approx -[(E_0^g - E_0^a) + \Omega\theta]$ [2,3]. The isotherm equation obtained in the case of mobile adsorption with lateral interactions is [2,3]

$$\theta = \frac{K_M P}{1 + K_M P} \tag{6.24a}$$

where

$$K_M = \left\{ \frac{Z_a^i}{Z_g^i} \right\} \left[\frac{b}{RT} \right] \exp \left(\frac{[(E_0^g - E_0^a) + \Phi\theta)]}{RT} \right) \qquad (6.24b)$$

or

$$K_M = K_0^M \exp \left(\frac{\Phi\theta}{RT} \right) \qquad (6.24c)$$

In addition, E_0^g is the reference energy state for the gas molecule, and E_0^a is the reference energy state for the adsorbed molecule in the homogeneous adsorption field inside the cavity or channel. Additionally, $\Phi = \alpha/b$ is a parameter characterizing lateral interactions, and b, the sorbate volume, is expressed in molar units.

Lastly, for mobile adsorption, the average adsorption field in the volume filled by the adsorbed molecules is $\xi(\theta) \approx -[(E_0^g - E_0^a) + \Phi\theta]$ [2,3]. It is evident, in the event that, $c \approx 0$ or $\alpha \approx 0$, Equations 5.23a and 5.24b reduce to a Langmuir-type (LT) adsorption isotherm equation

$$\theta = \frac{K_L P}{1 + K_L P} \qquad (6.25)$$

where depending on the case, that is, immobile or mobile, $K_L = K_0^I$ or $K_L = K_0^M$.

Equations 6.23a and 6.24a are of the Fowler–Guggenheim-type (FGT) adsorption isotherm equations describing volume filling rather than surface coverage. Similarly, Equation 6.25 is of the Langmuir Type (LT) adsorption isotherm equation, describing volume filling rather than surface coverage. This is also the situation for the adsorption isotherm Equation 6.21a, which reduces to an LT isotherm Equation 6.21b, describing volume filling rather than surface coverage. The discussion carried out so far shows that a variety of different approaches for describing the adsorption process in zeolites conducts to LT and FGT isotherms, describing adsorption as a volume filling effect [2,3]. As a result, Equations 6.21b, 6.23a, and 6.24a, obtained with totally different model descriptions of the adsorption phenomenon in microporous materials, all have an analogous mathematical form. Consequently, these adsorption isotherm equations can be useful in the evaluation of the micropore volume of micropore materials.

To show the utility of the obtained isotherm equations, an experimental test employing Ar adsorption at 87 K in the following commercial zeolites H-ZSM-5 (CBV-3020, $SiO_2/Al_2O_3 = 32$), H-ZSM-5(CBV5020, $SiO_2/Al_2O_3 = 53$), and H-ZSM-5 (CBV8020, $SiO_2/Al_2O_3 = 81$), all provided by the PQ Corporation, Malvern, PA, USA, was performed [3]. The experiments were carried out with a Micromeritics ASAP 2000 instrument for adsorption isotherm determination [3]. The experimental data were fitted with the linear form of the FGT-type isotherm equation

$$\ln \left(\frac{\theta}{1-\theta} \right) = \ln K + \frac{k\theta}{RT} \qquad (6.26)$$

which reduces to the LT-type isotherm equation for $k = 0$. It is evident from Figure 6.13 that the LT isotherm equation fits Ar adsorption in the range $0.01 < \Theta < 0.35$, for all the adsorption systems [3].

Since Equation 6.26, with $k = 0$, describes the adsorption of Ar at 87 K in the above referred zeolites, then the linear form of the LT isotherm

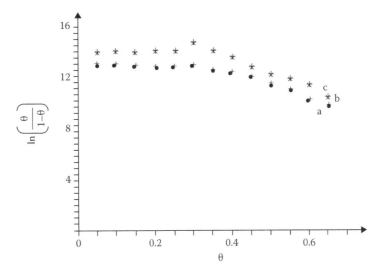

FIGURE 6.13 Plot of $\ln(\theta/1 - \theta)$ vs. θ, for the following commercial zeolites (a) H-ZSM-5 (CBV-3020), (b) H-ZSM-5(CBV5020), and (c) H-ZSM-5 (CBV8020).

TABLE 6.1
Micropore Volume, W^{Ar}, Measured with the Adsorption Isotherm of Ar at 87.3 K

Zeolite	W^{Ar} [cm^3/g]
H-ZSM-5 (CBV3020)	0.111
H-ZSM-5 (CBV5020)	0.112
H-ZSM-5 (CBV8020)	0.125

$$P = N_a \left(\frac{P}{n_a} \right) + \frac{1}{K} \tag{6.27}$$

allows a reliable measurement of the micropore volume (W^{Ar}) for the studied zeolites (see Table 6.1 [3]). The calculation of W^{Ar} is carried out with the equation $W^{Ar} = N_a\,b$, where N_a is determined with Equation 6.27. The value for the parameter b (volume occupied by 1 mol of adsorbed Ar molecules) was $b = 32.19\,cm^3/mol$.

The results reported in Table 6.1 closely agree with the micropore volume, W, reported for the zeolite ZSM-5 in the literature, which is $0.13\,cm^3/g$ [61].

6.8 ADSORPTION IN NANOPOROUS-ORDERED AND AMORPHOUS MATERIALS

6.8.1 MESOPOROUS MOLECULAR SIEVES

Recently, important efforts have been focused on obtaining materials with higher pore sizes [62], for example, the so-called mesoporous molecular sieves (MMS) [63–66].

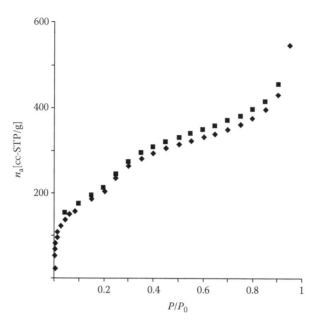

FIGURE 6.14　Adsorption isotherm of N_2 at 77 K on a mesoporous molecular sieve, MCM-41 (♦, adsorption branch; ■, desorption branch).

In Chapter 2, the structure of these materials and, in Chapter 3, the syntheses methods were described. In Figure 6.14, the adsorption isotherm of N_2 at 77 K on the mesoporous molecular sieve MCM-41 is shown [67]. The existence of capillary condensation is obvious from the isotherm. This fact implies the existence of pores in the mesopore range, that is, between 2 and 50 nm, which, in modern terms, is the nanoporous region [2]. Capillary condensation in mesopores is generally associated with a shift in the vapor–liquid coexistence in pores in comparison with the bulk fluid. That is, a fluid confined in a pore condenses at a pressure lower than the saturation pressure at a given temperature, given that the condensation pressure depends on the pore size and shape, and also on the strength of the interaction between the fluid and pore walls [2,4,5,41].

Specifically, pore condensation represents a confinement-induced shifted gas–liquid-phase transition [20]. This means that condensation takes place at a pressure, P, less than the saturation pressure, P_0, of the fluid [2,4,5]. The $x = P/P_0$ value, where pore condensation takes place, depends on the liquid-interfacial tension, the strength of the attractive interactions between the fluid and pore walls, the pore geometry, and the pore size [20].

Subsequently, it is assumed that for pores of a given shape and surface chemistry, there exists a one-to-one correspondence, between the condensation pressure and the pore diameter. Thus, adsorption isotherms contain unequivocal information about the PSD of the sample under analysis [2,20,41].

6.8.2　Amorphous Silica

The investigation of silica materials has been a field of huge interest for a long time, because this is an inert and very stable material [68–76]. In Chapter 2, the amorphous morphology of silica and, in Chapter 3, the method for silica synthesis were described.

In Figure 6.15, the adsorption isotherm of N_2 at 77 K on the silica 68bs1E [42], where the capillary condensation effect is obvious, is shown. Capillary condensation is normally characterized by a step in the adsorption isotherm. In materials with a uniform PSD, the capillary condensation step is remarkably sharp [20]. However, in practice, the hysteresis loop is seen in materials consisting of slit-like pores, cylindrical-like pores, and spherical pores, that is, ink-bottle pores [2,41]. The

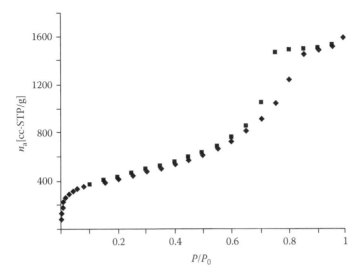

FIGURE 6.15 Adsorption isotherm of N_2 at 77 K on silica 68bs1E (♦, adsorption branch; ■, desorption branch).

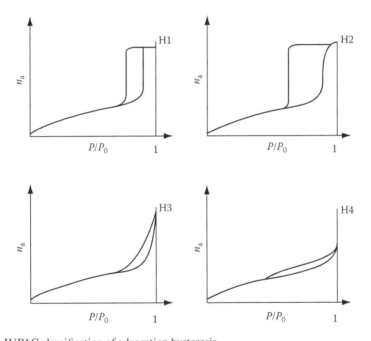

FIGURE 6.16 IUPAC classification of adsorption hysteresis.

hysteresis related with the silica material reported in Figure 6.15 is formed by suitably defined cylindrical-like pore channels or agglomerates of compacts of nearly homogeneous spheres [1,2,4,5,20], and is called the H1-type hysteresis (see Figure 6.16 [1]).

It was as well established that other materials that give rise to H2 hysteresis are commonly disordered, and their PSDs are not well defined [1,2,4,5,20] (see Figure 6.16 [1]). Systems showing H3-type hysteresis do not exhibit any limiting adsorption at a high relative pressure. This is seen in nonrigid aggregates of plate-like particles forming slit-shaped pores [1,2,4,5,20] (see Figure 6.16 [1]). Likewise, H4-type loops are commonly associated with thin slit pores, and also with pores in the micropore region [1,2,4,5,20] (see Figure 6.16 [1]).

The capillary condensation effect takes place in pores wider than approximately 5 nm; however, as the pore size decreases, the experimental hysteresis loop gradually narrows, and finally disappears for pores smaller than about 4 nm [41].

6.8.3 ADSORPTION IN ACTIVE CARBON AND CARBON NANOTUBES

In Chapters 2 and 3, the surface chemistry morphology and preparation methods of activated carbons and carbon nanotubes (CNT) were described.

Active carbon has exceptional adsorption properties because of its high surface area, its developed porosity, the wide variety of functional groups in its surface, and its relatively high mechanical strength [77–79]. As a result, active carbons are usually applied in numerous industrial procedures for the elimination of impurities from gases and liquids [77]. In this section, two examples of carbon-related adsorbents, that is, a Fisher powdered active carbon and a single-walled carbon nanotube (SWCNT), are presented [80].

As explained in Section 2.8, CNTs are another group of carbon-related materials that are extremely promising for different applications. CNTs possess excellent adsorption properties on account of their high surface area, their highly controlled PSD, and their high mechanical strength. As a result, CNTs are currently intensively studied as adsorbents [81].

In Figure 6.17a, the adsorption isotherm of N_2 at 77 K on a Fisher powdered active carbon is shown. In this isotherm, the capillary condensation effects are not pronounced, and the material, from an adsorption point of view, behaves similar to a zeolite. This is due to the highly developed micropore network present in this carbon and the few mesopores present in it. Additionally, in Figure 6.17b, the adsorption isotherm of N_2 at 77 K on a SWCNT is shown [80]. In the following sections, the data reported in Figure 6.17b are applied to illustrate the application of the BET method for the determination of the specific surface area, and the Saito–Foley and NLDFT methods for the determination of the PSD.

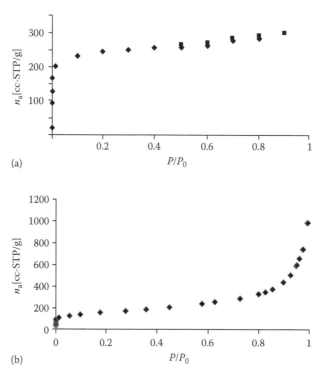

FIGURE 6.17 (a) Adsorption isotherm of N_2 at 77 K on an active carbon (♦, adsorption branch; ■, desorption branch) and (b) N_2 adsorption at 77 K on a SWCNT (♦, adsorption branch).

6.8.4 DETERMINATION OF THE SPECIFIC SURFACE AREA OF MATERIALS

The method as a rule used for the determination of the specific surface of a material is the Brunauer–Emmet–Teller (BET) method [2,4,5]. The BET theory of multilayer adsorption for the calculation of specific surface area, S, was originally developed by Brunauer, Emmett, and Teller [2,4,5]. The adsorption process, within the frame of the BET theory, is considered as a layer-by-layer process. In addition, an energetically homogeneous surface is assumed so that the adsorption field is the same in any site within the surface. Additionally, the adsorption process is considered to be immobile, that is, each molecule is adsorbed in a concrete adsorption site in the surface. Subsequently, the first layer of adsorbed molecules has an energy of interaction with the adsorption field, E_0^a, and a vertical interaction between molecules after the first layer, E_0^L, is explicitly analogous to the liquefaction heat of the adsorbate. Besides, adsorbed molecules do not interact laterally.

The deduction of the BET isotherm equation can be carried out by means of the grand canonical ensemble approach applying a methodology developed by Hill [12], which is exposed elsewhere [2]. The BET isotherm obtained is expressed as follows [2,4,5]

$$\frac{n_a}{N_m} = \frac{Cx}{(1-x+Cx)(1-x)} \tag{6.28a}$$

where
n_a is the amount adsorbed
N_m is the monolayer capacity
$x = P/P_0$
$C = K \exp\left(\dfrac{E_0^a - E_0^L}{RT}\right)$, where K is a constant

In order to apply real adsorption data to the BET isotherm equation, it is customary to use Equation 6.28a in the linear form

$$y = \frac{x}{n_a(1-x)} = \left(\frac{1}{N_m C}\right) + \left(\frac{C-1}{CN_m}\right)x = b + mx \tag{6.28b}$$

where $b = (1/N_m C)$, $m = (C - 1/N_m C)$, $y = x/n_a (1 - x)$, and $x = P/P_0$, in the region $0.05 < x < 0.4$ [2,4,5]. Considering the term $C - 1/C \approx 1$, then the slope m of the linear regression can be approximated to $m \approx 1/N_m$. As a result, the monolayer capacity N_m is determined, and the specific surface area can be calculated as

$$S = N_m N_A \sigma$$

where N_A is the Avogadro number, and σ is the cross-sectional area, that is, the average area occupied by each molecule in a completed monolayer, where $\sigma(N_2) = 0.162 \text{ nm}^2$ for N_2 at 77 K and $\sigma(Ar) = 0.138 \text{ nm}^2$ for argon at 87 K [2].

In the general case where the condition $C - 1/C \approx 1$ is not fulfilled, the parameters

$$b = \left(\frac{1}{N_m C}\right) \quad \text{and} \quad m = \left(\frac{C-1}{CN_m}\right)$$

must be calculated, formulating two equations with two unknowns that could be solved to get N_m and C. After this, S is calculated following the procedure previously explained. In Figure 6.18, the

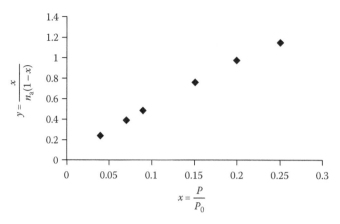

FIGURE 6.18 BET plot of a mesoporous molecular sieve MCM-41.

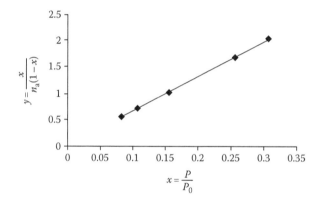

FIGURE 6.19 BET plot of a SWCNT.

BET plot ($0.04 < P/P_0 < 0.3$) for the adsorption of N_2 at 77 K in a silica material, explicitly, in the sample MCM-41 as previously described is shown (see Figure 6.14) [2]. The area measured for the MCM-41 sample tested was $S = 800\,m^2/g$.

The fitting process of the BET isotherm equation, as was persistently stated before, can be also carried out with the help of a nonlinear regression method [54].

In Figure 6.19, the BET plot ($0.04 < P/P_0 < 0.3$) for the adsorption of N_2 at 77 K in the SWCNT previously described (see Figure 6.17b) is shown. The area measured for this SWCNT sample was $S = 530\,m^2/g$.

The BET methodology must be very carefully applied in order to get correct results [2,4,5,20]. The methodology is really valuable in cases were the sorbates do not penetrate in the primary porosity, and, subsequently, the adsorption process occurs only in the outer surface. Consequently, the BET equation is really valid for the surface area analysis of nonporous and mesoporous materials consisting of pores of wide pore diameters. However, this is not truly valid in the case of microporous adsorbents [20], as long as the BET theory describes surface recovery, and adsorption in the primary porosity of zeolites is volume filling [2].

The BET method is not exact for the determination of the surface area of MMS of pore widths less than about 4 nm [20]. Given this, pore filling is observed at pressures very close to the pressure range where a monolayer–multilayer formation on the pore walls takes place, which may lead to an important overestimation of the monolayer capacity in case of a BET analysis [20]. However, the BET surface area is extensively taken as a reproducible parameter for the characterization of the surface of porous materials notwithstanding the fact that in some situations, it lacks of a precise physical meaning.

Another source of error in the use of the BET methodology is related to the surface chemistry of the sample under test. For instance, in the case of silica samples, the cross-sectional area of nitrogen on hydroxylated surfaces (σ (N_2)) [82], such as silica, is not always equal to $0.162 \, nm^2$, as is usually considered for the calculation of the BET surface area.

An additional cause of error in the application of the BET equation during adsorption experiments is the measurement of the adsorbent mass. As a result, the adsorbent mass must be very carefully measured with an analytical balance [4].

To conclude this section, it must be reiterated that following the discussion above, it is obvious that different causes exist for the spreading of the specific surface area measured in an adsorption experiment. Thus, it is usually estimated, by measuring repeatedly the tested samples, that the relative error in the BET surface area measurements of the adsorption parameters is normally around 20% [5]. For samples with very large surface areas, the relative error could be even 30% [2].

6.9 HOWARTH–KAWAZOE APPROACH FOR THE DESCRIPTION OF ADSORPTION IN MICROPOROUS MATERIALS FOR THE SLIT, CYLINDRICAL, AND SPHERICAL PORE GEOMETRIES

To make a model of an adsorption system, it is necessary to provide a description of the interaction field and the geometry of the pore system of the adsorbent [24]. The interaction fields were previously discussed in Section 6.4. On the other hand, in order to consider the effect of the pore geometry on adsorption, various models were worked out. In this regard, Horvath and Kawazoe [17] developed a methodology for the calculation of the micropore size distribution (MPSD) applying the slit potential model of Everett and Powl [45]. This methodology was later applied by Saito and Foley [18] to the case of the cylindrical pore geometry and by Cheng and Yang [19] to the case of the spherical pore geometry.

In the Horvath and Kawazoe (HK) method, the L–J (12–6) potential [2,13,17,45] was applied, where only the dispersion and repulsion interactions were included

$$\varepsilon(r) = K\varepsilon^* \left[\left(\frac{\sigma}{r} \right)^{12} - \left(\frac{\sigma}{r} \right)^6 \right]$$

where

$K = 4$ is a constant

ε^* is the minimum potential energy

n = 6 and m = 12 are the orders of the dispersion and repulsion terms, respectively

$\sigma = 0.858d$, where d is the minimum distance of separation between the interacting molecules

r is the actual distance of separation between the interacting molecules, for the case of the interaction of one adsorbate molecule with two infinite lattice planes separated by a distance, L, that is, the slit pore geometry (see Figure 2.21), which is the slit potential model of Everett and Powl [45]

$$E(z) = \frac{N_{AS} A_{AS}}{2\sigma^4} \left[\left(-\left(\frac{\sigma}{z} \right)^4 + \left(\frac{\sigma}{z} \right)^{10} \right) + \left(-\left(\frac{\sigma}{L-z} \right)^4 + \left(\frac{\sigma}{L-z} \right)^{10} \right) \right]$$

where

N_{AS} is the number of solid molecules/surface unit

L is the distance between the layers (Figure 2.21)

$\sigma = 0.858d$, where $d = (d_s + d_a)/2$, and d_s is the diameter of the adsorbent molecule d_a is the diameter of the adsorbate molecule

In addition, z is the internuclear distance between the adsorbate and adsorbent molecules, $(L - d_s)$ is the effective pore width, and A_{AS} is the dispersion constant, which takes into account the adsorbate–adsorbent interaction. The term A_{AS} is calculated with the help of the Kirkwood–Muller formula [8,13,17–19]

$$A_{AS} = \frac{6mc^2\alpha_S\alpha_A}{\left(\dfrac{\alpha_S}{\chi_S} + \dfrac{\alpha_A}{\chi_A}\right)}$$

where

m is the mass of an electron
c is the speed of light
α_A and α_S are the polarizabilities of the adsorbate and the adsorbent molecules, respectively
χ_A and χ_S are the magnetic susceptibilities of the adsorbate and the adsorbent, respectively

Horwath and Kawazoe proposed that the potential is increased by the adsorbate–adsorbate interaction, suggesting the following potential [17]

$$\Phi(z) = \frac{N_{AS}A_{AS} + N_{AA}A_{AA}}{2\sigma^4}\left[\left(-\left(\frac{\sigma}{z}\right)^4 + \left(\frac{\sigma}{z}\right)^{10}\right) + \left(-\left(\frac{\sigma}{L-z}\right)^4 + \left(\frac{\sigma}{L-z}\right)^{10}\right)\right] \quad (6.29)$$

where

N_{AA} is the number of adsorbed molecules/surface unit
L is the distance between the layers (Figure 2.21)
$\sigma = 0.858d$

Finally, A_{AA}, calculated with the help of the Kirkwood–Muller formula, is the constant characterizing the adsorbate–adsorbate interactions [2,13,17]:

$$A_{AA} = \frac{3mc^2\alpha_A\chi_A}{2}$$

The next step is to obtain the average interaction energy. This is made by volumetrically averaging the potential expressed by Equation 6.29, in this fashion [17]:

$$\xi(L) = N_A \int_d^{L-d} \frac{\Phi(z)dz}{(L-2d)} \quad (6.30a)$$

where N_A is the Avogadro number to get molar magnitudes. Integrating Equation 6.30 one obtains [9]

$$\xi(L) = \left(\frac{N_{AS}A_{AS} + N_{AA}A_{AA}}{2\sigma^4(L-2d)}\right)\left(\frac{\sigma^4}{3(L-d)^3} - \frac{\sigma^{10}}{9(L-d)^9} - \frac{\sigma^4}{3d^3} + \frac{\sigma^4}{9d^9}\right) \quad (6.30b)$$

where

$\xi(L)$ is the average potential in a given pore obtained by the integration across the effective pore width
$\Phi(z)$ is the adsorption field inside the slit pore

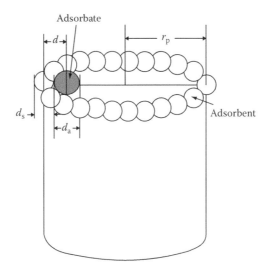

FIGURE 6.20 Cylindrical geometry description of the pore.

For the cylindrical geometry (see Figure 6.20), the interaction potential averaged over the cylinder allows us to get an approximate value for the adsorption field in the zeolite cavity or channel [18].

The adsorption process is, in this case, described with the help of a potential in between a perfect cylindrical pore of infinite length but finite radius, r_p [18]. The calculation is made with the help of a model similar to those developed by Horvath–Kawazoe for determining the MPSD [18], which includes only the van der Waals interactions, calculated with the help of the L–J potential. In order to calculate the contribution of the dispersion and repulsion energies, Everett and Powl [45] applied the L–J potential to the case of the interaction of one adsorbate molecule with an infinite cylindrical pore consisting of adsorbent molecules (see Figure 6.20), and obtained the following expression for the interaction of a molecule at a distance r to the pore wall [18]

$$E(r) = \frac{5}{2}\pi\varepsilon^* \left[\frac{21}{32}\left(\frac{d}{r_p}\right)^{10} \sum_{k=0}^{\infty} \alpha_k \left(\frac{r}{r_p}\right)^{2k} - \left(\frac{d}{r_p}\right)^4 \sum_{k=0}^{\infty} \beta_k \left(\frac{r}{r_p}\right)^{2k} \right] \tag{6.31}$$

where

$$\varepsilon^* = \frac{3}{10}\left(\frac{N_{AS}A_{AS} + N_{AA}A_{AA}}{2d^4} \right)$$

N_{AS}, d, d_s, d_a, A_{AS}, N_{AA}, and A_{AA} have already been defined, as explained in the description of the HK method.

And

$$\alpha_k = \left(\frac{\Gamma(-4.5)}{\Gamma(-4.5-k)\Gamma(k+1)} \right)^2$$

$$\beta_k = \left(\frac{\Gamma(-1.5)}{\Gamma(-1.5-k)\Gamma(k+1)} \right)^2$$

in which $\alpha_0 = \beta_0 = 1$. The next step is to obtain the linear average of the interaction energy, by averaging the potential expressed by Equation 6.31 as [18]

$$\xi(r_p) = N_A \frac{\displaystyle\int_0^{r_p-d} E(r)dr}{r_p - d} \tag{6.32a}$$

to get molar magnitudes. Then [18]

$$\xi(r_p) = \frac{3}{4}\pi N_A \left(\frac{N_{AS}A_{AS} + N_{AA}A_{AA}}{d^4}\right)\left(\sum_{k=0}^{\infty}\left[\frac{1}{2k+1}\left(1-\frac{d}{r_p}\right)^{2k}\left\{\frac{21}{32}\alpha_k\left(\frac{d}{r_p}\right)^{10} - \beta_k\left(\frac{d}{r_p}\right)^4\right\}\right]\right) \tag{6.32b}$$

where $\xi(r_p)$ is the average potential in a given pore obtained by the integration across the effective pore width.

For the spherical pore geometry (see Figure 6.21), the interaction between a single adsorbate molecule and the inside wall of the spherical pore cavity of radius R consisting of a single lattice plane is given by [19]

$$\Gamma_S(r) = \int_0^{\pi}\int_0^{2\pi} N_1\varepsilon(r,\theta)R^2(\sin\theta)d\theta d\phi$$

where

$$N_1 = 4\pi R^2 N_{AS}$$

is the number of atoms in the wall of the pore, $\varepsilon(r, \theta)$ is the L–J potential inside the spherical pore, r is the radial distance of the adsorbate molecule from the center of the cavity, θ is the polar angle, and ϕ is the azimuth angle. The integration of this expression gives

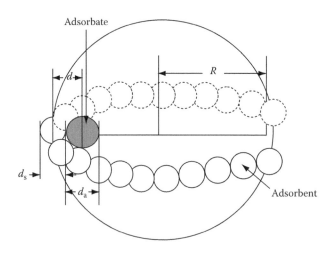

FIGURE 6.21 Spherical geometry description of the pore.

$$\Gamma_S(r) = 2N_1\varepsilon_{aA}^* \left[\begin{array}{l} -\left(\dfrac{d}{R}\right)^6 \left\{\dfrac{1}{4\left(\dfrac{r}{R}\right)}\right\} \left\{ \dfrac{1}{\left(1-\dfrac{r}{R}\right)^4} - \dfrac{1}{\left(1+\dfrac{r}{R}\right)^4} \right\} \\[3em] +\left(\dfrac{d}{R}\right)^{12} \left\{\dfrac{1}{10\left(\dfrac{r}{R}\right)}\right\} \left\{ \dfrac{1}{\left(1-\dfrac{r}{R}\right)^{10}} - \dfrac{1}{\left(1+\dfrac{r}{R}\right)^{10}} \right\} \end{array} \right]$$

Now, taking into account the adsorbate–adsorbate interaction

$$\Gamma(r) = 2\left(N_1\varepsilon_{aA}^* + N_2\varepsilon_{AA}^*\right) \left[\begin{array}{l} -\left(\dfrac{d}{R}\right)^6 \left\{\dfrac{1}{4\left(\dfrac{r}{R}\right)}\right\} \left\{ \dfrac{1}{\left(1-\dfrac{r}{R}\right)^4} - \dfrac{1}{\left(1+\dfrac{r}{R}\right)^4} \right\} \\[3em] +\left(\dfrac{d}{R}\right)^{12} \left\{\dfrac{1}{10\left(\dfrac{r}{R}\right)}\right\} \left\{ \dfrac{1}{\left(1-\dfrac{r}{R}\right)^{10}} - \dfrac{1}{\left(1+\dfrac{r}{R}\right)^{10}} \right\} \end{array} \right] \tag{6.33}$$

in which

$$N_2 = 4\pi(R-d)^2 N_{AA}$$

and

$$\varepsilon_{aA}^* = \frac{A_{aA}}{4d^6} \quad \text{and} \quad \varepsilon_{AA}^* = \frac{A_{AA}}{4d^6}$$

Applying the same approach developed by Howart and Kawazoe,

$$\xi(R) = N_A \frac{\displaystyle\int_0^{R-d} \Gamma(r)4\pi r^2 dr}{\displaystyle\int_0^{R-d} 4\pi r^2 dr}$$

The integration gives [19]

$$\xi(R) = \frac{6(N_1\varepsilon_{aA}^* + N_2\varepsilon_{AA}^*)}{(R-d)^3} \left[-\left(\frac{d}{R}\right)^6 \left(\frac{1}{12}T_1 + \frac{1}{8}T_2\right) + \left(\frac{d}{R}\right)^{12} \left(\frac{1}{90}T_3 + \frac{1}{80}T_4\right) \right] \tag{6.34}$$

where

$$T_1 = \frac{1}{\left(1 - \dfrac{R-d}{R}\right)^3} - \frac{1}{\left(1 + \dfrac{R-d}{R}\right)^3}$$

$$T_2 = \frac{1}{\left(1 + \dfrac{R-d}{R}\right)^2} - \frac{1}{\left(1 - \dfrac{R-d}{R}\right)^2}$$

$$T_3 = \frac{1}{\left(1 - \dfrac{R-d}{R}\right)^9} - \frac{1}{\left(1 + \dfrac{R-d}{R}\right)^9}$$

$$T_4 = \frac{1}{\left(1 + \dfrac{R-d}{R}\right)^8} - \frac{1}{\left(1 - \dfrac{R-d}{R}\right)^8}$$

As was previously stated, the molar integral change of free energy, at a given temperature, T, during adsorption is [2,13]

$$\Delta G^{ads} \approx \Delta H^{ads} \approx U_0 + P_a$$

Now, the average potentials for the action of the dispersion and repulsion forces for the three pore geometries $\xi(L)$, $\xi(r_p)$, and $\xi(R)$ can be represented by $\xi(\rho)$, where $\rho = L$, r_p or R. Then [24],

$$U_0 + P_a = \xi(\rho)$$

Besides, the free energy change upon adsorption can be calculated as follows [2,13]:

$$\Delta G^{ads} = RT \ln\left(\frac{P}{P_0}\right)$$

Consequently, for the three pore geometries [24]

$$RT \ln\left(\frac{P}{P_0}\right) = N_A \xi(\rho) \tag{6.35}$$

where N_A is the Avogadro number, and the energy is represented in molar units.

With the help of Equation 6.35 and Equation 6.30b for the H–K method, Equation 6.32b for the S–F method, and Equation 6.34 Cheng and Yang (Ch–Y) method, it is possible to calculate the MPSD for the slit pore geometry [17], for the cylindrical pore geometry, and for the spherical pore geometry [19], respectively. The original H–K method states that the relative pressure, $x = P/P_0$, required for the filling of micropores of a concrete size and shape is directly related to the adsorbate–adsorbent interaction energy [2,13]. This means that the micropores are progressively filled with an increase in adsorbate pressure. In the HK method, only pores with dimensions lower than a particular unique value will be filled for a given relative pressure of the adsorbate [2,13]. Following similar ideas, the S–F and Ch–Y methods allow the calculation of the pore size distribution in the micropore range at low pressures for cylindrical and spherical geometries [13,18,19].

In Figure 6.22a, the MPSD of the DAY zeolite calculated with the help of the SF method is reported. The DAY zeolite has an FAU framework, with a 12MR three-dimensional channel system with apertures of 7.4 Å (see Section 2.5.1) [83]. The obtained PSD for the DAY zeolite shows a clear maximum at about 7 Å. In Figure 6.22b, the S-F MPSD of the SWCNT is reported [80]. The calculated PSD for the SWCNT indicates that this material has a slightly heterogeneous distribution of pores with a clear, and highly populated, maximum at 13.5 Å [80]. The adsorption study, in both cases, was carried out using N_2 adsorption at 77 K isotherms, obtained with a Quantachrome Autosorb-1 equipment for adsorption isotherm determination by the volumetric method [21].

All these models, founded on the notion of intermolecular potential, have been applied by numerous authors not only merely to calculate PSDs but also to get information on other significant properties of the adsorption systems.

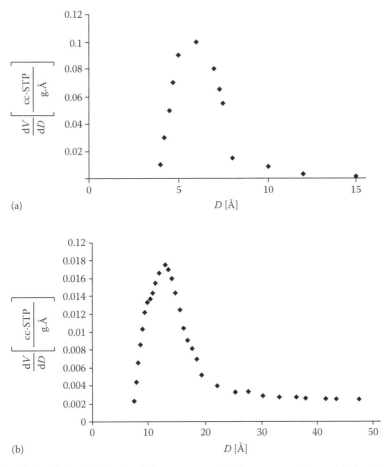

FIGURE 6.22 Saito–Foley MPSD (dn_a/dD_{ST} vs. pore width) corresponding to (a) the DAY zeolite and (b) a SWCNT.

6.10   ADSORPTION FROM LIQUID SOLUTIONS

6.10.1   Introduction

In this chapter, we have so far discussed the adsorption of gases in solids. This section gives a brief description of the adsorption process from liquid solutions. This adsorption process has its own peculiarities compared with gas–solid adsorption, since the fundamental principles and methodology are different in almost all aspects [2,4,5]. In the simplest situation, that is, a binary solution, the composition of the adsorbed phase is generally unknown. Additionally, adsorption in the liquid phase is affected by numerous factors, such as pH, type of adsorbent, solubility of adsorbate in the solvent, temperature, as well as adsorptive concentration [2,4,5,84]. This is why, independently of the industrial importance of adsorption from liquid phase, it is less studied than adsorption from the gas phase [2].

Nevertheless, liquid-phase adsorption fundamentally onto activated carbon [2,85], silica [2,86,87], zeolites [2,88,89], and resins [90] provides a feasible technique, and is one of the most extensively used technologies for removal of organic pollutants from industrial wastewater [2,8,84–90].

6.10.2   Isotherms for the Description of Adsorption from Liquid Phase

The interfacial layer is the inhomogeneous space region intermediate between two bulk phases in contact, and where properties are notably different from, but related to, the properties of the bulk phases (see Figure 6.1). Some of these properties are composition, molecular density, orientation or conformation, charge density, pressure tensor, and electron density [2]. The interfacial properties change in the direction normal to the surface (see Figure 6.1). Complex profiles of interfacial properties take place in the case of multicomponent systems with coexisting bulk phases where attractive/repulsive molecular interactions involve adsorption or depletion of one or several components.

The surface excess amount, or Gibbs adsorption (see Section 6.2.3), of a component i, that is, n_i^σ, is defined as the excess of the quantity of this component actually present in the system, in excess of that present in an ideal reference system of the same volume as the real system, and in which the bulk concentrations in the two phases stay uniform up to the GDS. Nevertheless, the discussion of this topic is difficult; on the other hand for the purposes of this book, it is enough to describe the practical methodology, in which the amount of solute adsorbed from the liquid phase is calculated by subtracting the remaining concentration after adsorption from the concentration at the beginning of the adsorption process.

The methodology is carried out as follows: first, it is required to determine the particle size of the adsorbent, adsorption temperature, pH of the solution, and contact time needed to achieve adsorption equilibrium [78,84,91]. Afterward, different amounts of the solid adsorbent, normally, few milligrams, to be precise 1–10 mg, are brought into contact with equivalent volumes of solution, as a general rule, 25–50 mL of the solution containing an established concentration of the substance to be adsorbed [78,88,91]. After this, all solutions are equilibrated, usually for 0.5–5 h, while being stirred. Once contact time has elapsed, the adsorbent is centrifuged and filtered. The resulting solutions are poured into vials, which are completely filled so the headspace is very low. Next, the lids are sealed with tape until the vials are analyzed by, for example, with a spectrophotometer [91]. The amount of solute adsorbed from the liquid phase is determined by subtracting the remaining analyzed concentration with the starting concentration [78,91]. Then, the amount adsorbed on the adsorbent is calculated from the initial liquid-phase concentration and equilibrium concentration with the help of the following simple equation [91]

$$q_1^e = \frac{V}{m_a}(C_1^0 - C_1^e)$$

where

C_1^0 and C_1^e are the initial and equilibrium concentrations, respectively

V is the volume of solution

m_a is the mass of adsorbent

At this point, it is feasible to correlate the liquid-phase adsorption equilibrium single component data, with the help of isotherm equations developed for gas-phase adsorption, since, in principle, it is feasible to extend these isotherms to liquid-phase adsorption by the simple replacement of adsorbate pressure by concentration [92]. These equations are the Langmuir, Freundlich, Sips, Toth, and Dubinin–Radushkevich equations [91–93]. Nevertheless, the Langmuir and Freudlich equations are the most extensively applied to correlate liquid-phase adsorption data. [2,87].

The Langmuir isotherm equation for the correlation of the liquid-phase adsorption equilibrium of a single component, can, in principle, as was previously stated, be extended to liquid-phase adsorption by the simple replacement of adsorbate pressure by concentration [2,87]:

$$q_1^e = \frac{q_m b_0 C_1^e}{1 + b_0 C_1^e}$$

and the Freundlich equation is given as

$$q_1^e = k(C_1^e)^{1/n}$$

where the experimental equilibrium data, expressed as q_1^e, is the amount of the solute component, 1, adsorbed per mass of adsorbent; this parameter is expressed in [mg/g] or [mmol/g]. C_1^e is the final solute concentration in solution, or the equilibrium concentration in [mg/L] or [mmol/L]. Besides, q_m and b_0 are the Langmuir parameters, and k and n are the Freundlich parameters.

The Sips equation is [91]

$$q_1^e = \frac{q_0(dC_1^e)^{1/S}}{1 + (dC_1^e)^{1/S}}$$

where q_0, d, and S are parameters of the Sips equation. The other three-parameter equation is the Toth equation [91,94]

$$q_1^e = \frac{q_0 b C_1^e}{1 + (bC_1^e)^{1/t}}$$

in which q_0, b, and t are the parameters.

The calculation of the parameters for the Langmuir equation is, in general, carried out by linearly regressing each set of experimental data using the following equation

$$y = C_1^e = q_m \left(\frac{C_1^e}{q_e}\right) + \frac{1}{b_0} = mx + b$$

which is a linear form of the Langmuir equation.

A standard regression analysis can be performed on each set of data by applying a commercial regression software [95,96]. The calculation can be also carried out by nonlinearly regressing each set of experimental data using directly the Langmuir equation for liquid-phase adsorption.

In the instance of the Freundlich equation, the parameters are calculated by linearly regressing each set of experimental data using the following equation [87]:

$$y = \log(q_e) = \log(k) + \left(\frac{1}{n}\right)\log(C_1^e) = mx + b$$

Also, the linear regression analysis can be carried out applying a commercial regression software, which can as well be used for the nonlinear regressing analysis of each set of experimental data using directly the Freundlich isotherm equation for liquid-phase adsorption [95,96].

Now, to conclude this section, it is necessary to affirm that any one of the equations described here to correlate the relation between the amount adsorbed, q_i^e, with the equilibrium concentration in solution, C_1^e, corresponds to a particular model for adsorption from solutions. That is, these should be considered as empirical isotherm equations [2].

6.11   DYNAMIC ADSORPTION: THE PLUG-FLOW ADSORPTION REACTOR

6.11.1   DYNAMIC ADSORPTION

To use adsorption as a unitary operation in industrial, pollution abatement, or energy production applications, in most cases, a reactor where a dynamic adsorption process will take place is packed with a concrete adsorbent. The adsorbents generally used for these applications are active carbons, zeolites and related materials, silica, mesoporous molecular sieves, alumina, titanium dioxide, magnesium oxide, clays, and pillared clays.

Dynamic adsorption is a mass-transfer problem that can be treated with complex mass-transfer models, where many parameters that have to be calculated by independent batch kinetic studies or estimated by appropriate correlations are required [97,98].

We discuss here the simplest application of dynamic adsorption, that is, the application of adsorbents to clean gas or liquid flows, by the removal of a low-concentrated impurity, with the help of a plug-flow adsorption reactor (PFAR) (see Figure 6.23). In this situation, during the operation of the PFAR, the change in concentration with time at the outlet of the reactor is represented by a break-through curve (see Figure 6.24) [99–106].

In Figure 6.24, C_0 is the initial concentration, C_e is the breakthrough concentration, V_e is the fed volume of the aqueous solution of the solute A to breakthrough, and V_b is the fed volume to saturation (the previous description is as well valid for a or gaseous mixture). This is a response curve where the relation between concentration and time at the outlet of the packed bed adsorption reactor is shown.

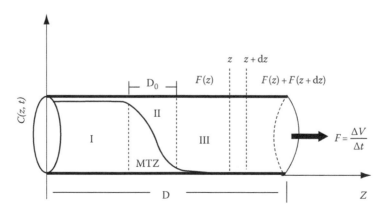

FIGURE 6.23   Packed-bed adsorption reactor.

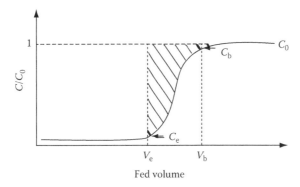

FIGURE 6.24 Breakthrough curve.

The volumetric flow rate passing through the reactor is given by

$$F = \frac{\Delta V}{\Delta t} = \frac{\text{volume}}{\text{time}}$$

where the passing fluid is an aqueous solution with an initial concentration C_0 [mass/volume] of a trace solute A, or a gas flow with a low concentration, C_0, of a gaseous impurity A [mass/volume].

The volume of the empty bed is $V_B = \varepsilon V$, where V is the bed volume and ε is the fraction of free volume in the bed, and the interstitial fluid velocity, u, is defined as [100]

$$u = \frac{F}{S} \tag{6.36}$$

Additionally, the contact, or residence, time of the fluid passing through the reactor, τ, is determined with the help of the equation [100]:

$$\tau = \frac{V_B}{F} \tag{6.37}$$

For the laboratory dynamic adsorption reactor to properly work, as a rule, it should satisfy the following points [2,78]:

1. *Residence time*: Since adsorption could be a slow process, the fluid contact or residence time should be long enough to allow molecular transport to the adsorption sites. Therefore, as a rule of thumb, it is possible to apply residence times around the following figures: $0.05\,\text{s} < \tau < 0.1\,\text{s}$ for gaseous dynamic adsorption and $0.5\,\text{s} < \tau < 1\,\text{s}$ for liquid-phase dynamic adsorption.
2. *Particle size*: If the particle size is small enough, it will lead to a significant pressure drop in the reactor. Accordingly, granular particles with big particle sizes should be packed in the reactor. The particular particle size depends on the size of the reactor, for this reason, as a rule of thumb, to construct the reactor the following approximate relation, $d_R/d_P \geq 10$, should be applied, where d_R is the reactor diameter and d_P is the particle size. For the laboratory testing of a material, which is the principal aim of this book, the relation $d_R/d_P \approx 10$ is a good choice.
3. *Reactor longitude*: Given that residence times are somewhat long, large facilities are sometimes required to reach the needed treatment capacities. Subsequently, in order to keep the

proportions of the reactor dimensions moderate, the next rule could be applied, $D/d_R \leq 10$, where D is the reactor length, and d_R is the reactor diameter. For the laboratory testing of a material, $D/d_R \approx 10$ is an excellent option.

Obviously, the above points are only approximate design criteria, which are exclusively justified for laboratory testing of materials and not for the design of industrial reactors, which is not the aim of this book.

6.11.2 PLUG-FLOW ADSORPTION REACTOR MODEL

The plug-flow model indicates that the fluid velocity profile is "plug shaped," that is, is uniform at all radial positions, fact which normally involves turbulent flow conditions, such that the fluid constituents are well-mixed [99]. Additionally, it is considered that the fixed-bed adsorption reactor is packed randomly with adsorbent particles that are fresh or have just been regenerated [103]. Moreover, in this adsorption separation process, a rate process and a thermodynamic equilibrium take place, where individual parts of the system react so fast that for practical purposes local equilibrium can be assumed [99]. Clearly, the adsorption process is supposed to be very fast relative to the convection and diffusion effects; consequently, local equilibrium will exist close to the adsorbent beads [2,103]. Further assumptions are that no chemical reactions takes place in the column and that only mass transfer by convection is important.

Finally, it must be pointed out that the adsorbent when it makes contact with a binary mixture, one component is selectively adsorbed by the solid adsorbent. In the flowing fluid, a trace of an adsorbable species is adsorbed from a relatively inert carrier. In addition, the heat effects can be ignored, as a result, isothermal conditions can be taken [9,103]. The flow is fed at the top of the bed at a constant flow rate, and under conditions such that mass-transfer resistance is insignificant [2,103].

The reactor has a cross-sectional area, S, column length, D, and adsorbent mass in the bed, M (see Figure 6.23). The adsorbent bed in the PFAR can be divided in three zones: I, the equilibrium zone; II, mass transfer zone (MTZ) with a length, D_0; and III, the unused zone [100,105,106]. In addition, the length of the MTZ, D_0, can be calculated with the following expression (see Figure 6.24) [106]:

$$D_0 = 2D \frac{V_b - V_e}{V_b + V_e} \tag{6.38}$$

Other parameters characterizing a PFAR are the column breakthrough capacity, B_C, the column saturation capacity, S_C, and the column efficiency, E, of the PFIEBR, which are calculated with the following equations [105]:

$$B_C = \frac{C_0 V_e}{M} \tag{6.39a}$$

$$S_C = \frac{C_0 V_b}{M} \tag{6.39b}$$

$$E = \frac{B_C}{S_C} \tag{6.39c}$$

At this point, it is necessary to state that the physical process of adsorption is so fast relative to other steps, such as diffusion within the solid. In this case, in and near the solid adsorbent, the general form for the equilibrium isotherm is [99]

$$q = KC^*$$ (6.40)

where
q is the equilibrium value of the adsorbate concentration, expressed as moles solute adsorbed per unit volume of the solid particle
C^* denotes the solute composition, in moles of solute per unit volume of fluid, which could exist at equilibrium
K is the linear partition coefficient

Taking into account all the preceding assumptions, the mass balance equation for the PFAR is

$$\text{in} - \text{out} = \text{accumulation}$$

which could be expressed mathematically as follows [103]

$$FC(z) - FC(z + dz) = \varepsilon \frac{\partial C}{\partial t} S \, dz + (1 - \varepsilon) \frac{\partial q}{\partial t} S \, dz$$

in which the first term is related to fluid flow, and the two terms on the other side of the equation are related to the accumulation in the fluid phase and the solid phase, respectively. Therefore, dividing by $S \, dz$, we get [8]

$$\frac{\partial C}{\partial t} + u \frac{\partial C}{\partial z} + \frac{1 - \varepsilon}{\varepsilon} \frac{\partial q}{\partial t} = 0$$ (6.41)

where
ε is the void fraction of the bed, that is, the volume between particles
$(1 - \varepsilon)$ denotes the fractional volume taken up by the solid
u is the interstitial velocity of the carrier fluid
t is the operating time
z is the distance from the inlet of the mobile phase
$C(z, t)$ is the flowing solute composition
q is the solute concentration in the stationary phase

To complete the required set of equations, it is necessary to incorporate the adsorption rate of the solute or contaminant, which can be described by the linear driving force model in terms of the overall liquid-phase mass-transfer coefficient [8,103,104]

$$\frac{\partial q}{\partial t} = k'(C - C^*)$$ (6.42)

where
C^* is the mobile phase concentration in equilibrium with the stationary phase concentration q
k' is the rate coefficient which is a lumped mass-transfer coefficient

where

$$k' = \frac{k_c a}{(1 - \varepsilon)}$$

in which
k_c is the mass-transfer coefficient per unit interfacial area
$k_c a$ is the mass-transfer coefficient per unit be volume
a is the total interfacial area per unit volume of packed column [99,103]

Equations 6.40 through 6.42, having three unknowns, q, C, and C^*, describe the system. Thus, it is possible to eliminate q, with the help of Equation 6.40, and then obtain [99]:

$$\frac{\partial C}{\partial t} + u \frac{\partial C}{\partial z} + \frac{1-\varepsilon}{\varepsilon} K \frac{\partial C}{\partial t} = 0 \tag{6.43a}$$

$$\frac{\partial C^*}{\partial t} = \frac{k_c a}{(1-\varepsilon)K}(C - C^*) \tag{6.43b}$$

The initial and boundary conditions associated with the above two partial differential equations (PDE) describing the operation of the PFAR are [99,103]

1. $C(z,0) = 0$ and $C^*(z,0) = 0$, initially clean interstitial fluid, for $0 \le z \le D$
2. $C(0,t) = C_0$, that is, constant composition at bed access

To calculate the breakthrough curve, analytical solutions of the system of Equations 6.43a and b can be obtained using the Laplace transform method [99]. It is then possible to express the solution of Equations 6.43a and b [99] as follows:

$$C(\xi,\tau) = C_0 u(\tau)\left[1 - \exp(-\tau)\int_0^\xi \exp(-\beta)I_0(2\sqrt{\beta\tau})d\beta\right] \tag{6.44}$$

where

I_0 is the modified Bessel function
$u(\tau)$ is the step function

$$\xi = \frac{k_c a}{\varepsilon}\left(\frac{z}{u}\right)$$

$$\tau = \frac{k_c a}{K(1-\varepsilon)}(\theta), \quad \text{where } \theta = t - \frac{z}{u}$$

β is only an integration mute variable

Now, applying the algorithm of the Bessel function [107,108] it is possible to represent Equation 6.44 as a series expansion [107–109]

$$C(\xi,\tau) = C_0 u(\tau)\left[1 - e^{-\tau}\xi + \left(\frac{e^{-\tau}}{2} - \frac{e^{-\tau}\tau}{2}\right)\xi^2 + \left(-\frac{e^{-\tau}}{6} + \frac{e^{-\tau}\tau}{3} - \frac{e^{-\tau}\tau^2}{12}\right)\xi^3 + O(\xi,\tau)^4 + \cdots\right] \tag{6.45}$$

which can be computed by the method described in Sections 5.7.7 and 5.9.2; that is, in Equation 6.45 the first three terms of the original expansion series are given, and the fourth term is given by

$$O(\xi,\tau)^4 = \left(\frac{e^{-\tau}}{24} - \frac{e^{-\tau}\tau}{8} + \frac{e^{-\tau}\tau^2}{16} - \frac{e^{-\tau}\tau^3}{144}\right)\xi^4$$

and the rest of the terms are not given. Then, cutting the series in Equation 6.45, and using only the first three or four terms, an approximation to Equation 6.44 is obtained, which could be numerically

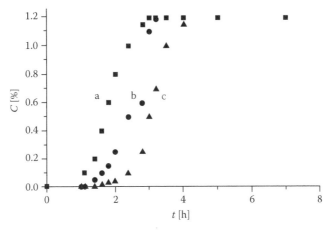

FIGURE 6.25 Breakthrough curves resulting from the dynamic adsorption of H_2O from a H_2O–air mixture by (a) HC, (b) Na-HC, and (c) Ca-HC.

fitted to experimental data with the help of a nonlinear regression method. The above-reported analytical solution of the simultaneous couple PDE describing the operation of the PFAR is an elegant example of the use of the Laplace transform methodology for the solution of the PDEs. More importantly, within the framework of the applications of adsorption, it is very useful when the relation between q, the equilibrium value of the adsorbate concentration, and C^*, the solute composition, is linear. On the other hand, numerical solutions offer more rigorous results because of a smaller amount of simplification is made, and represent a more flexible line of attack, since their use is not limited by the type of adsorption isotherm, or the initial and boundary conditions applied [2].

In Figure 6.25, in an application of a PFAR in a dynamic adsorption process of gas cleaning, three breakthrough curves describing the dynamic adsorption of H_2O from a H_2O–air mixture by a natural clinoptilolite, sample HC (see Table 4.1), and two homoionic clinoptilolites, Na-HC and Ca-HC, are shown [31,34,110].

The water concentration in the mixture prior to breakthrough was C_0 = 1.2 mg/L, which was reduced after the gas flow through the adsorption reactor to 0.01–0.03 mg/L [31,34,110]. The reactor had a cross-sectional area, S = 10.2 cm², column length, D = 6.9 cm, adsorbent mass in the bed, M = 30 g, the volume of the bed was 70 cm³, the volume free of adsorbent in the bed was about 35 cm³, and the volumetric flow rate was F = 7.7 cm³/s [31,34,110].

6.12 SOME CHEMICAL, SUSTAINABLE ENERGY, AND POLLUTION ABATEMENT APPLICATIONS OF NANOPOROUS ADSORBENTS

6.12.1 GAS SEPARATION AND CLEANING

Gas, or vapor molecules, after the degasification process, can go through the pore structure of crystalline and ordered nanoporous materials through a series of channels and/or cavities. Each layer of these channels and cavities is separated by a dense, gas-impermeable division, and within this adsorption space the molecules are subjected to force fields. The interaction with this adsorption field within the adsorption space is the base for the use of these materials in adsorption processes. Sorption operations used for separation processes imply molecular transfer from a gas or a liquid to the adsorbent pore network [2].

Zeolites A, -X, ZSM-5, chabazite, clinoptilolite, mordenite, and other nanoporous materials are used for removing H_2O, NH_3, NO, NO_2, SO_2, SH_2, CO_2, and other impurities from gas streams [2,27–37,110–114]. For instance, in gas cleaning, zeolites are normally used for the removal of H_2O,

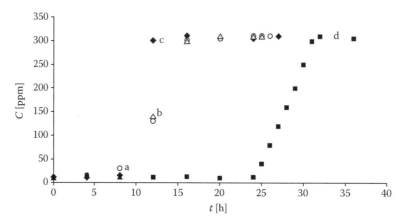

FIGURE 6.26 Breakthrough curves resulting from the dynamic adsorption of H_2O from $H_2O–CO_2$ by (a) Mg-CMT (○), (b) K-CMT (Δ), (c) Ca-CMT(♦), and Na-X (■).

SH$_2$, and CO$_2$ from sour natural gas streams [2,28,31,34]. They could be also used for drying of the CO$_2$ used in cryosurgery (see Figure 6.26 [34,110]); selective removal of NH$_3$ produced during the gasification of coal; removal of NH$_3$, SO$_2$, NO$_x$, and CO$_2$ from air; and for the selective adsorption of SH$_2$ from methane (CH$_4$) streams [31,34,110].

Among natural zeolites, clinoptilolite is more profusely distributed in the earth's core, and, consequently, the most applied in adsorption applications [31]. Clinoptilolite is a member of the heulandite family, with a molar ratio Si/Al > 4, and about 22 water molecules compose the unit cell, where Na, K, Ca, and Mg are the most common charge-balancing cations [31,115].

TABLE 6.2
Breakthrough Mass

Sample	B.M. [mg H_2O/cm³]
HC	39
Na-HC	44
Ca-HC	67
Na-X	96
Alumina	28

In Table 6.2, the breakthrough mass (BM) for air drying in the adsorption reactor filled with natural (HC sample) and homoionic sodium clinoptilolite (Na-HC), and homoionic calcium clinoptilite (Ca-HC) are reported [31]. The parameter, BM, was calculated with the help of Equation 6.39a, and using the breakthrough curves reported in Figure 6.25. But, the values are reported not in terms of adsorbent mass, but in terms of the reactor volume. The adsorbents applied for comparison were a synthetic Na-X zeolite provided by Laporte and an alumina (Al$_2$O$_3$) provided by Neobor [31]. The water concentration in the tested air prior to breakthrough was 1.2 mg/L, and it was reduced after the air flow through the adsorption reactor to 0.01–0.03 mg/L [31,34].

In Figure 6.26, the breakthrough curves resulting from the dynamic adsorption of H$_2$O from a H$_2$O–CO$_2$ mixture by the homoionic natural zeolites Mg-CMT, K-CMT, Ca-CMT (see Table 4.1), and the Na-X provided by Laporte are shown [34,110]. The water concentration in the mixture prior to breakthrough was $C_0 = 400$ ppm, which was reduced after the gas flow through the adsorption reactor to 10 ppm [34,110]. The reactor had a cross-sectional area, $S = 10.2$ cm², column length, $D = 6.9$ cm, adsorbent mass in the bed, $M = 30$ g, the volume of the bed was 70 cm³, the volume free of adsorbent in the bed was about 35 cm³, and the volumetric flow rate was $F = 7.7$ cm³/s [34,110].

The HEU framework of clinoptilolite is a two-dimensional micropore channel system, where the 10-membered ring (MR) channel A and the 8-MR channel B run parallel to each other and to the c-axis of the unit cell, while channel C (8-MR) is placed along the a-axis, intersecting both the A and B channels (see Figure 2.20) [83]. The elliptical shaped 8-MR and 10-MR that make up the channel system are nonplanar, and, consequently, cannot be simply dimensioned. The selectivity and uptake rate of gases by clinoptilolite, and also by other zeolites, are influenced by the type, number, and location of the charge-balancing cations residing in the clinoptilolite HEU framework channels [2,27,34,115]. Therefore, the positions adopted by exchangeable cations and the adsorbed molecules are interdependent, and the variations in the cation composition cause changes in the amount of adsorbed molecules as it is evident from Figures 6.21 and 6.22 [31,115]. In the case of the synthetic zeolite Na-A, it is possible to cation-exchange it with potassium and calcium, in order to get the 3A and 5A molecular sieves, respectively. Different studies [29,113,115,116] have revealed that the pore-opening size of zeolite and other molecular sieves can be controlled to fit desired applications by postsynthesis modification methods, such as internal or external surface modification by chemical reactions, preadsorption of polar molecules [116], chemical vapor deposition [117], or similar coating processes and thermal treatment [118].

It has also been revealed [29] that the pore size and the affinity of a zeolite structure can be modified by chemical treatment of the zeolite structure; the silane, borane, or disilane molecules are chemisorbed on the zeolite surface by reacting with the silanol groups of the zeolite [113]. Polar molecules, for example, water and amines presorbed in the zeolite can be used to modify the operation of the molecular sieve and the interaction toward adsorbate molecules of the zeolite [113].

In addition to zeolites, the MMS have been also applied in gas cleaning. In this regard, attention paid to this novel family of adsorbents is mostly because of the unique mesopore structure, which is not shared by any other families of adsorbents. On the other hand, the hydrophobic surface nature of the MMS [2] indicates that these materials are selective adsorbents for the removal of volatile organic compounds (VOCs) and other organic compounds contained in high-humidity gas streams or wastewater [119].

Therefore, MCM-41 is a possible adsorbent to substitute activated carbon for controlling VOCs [39,119,120]. But, the adsorption equilibrium of VOCs on MCM-41 frequently shows very low adsorption capacity in the low-concentration region, owing to its mesoporous structure, which considerably limits the application of MCM-41 as an adsorbent for low-concentration VOC removal [39].

The adsorption characteristics of MCM-41 for polar molecules greatly depend on the concentrations of surface silanol groups (SiOH) [121]. It has been demonstrated that different types of SiOH groups exist over MCM-41 surfaces, which can be qualitatively and quantitatively determined by a number of techniques [121].

In addition, the large pore volume, pore size flexibility, and structural variety of MCM-41 can be extensively used for the selective adsorption of a diversity of gases and liquids [39,40]. An extremely high sorption capacity for benzene has been demonstrated [40]. Widespread work has been carried out on the sorption properties of some adsorbates, such as nitrogen, argon, oxygen, water, benzene, cyclopentane, toluene, and carbon tetrachloride, as well as certain lower hydrocarbons and alcohols on MCM-41 [122].

It was also shown that the adsorptive capacity of the mesoporous materials is in excess of an order of magnitude superior than that of conventional porous adsorbent materials; thus, MCM-41 has the potential as a selective adsorbent in separation techniques, for example, high-performance liquid chromatography and supercritical fluid chromatography [122].

Additionally, the substitution of the surface hydroxyl groups in the pore wall with trimethylchlorosilane groups creates a more hydrophobic environment that substantially reduces the sorption capacity of polar molecules [123]. Siliceous MCM-41 samples were modified by silylation using trimethylchlorosilane (TMCS); the degree of silylation was found to linearly increase with increasing pre-outgassing temperature prior to silylation [123]. It was finally shown that surface modification of MCM-41 by silylation is an effective method in the development of selective adsorbents for the removal of organic compounds from streams or wastewater [123].

In addition to zeolites and MMS which are crystalline, ordered nanoporous materials and amorphous adsorbents, such as silica and active carbon, have been used in gas cleaning. The silica gel surface is generally terminated with OH groups bonded with a silicon atom, SiOH units, that is, silanols (see Figure 2.30). The concentration of OH groups at the surface is approximately $4–5 \times 10^{18}$ OH/m^2, and it is found to be almost independent of the synthesis conditions of porous silica [70]. These silanol groups are particularly reactive with H$_2$O and other polar molecules, such as SH$_2$, CO, N$_2$O, CH$_3$Cl, CH$_3$F, HCl, and NH$_3$, among others [124–126].

Dispersion and repulsion are the fundamental forces present during the adsorption of nonpolar molecules in silica; because the dipole moment of this molecule is null, the quadrupole moment is very low, and interactions with the hydroxyl groups do not exist. In the case of polar molecules, dispersion and repulsion interactions are present. But, specific interactions between the silica surface and the polar molecule, such as the dipole interaction, and, fundamentally, the interactions with the hydroxyl groups [124–126] are responsible for a more intense interaction of the silica surface with the polar molecules in comparison to nonpolar molecules [4].

For NH$_3$ adsorption, experimental evidence has shown that the main interaction mechanism is H bonding of Si-OH to the N atom in NH$_3$ [124]. Ammonia gas is a widely used chemical in industry, and it has to be removed to less than one ppm, for instance, from the gaseous effluents of ammonia fertilizer plants, urea plants, and other sources [127]. It is evident that silica is an excellent adsorbent of NH$_3$ [124–126]. Also, adsorption of ammonia on silica gel has received considerable attention recently, owing to its potential use in solar energy cooling cycles [128].

In the case of H$_2$O, CO, and N$_2$O, experimental evidence has demonstrated that the principal interaction mechanism is also H bonding of Si-OH, to the C end of CO, to the O end of N$_2$O, and to the O atom in, H$_2$O [124]. Because of these properties, silica gel could serve as an excellent adsorbent for water vapor and pollution gases. Particularly, zeolites and silica gels are the adsorbents mainly applied today as dryers [129]. An additional application of silica is hydrogen sulfide and carbon dioxide adsorptions [130,131].

In addition, activated carbons are the most extensively applied adsorbents for the elimination of contaminants from gaseous, aqueous, and nonaqueous streams [2,78]. In order to properly apply these adsorbents, like any other, equilibrium adsorption isotherms are used to calculate the optimum size of the adsorbent beds and their operating conditions [132,133].

For gas-phase applications, carbon adsorbents are regularly applied in the shape of hard granules, hard pellets, fiber, cloths, and monoliths since these prevent an extreme pressure drop [77,78].

In the majority of applications of activated carbons as adsorbents of vapors and gases, the contaminants are removed owing to the enhanced adsorption potential in the small pores of activated carbons [5,134].

VOCs are the most well-known air contaminants released by chemical, petrochemical, and other industries. Benzene, toluene, xylenes, hexane, cyclohexane, thiophene, diethylamine, acetone, and acetaldehyde are examples of VOCs [77,78]. Possibly, presently the most relevant technology for VOC control is adsorption on activated carbon [135–145]. It is a recognized technology, largely applied in industrial processes for the elimination and recovery of hydrocarbon vapors from gaseous streams [77,136]. Additionally, it offers several benefits over the others, that is, the opportunity of pure product retrieval for reuse, high removal efficiency at low inlet concentrations, and low fuel/energy costs [135].

One of the major uses of activated carbon is in the recovery of solvents from industrial process effluents. Dry cleaning, paints, adhesives, polymer manufacturing, and printing are some examples. Since, as a result of the highly volatile character of many solvents, they cannot be emitted directly into the atmosphere. Typical solvents recovered by active carbon are acetone, benzene, ethanol, ethyl ether, pentane, methylene chloride, tetrahydrofuran, toluene, xylene, chlorinated hydrocarbons, and other aromatic compounds [78]. Besides, automotive emissions make a large contribution to urban and global air pollution. Some VOCs and other air contaminants are emitted by automobiles through the exhaust system and also by the fuel system, and activated carbons are used to control these emissions [77,78].

6.12.2 HYDROGEN STORAGE

Currently, the problems related with the use of oil as an energy carrier are very clear. In contrast, hydrogen is an energy carrier with several benefits to its credit. However, the technology to begin the switch from a petroleum-based economy to a hydrogen-based economy will become accessible only in about 20 years. But, for this to take place, developing related hydrogen science and technology is needed. One of the principal uses of hydrogen as an energy carrier will be transportation. However, a concern with the application of hydrogen, on a large scale, for transportation exists, that is, it is extremely unstable. The fundamental components of the hydrogen energy economy are analogous to those in place for today's energy systems: production, purification, delivery, storage, and conversion [2,146–149]. In this regard, hydrogen storage is the confinement of hydrogen for delivery, and needs tanks for storing both gases and liquids at ambient and high pressures, and includes reversible and irreversible systems [146–148].

In the past two decades, the interest in the development of transportable reversible systems for high-capacity hydrogen storage has been growing [146,148]. New methods for storing hydrogen more efficiently could speed up the pace of the transition to the hydrogen economy radically [148]. In this sense, a very promising method of storage is by physical adsorption of molecular hydrogen on different materials, that is, carbon, silica, alumina, or zeolites [147]. But, to date the focus on hydrogen storage has mostly been limited to liquid hydrogen, and also on metal-hydride systems [146,148].

Nevertheless, higher-energy efficiency is attainable with systems in which hydrogen is concentrated by physical adsorption, above 70 K, using an appropriate adsorbent [147].

6.12.2.1 Hydrogen Storage in Zeolites

Recently, zeolites have been considered as good adsorbents for hydrogen adsorption [38,147,149–152]. In this regard, at cryogenic temperatures and at 1.5 MPa, a high gravimetric storage capacity of 2.19 wt % over Ca-exchanged X zeolites (Si/Al = 1.4) has been reported [152] (the structure of the FAU framework of zeolite X was explained in Section 2.5.1). Likewise, the total quantity of hydrogen adsorbed at 77 K and 0.1 MPa in zeolite L, ZSM-5, and ferrierite has been, in recent times, measured [147]. The structure of LTL zeolite L is composed of unidimensional 12-MR pores, parallel to the [001] crystallographic direction, of 0.71 nm diameter with lobes of 0.72 nm [83]. The ZSM-5 zeolite MFI framework comprises three-dimensional, interconnecting 10-ring pores of dimensions 0.55×0.51 nm and 0.53×0.56 nm [83]. Lastly, the organization of the FER framework of ferrierite consists of a two-dimensional pore network, containing cages with 8-MR windows of 0.35×0.48 nm and interconnecting 10-MR pores of 0.42×0.55 nm [83]. The amounts reported in literature for the hydrogen storage capacity at 77 K and 0.1 MPa in zeolite L was 0.52 wt %, for zeolite ZSM-5, 0.70 wt %, and for ferrierite, 0.57 wt % [147].

Besides, low-silica-type X zeolites (LSX), with silicon–aluminum ratio, Si/Al = 1, show the highest number of charge-compensating cations per unit cell among all materials with an FAU framework [38]. Therefore, it is an excellent adsorbent for gas separation and purification. In this sense, LSX zeolites completely exchanged with alkali-metal cations, Li^+, Na^+, and K^+, to get samples Li-LSX, Na-LSX, and K-LSX, were tested for their hydrogen storage capacities, at 77 K and near atmospheric pressures, obtaining the following amounts for the hydrogen storage capacity: 1.50 wt % for Li-LSX, 1.46 wt % for Na-LSX, and 1.33 wt % for K-LSX [38].

Hydrogen adsorption studies performed on H-SSZ-13 zeolite, at 77 K and 0.092 MPa, yielded a hydrogen capacity of 1.28 wt % [150]. This number is one of the highest hydrogen adsorption at 77 K and near atmospheric pressure for all zeolites that have been studied [38]. Zeolite H-SSZ-13 possesses a CHA-type framework [83], which is characterized by layers of double six-rings that are interrelated by four-rings, where the double six-ring layers stack in an ABC succession, conducting to a structure with an ordered array of barrel-shaped cages interconnected by eight-ring windows [150].

Hydrogen uptake at 77 K and 0.1 MPa have been studied in some aluminophosphate (AlPOs) molecular sieves [149]. These materials are produced when the Si, located in tetrahedral sites, is substituted by P [153]; in this case, a precise alternation of (AlO$_2$) and (PO$_2$) tetrahedral structures occurs. Consequently, the framework is electrically neutral, then, no extra-framework cations are needed for charge balance and the material has no acidity.

To conclude this section, it is necessary to state that experimental hydrogen adsorption studies on zeolites at room temperature are scarce. It has been reported that zeolites can store only very small amounts of hydrogen (<0.5 wt % at 60 bar) at room temperature [151]. In this regard, LSX zeolites, as well, fully exchanged with alkali-metal cations, that is, Li$^+$, Na$^+$, and K$^+$, to get samples Li-LSX, Na-LSX, and K-LSX, were tested for their hydrogen storage capacities at 298 K and 10 MPa, and the measured H$_2$ capacity was 0.6 wt % [38].

6.12.2.2 Hydrogen Storage in Mesoporous Molecular Sieves and Pillared Clays

The consideration given to the family of MMS adsorbents is mostly as a consequence of the unique ordered mesopore structure, which is not shared by other families of adsorbents [2]. Additionally, with regard to the pore structure, one of the most exceptional characteristics of these materials is the large BET surface area, usually in the range between 600 and 1300 m^2/g and pore volume generally greater than 0.6 cm^3/g.

In the case of pillared clays (PILCs), notwithstanding the fact that these materials were initially developed as catalysts, there have also been other investigations, where PILCs have been studied as potential sorbents, especially for gas separation applications. These materials have a developed micropore structure with a relatively large BET surface area, normally between 300 and 400 m^2/g and a micropore volume around 0.15 cm^3/g [154,155].

In this regard, as a result of their adsorption properties, MMS and PILCs could be convenient materials for hydrogen storage. But, there are very few reports on hydrogen adsorption in MMS [156] and pillared clays [9,157].

The obtained experimental results indicate that the hydrogen uptake at 77 K and 0.1 MPa in the silica mesoporous molecular sieve, MCM-41, is very low [147]. Nevertheless, MMS have been used as templates for the creation of ordered porous carbon with tailored pore sizes, which can be applied in hydrogen storage [158].

With respect to PILCs, as was previously stated, these materials are potentially good adsorbents, but these promises have not been, in general, accomplished [9,157], particularly in the case of hydrogen storage.

6.12.2.3 Hydrogen Storage in Silica

Reports in the literature regarding hydrogen storage in amorphous silica are scarce [159]. In addition, the hydrogen uptake at 77 K and 0.1 MPa in some silica samples show that the hydrogen adsorption capacity in amorphous silica is very low [147]. These negative results are related to the fact that dispersion and repulsion are the unique forces present during the adsorption of H$_2$ in silica [2]; Nevertheless, if the ammonia molecule, that is, NH$_3$, is applied in place of H$_2$ as the hydrogen carrier, in this situation, dispersion and repulsion interactions are present, but the dipole interaction and the specific interaction between the silica surface and the NH$_3$ molecule with the hydroxyl groups are present as well [124]. These interactions are the reason for a more intense interaction of the silica surface with the NH$_3$ molecule in contrast to the H$_2$ molecule [124,125].

Another advantage of ammonia as a hydrogen carrier is the fact that currently the fundamental hydrogen production processes, such as steam reforming, partial oxidation, and autothermal reforming of hydrocarbons, generate great amounts of CO$_x$ as byproducts. Consequently, one obvious alternative to CO$_x$-free hydrogen production is the catalytic decomposition of NH$_3$ [160],

$$NH_3 \leftrightarrow \tfrac{1}{2}N_2 + \tfrac{3}{2}H_2$$

an endothermic process where $\Delta H = 46$ kJ/mol, which is normally catalyzed by metals such as Al, Fe, Re, Rh, Ni, Pt, Ru, W, and Ir. This process, in addition to the previously mentioned advantage, has other benefits [161], since NH_3 is the second largest product produced by the chemical industry, has 30% additional energy per unit volume than liquid hydrogen, has the infrastructure for its transportation, distribution, storage, has wide applications, and has the flammability range for ammonia–air at STP narrower than that for hydrogen–air mixtures.

As a result, considering the benefits of NH_3, the author and collaborators [162] propose a more stable form of hydrogen supply, applying NH_3 as the hydrogen carrier, adsorbed on silica. However, various critical opinions at present exist, which state that the use of NH_3 for energy storage is not a good choice, given that it is harmful, and because of toxicity concerns. Moreover, it is considered that NH_3 storage, as a hydrogen carrier, is trivial, given that it is usually transported safely as a liquid, having both higher gravimetric and volumetric hydrogen densities over that of high surface area adsorbents.

In the author's view, if we use liquid NH_3 in vehicular and small-scale fuel cell use, we should be worried about its toxicity. However, if NH_3 is stabilized in a solid matrix like silica, the toxicity problems can be diminished. Because NH_3 possesses a characteristic strong odor, any leakage can be detected easily [162]. Besides, the adsorbent will steadily release it in the event of an escape or leakage. Another drawback is associated with the effective dissociation of NH_3 into hydrogen and nitrogen, given that standard NH_3 cracking reactors expend part of the produced hydrogen in the reactor operation. However, researchers at the Massachusetts Institute of Technology have created an efficient suspended-tube microreactor to generate hydrogen by thermally cracking ammonia [163,164].

In Figures 6.27 and 6.28, the semi-logarithmic plots of the amount adsorbed, n_a [cc-STP/g], versus $-\log [P/P_0]$ for the adsorption isotherms of N_2 at 77 K in the as-synthesized silica labeled 70bs2-a-s(25) and 70bs2-a-s(50), synthesized at 298 and 323 K respectively [162], and NH_3 adsorption isotherms at 300 K in the same silica material after 3 years of synthesis, that is, stable silica labeled 70bs2-s are shown [162]. The ammonia adsorption results indicates that at 300 K and 0.1 MPa the silica stored 2.0 wt % of hydrogen in the form of NH_3 (see Figure 6.28); but, at this point, the mesopore volume of the silica is still empty.

Subsequently, using the Gurvich rule [2], it is possible to obtain the relation $N_m = W/\overline{V}_{NH_3}$, which allows the calculation of the maximum amount of NH_3 that could be adsorbed in the stabilized silica samples. Then, the maximum amount of NH_3 that could be adsorbed in both samples is approximately $N_m = 40.1$ mmol/g. Since each molecule of ammonia contains three hydrogen atoms, 0.120 grams of hydrogen is stored in the form of NH_3 per gram of the silica adsorbent, that is, 0.120 g/g.

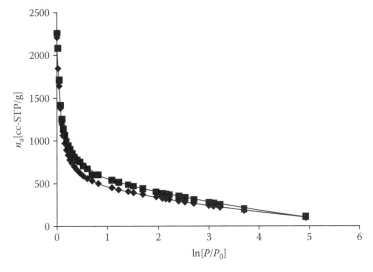

FIGURE 6.27   Semilogarithmic plot of the N_2 adsorption isotherm at 77 K on both samples 70bs2-a-s(25) and 70bs2-a-s(50).

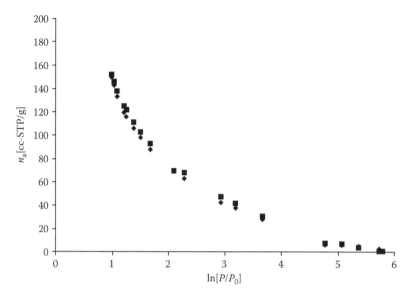

FIGURE 6.28   Semi-logarithmic plot of the NH_3 adsorption at 300 K on both samples 70bs2-s(25) and 70bs2-s(50).

Hence, the maximum amount of hydrogen stored in the adsorbent in the form of NH_3 will be about 11 wt %, a quantity higher than the established minimum figure of 6.5 wt %.

6.12.2.4   Hydrogen Storage in Carbon-Based Adsorbents

Active carbon (AC), carbon nanotubes, carbon fibers, and other carbonaceous adsorbents, owing to their low atomic weight, high surface area, pore volume, and suitable mechanical properties, are excellent options as adsorbents for hydrogen storage [81, 147,165–184].

Carbon adsorbs the undissociated hydrogen molecules by repulsion and dispersion forces at their pore network and surface [2]. However, since these binding forces are weak, the physical adsorption process at room temperature is virtually restrained by the thermal motion; thus, to store large quantities of hydrogen, the carbon samples have to be cooled [166]. However, a cryogenic hydrogen storage technique is, in the majority of cases, economically ineffective [173].

Hydrogen storage in carbon has been considered during the last few years on account of the existence of new carbon nanomaterials, such as fullerenes, superactivated carbons, carbon monoliths, carbon nanotubes, and carbon nanohorns [147,166,176–179], distinguished by their high adsorption capacities, hydrophobic nature, and high adsorption/desorption rates [170].

The adsorption of hydrogen on molecularly engineered carbon at 123 K has been studied. It has been reported that this material exhibited a capacity of 0.5 g of H_2/kg of carbon at 2 MPa pressure [174,175].

Metal-intercalated graphite has been studied as well for hydrogen storage [180]. It has been reported that up to 0.137 L (STP) of hydrogen per gram of carbon can be adsorbed between the layers of alkali-intercalated graphite [181,182].

A new type of carbonaceous porous material called mesocarbon microbeads (MCMBs) are microcarbon spheres produced by mesophase pitches. These materials have been principally used as fillers in paints, elastomers, and plastics to change the mechanical and electrical properties of materials [79]. Activated MCMBs (a-MCMBs) with high specific surface area greater than 3000 m²/g can be prepared. Many investigations demonstrate that a-MCMBs are expected to be more ordered in structure than the activated carbon fibers (ACF). ACFs are carbonaceous materials composed of piled, nanosized graphene layers, which shape an ordered structure of slit-like pores with many open edges. These materials are also capable of gas storage [166]. The calculations show that the adsorption amount of hydrogen in a-MCMB at 10 MPa can reach 3.2 wt % and

15 wt % at 298 K and 77 K, respectively, which are also higher values compared with other carbon materials [79].

Carbon nanotubes (CNTs), having nanosized tubular microstructures, are being studied extensively because of their potential ability to adsorb hydrogen [166,172,183,184]. In 1997, data on hydrogen adsorption on single-walled carbon nanotubes (SWCNTs) was first reported. Hydrogen storage was studied, applying a material containing 0.1–0.2 wt % of single-walled carbon nanotubes in addition to other materials. It has been reported that this sample adsorbed 5% of hydrogen at 273 K [172]. Multiwalled carbon nanotubes (MWNTs), with an average outer diameter of 5.1 nm, have been also tested for hydrogen storage measurements [168]. A scrutiny of the literature associated with adsorption studies on carbon nanotubes reveals that the reported hydrogen storage capacities of such materials differ over a wide range, that is, from 0.1 to 10 wt % [79,81]. This discrepancy in the experimental data is probably a result of the lack of reliable techniques for measurements of hydrogen uptake at high gas pressures [81]. Subsequently, despite the optimism raised from reports on the application of carbon nanotubes [172] and carbon nanofibers [165], as a result of additional research, the obtained results have become dubious [147]. Specifically, in a study with different carbonaceous materials [167], exceptionally huge amounts of hydrogen adsorption, such as those reported for carbon nanotubes [185] and carbon nanofibers [165,166], have not been measured in any of the tested samples. In fact, the highest amounts of hydrogen adsorption for a carbonaceous material was measured in anthracite (AC-KUA1) [167].

6.12.3 METHANE STORAGE IN ADSORBENTS

6.12.3.1 Introduction

Because of the problems of petroleum as an energy carrier, current universal consensus is that future alternative fuels must be nonpetroleum, due to energy security, environmental benefits, political stability, and other issues [186]. Air quality is a major public health problem, particularly in metropolitan areas; consequently, the use of nonpolluting fuels for transportation is strongly encouraged [187]. Natural gas, which primarily consists of methane, is a significant choice as a fuel having two key benefits in comparison with gasoline: low cost and clean burning characteristics. Subsequently, the interest in natural gas as a vehicular fuel has increased considerably, given that natural gas has significant benefits over standard fuels, from an environmental point of view and due to its natural abundance [77].

Undoubtedly, methane is cheap, and its abundant distribution worldwide is a highly valuable feature [186]. In addition, during methane combustion, low emission levels of ozone, unburned hydrocarbons, and SO_x and NO_x are observed [188]. Furthermore, the greenhouse effect is much less than that of classical liquid hydrocarbon fuels [187]. Finally, methane has the maximum possible hydrogen-to-carbon ratio and, thus, the highest energy per unit mass of all the other hydrocarbons [188].

Unfortunately, methane is supercritical at room temperature; consequently, it cannot be liquefied by compression above T_c, that is, cannot be stored at a density as high as other fuels. Subsequently, it has a lower heat of combustion per unit volume when compared with usual fuels [189]. Specifically, methane has an energy density about one-third that of gasoline for compressed natural gas (CNG) at 24.8 MPa [78]. In addition, the application of CNG has some disadvantages, for example, the high cost of the used cylinders [190]. Consequently, the storage of great amounts of methane in a given limited volume is a real challenge for its transportation and its use in gas-powered vehicles [187]. But, natural gas is principally stored as CNG at 20.7 MPa in pressure vessels requiring an expensive multistage compression [186]. Nevertheless, an attractive and alternative option to CNG is adsorbed natural gas (ANG), where the gas is stored as an adsorbed phase in a porous solid at a lower pressure. The U.S. Department of Energy (DOE) has defined, in 1993, the storage goal at 3.5 MPa as 150 v/v, that is, 150 STP (0.101 MPa, 298 K) liters of gas stored per liter vessel volume [191,192]. But, in recent times, this storage goal was modified to 180 v/v, so that the energy density of ANG becomes comparable to that of CNG [186].

Among the accessible adsorbents, activated carbons exhibit the largest adsorptive capacity for methane storage. High-capacity carbonaceous adsorbents has generated a great quantity of literature, an indication that activated carbons are very good adsorbents presenting the highest ANG energy densities [2,77–79,147,187,191,193,194]. In addition to carbon, IRMOF-6, which is a member of the family of metal-oxide clusters bridged by functionalized organic links and known as isoreticular metal-organic frameworks (IRMOFs), exhibits an extremely high capacity for methane storage [186]. An additional form of methane storage is by using a wet SBA-15 mesoporous molecular sieve [195]. As was previously stated, methane is more abundant in the form of clathrate hydrates than as natural gas [186]. Subsequently, the storage of natural gas in the form of a hydrate in a mesoporous material, such as the SBA-15 MMS, could be an excellent option to store natural gas in the presence of water [195].

6.12.3.2 Methane Storage in Carbonaceous Adsorbents

Nanoporous carbonaceous materials are appropriate containers for storing molecules under strong confinement [196–207]. In the case of the physical adsorption of methane with carbonaceous nanoporous materials, the attractive adsorption forces are only dispersive, therefore very weak; hence, efficient carbonaceous materials for methane storage should have the highest possible specific surface areas and pore volume. However, the factor area is just not enough to get an efficient material, that is, the sizes of the pores are significant, especially when adsorption of a supercritical gas is concerned [187]. In this sense, as previously discussed in the case of hydrogen storage, new carbon-based nanoporous materials, such as single-walled carbon nanotubes (SWNTs), wormlike single-walled carbon nanotubes (WLSWNTs), double-walled carbon nanotubes (DWNTs), single-walled carbon nanohorns (SWNHs), stacked-cup carbon nanofibers (SCCNT), doped fullerenes, wormlike graphitic carbon nanofibers, bamboo-type multiwalled carbon nanotubes, and ordered porous carbons, have been regarded as promising media for the efficient reversible storage of chemical species [199].

In relation to methane adsorption in active carbon, which as previously described is formed by slit pores (see Figure 2.21), several numerical simulations have revealed that the highest density of the adsorbed phase is achieved within slit pores of 1.12–1.14 nm of diameter [187,200]. For slit pores of a width, $L = 1.13$ nm (see Figure 2.21), two facing methane molecule monolayers may be inserted between pore walls [203–205].

Indeed, if the methane molecule is supposed to act similar to a spherical molecule, its diameter is 0.381 nm; hence, pores that have widths, L, of less than approximately 1 nm can house only one monolayer, while those narrower than around 0.4 nm cannot store methane at all [186]. If wider pores are considered, the attractive potential produced by the facing pore walls diminishes very fast with the value of the pore width, L, so that, if $L > 3$–4 nm, methane is weakly adsorbed, and its density is comparable to that of the gas phase in equilibrium with it. Subsequently, it is obvious that the maximum adsorption capacities are achieved with materials for which the volume of pores that have the pertinent width is the maximum [200,202]. As a result, active carbons having slit-shaped pores may be the best material for methane adsorption [187].

In active carbons, micropores are slit-shaped, as represented in Figure 2.21 [201]. In the image (Figure 2.21) of the slit pore are also described the meaning of the physical pore width, L and the accessible inner space, l [35,201]. The parameter L is in general measured with the help of the adsorption of N_2 at 77 K and Ar at 87 K [2] and the inner pore-wall spacing, l, is given by the relation [187,202]

$$l = L - 2r_C$$

in which r_C, the radius of a carbon atom, is assumed to be half of the normal separation of planes of graphite, that is, 0.17 nm. Accordingly, the most favorable pore width that has been discussed previously, $L \approx 1.13$ nm, is compatible with an ideal inner pore size of $l \approx 0.8$ nm, that is, just about the thickness of two methane molecules [203–205]. With this ideal value of l, the adsorbed phase has the maximum density at 3.5 MPa [205]. Activated carbons and activated carbon fibers synthesized from different raw materials and diverse activating agents covering a wide range of surface area and MPSDs have been studied [191]. In these studies, it has been established that the methane adsorption capacity of an adsorbent is interrelated with its micropore volume. Nonetheless, it has been demonstrated that the methane adsorption capacity is also related with its MPSD; as a result, activated carbons show larger methane adsorption capacities than activated carbon fibers with similar micropore volumes [191].

Methane adsorption in carbon nanotubes has also been investigated [208–210]. The use of carbon nanotubes in adsorption can be founded on the high intrinsic porosity of the lightweight carbon network, which is comprised of single-wall graphitic channels [211,212]. Methane storage has been investigated on isolated SWNTs by applying the density functional theory (DFT), and it was established that the excess gravimetric capacity reached 0.198 g CH_4/g C at room temperature and 4.0 MPa, when the reduced pore size of SWNT is 0.42 nm [210].

In addition, the adsorption of methane in single-walled nanohorns has been experimentally investigated. In this regard, the measured volumetric capacity reached 160 v/v [211]. Nanohorns, materials which are generated in a synthetic process without a metal catalyst, show a pure graphitic structure. These materials, contrary to carbon nanotubes which grow in a triangular lattice with thin intertubular gaps, self-assemble in spherical aggregates, with spaces between the neighboring nanohorns [211].

The study of methane adsorption on activated carbon fibers has demonstrated, as was previously explained, that these carbonaceous materials, because of their cylindrical morphology and smaller diameter, have higher packing density than activated carbons with similar micropore volumes [191]. Subsequently, the higher adsorption capacity for the powdered activated carbons against the higher packing density for the fibers helps both kinds of materials reach similar, maximum adsorption values [191].

To conclude this section, widespread consensus exist that the adsorbents with the highest methane capacity are the carbonaceous materials, especially activated carbons [192], where storage values up to 200 v/v have been reported [213].

6.12.4 WATER CLEANING

Because of the risk of environmental catastrophe due to the spilling of organic pollutants in water streams and increasing public concern, the establishment of limits on the suitable levels of particular contaminants in the environment exist. In this regard, adsorption methods are widely accepted as one of the successful processes to establish these limits.

Hydrophobic zeolites, as well as the all-silica zeolites or zeolites with a very small aluminum content, possess high capacity for adsorbing organic compounds dissolved in water. Some recent studies demonstrated that hydrophobic, dealuminated zeolites adsorbed organic compounds from water as effectively as activated carbon [2,37,88,89,214]. The hydrophobicity of zeolites is controlled basically by changing the Si/Al ratio in the framework by synthesis conditions and postsynthesis modification treatments [215].

The most-tested hydrophobic zeolite in adsorption of organic compounds from water solutions is silicalite-1 [89]. This material is a molecular sieve with an MFI structure composed of pure silica. The MFI framework has a 10-MR channel system with elliptical pores having diameters of 5.2 × 5.7 Å [83]. Additionally, other zeolites, as the all-silica β-zeolite [216] which possesses a three-dimensional, 12-numbered ring, interconnected channel system with pore diameters of 7.1 × 7.3 Å [83] have been used in the elimination of methyl tert-butyl ether (MTBE) from water solutions [88].

A recent study showed that hydrophobic, dealuminated mordenite adsorbed MTBE from water better than activated carbon. In this study, 5 mg of zeolite powders was equilibrated with 25 mL of aqueous solution containing 100 μg/L of MTBE for 15 min, and the dealuminated mordenite removed 96% of the MTBE [214]. Mordenite has a 12-membered ring zeolite with 6.5 × 7.0 Å pores [83].

However, activated carbons are the most extensively applied industrial adsorbents for the removal of pollutants from gaseous and aqueous and nonaqueous streams, because of their exceptionally powerful adsorption properties and their readily modifiable surface chemistry [217,218]. Carbon is the primarily applied adsorbent in the case of liquid–solid adsorption systems.

Phenol is one of the most significant compounds adsorbed by carbon from the liquid phase. Phenol is a fundamental structural part for a variety of synthetic organic compounds; subsequently, wastewater originating from many chemical plants and pesticide and dye manufacturing industries contain phenols [216]. Additionally, wastewater originating from industries like paper and pulp, resin manufacturing, gas and coke manufacturing, tanning, textile, plastic, rubber, pharmaceutical, and petroleum contains diverse types of phenols [219]. In addition to the phenols produced as a consequence of industrial activity, wastewaters contain phenols generated as a product of vegetation decay.

In view of the wide prevalence of phenols in different wastewaters and their toxicity to humans and animal life even at low concentrations, it is necessary to eliminate them before discharging wastewater into water bodies [85]. In this regard, numerous methods, such as oxidation with ozone/hydrogen peroxide, biological methods, membrane filtration, ion exchange, electrochemical oxidation, reverse osmosis, photocatalytic degradation, and adsorption, have been used for the removal of phenols but the adsorption process even now remains the best [219]. The adsorption of phenol and substituted phenols from aqueous solutions on activated carbons is one of the most studied of all liquid-phase applications of carbon adsorbents [2,85]. Besides, active carbons are also applied in the elimination of organic compounds from effluent water in petroleum refining, petrochemicals, metal extraction, detergent, margarine and soft fat manufacture, mineral extraction, food and beverage industries, pharmaceutical industries, and other industries [77,78,84]. However, carbon is expensive, and the regeneration of spent carbon is not easy because of considerable irreversible adsorption [220]; in addition, activated carbon is a flammable and hygroscopic material [221]. In this regard, the author and collaborators have studied the liquid-phase paranitrophenol (PNP) dynamic adsorption in a packed-bed adsorption reactor (PBAR), filled with well-characterized granulated active carbon (GAC, provided by Fisher Scientific, Pittsburgh, PA, USA, lot number 974914), and dealuminated Y zeolite (DAY-20F, Si/Al = 20, provided by Degussa AG, Dusseldorf, Germany), to compare the effectiveness of the dynamic adsorption process in the GAC and the DAY zeolites (see Figure 6.29) [222].

The PBAR was a laboratory made setup [222] comprising a paranitrophenol (PNP) container, in order to supply the PNP aqueous solution fluid to a high-performance liquid chromatography (HPLC) double piston KNAUER-K-501 pump, which feeds the PBAR. This setup was used for the measurement of the breakthrough curves, resulting from the dynamic adsorption of PNP in DAY and GAC adsorbent beds, using a Shimadzu UV-2401 PC, UV/VIS recording spectrophotometer to measure the input and output concentrations of the PNP aqueous solution fluid. The PBAR was built with a Phenomenex cylindrical stainless steel column with an internal diameter, $d = 4.6 \times 10^{-3}$ m, and a total length, $L = 0.15$ m. The reactor has a cross-sectional area, $S = 1.662 \times 10^{-5}$ m^2, bed length, $D = 4.6 \times 10^{-2}$ m, and adsorbent mass in the bed, $M = 0.5$ g [222].

In Figure 6.29, the breakthrough curves corresponding to the dynamic adsorption of PNP in the GAC (Figure 6.29a) and the DAY zeolites (Figure 6.29b), respectively are shown [222].

With the help of the breakthrough curves, and applying Equations 6.36, 6.38, and 6.39a to the breakthrough data, the operational parameters characterizing the PBAR were calculated (Table 6.3). Specifically, the interstitial fluid velocity, u, the breakthrough time, t_0, the breakthrough curve width expressed in time, Δt, the column breakthrough capacity, B_C, the length of the MTZ, D_0, for the different volumetric flows, F, were imposed with the help of the HPLC pump.

With the help of these results, it is possible to conclude that the reactor filled with the DAY zeolite operates more efficiently than those filled with the GAC zeolite [222]. In this case, the packed-bed adsorption reactor fulfill the conditions of a plug-flow adsorption reactior [222].

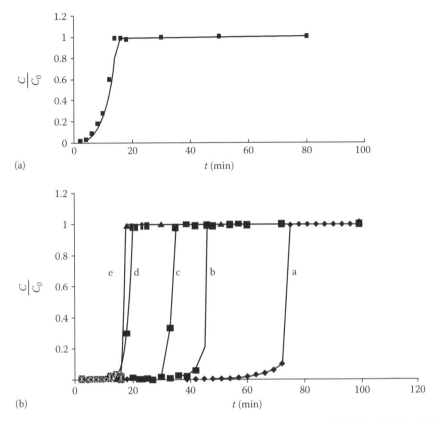

FIGURE 6.29 Breakthrough curves corresponding to the dynamic adsorption of PNP in the Fisher GAC (a) and the DAY (b: a, $F = 1$; b, $F = 1.5$; c, $F = 2$; d, $F = 3$; e, $F = 4\,mL/min$).

TABLE 6.3
Operational Parameters for the PBAR for the DAY Zeolite and the Fisher Granular Active Carbon

Adsorbent	F [mL/min]	u [mm/s]	t_0 [s]	Δt [s]	B_C [g/g]	D_0 [cm]
DAY	1	1.003	4200	170	0.19	0.18
DAY	1.5	1.505	2650	185	0.18	0.31
DAY	2	2.006	1800	190	0.17	0.43
DAY	3	3.009	980	180	0.18	0.55
DAY	4	4.012	970	220	0.18	0.85
GAC	4	4.012	351	400	0.07	2.4

6.13 POROUS POLYMERS AS ADSORBENTS

6.13.1 Porous and Coordination Polymers

We are interested in highly cross-linked, permanently porous polymers. These materials have a permanent porous structure, produced during their synthesis and preserved in the dry state, and are employed in a wide variety of applications as adsorbents [90,223–230].

In 1944, D'Aleleio obtained the well-known Amberlite, that is, polystyrene resins cross-linked with divinylbenzene (see Figure 2.45). These materials are widely applied as adsorbents because of their developed pore network. Their pore sizes are closely related to the percentage of divinylbenzene; that is, a 10% cross-linked polymer has smaller pores than a 2% one, because the supplementary amount of divinylbenzene forms further connections, creating a network where the pore sizes are typically less than 3 nm in diameter [223]. Consequently, cross-linked polymers are microporous and these small pores are produced among the spherical polymer particles during the cross-linking process.

In order to get mesopores and macropores in polymers, a new methodology was develop in the late 1950s by researchers at Rohm and Haas, Philadelphia, Pennsylvania and The Dow Chemical Company, Midland, Michigan. To get pores in the mesoporous region, the presence of a porogen agent is needed. The porogen is an inert organic compound that is a satisfactory solvent for the monomer but an inadequate solvent for the polymer in the copolymerization process [223]. Consequently, the solvent stays within the liquid globules of monomer mixture during suspension polymerization; subsequently, after polymerization, the solvent is removed, leaving behind a resin with permanent pores [224]. For example, porous poly(styrene–divinylbenzene) resins have been prepared by suspension polymerization of styrene–divinylbenzene (DVB) mixtures with DVB contents of 1–12 mol%, using 2-ethyl-hexan-1-ol as a porogen [225].

Other common examples of monomers used for the production of these resins are methacrylate and styrene where the cross-linkers are ethylene dimethacrylate (EDMA) and DVB, respectively [228]. Other high surface area nanoporous polymeric materials are Hayesep N, that is, poly(divinylbenzene-ethylenedimethacrylate); Hayesep S, that is, poly(divinylbenzene-4-vinylpyridine); Hypersol-Macronet MN200, which is a hypercrosslinked polystyrene; Hypersol-Macronet MN100, composed of an amine-functionalized hyper-crosslinked polystyrene; and Hypersol-Macronet MN500, a sulfonated hypercrosslinked polystyrene [229]. Another porous polymer is poly(glycidyl methacrylate–ethyleneglycol dimethacrylate) (see Figure 2.46), a material with an interconnected pore network, which permeates the extensively cross-linked polymcr matrix [230].

In addition, materials obtained by the addition of a known porosity inorganic matrix, for instance, silica gel or aluminum oxide, to the reacting mixture have a complex pore system, that is, the micropore network created during the cross-linking process and the mesopore network developed by the inorganic porosity agent [231]. Consequently, these porous polymers have a highly developed surface area [232,233]. A particular example of the previously described methodology is the cross-linked porous melamine–formaldehyde resin–silica system [232–234]. Here the polymer is produced by the condensation of formaldehyde with melamine and the fumed silica is applied as a template, forming the pore structure of the cross-linked polymer [232–234]. The produced polymer is modified by applying 2-methyl-1-butanol in order to differentiate its surface properties [233].

Ordered mesoporous polymers have been produced by filling with divinylbenzene, ethyleneglycol dimethacrylate, or a mixture of the two; the pores of a colloidal crystal are formed by silica spheres of 35 nm of diameter. Thereafter, the polymerization and subsequent dissolution of the silica template leave a polycrystalline network of interconnected pores [235].

In addition, the polymers of intrinsic microporosity (PIMs), such as phthalocyanine networks and the Co phthalocyanine network-PIM (CoPc20), display high specific surface area, as confirmed by the N_2 adsorption isotherm at 77 K, and by the adsorption of small organic probe molecules from aqueous solutions at 298 K [236]. This material is basically microporous with an increased concentration of effective nanopores.

Recently, the use of coordination polymers in the creation of novel porous materials to be applied in gas adsorption are being considered [237–247] (see Section 2.9.4). As was previously described, these materials, in comparison with standard crystalline microporous materials such as zeolites, are considered by some authors to be more promising, owing to the fact that their frameworks can be previously conceived and are more plastic on account of the diversity of coordination geometries of metal centers and the multifunctionality of the bridging organic parts [239]. That is, in addition to zeolites, this novel family of coordination polymers or metal-organic frameworks (MOFs) form zeolitic-like frameworks. The design and synthesis of MOFs have produced a number of structures that have been shown to possess effective gas and liquid adsorption properties.

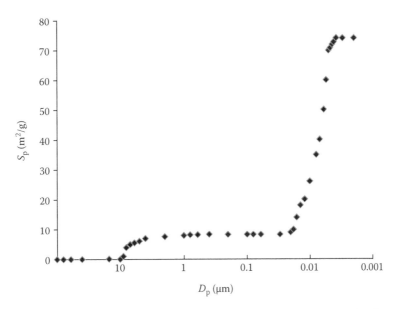

FIGURE 6.30 Cumulative pore area (S_p [m²/g]) vs. log D_p [μm] graphic plot for a sample of a poly-(HIPE).

Additionally, highly porous, open-cell polymers [poly-HIPE] have been prepared by polymerizing the monomers in the continuous phase of high internal phase emulsions (HIPE), using styrene as the monomer, 2-ethylhexyl acrylate as the cross-linking agent, divinylbenzene and ethylstyrene as comonomers, and toluene as the porogen added to the styrene [248,249]. The emulsifier for the cross-linked polystyrene poly(HIPE) was sorbitan monooleate and the emulsifier for the cross-linked poly(2-ethylhexyl acrylate poly(HIPE) was sorbitan monolaurate [248]. These materials are good adsorbents; this fact is evidenced in Figure 6.30 (in Figure 4.66 the differential intrusion of mercury (dV_p/dD_p) versus the log D_p profile for the same material is shown) where the cumulative pore area, S(m²/g), versus the log D_p(μm) profile of the highly porous open-cell polymer [poly-(HIPE)], prepared by polymerizing styrene (90%) and divinyl benzene (10%) (the poly-[HIPE] sample was provided by Dr. Michael Silverstein, Israel Institute of Technology, kindly) is shown.

The mercury porosimetry test (see Section 4.10) shown in Figure 6.30 was performed in a Micromeritics AutoPore IV-9500 automatic mercury porosimeter.* This study was carried out as follows: the porosimeter sample holder chamber was evacuated up to 5×10^{-5} Torr, the equilibration time was 10 s and the mercury intrusion pressure range was from 0.0037 to 414 MPa, that is, the pores size range was from 335.7 to 0.003 μm.

6.13.2 APPLICATIONS OF POROUS POLYMERS AND COORDINATION POLYMERS IN ADSORPTION PROCESSES

Polystyrenes cross-linked with divinylbenzene commercial resins, such as Lewatit VP-OC 1163 (Bayer AG, Leverkusen, Germany), Amberlite XAD-4 (Rohm & Haas, Frankfurt, Germany), and Serdolite PAD I, PAD II, and PAD III (Serva AG, Heidelberg, Germany), have been applied in the adsorption of phenol, chlorophenols, and dihydroxybenzenes from water solutions [90]. Besides, commercial polymethacrylate/divinylbenzene resins, for example, Supelcogel TPR-100

* The author thanks Roberto Cordero (TTC Analytical Service Corp., Caguas, PR), Michael L. Strickland (Micromeritics Instrument Corporation), and Micromeritics Analytical Services for the mercury porosimetry intrusion study of the poly-[HIPE] sample.

(Supelco, Deisenhofen, Germany) have been investigated in the adsorption of phenol, chlorophenols, and dihydroxybenzenes from water solutions [90].

CoPc20, previously mentioned, was compared with an activated carbon of similar specific surface area (Darco 20–40 mesh); it was shown that CoPc20 is more selective in its adsorption performance [236].

As was mentioned in the previous section, the coordination of metal ions to organic linkers, in order to construct open frameworks, has received a great deal of attention in the past decade, since it permits a more adaptable and logical design of such materials [238–247,250–254]. In this regard, the pioneering work of Yaghi and coworkers led to the successful synthesis of metal-organic frameworks with permanent porosities [242–244,250]. However, few metal-organic frameworks have been reported to possess permanent porosity in terms of reversible gas sorption behavior; however, the significant capacity of some of these materials in hydrogen adsorption allows some authors to predict their applications in hydrogen storage in the future [253]. As of now, they are very far from this goal and possibly the status quo will remain. In this regard, recently, metallo-organic framework materials with moderately high hydrogen sorption capacities at 78 K and ambient temperature under safe pressure, that is, up to 2 MPa [244] has been reported. MOF-5 showed 4.5 wt % hydrogen storage capacity at 78 K and moderate pressures [244,255].

In other similar structures, such as IRMOF-6 and IRMOF-8, the specific hydrogen uptake is approximately doubled and quadrupled, respectively, compared to MOF-5 at room temperature and 2.0 MPa pressure [244]. The hydrogen absorption capacity of these structures at room temperature is comparable to that of carbon nanotubes at cryogenic temperatures and can be fine-tuned by modifying the porosity of the structure with suitable linkers [244].

Other MOFs, which could be of interest in hydrogen storage, are the zeolitic imidazolate frameworks (ZIF), specifically, ZIF8, that show immense potential in hydrogen storage applications [241]. ZIF8 is a typical ZIF compound, with the following composition: $Zn(MeIM)_2$, where MeIM is 2-methylimidazolate [241]. This molecular sieve shows an SOD framework, that is, a sodalite zeolite-type structure, displaying a nanopore topology formed by 4-ring and 6-ring ZnN_4 clusters [241]. At high hydrogen loading, the ZIF8 structure is capable of holding up to $28 H_2$ molecules, which is equivalent to 4.2 wt % of hydrogen storage capacity in the form of highly symmetric novel three-dimensional interlinked H_2 nanoclusters [241].

Additionally, Prussian blue analogues, for example, $Co_3[Co(CN)_6]_2$, $Ni_3[Co(CN)_6]_2$, $Cu_3[Co(CN)_6]_2$ $Mn_3[Co(CN)_6]_2$, and $Zn_3[Co(CN)_6]_2$, show hydrogen storage capacities at 77 K and 890 Torr, ranging from 1.4 wt % in $Zn_3[Co(CN)_6]_2$ up to 1.8 wt % in $Cu_3[Co(CN)_6]_2$ [254].

Poly(HIPE) has been applied to remove trihalomethane: triboromomethane from aqueous solutions [249]. The well-known chlorination process practically eliminates bacteria, viruses, and parasites from drinking water supply, and, consequently, the illness produced by these microbes. But, chlorination produces byproducts named disinfection byproducts (DBP), which include a group of chemicals known as trihalomethanes, which includes four chemicals: chloroform, bromodichloromethane, dribromochloromethane, and tribromomethane [100]. These compounds can be carcinogenic [256]; consequently, the elimination of these organic compounds from water is normally performed by adsorption using granular activated carbon, air stripping, ozone oxidation, reverse osmosis, and pervaporation, among other methods [100,257].

Other type of porous polymers with adsorbent properties can be obtained from furfural $(C_5H_4O_2)$ (see Figure 2.47), an aldehyde that is obtained from agricultural products (see Section 2.9.3). Furfural reacts by means of the carbonyl (=CO) group and the furanic ring, when it is heated in the presence of acids to form thermosetting polymers [258–264]. In this regard, polymerizing furfural and furfural acetone solutions in the presence of a Brönsted acid catalyst leads to the formation of porous polymers, which are good adsorbents to remove ammonia from gaseous streams [262].

REFERENCES

1. K.S.W. Sing, D.H. Everett, R.A.W. Haul, L. Moscou, R.A. Pirotti, J. Rouquerol, and T. Siemieniewska, *Pure Appl. Chem.*, 57, 603 (1985).
2. R. Roque-Malherbe, *Adsorption and Diffusion in Nanoporous Materials*, CRC Press, Boca Raton, FL, 2007.
3. R. Roque-Malherbe, *Mic. Mes. Mater.*, 41, 227 (2000).
4. F. Rouquerol, J. Rouquerol, and K. Sing, *Adsorption by Powder Porous Solids*, Academic Press, New York, 1999.
5. S.J. Gregg and K.S.W. Sing, *Adsorption Surface Area and Porosity*, Academic Press, London, 1982.
6. S. Ross and J.P. Olivier, *On Physical Adsorption*, Wiley, New York, 1964.
7. A.W. Adamson and A.P. Gast, *Physical Chemistry of Surfaces* (6th edition), John Wiley & Sons, New York, 1997.
8. D.W. Ruthven, *Principles of Adsorption and Adsorption Processes*, Wiley, New York, 1984.
9. R.T. Yang, *Adsorbents: Fundamentals and Applications*, John Wiley & Sons, New York, 2003.
10. W. Rudzinski and D.H. Everett, *Adsorption of Gases in Heterogeneous Surfaces*, Academic Press, London, UK, 1992.
11. M.M. Dubinin, *Prog. Surf. Membr. Sci.*, 9, 1 (1975).
12. T.L. Hill, *Introduction to Statistical Thermodynamics*, Dover Publications Inc., New York, 1986.
13. S.U. Rege and R.T. Yang, in *Adsorption: Theory, Modeling and Analysis*, J. Toth, (editor), Marcel Dekker, New York, 2002, p. 175.
14. B.P. Bering, M.M. Dubinin, and V.V. Serpinskii, *J. Colloid Int. Sci.*, 38, 185 (1972).
15. R. Roque-Malherbe, *J. Thermal Anal.*, 32, 1361 (1987).
16. R. Roque-Malherbe, *Physical Adsorption of Gases*, ENPES-MES, Havana, Cuba, 1987.
17. G. Horvath and K. Kawazoe, *J. Chem. Eng. Jpn.*, 16, 470 (1983).
18. A. Saito and H.C. Foley, *AIChE J.*, 37, 429 (1991).
19. L.S. Cheng and R.T. Yang, *Chem. Eng. Sci.*, 49, 2599 (1994).
20. M. Thommes, in *Nanoporous Materials: Science and Engineering*, G.Q. Lu and X.S. Zhao (editors), Imperial College Press, London, UK, 2004, Chapter 11, p. 317.
21. AUTOSORB-1, Manual, Boynton Beach, FL, 2003.
22. Micromeritics, ASAP 2020. Description, Atlanta, GA, 1992.
23. R. Roque-Malherbe, Surface area and porosity characterization of porous polymers, in *Porous Polymers*, M.S. Silverstein, N. Cameron, and M. Hillmyer (editors), John Wiley & Sons Inc., New York (to be published).
24. R. Roque-Malherbe and F. Diaz-Castro, *J. Mol. Catal. A*, 280, 194 (2008).
25. R. Roque-Malherbe, *Physical Chemistry of Zeolites*, ENPES-MES, Havana, Cuba, 1988.
26. M.M. Dubinin, *Am. Chem. Soc. Symp. Ser.*, 40, 1 (1977).
27. D.W. Breck, *Zeolite Molecular Sieves*, John Wiley & Sons, New York, 1974.
28. R.M. Barrer, *Zeolites and Clay Minerals as Sorbents and Molecular Sieves*, Academic Press, London, 1978.
29. E.F. Vansant, *Pore Size Engineering in Zeolites*, John Wiley & Sons, New York, 1990.
30. H.G. Karge, M. Hunger, and H.K. Beyer, in *Catalysis and Zeolites*, J. Weitkamp and L. Puppe (editors), Springer, Berlin, 1999, p. 198.
31. R. Roque-Malherbe, L. Lemes-Fernandez, L. Lopez-Colado, C. de las Pozas, and A. Montes-Caraballal, in *Natural Zeolites '93 Conference Volume International Committee on Natural Zeolites*, D.W. Ming and F.A. Mumpton (editors), Brockport, New York, 1995, p. 299.
32. A. Corma, *Chem. Rev.*, 95, 559 (1995).
33. G.V. Tsitsisvili, T.G. Andronikashvili, G.N. Kirov, and L.D. Filizova, *Natural Zeolites*, Ellis Horwood, New York, 1992.
34. R. Roque-Malherbe, in *Handbook of Surfaces and Interfaces of Materials*, Vol. 5, H.S. Nalwa (editor), Academic Press, New York, 2001, Chapter 12, p. 495.
35. M. Guisnet and J.-P. Gilson (editors), *Zeolites for Cleaner Technologies*, Imperial College Press, London, UK, 2002.
36. P. Payra and P.K. Dutta, in *Handbook of Zeolite Science and Technology*, S.M. Auerbach, K.A. Corrado, and P.K. Dutta (editors), Marcel Dekker, Inc., New York, 2003, p. 1.
37. R. Roque-Malherbe and F. Marquez-Linares, *Facets-IUMRS J.*, 3, 8 (2004).
38. Y. Li and R.T. Yang, *J. Phys. Chem. B*, 110, 17175 (2006).

39. C. Nguyen, C.G. Sonwane, S.K. Bhatia, and D.D. Do, *Langmuir*, 14, 4950 (1998).
40. X.S. Zhao, Q. Ma, and G.Q. Lu, *Energy Fuels* 12, 1051 (1998).
41. A.V. Neimark and P.I. Ravikovitch, *Mic. Mes. Mat.*, 44–45, 697 (2001); P.I. Ravikovitch and A.V. Neimark, *J. Phys. Chem. B*, 105, 6817 (2001); *Langmuir*, 18, 1550 (2002).
42. F. Marquez-Linares and R. Roque-Malherbe, *J. Nanosci. Nanotech.*, 6, 1114 (2006).
43. I.G. Kaplan, *Intermolecular Interactions: Physical Picture, Computational Methods and Model Potentials*, John Wiley & Sons, New York, 2006.
44. P. Atkins, *Physical Chemistry* (6th edition), W.H. Freeman & Co., New York, 1998.
45. D.H. Everett and J.C. Powl, *J. Chem. Soc. Faraday Trans. I*, 72, 619 (1976).
46. J.D. Jackson, *Classical Electrodynamics* (2nd edition), John Wiley & Sons, New York, 1975.
47. C.F. Matta and R.J. Boyd, in *The Quantum Theory of Atoms in Molecules*, C.F. Matta and R.J. Boyd (editors), WILEY-VCH Verlag GmbH & Co. KGaA, Weinheim, Germany, 2007, p. 1.
48. A.D. Buckingham, P.W. Fowler, and J.M. Hutson, *Chem. Rev.*, 88, 963 (1988).
49. R. Roque-Malherbe, C. de las Pozas, and M. Hernandez-Velez, *Rev. Cubana Fis.* 5, 107 (1985).
50. E. Calvet and H. Prat, *Recent Progress in Microcalorimetry*, H.A. Skinner (editor), MacMillian, New York, 1963.
51. N.M. Avgul, B.C. Aristov, A.V. Kiseliov, and L.Ya. Kurdiukova, *Zh. Fiz. Xim.* 62, 2678 (1968).
52. W.W. Porterfield, *Inorganic Chemistry* (2nd edition), Academic Press, San Diego, CA, 1993.
53. C. Naccache and Y. Ben Taarit, *Acta Univ. Szegediensis, Acta Phys. Chem.*, 24, 23 (1978).
54. N.R. Draper and H. Smith, *Applied Regression Analysis* (3rd edition), Wiley, New York, 1998.
55. B.P. Bering and V.V. Serpinskii, *Ixv. Akad. Nauk, SSSR, Ser. Xim.*, 2427 (1974).
56. R. Roque-Malherbe, *KINAM*, 6, 35 (1984).
57. V.A. Bakaev, *Dokl. Akad. Nauk SSSR*, 167, 369 (1966).
58. D.M. Ruthven, *Nat. Phys. Sci.*, 232, 70 (1971).
59. M. Dupont-Pavlovskii, J. Barriol, and J. Bastick, *Coll. Int. CNRS*, No. 201 (Termochemie), 1972.
60. W. Schirmer, K. Fiedler, and H. Stach, *ACS, Symp. Ser.*, 40, 305 (1977).
61. R. Ravinshankar, T. Sen, V. Ramaswami, H.S. Soni, S. Ganapathy, and S. Sivansanker, *Stud. Surf. Sci. Catal. A*, 84, 331 (1994).
62. M.E. Davies, *Nature*, 417, 813 (2002).
63. T.J. Barton, L.M. Bull, G. Klemperer, D.A. Loy, B. McEnaney, M. Misono, P.A. Monson, G. Pez, G.W. Scherer, J.C. Vartulli, and O.M. Yaghi, *Chem. Mater.*, 11, 2633 (1999).
64. G.J.A.A. Soler-Illia, C. Sanchez, B. Lebeau, and J. Patarin, *Chem. Rev.*, 102, 4093 (2002).
65. C.T. Kresge, M.E. Leonowicz, W.J. Roth, J.C. Vartuli, and J.S. Beck, *Nature*, 359, 710 (1992).
66. J.S. Beck, J.C. Vartuli, W.J. Roth, M.E. Leonowicz, C.T. Kresge, K.D. Schmitt, C.T.-W. Chu, D.H. Olson, E.W. Sheppard, S.B. McCullen, J.B. Higgins, and J.L. Schlenker, *J. Am. Chem. Soc.*, 114, 10834 (1992).
67. R. Roque-Malherbe and F. Marquez-Linares, *Surf. Interface Anal.*, 37, 393 (2005).
68. R.K. Iler, *The Chemistry of Silica*, John Wiley & Sons, New York, 1979.
69. W. Stobe, A. Fink, and E. Bohn, *J. Colloids Interface Sci.*, 26, 62 (1968).
70. C.J. Brinker and G.W. Scherer, *Sol-Gel Science*, Academic Press, New York, 1990.
71. K. Unger and D. Kumar, in *Adsorption on Silica Surfaces*, E. Papirer (editor), Marcel Dekker, New York, 2000, p. 1.
72. C. Burda, X. Chen, R. Narayanan, and M.A. El-Sayed, *Chem. Rev.*, 105, 1025 (2005).
73. A.C. Pierre and G.M. Pajonk, *Chem. Rev.*, 102, 4243 (2002).
74. G.M.S. El Shaffey, in *Adsorption on Silica Surfaces*, E. Papirer (editor), Marcel Dekker, New York, 2000, p. 35.
75. J. Persello, in *Adsorption on Silica Surfaces*, E. Papirer (editor), Marcel Dekker, New York, 2000, p. 297.
76. A.P. Legrand, in *The Surface Properties of Silica*, A.P. Legrand (editor), John Wiley & Sons, New York, 1998, p. 1.
77. F. Derbyshire, M. Jagtoyen, R. Andrews, A. Rao, I. Martin-Guillon, and E.A. Grulke, in *Chemistry and Physics of Carbon*, Vol. 27, L.R. Radovic (editor), Marcel Dekker, New York, 2001, p. 1.
78. F. Rodriguez-Reinoso and A. Sepulveda-Escribano, in *Handbook of Surfaces and Interfaces of Materials*, Vol. 5, H.S. Nalwa (editor), Academic Press, New York, 2001, p. 309.
79. X. Shao, W. Wang, R. Xue, and Z. Shen, *J. Phys. Chem. B*, 108, 2970 (2004).
80. V. Lopez, F. Marquez-Linares, C. Morant, C. Domingo, E. Elizalde, F. Zamora, and R. Roque-Malherbe, *Nano Today* (submitted).
81. A. Lan and A. Mukasyan, *J. Phys. Chem. B*, 109, 16011 (2005).
82. A. Galarneau, D. Desplantier, R. Dutartre, and F. Di Renzo, *Mic. Mes. Mat.*, 27, 297 (1999).

83. Ch. Baerlocher, W.M. Meier, and D.H. Olson, *Atlas of Zeolite Framework Types*, Elsevier, Amsterdam, the Netherlands, 2001.

84. L.R. Radovic, C. Moreno-Castilla, and J. Rivera-Utrilla, in *Chemistry and Physics of Carbon*, Vol. 27, L.R. Radovic (editor), Marcel Dekker, New York, 2001, p. 227.

85. L.S. Colella, P.M. Armenante, D. Kafkewitz, S.J. Allen, and V. Balasundaram, *J. Chem. Eng. Data*, 43, 573 (1998).

86. J. Goworek, A. Derylo-Marczewska, and A. Borowka, *Langmuir*, 15, 6103 (1999).

87. D. Andrieux, J. Jestin, N. Kervarec, R. Pichon, M. Privat, and R. Olier, *Langmuir*, 20, 10591 (2004).

88. S. Li, V.A. Tuan, R. Noble, and J. Falcone, *Environ. Sci. Technol.*, 37, 4007 (2003).

89. S. Chempath, J.F.M. Denayer, K.M.A. De Meyer, G.V. Baron, and R.Q. Snurr, *Langmuir*, 20, 150 (2004).

90. K. Wagner and S. Schul, *J. Chem. Eng. Data*, 46, 322 (2001).

91. H. Hindarso, S. Ismadji, F. Wicaksana, Mudjijati, and N. Indraswati, *J. Chem. Eng. Data*, 46, 788 (2001).

92. C. Tien, *Adsorption Calculations and Modeling*, Butterworth, Boston, MA, 1994.

93. J. Oscik, *Adsorption*, Ellis Horwood, Chichester, UK, 1982.

94. J. Toth, in *Adsorption: Theory, Modeling and Analysis*, J. Toth (editor), Marcel Dekker, New York, 2002, p. 1.

95. PeakFit®, *Program System*, Seasolve Software Inc., Framingham, MA.

96. N.R. Draper and H. Smith, *Applied Regression Analysis*, John Wiley & Sons, New York, 1966.

97. A.J. Slaney and R. Bhamidimarri, *Water Sci. Technol.*, 38, 227 (1998).

98. A. Wolborska, *Chem. Eng. J.*, 37, 85 (1999).

99. R.G. Rice and D.D. Do, *Applied Mathematics and Modeling for Chemical Engineers*, John Wiley & Sons, New York, 1995.

100. R. Droste, *Theory and Practice of Water and Wastewater Treatment*, John Wiley & Sons, New York, 1997.

101. H. Scott-Fogler, *Elements of Chemical Reaction Engineering*, Prentice-Hall, Upper Saddle River, NJ, 1999.

102. F.G. Helfferich and G. Klein, *Multicomponent Chromatography: Theory of Interference*, Marcel Dekker, New York, 1970.

103. J.-M. Chern and Y.-W. Chien, *Ind. Eng. Chem. Res.*, 40, 3775 (2001).

104. T.K. Sherwood, R.L. Pigford, and C.R. Wilke, *Mass Transfer*, McGraw-Hill, New York, 1975.

105. M. Pansini, *Mineral. Deposita*, 31,653 (1996).

106. A.C. Michaels, *Ind. Eng. Chem.*, 44, 1922 (1952).

107. G.B. Arfken and H.J. Weber, *Mathematical Methods for Physicists* (5th edition), Harcourt/Academic Press, San Diego, CA, 2001

108. D.W. Hardy and C.L. Walker, *Doing Mathematics with Scientific Workplace and Scientific Notebook*, MacKisham Software Inc., Poulsbo, WA, 2003.

109. M.L. Boas, *Mathematical Methods in the Physical Sciences*, John Wiley & Sons, New York, 1966.

110. R. Roque-Malherbe, L. Lemes, M. Autie, and O. Herrera, *in Zeolite or the Nineties, Recent Research Reports, 8th International Zeolite Conference*, Amsterdam, July 1989, J.C. Hansen, L. Moscou, and M.F.M. Post (editors), IZA, 1989, p. 137.

111. E.J. Acosta, C.S. Carr, E.E. Simanek, and D.F. Shantz, *Adv. Mater.*, 16, 985 (2004).

112 M.A. Hernandez, L. Corona, A.I. Gonzalez, F. Rojas, V.H. Lara, and F. Silva, *Ind. Eng. Chem. Res.*, 44, 2908 (2005).

113. D. Chudasama, J. Sebastian, and R. V. Jasra, *Ind. Eng. Chem. Res.*, 44, 1780 (2005).

114. S. Sircar and A.L. Myers, in *Handbook of Zeolite Science and Technology*, S.M. Auerbach, K.A. Corrado, and P.K. Dutta (editors), Marcel Dekker, New York, 2003, p. 1063.

115. C. de las Pozas, R. López-Cordero, C. Díaz-Aguilas, M. Cora, and R. Roque-Malherbe, *J. Solid State Chem.*, 114, 108 (1995).

116. E.F. Vansant, in *Innovation in Zeolite Material Science*, P.J. Grobet (editor), Elsevier, Amsterdam, the Netherlands; *Stud., Surf. Sci. Catal.*, 37, 143 (1988).

117. S.M. Kuznicki, V.A. Bell, S. Nair, H.G. Hillhouse, R.M. Jacubinas, M.B. Carola, and B.H. Toby, *Nature*, 412, 720 (2001).

118. G. Giannetto, *Zeolitas*, Editorial Caracas, Caracas, Venezuela, 1990.

119. X. Hu, S. Qiao, X.S. Zhao, and G.Q. Lu, *Ind. Eng. Chem. Res.*, 40, 862 (2001).

120. X.S. Zhao, G.Q. Lu, and X. Hu, *Colloids Surf. A: Physicochemical Eng. Aspects*, 179, 261 (2001).

121. A. Cauvel, D. Brunel, F. Di Renzo, E. Garrone, and B. Fubini, *Langmuir*, 13, 2773 (1997).

122. P. Selvam, S.K. Bhatia, and C.G. Sonwane, *Ind. Eng. Chem. Res.*, 40, 3237 (2001).

123. X.S. Zhao and G.Q. Lu, *J. Phys. Chem. B*, 102, 1556 (1998).

124. E. Garrone and P. Ugliengo, *Langmuir*, 7, 1409 (1991).

125. B. Civalleri, E. Garrone, and P. Ugliengo, *Langmuir*, 9, 2712 (1993).
126. B. Fubini, V. Bolis, A. Cavenago, E. Garrone, and P. Ugliengo, *Langmuir*, 15, 5829 (1999).
127. J. Helmenin, J. Helenius, and E. Paatero, *J. Chem. Eng. Data*, 46, 391 (2001).
128. S.-L. Kuo, E.O. Pedram, and A.L. Hilnes, *J. Chem. Eng. Data*, 30, 330 (1985).
129. Z. Knez and Z. Novak, *J. Chem. Eng. Data*, 46, 858 (2001).
130. L. Zhou, L. Zhong, M. Yu, and Y. Zhou, *Ind. Eng. Chem. Res.*, 43, 1765 (2004).
131. H.Y. Huang, R.T. Yang, D. Chinn, and C.L. Munson, *Ind. Eng. Chem. Res.*, 42, 2427 (2003).
132. M.-B. Kim, Y.-K. Ryu, and C.-Ha Lee, *J. Chem. Eng. Data*, 50, 591 (2005).
133. R.T. Yang, *Gas Separation by Adsorption Process*, Butterworth, Boston, MA, 1987.
134. R.C. Bansal, J.B. Donnet, and F. Stoeckli, *Active Carbon*, Marcel Dekker, New York, 1988.
135. J. Benkhedda, J.N. Jaubert, D. Barth, and L. Perrin, *J. Chem. Eng. Data*, 45, 650 (2000).
136. G. Kimber, M. Jagtoyen, Y. Fei, and F. Derbyshire, *Gas Sep. Purif.*, 10, 131 (1996).
137. E.N. Rubby and L.A. Carroll, *Chem. Eng. Prog.*, 28 (1993).
138. M.H. Stenzel, *Chem. Eng. Prog.*, 36 (1993).
139. D.P. Valenzuela and A.L. Myers, *Adsorption Equilibrium Data Handbook*, Prentice Hall, Englewood Cliffs, NJ, 1989.
140. A.L. Myers, C. Minka, and Ou, Y. D. *AIChE J.*, 28, 97 (1982).
141. P.G. Hall and R.T. Williams, *J. Colloid Interface Sci.*, 113, 301 (1986).
142. T. Nabarawy, N.S. Petro, and S. Abdel-Aziz, *Adsorpt. Sci. Technol.*, 15, 47 (1997).
143. J.H. Yun and D.K. Choi, *J. Chem. Eng. Data*, 42, 894 (1997).
144. K.P. Gadkaree, *Carbon*, 36 981 (1998).
145. J. Benkhedda, J.N. Jaubert, D. Barth, J. Perrin, and M. Bailly, *J. Chem. Thermodyn.*, 32, 401 (2000).
146. W. Grochala and P.E. Edwards, *Chem. Rev.*, 104, 1283 (2004).
147. M.G. Nijkamp, J.E.M.J. Raymakers, A.J. van Dillen, and K.P. de Jong, *Appl. Phys. A*, 72, 619 (2001).
148. G.W. Crabtree, M.S. Dresselhaus, and M.V. Buchanan, *Phys. Today*, 57 (12), 39 (2004).
149. S.H. Jhung, H.-K. Kim, J.W. Yoon, and J.-S. Chang, *J. Phys. Chem. B*, 110, 9371 (2006).
150. A. Zecchina, S. Bordiga, J.G. Vitillo, G. Ricchiardi, C. Lamberti, G. Spoto, M. Bjrgen, and K.P. Lillerud, *J. Am. Chem. Soc.*, 127, 6361 (2005).
151. S.B. Kayiran and F.L. Darkrim, *Surf. Interface Anal.*, 34, 100 (2002).
152. H.W. Langmi, D. Book, A. Walton, S.R. Johnson, M.M. Al-Mamouri, J.D. Speight, P.P. Edwards, I.R. Harris, and P.A. Anderson, *J. Alloys Compd.*, 404–406, 637 (2005).
153. J.A. Martens and P.A. Jacobs, in *Catalysis and Zeolites. Fundamentals and Applications*, J. Weitkamp and L. Puppe (editors), Springer-Verlag, Berlin, 1999, p. 53.
154. N.D. Hutson, D.J. Gualdoni, and R.T. Yang, *Chem. Mater.*, 10, 3707 (1998).
155. A. Gil, G. Guiu, P. Grange, and M. Montes, *J. Phys. Chem.*, 99, 301 (1995).
156. K.J. Edler, P.R. Reynolds, P.J. Branton, F.R. Trouw, and J.W. White, *J. Chem. Soc. Faraday Trans.*, 93, 1667 (1997).
157. R.T. Yang, in *Nanostructured Materials*, J.Y. Ying (editor), Elsevier, Amsterdam, the Netherlands, 2001. p. 80.
158. B. Sakintuna and Y. Yurum, *Ind. Eng. Chem. Res.*, 44, 2893 (2005).
159. J.J. Petrovic, *Advanced Concepts for Hydrogen Storage*, Materials Science and Technology Division, Los Alamos National Laboratory, DOE Hydrogen Storage Workshop, Argonne National Laboratory, August 14–15, 2002.
160. C. Liang, W. Li, Z. Wei, Q. Xin, and C. Li, *Ind. Eng. Chem. Res.*, 39, 3694 (2000).
161. A. Raissi, *Hydrogen, Fuel Cells, and Infrastructure Technologies*, DoE, FY, 2002, Progress Report.
162. R. Roque-Malherbe, F. Marquez-Linares, W. Del Valle, and M. Thommes, *J. Nanosci. Nanotech.*, 8, 5993 (2008).
163. L.R. Arana, S.B. Schaevitz, A.J. Franz, M.A. Schmidt, and K.F. Jensen, *JMEMS*, 12, 600 (2003).
164. D.J. Quiram, I.-M. Hsing, A.J. Franz, K.F. Jensen, and M.A. Schmidt, *Chem. Eng. Sci.*, 55, 3065 (2000).
165. A. Chambers, C. Park, R.T.K. Baker, and N.M. Rodriguez, *J. Phys. Chem. B*, 102, 4253 (1998).
166. M. Rzepka, P. Lamp, and M.A. de la Casa-Lillo, *J. Phys. Chem. B*, 102, 10884 (1998).
167. M.A. de la Casa-Lillo, F. Lamari-Darkrim, D. Cazorla-Amoros, and A. Linares-Solano, *J. Phys. Chem. B*, 106, 10930 (2002).
168. P. Hou, Q. Yang, S. Bai, S. Xu, M. Liu, and H. Cheng, *J. Phys. Chem. B*, 106, 963 (2002).
169. B. Fang, H. Zhou, and I. Honma, *J. Phys. Chem. B*, 110, 4875 (2006).
170. P. Kowalczyk, H. Tanaka, R. Hołyst, K. Kaneko, T. Ohmori, and J. Miyamoto, *J. Phys. Chem. B*, 109, 17174 (2005).
171. Z. Yang, Y. Xia, and R. Mokaya, *J. Am. Chem. Soc.*, 129, 1673 (2007).
172. A.C. Dillon, K.M. Jones, T.A. Bekkedahl, C.H. Kiang, D.S. Bethune, and M.J. Heben, *Nature*, 387, 377 (1997).

173. S. Hynek, W. Fuller, and J. Bentley, *Int. J. Hydrogen Energy*, 22, 601 (1997).
174. K.A.G. Amankwah and J.A. Schwarz, *Int. J. Hydrogen Energy*, 14, 437 (1988).
175. J. Jagiello, T.J. Bandosz, K. Putyera, and J.A. Schwarz, *J. Chem. Soc., Faraday Trans.*, 91, 2929 (1995).
176. K. Murata, K. Kaneko, H. Kanoh, D. Kasuya, K. Takahashi, F. Kokai, M. Yudasaka, and S. Iijima, *J. Phys. Chem. B*, 106, 11132 (2002).
177. A. Anson, M.A. Callejas, A.M. Benito, W.K. Maser, M.T. Izquierdo, B. Rubio, J. Jagiello, M. Thommes, J.B. Parra, and M.T. Martinez, *Carbon*, 42, 1237 (2004).
178. A. Zutel, P. Sudan, P. Mauron, and P. Wenger, *Appl. Phys. A*, 78, 941 (2004).
179. M. Shiraishi, T. Takenobu, H. Kataura, and M. Ata, *Appl. Phys. A*, 78, 947 (2004).
180. N. Akuzawa, Y. Amari, and T. Nakajima, *J. Mater. Res.*, 5, 2849 (1990).
181. K. Watanabe, M. Soma, T. Ohishi, and K. Tamaru, *Nature*, 233, 160 (1971).
182. P. Lagrange, A. Metror, and A. Herold, *Comp. Rend.*, 275, 160 (1971).
183. M. Ritschel, M. Uhlemann, O. Gutfleisch, A. Leonhardt, A. Graff, C. Taschner, and J. Fink, *Appl. Phys. Lett.*, 80, 2985 (2002).
184. G.G. Tibbetts, G.P. Meisner, and C.H. Olk, *Carbon*, 39, 2291 (2001).
185. C. Liu, Y.Y. Fan, M. Liu, H. Cong, H.M. Chen, and M.S. Dresselhaus, *Science*, 286, 1127 (1999).
186. T. Duren, L. Sarkisov, O. M. Yaghi, and R. Q. Snurr, *Langmuir*, 20, 2683 (2004).
187. A. Celzard and V. Fierro, *Energy Fuels*, 19, 573 (2005).
188. D. Lozano-Castello, J. Alcañiz-Monge, M.A. de la Casa-Lillo, D. Cazorla-Amoros, and A. Linares-Solano, *Fuel*, 81, 1777 (2002).
189. J.J. Eberhardt, *Gaseous Fuels in Transportations Prospects and Promise*; Presented at the Gas Storage Workshop, Kingston, ON, Canada, July 10–12, 2001.
190. N.D. Parkyns and D.F. Quinn, in *Porosity in Carbons: Characterization and Applications*, J.W. Patrick (editor), Edward Arnold, London, 1995, p. 302.
191. D. Lozano-Castello, D. Cazorla-Amoros, and A. Linares-Solano, *Energy Fuels*, 16, 1321 (2002).
192. J. Wegrzyn and M. Gurevich, *Appl. Energy*, 55, 71 (1996).
193. D. Cao, X. Zhang, J. Chen, W. Wang, and J. Yun, *J. Phys. Chem. B*, 107, 13286 (2003).
194. J.A.F. MacDonald and D.F. Quinn, *Fuel*, 77, 61 (1998).
195. Q. Huo, R. Leon, P. Petroff, and G.D. Stucky, *Science*, 268, 1324 (1995).
196. G.E. Gadd, M. Blackford, S. Moricca, N. Webb, P.J. Evans, P.J.A.M. Smith, G. Jacobsen, S. Leung, A. Day, and Q. Hua, *Science*, 277, 933 (1997).
197. L. Schlapbach and Züttel, *Nature*, 414, 353 (2005).
198. S.K. Bhatia and A.L. Myers, *Langmuir*, 22, 1688 (2006).
199. P. Kowalczyk, L. Brualla, A. Zywocinski, and S. K. Bhatia, *J. Phys. Chem. C*, 111, 5250 (2007).
200. Z. Tan and K.E. Gubbins, *J. Phys. Chem.*, 94, 6061 (1990).
201. C. Lastoskie, K.E. Gubbins, and N. Quirke, *J. Phys. Chem.*, 97, 4786 (1993).
202. P.N. Aukett, N. Quirke, S. Riddiford, and S.R. Tennisson, *Carbon*, 30, 913 (1992).
203. J.-W. Lee, M.S. Balathanigaimani, H.-C. Kang, W.-G. Shim, C. Kim, and H. Moon, *J. Chem. Eng. Data*, 52, 66 (2007).
204. K.A. Sosin and D.F. Quinn, *J. Porous Mater.*, 1, 111 (1995).
205. J. Sun, T.D. Jarvi, L.F. Conopask, S. Satyapal, M.J. Rood, and M. Rostam-Abadi, *Energy Fuels*, 15, 1241 (2001).
206. D. Lozano-Castello, D. Cazorla-Amoros, A. Linares-Solano, and D.F. Quinn, *Carbon*, 40, 988 (2002).
207. B.U. Choi, D.K. Choi, Y.W. Lee, and B.K. Lee, *J. Chem. Eng. Data*, 48, 603 (2003).
208. X.R. Zhang and W.C. Wang, *Fluid Phase Equilib.*, 194, 288 (2002).
209. M. Muris, N. Dufau, M. Bienfait, N. Dupont-Pavlovsky, Y. Grillet, and J.P. Palmari, *Langmuir*, 16, 7019 (2000).
210. H. Tanaka, M. El-Merraoui, W.A. Steele, and K. Kaneko, K. *Chem. Phys. Lett.*, 352, 334 (2002).
211. E. Bekyarova, K. Murata, M. Yudasaka, D. Kasuya, S. Iijima, H. Tanaka, H. Kahoh, and K. Kaneko, *J. Phys. Chem. B*, 107, 4682 (2003).
212. S. Iijima, *Nature*, 354, 56 (1991).
213. V.C. Menon and S. Komarneni, *J. Porous Mater.*, 5, 43 (1998).
214. M.A. Anderson, *Environ. Sci. Technol.*, 34, 725 (2000).
215. M.L. Occelli and H. Kessler (editors), *Synthesis of Porous Materials*, Marcel Dekker, New York, 1997.
216. M.A. Camblor, A. Corma, and S. Valencia, *J. Chem. Soc. Chem. Commun.*, 2365 (1996).
217. J.K. Brennan, T. J. Bandosz, K. T. Thomson, and K. E. Gubbins, *Colloids Surf. A*, 187–188, 539 (2001).
218. C. Leon-Leon and L. Radovic, in *Chemistry and Physics of Carbon*, Vol. 24, P. Thrower (editor), Marcel Dekker, New York, 1994.
219. A. Jain, V.K. Gupta, S. Jain, and Suhas, *Environ. Sci. Technol.*, 38, 1195 (2004).

220. A. Garcia, J. Silva, L. Ferreira, A. Leitao, and A. Rodrigues, *Ind. Eng. Chem. Res.*, 41, 6165 (2002).
221. T. El Brihi, J.N. Jaubert, D. Barth, and L. Perrin, *J. Chem. Eng. Data*, 47, 1553 (2002).
222. C. Muñiz-Lopez, J. Duconge, and R. Roque-Malherbe, *J. Colloid Interface Sci.*, 329, 11 (2009).
223. J.R. Benson, *Am. Lab.*, April, 2003.
224. M.I. Abrams and J.R. Millar, *React. Funct. Polym.*, 35, 7 (1997).
225. S.M. Howdle, K. Jerábek, V. Leocorbo, P.C. Marr, and D.C. Sherrington, *Polymer*, 41, 7273 (2000).
226. D.C., Sherrington, *Chem. Commun.*, 2275 (1998).
227. F. Schuth, K.S.W. Sing, and J. Weitkamp (editors), *Handbook of Porous Solids*, John Wiley & Sons, New York, 2002.
228. A.K. Hebb, K. Senoo, R. Bhat, and A.I. Cooper, *Chem. Mater.*, 15, 2061 (2003).
229. J. Germain, J. Hradil, J.M.J. Frechet, and F. Svec, *Chem. Mater.*, 18, 4430 (2006).
230. N.S. Pujari, A.R. Vishwakarma, T.S. Pathak, A.M. Kotha, and S. Ponrathnam, *Bull. Mater. Sci.*, 27, 529 (2004).
231. B. Feibush and N.-H. Li, US Patent No. 4,933,372 (1990).
232. A. Deryło-Marczewska and J. Goworek, *Langmuir*, 17, 6518 (2001).
233. A. Deryło-Marczewska, J. Goworek, S. Pikus, E. Kobylas, and W. Zgrajka, *Langmuir*, 18, 7538 (2002).
234. A. Deryło-Marczewska, J. Goworek, R. Kusak, and W. Zgrajka, *Appl. Surf. Sci.*, 195, 117 (2002).
235. S.A. Johnson, P.J. Ollivier, and T.E. Mallouk, *Science*, 283, 963 (1999).
236. A.V. Maffei, P.M. Budd, and N.B. McKeown, *Langmuir*, 22, 4225 (2006).
237. W. Mori, F. Inoue, K. Yoshida, H. Nakayama, and S. Takamizawa, *Chem. Lett.*, 1219 (1997).
238. K. Seki, S. Takamizawa, and W. Mori, *Chem. Lett.*, 122 (2001).
239. K. Seki and W. Mori, *J. Phys. Chem. B*, 106, 1389 (2002).
240. K. Seki, S. Takamizawa, and W. Mori, *Chem. Lett.*, 332 (2001).
241. H. Wu, W. Zhou, and T. Yildirim, *J. Amer. Chem. Soc.*, 129, 5314 (2007).
242. B. Chen, M. Eddaoudi, S.T. Hyde, M. O'Keeffe, and O.M. Yaghi, *Science*, 291, 1021 (2001).
243. N.L. Rosi, J. Kim, M. Eddaoudi, B. Chem, M. O'Keeffe, and O.M. Yaghi, *J. Am. Chem. Soc.*, 127, 1504 (2005).
244. N. Zabukovec-Logar and V. Kaučič, *Acta Chim. Slov.*, 53, 117 (2006).
245. M. Jacoby, *Chem. Chem. Eng. News*, 83, 43 (2005).
246. B. Chen, M. Eddaoudi, T. Reineke, J.W. Kampf, M. O'Keeffe, and O.M. Yaghi, *J. Am. Chem. Soc.*, 122, 11559 (2000).
247. M. Eddaoudi, H. Li, and O.M. Yaghi, *J. Am. Chem. Soc.*, 122, 1391 (2000).
248. H. Tai, A.Y. Sergienko, and M.S. Silverstein, *Polym. Eng. Sci.*, 41, 1540 (2001); *Polymer*, 42, 4473 (2001).
249. A.Y. Sergienko, H. Tai, M. Narkis, and M.S. Silverstein, *J. App. Polym. Sci.*, 94, 2233 (2004).
250. J.L.C. Rowsell, A.R. Millward, K.S. Park, and O.M. Yaghi, *J. Am. Chem. Soc.*, 126, 5666 (2004).
251. B. Panella and M. Hirscher, *Adv. Mater.*, 17, 538 (2005).
252. U. Mueller, M. Shubert, F. Teich, H. Puetter, K. Schierle-Arndt, and J. Pastre, *J. Mater. Chem.*, 16, 626 (2006).
253. D.N. Dybtsev, H. Chun, S.H. Yoon, D. Kim, and K. Kim, *J. Am. Chem. Soc.*, 126, 32 (2004).
254. S.S. Kaye and J.R. Long, *J. Am. Chem. Soc.*, 127, 6506 (2005).
255. N.L. Rosi, J. Eckert, M. Eddaoudi, D.T. Vodak, J. Kim, M. O'Keefee, and O.M. Yaghi, *Science*, 300, 1127 (2003).
256. I.-Y.R. Yeh, *Cancer Risk Assessment of Trihalomethanes in Drinking Water*, VDM Verlag, Saarbrucken, Germany, 2008.
257. J. DeZuane, *Handbook of Drinking Water Quality*, John Wiley & Sons, New York, 1997.
258. A. Dunlop and F.N. Peters, *The Furans*, American Chemical Society Monograph Series, Washington, DC, 1958.
259. R. Sanchez, PhD Dissertation, National Center for Scientific Research, Havana, Cuba, 1988.
260. R. Roque-Malherbe, J. Onate-Martinez, and E. Navarro, *J. Mat. Sci. Lett.*, 12, 1037 (1993).
261. R. Sanchez and C. Hernandez, *Eur. Polym. J.*, 30, 51 (1994).
262. R. Sanchez, C. Hernandez, R. Roque-Malherbe, and H. Campaña, Cuban Patent Certificate No. 21644 A1, Cuban Office for the Industrial Property, Havana, Cuba (1987).
263. T. Budinova, D. Savova, N. Petrov, M. Razvigorova, V. Minkova, N. Ciliz, E. Apak, and E. Ekinci, *Ind. Eng. Chem. Res.*, 42, 2223 (2003).
264. I.V. Kaminskii, N.V. Ungurean, and V.I. Ilinskii, *Plat. Massy*, 12, 9 (1960).

7 Ion Exchange

7.1 INTRODUCTION

Ion exchange, in general, describes the process of interchange of ions between an ion-exchange material and a liquid electrolyte solution [1–5]. It is an efficient and flexible method of conditioning the feed water for different processes, wastewater treatment, and as well for the modification of materials which possess an ion-exchange property. Among the best-known inorganic compounds with ion-exchange behavior are zeolites, both natural and synthetic [1–4]. However, other inorganic materials, for example, clays, zirconium phosphates, ferrocyanides, titanates, apatites, hydrotalcites [3], and many coals also have ion-exchange properties [6]. Besides, polymeric resins can be considered as the most important industrial ion-exchange material currently applied in industry [4].

Within the framework of this book, we do not discuss all the existing inorganic and organic ionic exchangers; instead, we concentrate on one archetypal crystalline ionic exchanger, that is, aluminosilicate zeolites [1–3,5]. The description of the theory and practice of ion exchange in zeolites will bring about a good understanding of the process of ionic exchange in inorganic crystalline ion exchangers. Toward the end of the chapter, examples of ion exchange in other significant inorganic ion exchangers such as hydrotalcites, titanates, zirconium phosphates [3], and polymeric resins are discussed [4].

7.2 ALUMINOSILICATE ZEOLITE ION EXCHANGERS

Aluminosilicate zeolites are a key group of crystalline inorganic ion exchangers that are useful in industrial, agricultural, and environmental applications [5], extending from wastewater cleaning [7–11], agriculture [12,13], aquaculture [12], detergency [14–16], to the removal of radioactive nuclei from nuclear wastewaters [17–21], and other possible applications owing to their ion-exchange properties [22–25]. On the other hand, ionic exchange is possibly the most important method currently used to modify zeolites [21].

It was Thomson [26], in 1845, who first performed a scientific study of ion exchange in soils. Almost, at the same time, Way [27] showed that the effect was related to aluminosilicates present in soils. Later, Lemberg and Wiegner [28,29] identified zeolites, clays, and other minerals as the aluminosilicates responsible for ion exchange in soils. Moreover, Eichorn [30], for the first time, studied ion exchange quantitatively, showing the reversibility of the reaction in some natural zeolites, such as chabazite and natrolite.

The advances in the study of ion exchange triggered the industrial application of zeolites in water softening. However, in 1935, it was discovered that by grounding phonographic disks, a powder is obtained, which was a good ion-exchange material. This finding incited Adams and Holmer to develop ion-exchange resins whose ion-exchange properties were better than those of zeolites [25]. These first resins were similar to Bakelite and were synthesized by condensing polyhydric phenols or phenolsulfonic acids with formaldehyde. Later, other types of polymeric resins and ion exchangers based on cross-linked polystyrene derivatives that show selectivity between ions were developed. Common examples of a material extensively applied for cation exchange are resins prepared by sulfonating styrene/divinyl benzene copolymer beads [4]. These resins are frequently applied in industrial ion-exchange processes. Therefore, in contrast to ion-exchange resins, aluminosilicate zeolites have found relatively limited use as ion exchangers in the recent years.

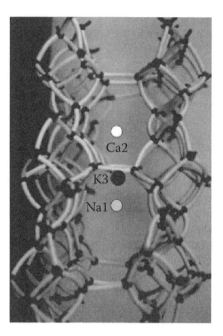

FIGURE 7.1 Schematic representation of the HEU framework structure of clinoptilolite along the crystallographic direction [001].

Usually, zeolites are used because of economic considerations, or in those cases where a high thermal or radiation flux prevent the application of ion-exchange resins [5,14]. Besides, zeolites, in comparison with ion-exchange resins, have a high selectivity toward certain metal cations [8] as for example, ammonium in the case of clinoptilolite and zeolite Na-A for water softening [5]. Consequently, zeolites are widely applied as water softeners in detergency [8,14–16], where the comparative cheapness of Na-A zeolite makes it an appealing alternative in use-and-dispose [14]. Another important use of zeolites is in the elimination and the storage of radioactive nuclei [5,17–20]. In this case, the significant resistance of various zeolites, fundamentally natural zeolites [5], to radiation and high temperatures makes these materials a good alternative [14].

Figure 7.1 shows a schematic illustration of the HEU framework structure of clinoptilolite, all along the crystallographic direction [001] [31]. In this illustration, the 10-member ring (MR) channel and the two 8-member ring (MR) channels are shown. In addition, the small circles tagged Na1, Ca2, and K3 represent the position of extraframework cations corresponding to this zeolite [32]. The HEU-type framework is congruent with the natural zeolites, heulandite and clinoptilolite [5,32], and the synthetic zeolite, LZ-219 [31].

The structure of clinoptilolite (HEU-type framework, [31]) shows three channels: one 8-MR channel along [100], with a window of access of $2.6\,\text{Å} \times 4.7\,\text{Å}$; and two parallel channels along [001] that include one 8-MR with a window of access of $3.3\,\text{Å} \times 4.6\,\text{Å}$, and a 10-MR with a window of access of $3.0\,\text{Å} \times 7.6\,\text{Å}$ (see Figure 7.1) [31,32].

7.3   SOME DEFINITIONS AND TERMS

In ion exchange, the solid, that is, the ion exchanger, is a matrix of fixed ions named the co-ions, forming a charged framework, whose charge is balanced by mobile ions, located in definite sites of the channels or cavities conforming to the solid matrix, named counterions [25]. Then, ion exchange is an effect or reaction where ions included in an electrolytic solution react or exchange with the counterions included in the ion exchanger in contact with the electrolytic solution, and where the reaction continues till it reaches equilibrium [6,25].

Ion exchange in aluminosilicate zeolites (or another ion exchanger) is a reaction where two particles take part: the extra-framework cations present in the zeolite and the ions dissolved in the electrolytic solution. The ion-exchange reaction in aluminosilicate zeolites is represented by [21,24]:

$$z_B A\{z_A^+\} + BZ \leftrightarrow z_A B\{z_B^+\} + AZ \tag{7.1}$$

where

$z_A e^+$ and $z_B e^+$ are the charges of cations A and B, respectively
$A\{z_A^+\}$ and $B\{z_B^+\}$ describe the cations A and B in the solution
AZ and BZ are the cations A and B in the zeolite

Ion-exchange equilibrium data are reported in the form of isotherms with the help of the equivalent ionic fraction in the solution (Figure 7.2):

$$X_A = \frac{z_A m_A}{z_A m_A + z_B m_B} \quad \text{and} \quad X_B = \frac{z_B m_B}{z_A m_A + z_B m_B} \tag{7.2a}$$

In the zeolite,

$$\overline{X}_A = \frac{z_A \overline{m}_A}{z_A \overline{m}_A + z_B \overline{m}_B} \quad \text{and} \quad \overline{X}_B = \frac{z_B \overline{m}_B}{z_A \overline{m}_A + z_B \overline{m}_B} \tag{7.2b}$$

where

m_A and m_B are the molalilities of A and B in the solution
$\overline{m}_A$ and $\overline{m}_B$ are the quantity of moles of A and B per gram of the zeolite

The total exchange capacity (TEC) of an aluminosilicate zeolite is a function of the framework composition that is, the Si/Al ratio. It is easy to obtain a numerical relation between the TEC, C, in milliequivalent per gram, and the number of Al atoms per framework unit cell, N^{Al} [5]

$$C = \frac{N^{Al}}{N_{Av} \rho V_c} \tag{7.3}$$

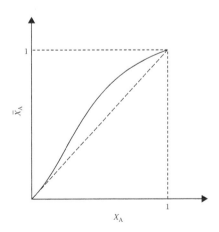

FIGURE 7.2 Graphic representation of an ion-exchange isotherm.

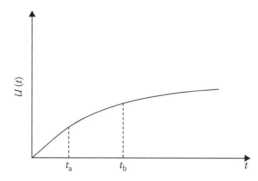

FIGURE 7.3   Ion-exchange kinetics.

where

N_{Av} is the Avogadro number
ρ is the zeolite density
V_c is the volume of the framework unit cell

The relation follows from N^{Al}/N_{Av}, which is the total number of equivalents of exchangeable cations per unit cell, and ρV_c, the mass of the unit cell.

Kinetic aspects are significant in ionic exchange in zeolites. The kinetics of ion exchange is represented with a plot of $U(t)$ versus. time t (Figure 7.3), in which

$$U(t) = \frac{Q_B(t) - Q_B(t)}{Q_B(0) - Q_B(\infty)} = \frac{Q_A(t)}{Q_A(\infty)}$$
(7.4)

where

$Q_B(0)$ is the initial magnitude of cation B in the zeolite
$Q_B(t)$ is the magnitude of cation B at time t
$Q_B(\infty)$ is the equilibrium magnitude
$Q_A(t)$ is the magnitude of cation A in the zeolite at time t
$Q_A(\infty)$ is the equilibrium magnitude

It is generally accepted that the ionic exchange process in zeolites is described by three steps [23–25] (see Figure 7.3): (1) interdiffusion in the adhering liquid thin layer ($0 < t < t_a$); (2) intermediate step, where interdiffusion in the liquid thin layer and crystalline interdiffusion are both present ($0 < t < t_b$); and (3) interdiffusion of A and B in the zeolite crystals ($t > t_b$).

7.4   THERMODYNAMICS OF ION EXCHANGE

The thermodynamic equilibrium constant for an equilibrium ion-exchange process, described by Equation 7.1, can be expressed as follows [14]

$$K_B^A = \frac{(\bar{a}_A)^{z_B} (a_B)^{z_A}}{(\bar{a}_B)^{z_A} (a_A)^{z_B}}$$
(7.5)

in which

$\bar{a}$ means activity in the zeolite for both components A and B
a is the activity in the solution phase for both components A and B

In addition,

$$\Delta G^0 = \frac{RT}{z_A - z_B} \ln K_B^A \tag{7.6}$$

is the standard variation of the Gibbs free energy during the ion-exchange reaction.

In addition to the representation of concentration with the help of the equivalent ionic fraction (Equations 7.2a and b), there are other forms of expressing the concentration in ion-exchange experiments such as the cationic mole fraction defined as [33]

$$M_A = \frac{m_A}{m_A + m_B} \quad \text{and} \quad M_B = \frac{m_B}{m_A + m_B} \tag{7.7}$$

Using the above definitions, it is possible to express the activities as follows [33]:

$$a_A = f_A M_A \quad \text{and} \quad a_B = f_B M_B \tag{7.8}$$

Then, finally, it is possible to write the thermodynamic equilibrium constant for an equilibrium ion-exchange process expressed by Equation 7.5 with the help of the following equation

$$K_B^A = K_V \frac{(f_A)^{z_B}}{(f_B)^{z_A}} \tag{7.9}$$

where
f_A and f_B are the rational activity coefficients of ions A and B in solution
K_V is the Vanselow-corrected selectivity quotient

Nevertheless, the thermodynamic formulation normally used for zeolites is the one developed by Gaines and Thomas [34], who expressed the equilibrium constant (Equation 7.5) as follows.

$$K_B^A = \frac{(\bar{X}_A g_A)^{z_B} (\gamma_B c_B)^{z_A}}{(\bar{X}_B g_B)^{z_A} (\gamma_A c_A)^{z_B}} = K_C \frac{(g_A)^{z_B}}{(g_B)^{z_A}} \tag{7.10}$$

where
$\bar{X}_A$ and $\bar{X}_B$ are the equivalent fractions of exchangeable ions A and B in the zeolite, respectively
a_A and a_B are the corresponding ion activities in the electrolytic solution
c_A and c_B in [mol/L] are the corresponding concentrations
γ_A and γ_B are the corresponding activity coefficients for the cations in solution and g_A and g_B are the activity coefficients for the cations in the zeolite
K_C is the Kielland coefficient

The ratio of ion activity coefficients, $(\gamma_B)^{z_A}/(\gamma_A)^{z_B}$, in a solution is normally calculated in the electrolytic solution of A and B by a method developed by Fletcher and Townsend [35]. Then, K_C can be evaluated.

Within the framework of the method of Gaines and Thomas, Equation 7.10 is differentiated, and then the Gibbs–Duhem equation is applied to obtain the equilibrium constant:

$$\ln K_B^A = (z_B - z_A) + \int_0^1 \ln K_C d\bar{X}_A \tag{7.11}$$

The method for the evaluation of K_B^A uses the so-called Kielland plots, that is, a plot of $\ln K_C$ versus $\bar{X}_A$, which is practically feasible, since a_A and a_B, the ion activities in the electrolytic solution, and $\bar{X}_A$ and $\bar{X}_B$, the equivalent fractions of exchangeable ions A and B in the zeolite, can be experimentally measured. On the contrary, $\bar{a}_A$ and $\bar{a}_B$ cannot be measured.

Barrer and Townsend [36] consider that Equation 7.11 is not an exact expression, since the presence of water as a third component is not considered. Considering water as a third component, we obtain [36]:

$$\ln{}^{W}K_B^A = \int_0^1 \ln K_C d\,\overline{X}_A \qquad (7.11a)$$

where ${}^{W}K_B^A$ is the equilibrium constant when water is considered as a third component in the equilibrium.

7.5 RULES GOVERNING THE ION-EXCHANGE EQUILIBRIUM IN ZEOLITES

7.5.1 REGULAR SYSTEMS

In these systems, the ion- exchange process obeys the mass action law (see Figure 7.4a and d) [24]

$$\alpha_B^A = \frac{(\overline{X}_A)(X_B)}{(\overline{X}_B)(X_A)}$$

where $\alpha_B^A x$ is the affinity coefficient constant. This is a result of the interaction between the cations A and B within the zeolite framework, independent of the concentration.

7.5.2 SPACE LIMITATIONS AND MOLECULAR SIEVING

In some cases, the exchanging cation is so large that, independently of the fact that it is still enough charge to compensate, there is not enough space to accommodate the exchanging cation (Figure 7.4b). For example, during the exchange of tetramethylammonium $[N(CH_3)_4^+]$ in faujasite, it is possible to

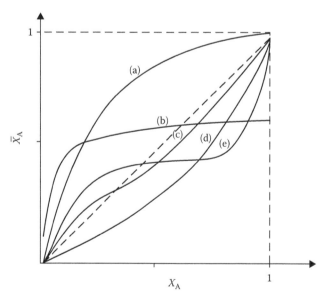

FIGURE 7.4 Characteristic ion-exchange isotherms of zeolites: (a) Regular system with selectivity for cation A, (b) incomplete ion exchange of cation A, (c) irregular system with selectivity reversal, (d) regular system with selectivity for cation B, and (e) irregular system with selectivity reversal.

accommodate merely 30 molecules of $N(CH_3)_4^+$, because of space limitations in the faujasite cage; however, 90 $N(CH_3)_4^+$ molecules are necessary to compensate for the framework charge [24].

In other cases, because of molecular sieving, the cation is not accessible to the zeolite and there is no exchange, and in other cases the ion exchange is partial because the cation do not have access to some channel systems or cavities of the zeolite [14].

7.5.3 IRREGULAR SYSTEMS

When the selectivity changes, that is, when α_B^A changes with the composition, the ion-exchange isotherms show a sigmoid-type form (Figure 7.4c and e) [23].

7.5.4 SYSTEMS WITH PHASE TRANSFORMATIONS

Some zeolites undergo a discontinuous change in the framework structure during ion exchange. A characteristic example is analcite [24]:

$$Na\text{-Analcite} + K^+ \leftrightarrow K\text{-Leucite} + Na^+$$

This means that Na-analcite when exchanged with K^+, becomes K-leucite, that is, as soon as the quantity of K^+ surpasses a definite value, the crystal lattice experiences a structural transformation. On the other hand, K-leucite when exchanged with Na^+ becomes Na-analcite, that is, when the amount of Na^+ exceeds a specific quantity, the crystal lattice goes through a structural transformation.

7.5.5 ELECTROSELECTIVITY

In an ion-exchange process, ions of the same valencies are involved, and there is no change in selectivity with concentration. However, when the exchanged ions are of different valencies, the zeolite then prefers the ion with a higher valence when the concentration is lowered [23]. This effect takes place because the zeolite needs to compensate for the framework charge. Subsequently, in a system where a two-valence cation and a one-valence cation are exchanged, the zeolite needs only one cation to compensate for the two charges, if the two-valence cation is exchanged. Consequently, the zeolite will prefer to exchange the two-valence cation than the one-valence cation.

In Figures 7.5 and 7.6, two examples of electroselectivity for the following ion-exchange process are shown [5,37]:

$$2NH_4^+ + Ni\text{-HEU} \leftrightarrow Ni^{2+} + NH_4\text{-HEU}$$

The experiment shows that at high concentrations, the zeolite is selective for the cation A, that is, NH_4^+, and at low concentrations, the zeolite is selective for cation B, that is, Ni^{2+}. The maximum concentration of NH_4^+ in the electrolytic solution was 2 M (Figure 7.5), and the maximum concentration of NH_4^+ in the solution was 0.01 M (Figure 7.6) [5,37].

7.5.6 EFFECT OF pH OF THE ELECTROLYTIC SOLUTION ON THE ION-EXCHANGE PROCESS

Zeolites behave as a "salt" and experience alkaline hydrolysis. That is, when a zeolite contacts water, the solution turns basic, that is, pH >7; this means that a proton exchange has occurred. It is necessary then to take into account proton concentration as a third component in binary ion exchange [37,38].

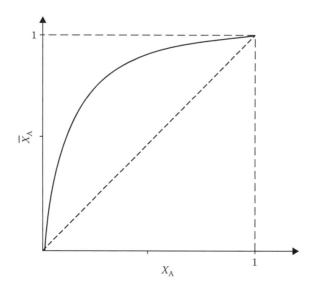

FIGURE 7.5 Electroselectivity in the NH_4^+-Ni-clinoptilolite system where the concentration of NH_4^+ in the electrolytic solution was 2 M.

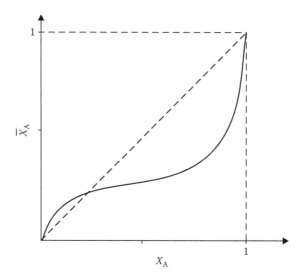

FIGURE 7.6 Electroselectivity in the NH_4^+-Ni-clinoptilolite system where the concentration of NH_4^+ in the electrolytic solution was 0.01 M.

7.6 ION-EXCHANGE HEAT

7.6.1 Ion-Exchange Heat Measurement

For an ion-exchange process in a two-component system (Equation 7.1), it is possible to measure the ion-exchange heat in a flow microcalorimeter. The author and collaborators measured ion-exchange heat [39,40] with the help of a Wadso and Monk type isothermic flow microcalorimeter [41], LKB 2277 [39,40]. Figure 7.7 schematically describes the initial state of the thermodynamic system releasing energy in this experiment. The total energy of this initial state is given by [40]

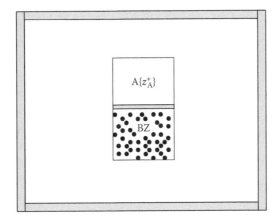

FIGURE 7.7 Thermostat of the heat-flow microcalorimeter showing the sample cell with the solution and the ion exchanger separated by a membrane to prevent ion exchange and to guarantee the initial state of the experiment.

$$E_i^T = n_A^i E_A^1 + n_B^i \bar{E}_B^1 + E_1 + \varepsilon_i^z + \varepsilon_i^s \tag{7.12}$$

where

n_A^i are the moles of cation A in the initial solution
n_B^i are the moles of cation B in the zeolite previous to the exchange
E_A^1 is the partial molar energy of cation A in the solution
$\bar{E}_B^1$ is the partial molar energy of cation B in the zeolite in the initial state
E_1 is the thermostat energy
ε_i^z and ε_i^s are the energies of the liquid and solid solvents, respectively

In the final state, that is, after the exchange, n_B moles of the cation B pass from the solid to the solution, and $\bar{n}_A$ moles pass from the solution to the zeolite.
Consequently [40],

$$E_f^T = \left(n_A^i - \bar{n}_A\right)E_A^2 + \bar{n}_A \bar{E}_A^2 + n_B \bar{E}_B^2 + \left(n_B^i - n_B\right)E_B^2 + E_1 + E_2 + \varepsilon_f^z + \varepsilon_f^s \tag{7.13}$$

By definition, the thermodynamic system defined in Figure 7.7 is an adiabatic system, then $E_f^T = E_i^T$. Therefore, combining Equations 7.12 and 7.13, and making the following approximations [40]

$$E_A^1 \approx E_A^2 = E_A; \quad \bar{E}_A^1 \approx \bar{E}_A^2 = \bar{E}_A; \quad E_B^1 \approx E_B^2 = E_B; \quad \bar{E}_B^1 \approx \bar{E}_B^2 = \bar{E}_B$$

and

$$\varepsilon_i^z \approx \varepsilon_f^z \quad \text{and} \quad \varepsilon_i^s \approx \varepsilon_f^s$$

The integral heat of ion exchange can be expressed as follows [39,40]

$$Q = E_2 - E_1 = \bar{n}_A(E_A - \bar{E}_A) - n_B(E_B - \bar{E}_B) \tag{7.14}$$

Now, if we use the equation of charge balance

$$z_A \bar{n}_A = z_B n_B$$

it is possible to write Equation 7.14 as follows [39,40]:

$$Q = n_B \left[\left\{ \bar{E}_B - \left(\frac{z_B}{z_A} \right) \bar{E}_A \right\} - \left\{ E_B - \left(\frac{z_B}{z_A} \right) E_A \right\} \right]$$

Besides, it is possible to define the differential heat of ion exchange [39] as

$$Q_d = \frac{Q}{n_B} = \left\{ \bar{E}_B - \left(\frac{z_B}{z_A} \right) \bar{E}_A \right\} - \left\{ E_B - \left(\frac{z_B}{z_A} \right) E_A \right\} \qquad (7.15)$$

Now, it is possible to make the following approximation [39]:

$$H_a^B - \left(\frac{z_B}{z_A} \right) H_a^A \approx E_B - \left(\frac{z_B}{z_A} \right) E_A \qquad (7.16)$$

where H_a^B and H_a^A are the hydration heats of cation A and B, respectively. Then

$$Q_d = \frac{Q}{n_B} = \Delta H - DH \qquad (7.17)$$

where the term $\Delta H = \left\{ \bar{E}_B - \left(\frac{z_B}{z_A} \right) \bar{E}_A \right\}$, that is, the heat evolved in the zeolite is evidently related with the electrostatic interaction between the exchanged cations and the zeolite framework. On the other hand, the term $DH = \left\{ E_B - \left(\frac{z_B}{z_A} \right) E_A \right\}$, which is the heat evolved in the solution during the ion-exchange process, as was previously stated, is related with H_a^B and H_a^A, that is, the hydration heats of cation A and B, respectively. Therefore, if $Q_d > 0$ [39], the process is exothermic and cation A is selectively exchanged in relation to cation B for the exchange process described by Equation 7.1.

Table 7.1 [39] shows the differential heats of ion exchange, Q_d [kJ/mol], measured by means of an LKB 2277 heat-flow isothermal microcalorimeter for the ion-exchange reaction

$$\text{ACl}_n + \text{B-Zeolite} \leftrightarrow \text{BCl}_m + \text{A-Zeolite}$$

where ACl_n were 3 M solutions of NaCl, KCl, and NH$_4$Cl for the univalent cations and CaCl$_2$, MgCl$_2$, and NiCl$_2$ for the divalent cations. The zeolites were homoionic mordenites, labeled B-MP, and homoionic clinoptilolites, labeled B-HC, where B was Na$^+$, K$^+$, NH$_4^+$, Ca^{2+}, Mg^{2+}, and Ni^{2+} [39].

From the results reported in Table 7.1, it is possible to conclude that in the case of heulandite and mordenite zeolites with a relatively high Si/Al ratio (see Table 7.1), the selectivity sequence K > NH$_4$ > Na > Ca is related to the hydration heats (see Table 7.2 [43]) with the exception of the pairs NH$_4$–K and Mg–Ca. These conclusions were verified by independent calorimetric studies carried out by other authors [42].

TABLE 7.1
Differential Heats of Ion Exchange, Q_d [kJ/mol]

ACl + B-Zeolite	Q_d [kJ/mol]	n_B [mmol]
NaCl + K-MP	−2.73	0.82
NaCl + K-HC	−3.35	0.51
$CaCl_2$ + K-MP	−4.40	0.52
$CaCl_2$ + K-HC	−3.85	0.62
$MgCl_2$ + Na-MP	−3.81	0.54
$MgCl_2$ + Na-HC	−3.88	0.50
$MgCl_2$ + K-MP	−1.84	0.44
$MgCl_2$ + K-HC	−2.86	0.36
$CaCl_2$ + Na-MP	−2.03	0.60
$CaCl_2$ + Na-HC	−2.12	0.60
NaCl + Ca-MP	5.30	0.53
NaCl + Ca-HC	5.60	0.47
NH_4Cl + Ca-MP	11.0	0.71
NaCl + Mg-HC	32	0.22
KCl + Mg-HC	35	0.46
$MgCl_2$ + Ca-MP	2.25	0.20
NH_4Cl + Na-MP	0.20	1.39
NH_4Cl + K-MP	−1.30	1.02
NH_4Cl + Ni-HC	34	0.34
$NiCl_2$ + NH_4-HC	−3.61	0.20

Note: $Q_d > 0$ for exothermic processes.

TABLE 7.2
Hydration Enthalpies ($-H_a$) vs. Cationic Radius (R) for Some Divalent Cations

Cation	R [Å]	H_a [kJ/mol]
Be^{2+}	0.45	2484
Ni^{2+}	0.70	2096
Co^{2+}	0.70	2010
Cu^2	0.73	2099
Cd^{2+}	0.95	1809
Ca^{2+}	1.00	1579
Sr^{2+}	1.18	1446
Pb^2	1.19	1485
Ba^{2+}	1.35	1309

7.7 ION-EXCHANGE SELECTIVITY IN ZEOLITES

Using the results obtained in the previous section, one can state that as a rule, for the ion-exchange reaction (Equation 7.1) in the case of zeolites with high Si/Al ratio, that is, zeolites with low TEC, ΔH is not very important. Because the aluminum content in the zeolite is low, the negative charge in the framework is low, and, consequently, the framework electrostatic interaction with the exchanging ions is not significant. In this case, A cations with low hydration heats are selectively exchanged by the zeolite [5,39,40], because of the considerable contribution of DH to the differential heat of ion exchange, Q_d. On the other hand, using the results obtained in the previous section one can state that as a rule, for zeolites with low Si/Al ratio, that is, zeolites with high TEC, ΔH is very important as a consequence of the high aluminum content in the zeolite, which creates a high negative charge in the framework, and, as a result, a high framework electrostatic interaction with the exchanging cations. Consequently, A cations with high hydration heats are selectively exchanged by the zeolite [5,39,40], on account of the considerable contribution of ΔH to the differential heat of ion exchange, Q_d.

Sherry, in the last few years, has been generalizing all knowledge about selectivity during ion exchange in zeolites; this author has summarized the selectivity rules as follows [21]: every zeolite has preference for Na^+ instead of Li^+ and NH_4^+ instead of Na^+; zeolites with a low Si/Al ratio have preference for Ca^{2+} and zeolites with a high Si/Al ratio have preference for alkaline cations; zeolites are selective for polarizable cations; and the electroselectivity, molecular sieving, and space limitations rules are valid.

7.8   ION-EXCHANGE KINETICS

In Section 7.3, it was stated that the kinetics of ion exchange is normally represented by the help of the plot of $U(t)$ versus time t (see Figure 7.3). It was also acknowledged that the ion-exchange process in zeolites is described in three steps (see Figure 7.3). Here, we study separately these processes [23–25,44–46].

7.8.1   Interdiffusion in the Adhering Liquid Thin Layer as the Limiting Step

It is possible that the rate-determining process in the kinetics of ion exchange is the film diffusion. Consider a spherical particle encircled by an aqueous solution sphere (see Figure 7.8), in which the zeolite is homoionic at $t = 0$, the electrolytic solution has a very high volume (i.e., $C_2^A \approx$ constant), and the diffusion is stationary and one-dimensional in a direction perpendicular to the zeolite surface [44]. Then, under the conditions discussed above, it is possible to calculate the exchange flux of cation B, that is, J_C^B, as follows [23]:

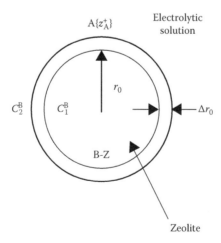

FIGURE 7.8   Liquid thin layer adhering to a zeolite (B-Z) spherical grain, where r_0, sphere radius; Δr_0, film thickness; and $\Delta C_B = C_1^B - C_2^B$ = constant; in which C_1^B and C_2^B are the concentrations of B at the solid–thin layer and at the electrolyte solution, respectively.

$$J_C^B = D_C \frac{\Delta C_B}{\Delta r_0} \tag{7.18}$$

where D_C is the diffusion coefficient in the adhering liquid thin layer.

The mass balance equation in this case can be written as follows [44]

$$-\frac{d\overline{C}_B V_P}{dt} = -\frac{d\left[\overline{C}_B\left(\frac{4\pi r_0^3}{3}\right)\right]}{dt} = 4\pi r_0^2 J_C^B = A_P J_C^B \tag{7.19}$$

where

A_P and V_P are the area and volume of the spherical zeolite grain of radius r_0, respectively

$\overline{C}_B$ is the concentration of B in the zeolite grain

Now, since the volume of the solution is very high, then $C_2^B \approx 0$:

$$-\frac{d\overline{C}_B}{dt} = \left(\frac{3}{r_0}\right)\left(\frac{D_C}{\Delta r_0}\right)C_1^B \tag{7.20}$$

The distribution coefficient between the liquid and the solid phases can be defined as follows [44]:

$$\frac{\overline{C}_B}{C_1^B} = K_B \tag{7.21}$$

Consequently, substituting Equation 7.21 in Equation 7.20, we obtain:

$$\frac{1}{\overline{C}_B}\frac{d\overline{C}_B}{dt} = -\left(\frac{3D_C}{r_0 \Delta r_0 K_B}\right) \tag{7.22}$$

Then, integrating Equation 7.22, it is shown that

$$U(t) = 1 - \exp(-Rt) \tag{7.23a}$$

where

$$R = \left(\frac{3D_C}{r_0 \Delta r_0 K_B}\right) \tag{7.23b}$$

In zeolites in zone I of the ion-exchange kinetics (see Figure 7.3):

$$U(t) \approx Rt$$

This expression is obtained from Equation 7.23a for $t \to 0$. Consequently, in practice, to obtain R, we need to calculate (see Equation 7.4):

$$R = \lim_{t \to 0}\left(\frac{dU(t)}{dt}\right) = \lim_{t \to 0}\left[\left(\frac{1}{Q_A(\infty)}\right)_{t \to 0}\left(\frac{dQ_A(t)}{dt}\right)\right] \tag{7.24}$$

7.8.2 INTERDIFFUSION OF A AND B IN ZEOLITE CRYSTALS AS THE LIMITING STEP

The limiting step in the kinetics of ion exchange in the zeolite is the interdiffusion of the electrolyte ions $A\{z_A^+\}$ and ions of the species B [24]. In the case where the solid ion–exchanger particle is spherical (see Figure 7.9) and the particle diffusion control is the rate-determining process, then Fick's second law equation in spherical coordinates is [47]

$$\frac{\partial \overline{C}_B}{\partial t} = D_e \left[\frac{\partial^2 \overline{C}_B}{\partial r^2} + \frac{2}{r_0} \left(\frac{\partial \overline{C}_B}{\partial r} \right) \right] \tag{7.25}$$

If the electrolytic solution has a very high volume and, consequently, the concentration of the cation A is practically constant throughout the ion-exchange process, the experimental setup exhibits the following initial and boundary conditions [44]:

$$\overline{C}_B(r,0) = \overline{C}_B^0 \quad \text{for} : 0 < r < r_0 \tag{7.26a}$$

and

$$\overline{C}_B(r_0,t) = \overline{C}_B(\infty) \tag{7.26b}$$

Then, the solution of Equation 7.25 is given by [44–46]:

$$U(t) = 1 - \frac{6}{\pi^2} \sum_{n=1}^{\infty} \frac{1}{n^2} \exp\left(-\frac{D_e t \pi^2 n^2}{r_0^2} \right) \tag{7.27}$$

For the calculation of the effective diffusion coefficient, D_e, a procedure where the series in Equation 7.27 is truncated can be applied, and taking the first four terms, that is,

$$U(t) \approx 1 - \frac{6}{\pi^2} \sum_{n=1}^{4} \frac{1}{n^2} \exp\left(-\frac{D_e t \pi^2 n^2}{r_0^2} \right) \tag{7.28}$$

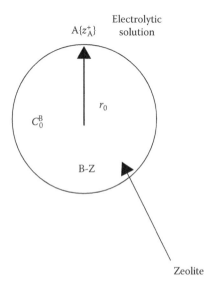

FIGURE 7.9 Interdiffusion in the zeolite spherical grain.

TABLE 7.3
Parameters Characterizing the Exchange of NH_4^+ and Ni^{2+} in the CMT-C Sample

Cation	R [s^{-1}]	D_e [m^2/s] $\times$ 10^{16}	TEC [mequiv/g]
NH_4^+	0.021	0.012	2.04
Ni^{2+}	0.013	6.4	1.95

which is numerically fitted to the experimental data. The fitting process is normally carried out with a nonlinear regression analysis software based on a least-square procedure, which allows calculation of the best-fitting parameters, that is, the numerical values of the relation between the effective diffusion coefficient and the square of the particle radius, D_e/r_0^2, and the standard errors (See Sections 5.7.7 and 5.9.2). An approximate solution, which fits the whole range $0 \leq U(t) \leq 1$, is given by [46]

$$U(t) \approx \left(1 - \exp\left[-\frac{D_e t \pi^2}{r_0^2}\right]\right)^{\frac{1}{2}} = \left(1 - \exp[-Bt]\right)^{\frac{1}{2}}$$

where

$$B = \frac{D_e \pi^2}{r_0^2}.$$

7.8.3 EXPERIMENTAL RESULTS

The reported values of the diffusion coefficients of cations in zeolites are around 10^{-13}–10^{-15} [m^2/s] [5,37,48], and the factors which affect the velocity of cationic exchange are

- Movement of monovalent cations—Monovalent cations move faster in zeolites and have lower activation energies than divalent cations [5,23].
- Structure of the zeolite or, more specifically, the dimensions of zeolite channels and cavities—Cations move faster in mordenite than in clinoptilolite and faster in clinoptilolite than in erionite [5,48].
- Cationic radius—Cation-exchange velocity decreases with the increase of the cationic radius [23].
- Concentration of the exchange solution—The velocity of the ionic exchange increases with increments of the solution concentration [48].

Here, we report kinetic data measured during the ion-exchange process in the NH_4^+–Ni^{2+} system in the sample CMT-C [5,37]. CMTC-C was obtained from deposits located in Tasajeras, Cuba (see Table 4.1) [5]. The parameters characterizing the kinetics of ion exchange of NH_4^+ and Ni^{2+} in the sample CMT-C [5,37], that is, R, D_e and the TEC, are reported in Table 7.3.

7.9 PLUG-FLOW ION-EXCHANGE BED REACTORS

7.9.1 INTRODUCTION

The plug-flow model signifies that the fluid velocity profile is plug shaped, and is uniform at all radial positions, as explained earlier for the plug-flow adsorption reactor [38,49–53] (see Section 6.11.2). The fixed-bed ion-exchange reactor is packed randomly with particles from a solid ion exchanger that are clean, or have just been regenerated. The ion-exchange process is supposed to be very fast relative to the convection and diffusion effects; subsequently, local equilibrium will exist close to the ion-exchange beads. Further assumptions are that no chemical reactions occur in the column, and that only mass transfer by convection is significant.

The operation of an ion exchanger in dynamic conditions is characterized by the TEC, the selectivity and the kinetics of the ion-exchange process, and also by the mechanical properties of the exchanger materials. The zeolite bed in the plug-flow ion-exchange bed reactors (PFIEBR) (Figure 7.10) can be divided in three zones: (I) the equilibrium zone, (II) mass transfer zone (MTZ) with a length D_0, and (III) the unused zone [25,49]. The PFIER has a cross sectional area, S, column length, D, and ion-exchanger mass in the bed, M (see Figure 7.10).

This PFIEBR operates in a steady-state regime and a volumetric flow rate [49]

$$F = \frac{\Delta V}{\Delta t} = \frac{\text{Volume}}{\text{Time}}$$

of an aqueous solution with an initial concentration, C_A^0, flows through it, (see Equation 7.1). The output of the operation of the PFIER is a breakthrough curve [25,49,51]. Figure 7.11 shows a breakthrough curve, where C_A^0 is the initial concentration, C_A^e is the breakthrough concentration, V_e is the fed volume of the aqueous solution of the electrolyte solution to breakthrough, and V_b is the fed volume to saturation. This is a response curve where the relation between concentration at the exit of the packed-bed ion-exchange reactor and time is described.

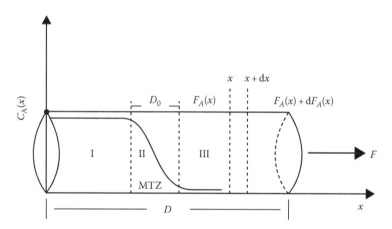

FIGURE 7.10   Schematic view of the PFIEBR.

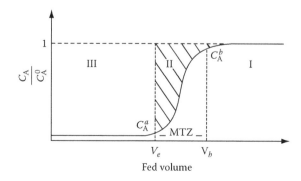

FIGURE 7.11   Breakthrough curve.

The length of the MTZ, D_0, is defined by the following expression (see Figure 7.11) [52]:

$$D_0 = 2D \left[\frac{V_b - V_e}{V_b + V_e} \right] \tag{7.29}$$

The volume of the empty bed is $V_B = \varepsilon V$, where V is the bed volume, and ε is the fraction of the free volume in the bed, and the interstitial fluid velocity, u, is defined as [49,50]

$$u = \frac{F}{S} \tag{7.30}$$

The contact or residence time of the fluid, τ, passing though the reactor is calculated with the equation:

$$\tau = \frac{V_B}{F} \tag{7.31}$$

Other parameters characterizing a PFIER are the column breakthrough capacity, B_C, the column saturation capacity, S_C, and the column efficiency, E, of the PFIEBR, which are calculated with the following equations [51]:

$$B_C = \frac{C_A^0 V_e}{M}, \quad S_C = \frac{C_A^0 V_b}{M}, \quad \text{and} \quad E = \frac{B_C}{S_C} \tag{7.32}$$

7.9.2 PARAMETERS FOR THE DESIGN OF A LABORATORY PFIER

For the dynamic ion-exchange reactor to work properly, it should satisfy the following features, as described in Section 6.11.1 for the case of adsorption.

Residence time: Since ion exchange is a slow process, the fluid contact or residence time should be long enough for the molecular transport to the ion-exchange sites. Therefore, it is possible to tryout residence times around the following figures $1\,s < \tau < 10\,s$.

Particle size: Granular particles with large particle sizes should be used, since if the particle size is small, a pressure drop inside the reactor will be generated. As in the case of adsorption, the particle size depends on the dimension of the reactor. A good practice is to construct the reactor following the approximate relation $d_R/d_P \geq 10$, where d_R is the reactor diameter and d_P is the particle size. For testing a material in the laboratory, $d_R/d_P \approx 10$ is a good choice.

Reactor longitude: Since the applied residence times are long, huge facilities are required to attain, in some cases, the required capacities. Afterward, to maintain the proportions in the reactor dimensions, the following rule can be applied $D/d_R \leq 10$, where D is the reactor length and d_R is the reactor diameter. For testing an ion-exchange material in the laboratory, $D/d_R \approx 10$ is a good option.

These features are only approximate design criteria, which are exclusively used for laboratory tests of ion exchangers.

7.10 CHEMICAL AND POLLUTION ABATEMENT APPLICATIONS OF ION EXCHANGE IN ZEOLITES

7.10.1 INTRODUCTION

Aluminosilicate zeolites are microporous crystalline materials recognized as excellent ion exchangers [1–3,5,20,21,51,53–85]. Particularly, natural zeolites have been of interest to geologists, chemists, engineers, veterinarians, medical doctors, physicists, and other specialists in different fields on account of their remarkable properties. However, the rapid advance in the use of synthetic zeolites has overtaken the widespread commercial utilization of natural zeolites [5]. However, the use of natural zeolites is unavoidable in the future in order to deal with several environmental problems, because of its low price and widespread availability in many countries [5].

Ion exchange is perhaps the most useful attribute of natural zeolites. It is important that this effect is used from the viewpoint of zeolite modification [5], as was previously discussed (see Chapters 2 and 3), and on the other hand, ionic exchange is employed in industrial, agricultural, aquacultural, and pollution abatement applications of natural zeolites. The applications of ion exchange in natural zeolites, after the rediscovery of these minerals in the late 1950s, started with the studies carried out by Ames [7] and coworkers. They discovered that chabazite, clinoptilolite, and mordenite are excellent exchangers for Cs^+, Sr^{2+}, Rb^+, K^+, and NH_4^+. On the basis of this knowledge and applying clinoptilolite, chabazite, and phillipsite, different processes for the treatment of municipal wastewater, purification of radioactive wastewater, applications in agriculture, processing of wastewater with high chemical oxygen demand (COD), animal nourishment, aquaculture, and the removal of heavy metals from wastewater have been developed [5]. However, the major use of ion exchange in zeolites today is the application of zeolite A in detergents for the softening of laundry water [54,55].

7.10.2 HEAVY METAL REMOVAL FROM WASTEWATER

Heavy metals (Pb, Cd, Co, Cr, Zn, Ni, Fe, Hg, Mn, Sn, Ti, W, V, and Zr) in water and wastewater turn out to be a serious environmental problem on account of the highly toxic effect of these metals on living organisms and because they are prohibited in some manufacturing operations [5,38,58–68]. The common processes for the elimination of heavy metals from wastewater are by its precipitation with lime or soda [56]. However, this procedure generates large amounts of sludge, resulting in unacceptably high levels of residual metal salts and requiring long settling times [58]. The ion-exchange operation with low-cost minerals such as natural zeolites can be an adequate solution in many cases [5,58–68]. Since the exchange reaction is very fast, specific, and characterized by stoichiometry, the pH of the original wastewater remains unaffected or moves toward neutrality after the treatment with zeolites, and the regeneration of the exhausted bed is relatively easy [58].

The application of natural zeolites in heavy metal removal is described now. The methodology consists of the development of a process for heavy metal removal from wastewater using dynamic ion exchange in natural zeolite columns [38,53].

For designing a canister system for heavy metal removal (see Figure 7.12) [38,53], a simple phenomenological description of dynamic ion exchange in zeolite bed reactors was worked out, which allows for the design of modular canister ion-exchange bed reactors for applications in heavy metal removal from wastewater.

FIGURE 7.12 Laboratory assembled modular canister setup composed of (a) ionic exchange column, (b) pump, (c) inlet valve, (d) elbow, (e) T connection, and (f) column valve.

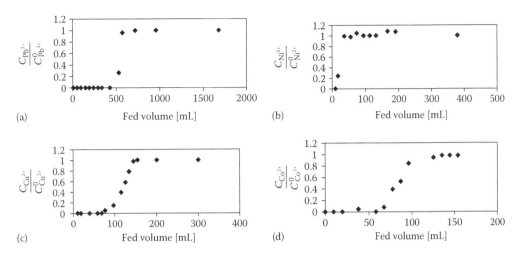

FIGURE 7.13 Experimental breakthrough curve of aqueous solutions of (a) $Pb(NO_3)_2$, (b) $Ni(NO_3)_2 \cdot 6H_2O$, (c) $Cu(NO_3)_2 \cdot 2.5\ H_2O$, and (d) $Co(NO_3)_2 \cdot 6H_2O$.

The experimental measurement of some of the parameters that describe the operation of a real PFIEBR consisting of natural sodium clinoptilolite ion-exchange columns for removing Pb^{2+}, Ni^{2+}, Co^{2+} and Cu^{2+} from water solutions (see Figure 7.13a through d) was carried out.

The natural zeolite tested, sample SW, mined from deposits at Sweetwater, Wyoming, was provided by ZeoponiX Inc., Louisville, Colorado [86]. The natural zeolite sample SW was studied in the homoionic form. This term denotes that the zeolite contains a prevalent cation exchanged in the cationic sites of its channels and cavities [86]. The sample SW was refluxed (at 373 K) five times, each for a 4 h period, in a 2 M NaCl solution to produce the sample Na-SW, that is, the Na-clinoptilolite ion exchanger [5,53]. In Table 7.4, the chemical composition of the sample determined with EDAX in a JEOL Model 5800LV SEM is reported. The direct experimental evaluation of the total cation-exchange capacity TCEC(SW-d) of the natural CSW clinoptilolite sample was carried out with the help of the following methodology: 1 g of the natural zeolite sample SW was refluxed (at 373 K) once for 6 h in 1 L of a 2 M NH_4Cl solution to produce a homoionic sample, NH_4-SW [5,53].

The mineralogical phase composition of the sample SW [86] (in wt %) is 90% ± 5% clinoptilolite and 10% ± 5% others, which include montmorillonite (2–10 wt %), quartz (1–5 wt %), calcite (1–6 wt %), feldspars (0–1 wt %), magnetite (0–1 wt %), and volcanic glass (3–6 wt %). Employing this sample and a pure clinoptilolite, whose TCEC fluctuates between 2.0–2.2 mequiv/g depending on the Si/Al relation of the clinoptilolite monocrystal, it is possible to indirectly evaluate the total cation-exchange capacity of the sample SW as follows:

$$\begin{aligned} TCEC(SW\text{-}i) &\approx [TCEC\ of\ a\ clinoptilolite\ monocrystal] \\ &\quad \times [sample\ mineralogical\ phase\ composition] \\ &= (2.1 \pm 0.1) \times 0.9\ mequiv/g = 1.9 \pm 0.1\ mequiv/g \end{aligned}$$

The direct experimental evaluation of the TCEC of the natural SW clinoptilolite sample was carried out with the help of a methodology previously described, which resulted in the measurement of

TABLE 7.4
Chemical Composition (in wt %) of the Na-CSW Sample

Sample	O	Si	Al	Fe	Ca	Mg	Na	K
Na-CSW	45.25	37.48	9.25	1.06	0.05	0.58	5.46	0.90

the cationic composition of the sample SW in milliequivalent per gram of Na+, K+, Ca²⁺, and Mg²⁺ present in the mineral. The obtained results are reported in Table 7.3, resulting in a TCEC (SW-d) value of 2.0 ± 0.1 mequiv/g [53].

The experimental measurement of some of the operational parameters for the PFIER was carried out with the help of SPECTRUM disposable, cylindrical polystyrene minicolumns with an internal diameter (d) of 0.732 ± 0.001 cm, which implies a cross sectional area (S) of 0.421 ± 0.002 cm², and a total length of 7 cm [38,53]. The columns were prepared with a bed length (D) of 4.20 ± 0.02 cm filled with crushed and sieved Na-SW zeolite. The free bed volume was $V_B \approx 0.8 \pm 0.1$ cm³. In the experimental PFIER, volumetric flows of dilute aqueous solutions of Pb(NO₃)₂, Cu(NO₃)₂, Co(NO₃)₂, and Ni(NO₃)₂ with initial concentrations that are reported below were passed [38,53]. The used salts were pure per the analysis products provided by Fisher.

Figure 7.13a shows the experimental breakthrough curve obtained during the operation of the tested PFIEBR filled with a mass (M) of 1.5 g of homoionic Na-clinoptilolite, with a grain size, ϕ, of 0.6–0.8 mm, a flow rate (F) of 0.8 [cm³/min] of aqueous solutions of Pb(NO₃)₂ with a Pb²⁺ initial concentration, $C^0_{Pb^{2+}} = 0.45$ [mg/cm³][53]. In Figure 7.13b, the experimental breakthrough curve obtained during the operation of the tested PFIEBR filled with a mass of $M = 1.5$ g of Na-clinoptilolite with a grain size, ϕ, of 0.6–0.8 mm, a flow rate of $F = 0.32$ [cm³/min] of aqueous solutions of Ni(NO₃)₂•6H₂O with a Ni²⁺ initial concentration, $C^0_{Ni^{2+}} = 0.65$ [mg/cm³], is reported [53]. In Figure 7.13c, the experimental breakthrough curve achieved during the performance of the tested PFIEBR loaded with 1.5 g of Na-clinoptilolite with a grain size, ϕ, of 0.6–0.8 mm, a flow rate of $F = 0.32$ [cm³/min] of aqueous solutions of Cu(NO₃)₂•2.5H₂O with a Cu²⁺ initial concentration, $C^0_{Cu^{2+}} = 0.35$ [mg/cm³], is reported [53]. In Figure 7.13d, the experimental breakthrough curve acquired throughout the action of the tested bed reactor filled with 1.5 g of Na-clinoptilolite with a grain size, ϕ, of 0.6–0.8 mm, a flow rate of $F = 0.32$ [cm³/min] of aqueous solutions of Co(NO₃)₂•6H₂O with a Co²⁺ initial concentration, $C^0_{Co^{2+}} = 0.40$ [mg/cm³], is shown [53].

In Table 7.5, the operational parameters of the tested PFIEBR, where τ was calculated with Equation 7.31, and B_C, D_0, and E were calculated with the help of the experimentally obtained breakthrough curves using Equation 7.32 and the established operational parameters F and ϕ, are shown.

In order to perform the design of a real canister system for heavy metal removal, a simple phenomenological description of dynamic ionic exchange in zeolite bed reactors is provided [38]. In this sense, the mass balance equation for the PFIEBR is [49]

TABLE 7.5
Operational Parameters τ, B_C, D_0, E, F, and ϕ of the Tested PFIEBR

Cation	τ [s]	B_C [mequiv/g]	D_0 [cm]	E [%]	F [cm³/min]	ϕ [mm]
Pb²⁺	75	0.42	3.5	32	0.80	0.8–2
Pb²⁺	60	1.25	1.0	75	0.80	0.6–0.8
Ni²⁺	<60	0.00	>3.7	0	0.80	0.8–2
Ni²⁺	148	0.30	3.3	40	0.32	0.6–0.8
Ni²⁺	60	0.27	3.4	35	0.80	0.6–0.8
Co²⁺	148	0.27	2.3	51	0.32	0.6–0.8
Cu²⁺	148	0.57	2.1	52	0.32	0.6–0.8
Cu²⁺	75	0.45	2.4	50	0.80	0.6–0.8

Note: The errors in τ, B_C, D_0, E, and F are ± 10 s, 0.05 mequiv/g, ± 0.3 cm, $\pm 5\%$, and 0.02 cm³/ min, respectively.

$$IN - OUT = ACCUMULATION$$

This means that the transported cations, (A^{zA+}; see Equation 7.1), into the reactor {IN} minus the transported cations outside the reactor, {OUT}, is equal to the accumulation of cations of the atom A (AZ; see Equation 7.1) in the zeolite bed, {ACCUMULATION}. In Figures 7.10 and 7.11, the concentration profile, {$C(x)$}, of A^{zA+} in an aqueous solution through the PFIEBR is shown.

The accumulation rate of cations of atom A in the zeolite grains can be, in general, calculated with the help of the following relation [49,50]

$$r_A = k[C(x)]^n \tag{7.33}$$

where

r_A [mass/volume − time] is the rate of the reaction producing the accumulation of the cations of atom A in the zeolite

k is the rate coefficient

If we define (see Figure 7.10) [49,50]

$$F_A(x) = C(x)F \tag{7.34}$$

it is possible to express numerically the mass balance equation by (see Figure 7.10) [49,50]:

$$F_A(x) - [F_A(x) + dF_A(x)] = r_A dV = r_A S dx \tag{7.35}$$

Now, to explain the operation of the PFIEBR, it is proposed that the interdiffusion in the adhering liquid thin layer is the rate-determining step, then, it is possible to consider that $n = 1$ in Equation 7.33, since for this transport process the diffusion rate, k, is proportional to concentration [38]. This approach is based on the assumption that states that the rate-determining process during the dynamic ionic exchange in zeolite columns determines the diffusion in the zeolite secondary porosity, that is, the transport process in the macro- and mesoporosities formed by the matrix inserted between zeolite crystals and the diffusion in the zeolite primary porosity, that is, in the cavities and channels which constitute the zeolite framework [38]. This fact is experimentally justified later. With the help of Equations 7.33 through 7.35, we obtain:

$$-FdC(x) = SkC(x)dx \tag{7.36}$$

$$-\frac{dC(x)}{C(x)} = \frac{kS}{F}dx \tag{7.37}$$

With the help of the definition of the interstitial fluid velocity, $u = \dfrac{F}{S}$, Equation 7.37 can be written as follows

$$-\frac{dC(x)}{C(x)} = \frac{k}{u}dx \tag{7.38}$$

and Equation 7.38 can be integrated in the MTZ as follows

$$\int_{C_A^0}^{C_A^e} -\frac{dC(x)}{C(x)} = \int_0^{D_0} \frac{k}{u}dx \tag{7.39}$$

resulting in

$$D_0 = \frac{u}{k} \ln\left(\frac{C_A^0}{C_A^e}\right)$$ (7.40)

Consequently, combining Equations 7.29 and 7.40, we get

$$2D\left[\frac{V_b - V_e}{V_b + V_e}\right] = \frac{u}{k} \ln\left(\frac{C_A^0}{C_A^e}\right)$$ (7.41)

which is the equation that we propose for calculating the kinetic parameter, k, that defines this process, using data obtained from the experimental breakthrough curves (Figure 7.13) for the dynamic ionic exchange in a zeolite bed placed in a PFIEBR (see Table 7.6). As it is obvious from Table 7.6, $k = 0.018 \pm 0.003\,\text{s}^{-1}$ for Ni^{2+}, a value similar to that determined for the parameter R (see Table 7.3). This fact supports our previous assumption that the interdiffusion in the adhering liquid thin layer is the rate-determining step in the present case [38].

Our previous results suggest that the order of selectivity for the studied dynamic exchange processes is $Pb^{2+} > Cu^2 > Co^{2+} > {}^+Ni^{2+}$. Therefore, the longest length of the MTZ can be calculated by substituting the numerical value of k for Ni^{2+}, which is the less selectively exchanged cation in the studied set, in Equation 7.41, as follows:

$$D_0 \approx \frac{u}{0.018} \ln 100$$ (7.42)

and with the help of a very simple calculation, it is possible to get the following semiempirical equation

$$D_0 \approx Cu \ [\text{m}]$$ (7.43)

where

$$C = [260 \pm 100] \ [\text{s}]$$ (7.44)

TABLE 7.6

Operational Parameter k and the Previously Reported Parameters F and ϕ of the Tested PFIEBR

Cation	k [s⁻¹]	F [cm³/min]	ϕ [mm]
Pb^{2+}	0.04	0.80	0.8–2
Pb^{2+}	0.14	0.80	0.6–0.8
Ni^{2+}	—	0.80	0.8–2
Ni^{2+}	0.018	0.32	0.6–0.8
Ni^{2+}	0.040	0.80	0.6–0.8
Co^{2+}	0.025	0.32	0.6–0.8
Cu^{2+}	0.027	0.32	0.6–0.8
Cu^{2+}	0.058	0.8	0.6–0.8

Note: The error in k is $\pm 0.003\,\text{s}^{-1}$.

The method to calculate the dimensions of the modular canister, or a set of modular canisters to be used, is to employ Equations 7.42 and 7.43 to calculate the maximum D_0, and also the previously estimated operational parameters of the PFIEBR, that is, $\tau \approx 60\,s$, and $\phi \leq 0.1\,d$, where, ϕ is the ion-exchanger particle diameter and d is the reactor internal diameter and our knowledge about the total cation-exchange capacity of the packed natural zeolite.

This method can be used, with a relatively high degree of generality, for the design of modular canister ionic exchange bed reactors packed with natural zeolites, such as clinoptilolite, chabazite, phillipsite, mordenite, and gismondine in sodium form, since Ni^{2+} (due to thermodynamic reasons) is one of the less selectively exchanged heavy metal cations in natural zeolites. As previously stated, cations with low hydration heats are selectively exchanged by zeolites [5,38,39].

7.10.3 Recovery of Ni^{2+} from the Waste Liquors of a Nickel Production Plant

For the recovery of Ni^{2+} from the waste liquors (WL) of the nickel plant located in Nicaro, Cuba, the following methodology was proposed [5,37,60] (see Figure 7.14): carbonate–ammonium liquors (fresh liquors [FL]) are used for the exchange of the natural zeolite to the ammonium form (first cycle), and for the activation of the exhausted natural zeolite after the removal of nickel from the WL previously passed through the zeolite column. The exchanged nickel is recovered and the obtained NH_4-zeolite is heated to get H-zeolite, thereby recovering the ammonium. The process is thus repeated cyclically.

7.10.4 Municipal Wastewater Treatment

Sewage treatment has both ecological and economical importance. Ammonium and phosphate ions are the primary inorganic contaminants in municipal wastewater and these ionic compounds are responsible for the phenomenon of eutrophication. In common sewage treatment, ammonium is removed by biological denitrification and phosphates are precipitated with the help of aluminum and iron salts [56]. However, the use of low-cost natural zeolites has been found promising for

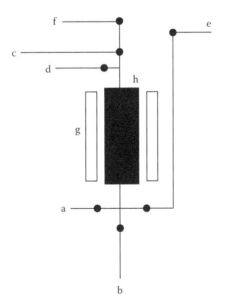

FIGURE 7.14 Sketch of a plant proposed for the treatment of the waste liquors (WL) of a nickel production plant at Nicaro, Holguin, and Cuba. (a) Fresh liquor (FL), (b) air supply, (c) waste liquor (WL), (d) air + NH_3, (e) waste liquor, (f) FL + Ni^{2+} + Co^{2+}, (g) heaters, and (h) zeolite columns.

ammonium removal, based on the high selectivity of chabazite, clinoptilolite, and mordenite for NH_4^+ in comparison with organic ion- exchange resins [5,7,9–11]. In this regard, in a dilute electrolyte solution, such as municipal wastewater, NH_4^+ breaks in an organic resin column before Ca^{2+} and Mg^{2+}; for a natural, zeolite this effect is reverse [7] and it reduces the amount of regenerating solution in the case of natural zeolites. Furthermore, zeolites make the disposal of spent regenerate easier.

Because of the properties of natural zeolites as ion exchangers, these materials have been considered as an alternative technology for ammonium removal from municipal wastewater, and large-scale facilities were constructed to accomplish this goal [5,10,11], for example, Tahoe-Truckee, California (22,500 m³/day); Upper Occoquan, Virginia (85,000 m³/day); Denver, Colorado (8,300 m³/day); Manfredoni, Italia (11,000 m³/day), South Lyon, Michigan [11], and Budapest, Hungary [10]. In the plants constructed by Hungarian scientists (Figure 7.15), clinoptilolite is used as a flocculent before the sewage enters the aerator (between the items b, c and d, f; see Figure 7.15; and later as ammonium ionic exchanger (in Figure 7.15)) for the removal of ammonium. For the purpose of comparison, the plant is duplicated up to the points (e) and (f) in Figure 7.15, where in one of the branches, the zeolite column and the rest of the equipment for ammonium recovery for fertilizer production are added [10].

7.10.5 RADIOACTIVE WASTEWATER TREATMENT

Another ion-exchange application of natural zeolites is the removal of radioactive ions from wastewater [7,17,18,20,57,70,71]. Chabazite, clinoptilolite, and mordenite selectively exchange radioactive Cs^+ and Sr^{2+} from solutions [5,7,17,18,20,57,70,71]. In addition, the high temperature, because of the activity of these radionuclides, and the effect of gamma radiation do not affect the performance of natural zeolites, which is the case for organic ion-exchange resins [72].

Besides, radioactive cobalt is a common radionuclide in liquid wastes from nuclear facilities, and natural erionite is a good exchanger for $^{60}Co^{2+}$. Studies carried out with this material reveal its possibilities in the elimination of radioactive cobalt from solutions [73]. The exchange of $^{232}Th^{4+}$ in natural clinoptilolite and mordenite from liquid solutions has also been studied [74]. All these peculiarities of natural zeolites make it suitable to be exploited as natural barriers for the migration of radionuclides and, consequently, natural zeolite deposits can be potential sites for a radioactive waste repository [19]. These materials have also been employed for the removal of radionuclides from polluted areas in places where nuclear power station accidents have occurred or where

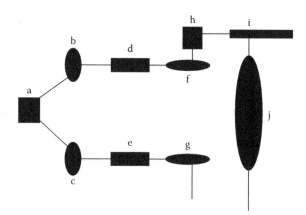

FIGURE 7.15 Graphic representation of a plant for sewage treatment using clinoptilolite as the ionic exchanger and flocculent. The plant is composed of (a) buffer tank, (b, c) primary settling tank, (d, e) aerators, (f, g) secondary settling tanks, (h) clinoptilolite ion exchanger, (i) stripper, and (j) absorber.

nuclear weapons are located, such as Chernobyl in Ukraine, Chelyabinsk in Russia, Semipalatinsk in Kazakhstan, and Kranoyarsk in Siberia [20].

One more important application is the eradication of radionuclides from animal products such as milk and meat produced by farm animals exposed to the contamination of radionuclides [69,70], and the lowering of the transference of radionuclides from soils to plants raised in contaminated areas [71].

7.10.6 CATALYTIC EFFECT OF PROTON EXCHANGE IN NATURAL ZEOLITES IN BIOGAS PRODUCTION DURING ANAEROBIC DIGESTION

In the absence of oxygen, in a closed reactor, anaerobic bacteria ferment organic matter into methane and carbon dioxide, a mixture named biogas. Biogas is produced through the activity of common methane bacteria that cause anaerobic degradation over a broad temperature range from 10°C to over 100°C.

It is well known that zeolites neutralize biological media by proton exchange [5,75]. This can be very useful in anaerobic wastewater treatment on account of the lowering of the pH of the reactor feed, owing to the proton production of the anaerobic bacteria metabolism during anaerobic digestion [5,75]. Zeolites used as packing material in anaerobic filters used in the purification of wastewater with a high volumetric organic loading can be effective in the decrease of the chemical oxygen demand and acidity, intensifying the biogas production [5,75]. In order to prove this hypothesis, a cylindrical anaerobic reactor with height, $h = 1.1$ m, and diameter, $d = 0.095$ m [75], was assembled. The reactor was packed with rod-shaped PVC pellets, ceramic rings, and $0.01 \times 0.01 \times 0.02$ m^3 zeolite gravel. The reactor free volume was 80% for the PVC pellets, 83% for the ceramic rings, and 70% for the zeolite gravel. Subsequently, the reactor was fed with baker's yeast wastewater having the following characteristics: COD 20 [g/L]; BOD 10 [g/L]; acidity 1.25 [g/L]; alkalinity 2.95 [g/L]; pH = 5.5 and a COD:N:P relation of 360:9:1 at residence times of 1, 3, 5, and 6 days for a 1-year measuring input and output of COD, BOD, acidity, alkalinity, and daily biogas production [75], where the daily biogas production was measured in $\dfrac{\text{Volume of biogas}}{\text{Reactor volume/day}}$.

Figure 7.16 [5,75] shows the effect of the volumetric organic loading, that is, B_v, expressed in kilograms of COD per cubic meter of reactor (m^3) day (d), that is, kg/m^{3}day, in the percentage of COD removal, E [%], where

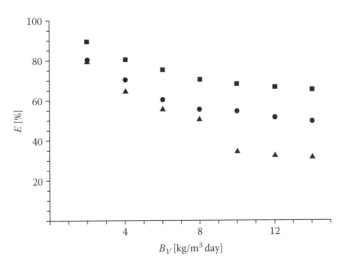

FIGURE 7.16   Plot of E [%] vs. B_V for PVC pellets (▲), ceramic rings (●), and CMT zeolite gravel (■).

$$E\,[\%] = \frac{COD_{Input} - COD_{Output}}{COD_{Input}} \times 100$$

for PVC pellets, ceramic rings, and zeolite gravel with a particle diameter, φ, in the range $10 < \varphi < 20\,mm$, of the natural zeolite CMT [75] (see Table 4.1). Table 7.7 shows the daily biogas production for a reactor packed with PVC pellets, ceramic rings, and zeolite gravel [75].

As shown in Figure 7.16 and Table 7.7, the effect of proton exchange and biofilm fixing by the zeolite radically increased the process efficiency and biogas production [75].

7.10.7 Zeolite Na-A as Detergent Builder

A builder is a compound that removes calcium and magnesium ions normally present in water, and, as a result, reduces the concentration of surfactants required to carry out the detergent action. Currently, the builder mainly used in practice is sodium tripolyphosphate. However, phosphates are plant nutrients; and provoke eutrophication in lakes and streams which receive municipal wastewater contaminated with detergent residuals. Consequently, the use of phosphates in detergents has been restricted.

In agreement with Löwenstein's rule, in a zeolite not more than half the atoms in the crystal lattice can be Al atoms, then, the maximum Si/Al ratio for an aluminosilicate zeolite is unity. In this case, the zeolite will have the maximum ion-exchange capacity. Zeolite Na-A has an Si/Al ratio of 1. Zeolite Na-A is a good candidate that can be applied as a detergent builder. In fact, Na-A zeolite is broadly applied as a detergent builder [54,55,76–78], because of its high exchange capacity for Ca^{2+} and Mg^{2+}, fast exchange kinetics for Ca^{2+} and Mg^{2+} [76], small particle size, and cubic shape of the crystals [78] with rounded corners and edges, which are easily removed on rinsing.

7.10.8 Aquaculture

The presence of NH_4^+ has been observed in a variety of agricultural, domestic, and industrial wastewaters, and in effluents from aquacultural activity [5,79]. The existence of a huge quantity of NH_4^+ in these wastewaters can produce eutrophication in the receiving water body [5,49,79]. Frequently, the NH_4^+ concentration in an aquacultural fishpond increases beyond acceptable levels necessary for a typical aquacultural activity. In this case, NH_4^+ elimination from the wastewaters is desirable to maintain appropriate water quality for a normal aquacultural life [79].

Several techniques for NH_4^+ removal exists, such as stripping, activated carbon adsorption, ozonation, and nitrification [49,79]. However, notwithstanding the fact that all these methods are effective, as well as having some setbacks, methods capable of effective NH_4^+ elimination from wastewaters are highly desirable [5,79].

Chabazite, clinoptilolite, and mordenite are selective for NH_4^+ exchange. The use of clinoptilolite filters has proven effective in removing NH_4^+ from hatchery effluent waters [5,80–85]. However, the addition of a filtering system can be expensive; hence, natural zeolite can be added to the feed of

TABLE 7.7
Daily Biogas Production for the Different Supports at Different Residence Times

Residence Time [days]	1	3	5	6
PVC pellets	2.7	1.8	1.3	1.3
Ceramic rings	4.0	3.1	1.4	1.2
CMT gravel	5.1	3.4	1.4	1.3

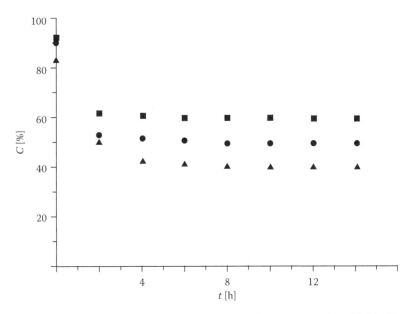

FIGURE 7.17 Experimental test of the reduction of the ammonium concentration C[%] in 5 L vessels, containing 500 g of young *Tilapia* fish, to which 40% (▲), 20% (●), and 10% (■) of the zeolite sample CMT were added. The concentration C[%] was measured using for comparisson a container with 0% of zeolite added where the ammonium concentration was taken as $C = 100\%$.

hatchery-reared fish, and the zeolite, after passing through the gastrointestinal tract, will reduce NH_4^+ levels [84].

Another method for the reduction of NH_4^+ levels in fish culture installations is the addition of clinoptilolite directly to fish culture tanks [5,85]. Figure 7.17 shows a decrease in the ammonium concentration, C[%], versus time in a 5 L vessel containing 500 g of *Tilapia* juvenile fishes, where 40%, 20%, and 10% of the zeolite sample CMT were added, and where the percent (%) decrease is represented by [85]

$$\% = \left(\frac{\text{Mass of zeolite}}{\text{Mass of fish}} \right) \times 100$$

The concentration was measured by comparing against a container (% = 0) having no zeolite, where the ammonium concentration in this container was taken as C [%] = 100%. A similar test was carried out in the aquaculture experimental station in Manzanillo, Cuba [85]. Even though natural zeolites are inexpensive, it has limited NH_4^+ removal capacity and cannot be easily regenerated when exhausted; therefore, synthetic ion-exchange resins have been proposed for NH_4^+ removal [79].

7.11 APPLICATIONS OF OTHER CRYSTALLINE INORGANIC ION EXCHANGERS

This section describes the applications of some important crystalline inorganic ion exchangers, such as the hydrotalcites, titanates, and zirconium phosphates [3,87–111].

7.11.1 HYDROTALCITES

The hydrotalcite clays and their synthetic analogues are a large class of anion exchangers [3,87–90]. The structure of hydrotalcite (see Section 2.6.2) can be envisaged as being made of brucite-type octahedral layers, where the $M^{(III)}$ cations partially substitute for $M^{(II)}$ cations [88]; this replacement generates positively charged [M(II)/M(III)/OH] layers, which are compensated for by anions located between them [87]. That is, the hydrotalcite structure is composed of single-layer metal hydroxide sheets alternating with interlayers that enclose anions and water molecules; consequently, the hydroxide sheets build on a positive charge, which is compensated by the interlayer and surface anions [87,88]. As a result, these materials possess large anion-exchange capacities [89,90,105].

A great variety of anions have been incorporated in the interlamellar spaces. It was established that CO_3^{2-} is strongly exchanged in alkaline solution [3]. The ion exchange of iodine in anionic form, that is, I^-, which is an important contaminant, has been studied, and it was shown that it is acceptably exchanged in hydrotalcites [90]. A Li/Al-layered double hydroxide, that is, $[LiAl_2(OH)_6]Cl \cdot 2H_2O$, has been synthesized as well, which has a rhombohedral-layer stacking sequence. The rhombohedral, $[LiAl_2(OH)_6]Cl \cdot 2H_2O$, has been found to easily experience anion-exchange reactions with a range of dicarboxylate anions [105].

7.11.2 Sodium Titanates

As described in Section 2.6.3.2, sodium titanates are built from three-TiO_6-octahedra-sharing edges, which are mutually connected through their corners to other similar chains of octahedra, forming, layers where the sodium ions are placed in the middle of the layers [3].

The ion-exchange properties of this material are usually controlled by the interplanar distance, which can be modified to be large enough to permit the diffusion of cations in their hydrated forms into the structure [91].

Nuclear waste produced from fuel reprocessing operations requires treatment to remove ^{90}Sr before discarding the low-level waste. It has been demonstrated that sodium titanate materials display excellent removal kinetics and capacity for ^{90}Sr [91–94]. In particular, the ion exchange of Sr^{2+} was studied in the sodium titanate, $Na_4Ti_9O_{20} \cdot xH_2O$, yielding a strontium ion-exchange capacity of 5.30 [mequiv/g], which corresponds well with the theoretical value [92]. Besides, it was as well shown that sodium nonatitanate, that is, $Na_4Ti_9O_{20}$, is a highly selective inorganic ion exchanger for strontium [93]. It can be regenerated, is radiation resistant, and has a high cation-exchange capacity [93].

7.11.3 Titanium Silicates

The original [95] synthetic inclusion of Ti in a microporous crystalline framework was reported for the MFI-type [31] framework. In this case, a zeolite-like material, called TS-1, where the $Ti^{(IV)}$ atoms are tetrahedrally coordinated, was obtained [96]. Microporous titanosilicates with a framework different to those encountered in zeolites have a three-dimensional structure composed of SiO_4 tetrahedra and TiO_6 octahedra by corner-sharing oxygen atoms, where the presence of each tetravalent Ti atom in an octahedron generates two negative charges, which are balanced by alkaline exchangeable cations [97].

The first materials discovered which belong to this novel group of titanosilicates, having numerous sixfold-coordinated Ti sites in a microporous crystalline array, was ETS-4, which is an analogue of the mineral zorite while the other, ETS-10, is topologically comparable to zeolite β [96].

Microporous titanosilicate compounds have received significant consideration as ion exchangers ever since the introduction of ETS-4 and ETS-10 materials in 1989 [91,97–100]. Some of these titanosilicates have proven to be good ion exchangers in neutral and alkaline solutions. For example, a potassium titanosilicate analogue of the mineral pharmacosiderite has been evaluated for the removal of cesium and strontium from various aqueous waste streams [91,101,102].

The ion-exchange properties of $M_2 - Ti_2O_3SiO_4 \cdot nH_2O$ (where M = H and Na) for $^{137}Cs^+$ and $^{89}Sr^{2+}$ were studied in the presence of 0.01–6 M, $NaNO_3$, $CaCl_2$, $NaOH$, and HNO_3 solutions, and it was shown that the sodium titanium silicate was an effective Cs^+ and Sr^{2+} ion exchanger [103].

7.11.4 ZIRCONIUM PHOSPHATES

As discussed in Section 2.6.4, this group of compounds consists of phosphates of both group IVA (group 14) and group IVB (group 4) elements. Crystalline zirconium phosphate exists in several polymorphic forms, but the most stable is α-ZrP. This zirconium phosphate has a layered structure (see Figure 2.27), where each layer is composed of zirconium atoms arranged practically in a plane and bridged by phosphate groups situated above and below the plane [108,112]. Adjacent layers are assembled in such a way so as to generate zeolitic cavities with slightly constrained accesses into the cavity [111]. The interlayer distance is 7.6 Å, and the forces holding the layers together are very weak, being either long hydrogen bonds or van der Waals forces.

The zirconium phosphate, α-ZrP, behaves as an ion exchanger where both hydrogen ions of the orthophosphate groups are exchangeable with sodium, potassium, and ammonium in two stages [109]. It has an exchange capacity of 6.64 mequiv/g [111].

In general, all of the zirconium phosphate phases have been shown to have interesting ion-exchange properties [107], where the exchange occurs in the crystals by diffusion of ions from the outer surface inward to the advancing phase boundary. During this process, the layers expand to house not only the ions but also the water of hydration [3]. In fact, there are numerous reports on the ion-exchange properties of different phases of $H_2Zr(PO_4)_2$ with alkali and alkaline earth metal ions in an aqueous solution [107–109].

7.12 ION-EXCHANGE POLYMERIC RESINS

Ion-exchange polymeric resins are the most important types of exchangers currently in use [113–123]. The first, totally organic ion-exchange resin was synthesized in 1935 by Adams and Holmer, when they produced a phenol–formaldehyde cation-exchange resin and an amine–formaldehyde anion-exchange resin, both obtained with the help of condensation polymerization reactions [113]. In 1944, D'Alelio synthesized styrene-based polymeric resins, which could be modified to obtain both cationic- and anionic-exchange resins. The majority of the resins commercially applied currently are of this type, for example, Amberlite IR-20, Lewatit S-100, Permutit Q, Duolite, C-20, Dowex-50, and Nalcite HCR.

7.12.1 GENERAL CHARACTERISTICS OF ION-EXCHANGE RESINS

In the majority of cases, ion-exchange resins consist of a matrix of an insoluble cross-linked polymer that includes fixed anionic or cationic groups which balance the negative and positive excess charges of the polymer frame, respectively. Commonly applied ion-exchange beads in ion-exchange operations are three-dimensional polymeric networks to which ion-exchange groups are bonded, and where the networks, along with the attached ions, which provide either cationic- or anionic-exchange sites, maintain the structural stability of the material [114].

In the case of cationic resins, the polymeric networks are generally attached to the following groups [4,6,23]: $-SO_3^-, -COO^-, -PO_3^{2-}$, and in some cases, $-AsO_3^-$; on the other hand, in the case of anionic resins, the attached groups are $-NH_3^+, =NH_2^+, \equiv N^+$, and in some cases $\equiv S^+$.

For example, the typical styrene-based matrix is prepared by suspension polymerization, using styrene with varying ratios of divinylbenzene as the reactants. Both reactants are added to the reactor, with equal quantities of water, and mixed with a surfactant; then, an agitation process forms globules, and when the globules break into droplets of approximately 1 mm, polymerization is started by adding benzoyl peroxide to obtain plastic beads [115]. The degree of cross-linking is measured with the help of the molar amounts of pure divinylbenzene in the reactant mixture, which could be as small as 0.25 mol% or up to 25 mol% [4,113].

To manufacture cationic materials, the resin beads are swollen with an organic polar solvent to prevent cracking during sulfonation, and then sulfonated with hot and concentrated sulfuric acid (see Figure 7.18) [116–118]. With this method, a strong-acid cation exchanger can be produced [6].

On the other hand, to obtain the anionic resin, the beads are chloromethylated with chlormethoxymethane (CH_3OCH_2Cl) to attach a chloromethyl group ($-CH_2Cl$) to the ethyl benzene nuclei, and, subsequently, the chloromethylated plastic bead is aminated with, for example, trimethylamine to obtain $RCH_2N(CH_3)_3^+Cl^-$, in order to obtain an anionic-exchange resin (see Figure 7.19) [113,117,118]. With this method, a strong-base anion exchanger can be produced, for example, the quaternary ammonium resin ($RCH_2N(CH_3)_3^+Cl^-$) [6].

The weak-base anion-exchange resins, that is, the secondary RCH_2NHR' and tertiary amine RCH_2NR_2', are produced by reacting the chloromethylated resin with lower substituted amines or with ammonia [124].

Ethylbenzene is not the only monomer that can be cross-linked with divinylbenzene to get a cross-linked polymer. By using the addition polymerization process, if methyl propionic acid (see Figure 7.20) and divinylbenzene are cross-linked, it is possible to obtain methacrylic divinylbenzene, a weakly acidic cationic resin (Figure 7.21) [118].

Ion-exchange resins are produced as gelular resins, which are homogeneous cross-linked polymers, or macroporous beads which have pores produced with a porogen, to allow ion transport [113,119].

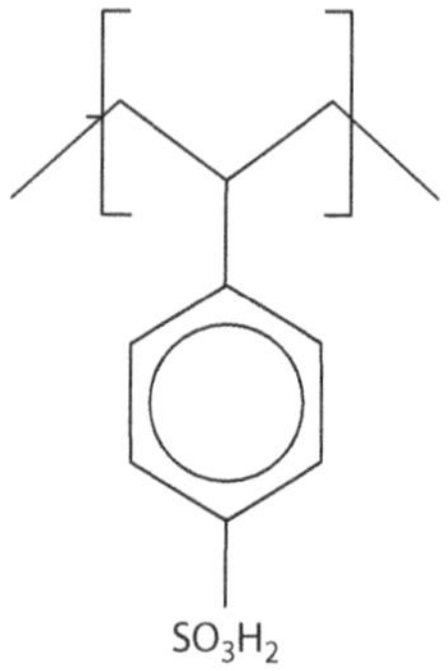

7.12.2 Ion-Exchange Resin Swelling

Swelling is the expansion of the ion-exchange material during solvent adsorption [23]. During the ion-exchange process, inorganic ion exchangers, such as zeolites, do not swell; however, organic polymeric resins experience this phenomenon. Dry resins swell in aqueous media mainly on account of the hydration of the fixed ions and counterions, and the osmotic pressure due to the latter [125]. Then, when the resins swell, their framework is expanded and exercise pressure on the internal pore [4,123]. The structure of cross-linked ion-exchange beads consists of three-dimensional flexible, random pore networks, which can swell to a certain extent [4,6]. The pore framework of the resins can be microporous when porogens are not added during their synthesis; if porogens are added, the pore structures contain the micropores of the cross-linked resin as well as the macro- and mesopores (see Figure 7.22) [6,124].

The first model which took into account the swelling properties of the polymeric resins was Gregor's model, which is based on the concept of inter-phase distribution [125] and considers the matrix of the resin as a network of elastic springs [4]. When the resin swells, the network is expanded and exerts pressure on the internal pore (see Figure 7.23); thereafter, the swelling pressure developed in the resin influences the ion-exchange equilibrium.

The real swelling behavior of cross-linked ion-exchange resins immersed in electrolyte solutions of varying concentrations denote that these systems can be considered as a charged

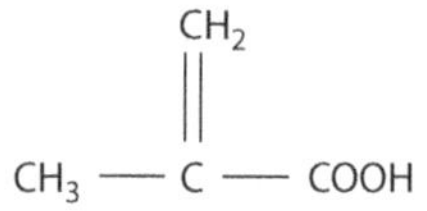

FIGURE 7.18 Cationic resin.

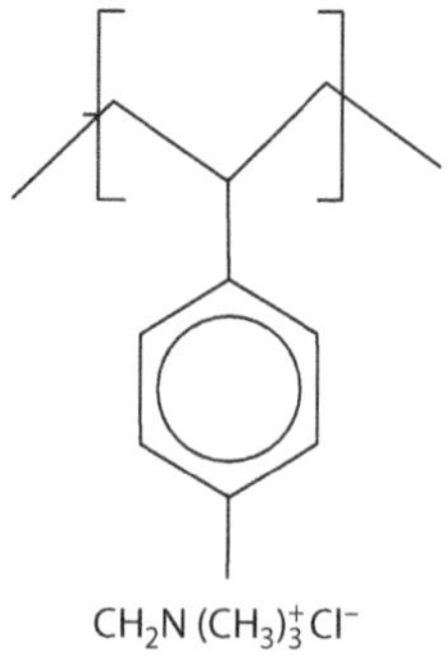

FIGURE 7.19 Anionic resin.

FIGURE 7.20 Methyl propionic acid.

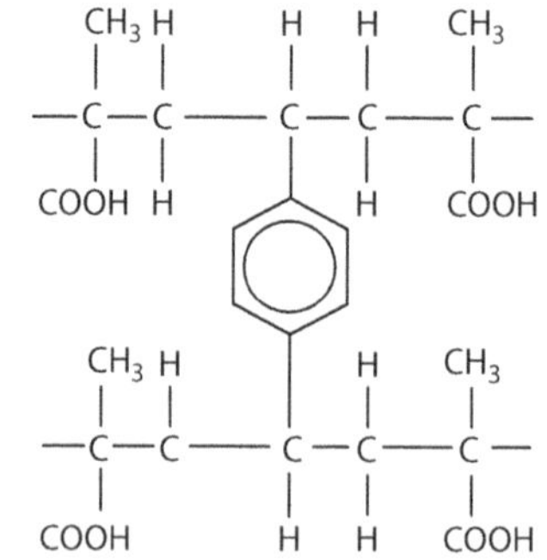

FIGURE 7.21 Methacrylic divinylbenzene cationic resin.

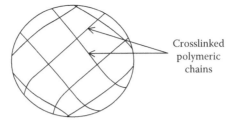

FIGURE 7.22 Cross-linked bead structure.

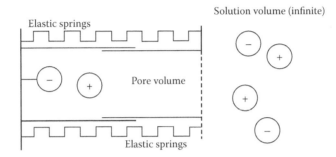

FIGURE 7.23 Gregor model.

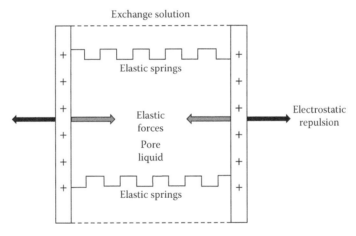

FIGURE 7.24 Lazare–Gregor model.

polymeric network, and as an internal solution consisting of interstitial water, counterions, and diffusible electrolyte in equilibrium across a thermodynamic phase boundary with the solution [126]. In the first model, developed by Gregor, only mechanical forces were included that considered their effect to be the single cause of the swelling pressure on the ion-exchange equilibrium. However, it is obvious that swelling is a balance between opposite forces, that is, the tendency of charged particles inside the resin to become surrounded with solvent molecules, to form salvation shells, and, thereafter, extend the resin framework and expand the matrix resistance [4,125,126].

In order to incorporate both tendencies, Lazare, Sundheim, and Gregor developed an improved model, where the elastic and electrostatic interactions were included. In this model, the resin was regarded as a set of charged planar capacitors, with the plates interconnected by elastic springs (Figure 7.24) [126]. The balance between forces is attained when the elastic forces provided by the polymeric resin stabilize the dissolution propensity.

Swelling can also be understood as the internal solution of the ion-exchange resin, and then the system can be treated as a concentrated solution (the polymeric resin) in osmotic equilibrium with a more diluted solution (the exchange solution) [4].

With regard to the previous arguments, some rules can be proposed that qualitatively describe the swelling equilibrium [4,6,113]: first, it is obvious that highly cross-linked polymers possess less ability to swell than less cross-linked ones because they are more inflexible; second, polar solvents are better swelling agents since their molecules interact powerfully with the charged particles included in the resin; third, resins with high TEC have more charged particles and consequently a higher propensity to swell; finally, the character of the coions and counterions that interact with the solvent modify the swelling process in a complex form. Consequently, the higher the valence of the ion, the higher the ability to be removed by a resin; and within the same valence, the lower the hydration enthalpy $(-H_a)$ of the ion, the higher the affinity of the resin for this ion [6,113].

7.12.3 Applications of Ion-Exchange Polymeric Resins

These materials are widely applied for hardness removal, and demineralization and purification of water, since they have high ion-exchange capacities even at low concentrations of the ion to be removed in solution, and are stable and easily regenerated.

Removal of hardness from water is a process applied in diverse systems, including laundries, water treatment plants, chemical industries, and power plants; the two generally applied methods for hardness elimination are lime softening and ion exchange, but ion-exchange processes generate less waste [5,49,56,127]. For hardness removal, weak-acid ion-exchange polymeric resin beads and fibers containing carboxylate ($-COOH$) functional groups are applied [127].

For demineralization, a set of strong-acid and weak-acid cation-exchange resin columns, and strong-base anion-exchange and weak-base ion-exchange resin columns have been applied [6].

We have previously described ammonia removal with zeolites. However, the development of new methods capable of effective and total ammonia removal from wastewater are extremely attractive; in this regard, the ion-exchange removal of ammonia with a Dowex HCR-S ion-exchange resin has been carried out [79].

Laboratory studies have demonstrated the possibility of using ion-exchange resins (Duolite GT-73, Amberlite IRC-748 Rohm and Haas, Philadelphia, PA, USA and two solvent-impregnated resins prepared from Amberlite IRA-96 Rohm and Haas) in permeable reactive barriers for the remediation of groundwater contaminated with heavy and transition metals, since all four resins were able to reduce cadmium, lead, and copper concentrations [120]. Besides, the study of the exchange equilibria between H-Amberlite IR-120 and aqueous solutions (0.1 equiv/L) of $Pb(NO_3)_2$;$Ni(NO_3)_2$ and $Cr(NO_3)_3$ at pH <3, at 283, 303, and 323 K, has shown the following order of selectivity: Cr^{3+} > Pb^{2+} > Ni^{2+}, at all temperatures, in agreement with the rule of higher selectivity at a higher counterion charge [121]. As well, a Hypersol–Macronet polymer, MN-600, and the weakly acidic acrylic ion-exchange resin, C-104E, containing mainly carboxylic functionality, have been applied for trace heavy metal removal [128]. Macronet is an hyper-crosslinked polymer based on a spherical styrene divinylbenzene copolymer, which is cross-linked when the polymer is in the swollen state [129], to obtain a stretched and rigid polymeric network with low density of chain packing owing to the large number of bridges, showing new types of porosity with high surface areas (1000–1900 m^2/g) [127]. Also, heavy metal recovery in the electroplating industry is a common process, for example, Ni^{2+} and Cr^{3+} can be recovered from plating wastes [6]. Additionally, the ion-exchange equilibria of Cu^{2+}, Cd^{2+}, Zn^{2+}, and Na^+ in aqueous solution at 283 and 303 K in the strong acid resin, Amberlite IR-120, have been studied in order to evaluate the possibility of applying ion exchange to remove heavy metal ions from industrial aqueous streams [130].

Anion-exchange resins, particularly quaternary ammonium anion-exchange resins, are extensively applied to recover radionuclides from different process waste streams; for instance, Pu(IV) and Th(IV) can be successfully recovered from uranyl ions using quaternary ammonium–based anionic resins in concentrated nitric acid [131–133].

Amino acids are of significant value in the food and pharmaceutical industries; in this regard, for the recovery and purification of amino acids from the products of these operations, ion-exchange resins are broadly applied. On account of their amphoteric nature, amino acid charges vary with pH, allowing their fixation to or elution from resins [134].

Perchlorate is a common contaminant of groundwater and surface water, is found in diary milk and breast milk, tobacco, leafy vegetables, fruit, plant species, and bottled water [135]. The effect of resin cross-linking, porosity, and functional groups on perchlorate selectivity has been studied, and it was shown that many hydrophobic resins demonstrate perchlorate affinity [136]. On the basis of the available information, most polystyrene-based resins are capable of acceptable perchlorate capacity; however, the regeneration effectiveness of perchlorate-loaded styrenic resins is normally, limited, in contrast to acrylic resins which are generally easier to regenerate even though their perchlorate adsorption capacity is low [135]. New methods for perchlorate removal are being explored, for example, using a bifunctional resin (A-530E), a class of polymeric ligand exchangers (PLEs), and an ion-exchange fiber (IXF) [135].

Ion-exchange materials coated on to glass fiber substrates have several advantages when compared with the usual ion-exchange beads, for instance, simpler syntheses techniques, different product forms such as papers and fabrics, increased rate of both reaction and regeneration, and physical and mechanical stabilities [122]. Cation-exchange fibers can be prepared by coating glass fiber substrates with a polystyrene/divinylbenzene oligomer, curing, and, finally, sulfonating. For the obtained materials, the contact efficiencies were significantly enhanced compared with the conventional beads, on account of the higher surface-to-volume ratio and shorter diffusion path lengths [114].

REFERENCES

1. D.W. Breck, *Zeolite Molecular Sieves*, John Wiley & Sons, New York, 1974.
2. R.M. Barrer, *Hydrothermal Chemistry of Zeolites*, Academic Press, London, 1982.
3. A. Cleafield, *Chem. Rev.*, 88, 125 (1988).
4. A.A. Zagorodni, *Ion Exchange Materials: Properties and Applications*, Elsevier Science & Technology Books, Amsterdam, the Netherlands, 2006.
5. R. Roque-Malherbe, in *Handbook of Surfaces and Interfaces of Materials*, Vol. 5, H.S. Nalwa, (editor), Academic Press, New York, 2001, p. 495.
6. A.M. Wachinski and J.E. Etzel, *Environmental Ion Exchange*, CRC Press, Boca Raton, FL, 1997.
7. B.W. Mercer and L.L. Ames, in *Natural Zeolites, Occurrence, Properties, Use*, L.B. Sand and F.A. Mumpton, (editors), Pergamon Press, New York, 1978, p. 451.
8. M. Kuronen, R. Harjula, J. Jernstrom, M. Vestenius, and J. Lehto, *Phys. Chem. Chem. Phys.*, 2, 2655 (2000).
9. J. Benitez, E. Sanchez, L. Travieso, and R. Roque-Malherbe, *International Conference on Advanced Wastewater Treatment and Reclamation*, Cracow, Poland, September 25–27, 1989.
10. D. Kallo, D., in *Natural Zeolites 93 Conference Volume*, D.W. Ming and F.A. Mumpton, (editors), International Committee on Natural Zeolites, Brockport, NY, 1995, p. 341.
11. L. Liberti, A. Lopez, V. Amicarelli, and G. Boghetich, in *Natural Zeolites 93 Conference Volume*, D.W. Ming and F.A. Mumpton, (editors), International Committee on Natural Zeolites, Brockport, NY, 1995, p. 351.
12. W.G. Pond and F.A. Mumpton, (editors), *Zeo-Agriculture: Use of Natural Zeolites in Agriculture and Aquaculture*, Westview, Boulder, CO, 1984.
13. S. Ganev, *Modern Soil Chemistry*, Nauka i Iskustvo, Sofia, 1990.
14. R.P. Townsend, *Pure Appl. Chem.*, 58, 1359 (1986).
15. A. Dyer, *An Introduction to Zeolite Molecular Sieves*, John Wiley & Sons, New York, 1988.
16. M.J. Schwuger and M. Liphard, *Stud. Surf. Sci. Catal.*, 46, 673 (1989).
17. S.M. Robinson, W.D. Arnold, and C.H. Byers, *ACS Symp. Ser.*, 468, 133 (1991).
18. H. Fagghibian, M.M. Ghannadi, and H. Kazeman, *App. Radiat. Isotopes*, 50, 655 (1999).
19. D.T. Vanimam and D.L. Bish in *Natural Zeolites 93 Conference Volume*, D.W. Ming and F.A. Mumpton, (editors), International Committee on Natural Zeolites, Brockport, NY, 1995, p. 533.
20. N.F. Chelichev, in *Natural Zeolites 93 Conference Volume*, D.W. Ming and F.A. Mumpton, (editors), International Committee on Natural Zeolites, Brockport, NY, 1995, p. 525.

21. H.S. Sherry, in *Handbook of Zeolite Science and Technology*, S.M. Auerbach, K.A. Corrado, and P.K. Dutta, (editors), Marcel Dekker, New York, 2003, p. 1007.
22. F. Pepe, D. Caputo, and C. Colella, *Ind. Eng. Chem. Res.*, 42, 1093 (2003).
23. F. Helferich, *Ion Exchange*, McGraw-Hill, New York, 1962.
24. R.M. Barrer, in *Natural Zeolites, Occurrence, Properties, Use*, L.B. Sand and F.A. Mumpton, (editors), Pergamon Press, New York, 1978, p. 385; *Pure Appl. Chem.*, 51, 1091 (1979); 52, 1980 (1980).
25. J.D. Seader and E.J. Henley, *Separation Process Principles*, John Wiley & Sons, New York, 1998.
26. H.S. Thompson, *J. Royal Agric. Soc. Engl.*, 11, 68 (1850).
27. J.T. Way, *J. Royal Agric. Soc. Engl.*, 11, 313 (1850).
28. J. Lemberg, *Z. Deut. Geol. Ges.*, 22, 335 (1870).
29. G. Wiegner, *J. Landwirtsch*, 60, 111 (1912).
30. H. Eichorn, *Poggendorf Ann. Phys. Chem.*, 105, 126 (1858).
31. Ch. Baerlocher, W.M. Meier, and D.M. Olson, *Atlas of Zeolite Framework Types*, (5th edition), Elsevier, Amsterdam, the Netherlands, 2001.
32. K. Koyama and T. Takeuchi, *Z. Kristallogr.*, 145, 216 (1977).
33. A.P. Vanselow, *J. Amer. Chem. Soc.*, 54, 1307 (1932).
34. G.L. Gaines and H.C. Thomas, *J. Chem. Phys.*, 21, 714 (1953).
35. P. Fletcher and R. P. Townsend, *J. Chem. Soc. Faraday Trans. 1*, 77, 2077 (1981).
36. R.M. Barrer and R.P. Townsend, *Zeolites*, 5, 287 (1985).
37. G. Rodriguez-Fuentes, PhD thesis, National Center for Scientific Research, Havana, Cuba, 1987.
38. R. Roque-Malherbe, W. del Valle, J. Ducongé, and E. Toledo, *Int. J. Env. Pollut.*, 31, 292 (2007).
39. R. Roque-Malherbe, *Chemical Physics of Zeolites*, Publishing Department of the Ministry of Higher Education, Havana, Cuba, 1988.
40. R. Roque-Malherbe, A. Berazain, and J.A. del Rosario, *J. Therm. Anal.*, 32, 949 (1987).
41. P. Backman, M. Bastos, L.-E. Briggner, S. Hagg, D. Hallen, P. Lonnbro, S.-O. Nilsson et al., *Pure Appl. Chem.*, 66, 375 (1994).
42. N. Petrova, L. Filizova, and G.N. Kirov, in *Natural Zeolites '93 Conference Volume*, D.W. Ming and F.A. Mumpton, (editors), International Committee on Natural Zeolites, Brockport, NY, 1995, p. 281.
43. C.D. Schoeffer, Jr., C.A. Strausser, N.W. Thomsen, and C.H. Yoder, *Data for General Inorganic, Organic and Physical Chemistry*, (2001). http://wulfenite.fandm.edu/Data20%/Table_15.html
44. G.E. Boyd, A.W. Adamson, and L.S. Myers, *J. Amer. Chem. Soc.*, 69, 2836 (1947).
45. R.M. Barrer, R.F. Barthelemew, and L. Rees, *J. Phys. Chem. Solids*, 24, 51 (1963).
46. T. Vermeulen, *Ind. Eng. Chem.*, 45, 1664, 1953.
47. J. Crank, *The Mathematics of Diffusion of Second Edition*, Oxford Science Publications, Oxford, 1975.
48. N.F. Chelishchev, B.G. Berenshtein, and V.F. Volodin, *Tseolti Novii Tip Mineralnogo Siria*, Nedra, Moscow, 1987.
49. R. Droste, *Theory and Practice of Water and Wastewater Treatment*, John Wiley & Sons, Hoboken, NJ, 1997.
50. H. Scott-Fogler, *Elements of Chemical Reaction Engineering*, Prentice-Hall, New Jersey, 1999.
51. M. Pansini, *Miner. Deposita*, 31, 563 (1996).
52. A.C. Michaels, *Ind. Eng. Chem.*, 44, 1922 (1952).
53. R. Roque-Malherbe, W. del Valle, N. Planas, K. Gómez, D. Ledes, L. Garay, and J. Ducongé, *Zeolites '02 Book of Abstracts*, P. Misaelides, (editor), 7th International Conference on the Occurrence, Properties and Use of Natural Zeolites, Thessaloniki, Greece, June 3–7, 2002, p. 316.
54. J. Kecht, B. Mihailova, K. Karaghiosoff, S. Mintova, and T. Bein, *Langmuir*, 20, 5271 (2004).
55. A.J. Celestian, J.B. Parise, C. Goodell, A. Tripathi, and J. Hanson, *Chem. Mater.*, 16, 2244 (2004).
56. Drew Chemical Corporation, *Principles of Industrial Water Treatment*, Drew Chemical Corporation, Boonton, NJ, 1985.
57. T.-J. Liang and Y-Ch. Tsai, *Appl. Radiat. Isot.*, 40, 7 (1995).
58. C. Colella, in *Natural Zeolites 93 Conference Volume*, D.W. Ming and F.A. Mumpton, (editors), International Committee on Natural Zeolites, Brockport, NY, 1995, p. 363.
59. M.J Semmens, *Natural Zeolites, Occurrence, Properties, Use*, L.B. Sand and F.A. Mumpton, (editors), Pergamon Press, New York, 1978, p. 517.
60. R. Roque-Malherbe, A. Picart, C. Diaz, and G. Rodriguez, Patent Certificate 21055, ONIITEM, Havana, Cuba, 1986.
61. V.J. Inglezakis, H.D. Loizidou, and H.D. Grigoropoulu, *J. Colloid Int. Sci.*, 215, 54 (1999).
62. M. Loizidou, K.J. Haralambous, A. Lokauttos, and D. Dimitrakopoulo, *J. Env. Sci. Heal. A.* A27, 1759 (1992).

63. P. Papakrishtou, K.J. Haralambous, M. Loizidou, and N. Syrellis, *J. Env. Sci. Heal. A*. A28, 135 (1993).
64. E. Malliou, M. Loizidou, and N. Spyrellis, *Sci. Total Environ.*, 149, 139 (1994).
65. J.E. García, J.S. Notario del Pino, and M.M. Gonzales Martín, *Appl. Clay Sci.*, 9, 239 (1994).
66. V. Albino, R. Cioffi, M. Pansini, and C. Colella, *Env. Tech.*, 16, 147 (1995).
67. C. Colella and M. Pansini, *ACS Symp. Ser.*, 368, 500 (1988).
68. C. Colella, M. de Gennaro, A. Langella, and M. Pansini, in *Natural Zeolites 93 Conference Volume*, D.W. Ming and F.A. Mumpton, (editors), International Committee on Natural Zeolites, Brockport, NY, 1995, p. 377.
69. M. Vavrova, O. Musatovova and S. Bartha, *Isotopenpraxis*, 27, 7 (1991).
70. O. Musatovova, M. Vavrova, A. Mitro, and A. Bartha, in *Zeolite '93* 4th International Conference on the Occurrence, Properties and Utilization of Natural Zeolites, Program and Abstracts, International Committee on Natural Zeolites, Boise, Idaho, 1993, p. 145.
71. G.D. Gradev, I.G. Stefanova, A.G. Milusheva, and A.P. Naidenov, in *Zeolite '93* 4th International Conference on the Occurrence, Properties and Utilization of Natural Zeolites, Program and Abstracts, International Committee on Natural Zeolites, Boise, Idaho, 1993, p. 111.
72. J.D. Sherman, *NATO-ASI Series E*, 583 (1984).
73. L.M. Carrera, S. Gomez, P. Bosh, and S, Bulbalian, *Zeolites*, 13, 622 (1993).
74. C. Constantinodopoulo, M. Loizidou, Z. Loizou, and N. Sperillis, *J. Radioan. Nucl. Chem. Ar.*, 178, 143 (1994).
75. E. Sanchez and R. Roque-Malherbe, *Biotechnol. Lett.*, 9, 671 (1987).
76. D. Drummond, A. De Jonge, and L.V.C. Rees, *J. Phys. Chem.*, 87, 1967 (1983).
77. B.H. Wlers, R.J. Grosse, and W.A. Cilley, *Environ. Sci. Technol.*, 16, 617 (1982).
78. C. de las Pozas, C. Becquer, M. Carreras-Gracial, L. López-Colado, T. Marquez, and R. Roque-Malherbe, in *Zeolites '91 Full Papers Volume*, G. Rodríguez and J.A. González, (editors), International Conference Center Press, Havana, Cuba, 1992, p. 278.
79. S.H. Lin and C.L. Wu, *Ind. Eng. Chem. Res.*, 35, 553 (1996).
80. F.A. Mumpton, in *Natural Zeolites, Occurrence, Properties, Use*, L.B. Sand and F.A. Mumpton, (editors), Pergamon Press, New York, 1978, p. 3; Natural Zeolites: Occurence, Properties, Uses of Natural Zeolites, D. Kallo and H.S. Sherry, (editors), Akademiai Kiado, Budapest, 1988, p. 333.
81. G.V. Tsisihsvili, T.G Andronikashvili, G.N. Kirov, and L.D. Filizova, *Natural Zeolites*, Ellis Horwood, New York, 1992.
82. P.W. Jonhson and J.M. Sieburth, *Aquaculture*, 4, 61 (1974).
83. M.J. Semmens, in *Zeo-Agriculture: Use of Natural Zeolites in Agriculture and Aquaculture*, W.G. Pond and F.A. Mumpton, (editors), Westview, Boulder, CO, 1984, p. 45.
84. D.A. Edssall and C.H. Smith, *Prog. Fish Cult.*, 51, 98 (1989).
85. C. Becker, C. de las Pozas, E. Pullé, J. Fajardo, and R. Roque-Malherbe, *Rev. CNIC. Ciencias Químicas*, 21, 223 (1990).
86. R. Roque-Malherbe, W. del Valle, F. Marquez, J. Duconge, and M.F.A. Goosen, *Sep. Sci. Technol.*, 41, 73 (2006).
87. R.P. Bontchev, S. Liu, J.L. Krumhansl, J. Voigt, and T.M. Nenoff, *Chem. Mater.*, 15, 3669 (2003).
88. M. Bellotto, B. Rebours, O. Clause, J. Lynch, D. Bazin, and E. Elkaim, *J. Phys. Chem.*, 100, 8527 (1996).
89. S. Miyata, *Clay Clay Miner.*, 31, 305 (1983).
90. H. Curtius and Z. Kattilparampil, *Clay Clay Miner.*, 40, 455 (2005).
91. T. Möller, Selective crystalline inorganic materials as ion exchangers in the treatment of nuclear waste solutions, PhD dissertation, Faculty of Science, University of Helsinki, Helsinki, 2002.
92. J. Lehto and A. Clearfield, *J. Radioanalyt. Nuc. Chem.*, 118, 1588 (1987).
93. S.F. Yates and P. Sylvester, *Sep. Sci. Technol.*, 36, 867 (2001).
94. D.T. Hobbs, M.R. Poirier, M.J. Barnes, M.E. Stallings, and M.D. Nyman, WM'06 Conference, Tucson, AZ. WSRC-MS-2005-00477, February 26–March 2, 2006.
95. M. Taramasso, G. Perego, and B. Notari, US Patent 4,410,501 (1983).
96. M.E. Grillo and J. Carrazza, *J. Phys. Chem.*, 100, 12261 (1996).
97. J. Rocha and Z. Lin, *Rev. Min. Geochem.*, 57, 173 (2005).
98. D.M. Poojary, R.A. Cahill, and A. Clearfield, *Chem. Mater.*, 6, 2364 (1994).
99. V. Valtchev, J.-L. Paillaud, S. Mintova, and H. Kessler, *Mic. Mes. Mat.*, 32, 287 (1999).
100. J. Rocha and M.W. Amerson, *J. Eur. Inorg. Chem.*, 5, 801 (2000).
101. E. Behrens and A. Clearfield, *Mic. Mat.*, 11, 65 (1997).
102. E.A. Behrens P. Sylvester, and A. Clearfield, *Environ. Sci. Technol.*, 32, 101 (1998).

103. S. Solbra, N. Allison, S. Waite, S. Mikhalovsky, A. Bortun, L. Bortun, and A. Clearfield, *Env. Sci. Technol.*, 35, 626 (2001).

104. M. Jobbagy and A.E. Regazzoni, *J. Phys. Chem. B*, 109, 389 (2005).

105. A.M. Fogg, A.J. Freij, and G.M. Parkinson, *Chem. Mater.*, 14, 232 (2002).

106. D.A. Burwell and M.E. Thompson, in *Supramolecular Architecture*, T. Bein, (editor), American Chemical Society Symposium Series, Washington, DC, 1992, p. 166.

107. A.E. Gash, P.K. Dorhout, and S. H. Strauss, *Inorg. Chem.*, 39, 5538 (2000).

108. A. Clearfield and J. Troup, *J. Phys. Chem.*, 74, 314 (1970).

109. A. Clearfield, L. Kullberg, and A. Okarsson, *J. Phys. Chem.*, 78, 1150 (1974).

110. A. Clearfield and J. Troup, *J. Phys. Chem.*, 77, 243 (1977).

111. A. Clearfield, W.L. Duax, A.S. Medina, G.D. Smith, and J. R. Thomas, *J. Phys. Chem.*, 73, 3424 (1969).

112. A. Marti and J. Colon, *Inorg. Chem.*, 42, 2830 (2003).

113. F.J. DeSilva, 25th Annual Water Quality Association Conference, Fort Worth, TX, March 1999.

114. J. Economy and L. Dominguez, *Ind. Eng. Chem. Res.*, 41, 6436 (2002).

115. S.B. Brijmohan, S. Swier, R.A. Weiss, and M.T. Shaw, *Ind. Eng. Chem. Res.*, 44, 8039 (2005).

116. K. Dorfner, *Ion Exchangers*, Walter de Gruyter, Berlin, 1991.

117. S.D. Alexandratos and D.W. Crick, *Ind. Eng. Chem. Res.*, 35, 635 (1996).

118. C.H. Harland, *Ion Exchange. Theory and Practice* (2nd edition), Royal Society of Chemistry, London, UK, 1994.

119. I.M. Abrams and J.R. Millar, *Reactive Funct. Polym.*, 35, 7 (1997).

120. M.Y. Vilensky, B.N. Berkowitz, and A. Warchawsky, *Environ. Sci. Technol.*, 36, 1851 (2002).

121. M. Carmona, J. Warchoł, A. de Lucas, and J.F. Rodriguez, *J. Chem. Eng. Data*, 53, 1325 (2008).

122. L. Dominguez, and K. Benak, *J. Polym. Adv. Technol.*, 12, 197 (2001).

123. G.V. Samsonov and V.A. Passchnik, *Russ. Chem. Rev.*, 38, 547 (1969).

124. J.S. Fritz and D.T. Djerde, *Ion Chromatography* (3rd edition), Wiley-VHC, Weinheim, Germany, 2000.

125. H.P. Gregor, *J. Amer. Chem. Soc.*, 70, 1293 (1948); 73, 642 (1951).

126. L. Lazare, B.R. Sundheim, and H.P. Gregor, *J. Phys. Chem.*, 60, 641 (1956).

127. J.E. Greenleaf and A. Sengupta, *Environ. Sci. Technol.*, 40, 370 (2006).

128. B. Saha and M. Streat, *Ind. Eng. Chem. Res.*, 44, 8671 (2005).

129. V.A. Davankov and M.P. Tsyurupa, *Pure Appl. Chem.*, 61, 1881 (1989).

130. J.L. Valverde, A. de Lucas, M. Gonzalez, and J.F. Rodriguez, *J. Chem. Eng. Data*, 46, 1404 (2001).

131. S.F. Marsh, G.D. Jarvinen, J.S. Kim, J. Nam, and R.A. Bartsch, *React. Func. Polym.*, 35, 75 (1997).

132. S.F. Marsh, G.D. Jarvinen, R.A. Bartsch J. Nam, and M.E. Barr, *J. Radioanal. Nucl. Chem.*, 235, 37 (1998).

133. Y.H. Ju, O.F. Webb, S. Dai, J.S. Lin, and C.E. Barnes, *Ind. Eng. Chem. Res.*, 39, 550 (2000).

134. A. Zammouri, S. Chanel, L. Muhr, and G. Grevillot, *Ind. Eng. Chem. Res.*, 39 1397 (2000).

135. Z. Xiong, D. Zhao, and W.F. Harper, *Ind. Eng. Chem. Res.*, 46, 9213 (2007).

136. A.R. Tripp and D.A. Clifford, in *Perchlorate in the Environment*, E.T. Urbansky, (editor), Kluwer Academic-Plenum Press, New York, 2000, p. 123; *J. Am. Water Works Assoc.*, 98, 105 (2006).

8 Solid-State Electrochemistry

8.1 INTRODUCTION

The world's energy production based on the combustion of fossil fuels is having a serious impact on the ecology worldwide, besides causing economic and political problems [1–4]. Therefore, the electrochemical generation of energy is a viable alternative option, given that this approach is more sustainable and environmentally friendly than fossil fuel combustion [5–8]. In this chapter, a brief review of solid-state electrochemistry emphasizing on solid oxide fuel cells and, to a certain extent, polymer electrolyte fuel cells, is given.

8.1.1 BATTERIES AND FUEL CELLS

Electrochemical energy storage and conversion systems described in this chapter comprise batteries and fuel cells [6–11]. In both systems, the energy-supplying processes occur at the phase boundary of the electrode–electrolyte interface; moreover, the electron and ion transports are separate [6,8]. Figures 8.1 and 8.2 schematically illustrate the electron and ion conductions in both the electrodes and the electrolyte in Daniel and fuel cells. The production of electrical energy by the conversion of chemical energy by means of an oxidation reaction at the anode and a reduction reaction at the cathode is also described.

Batteries and fuel cells are differentiated based on the sites of energy storage and conversion. Batteries are closed systems, where the anode and the cathode are the charge-transfer media, in which oxidation and reduction reactions take place over an active mass. In the Daniel cell, dissolving (Zn) and depositing (Cu) energy storage and conversion occur in the same compartment. On the other hand, fuel cells are open systems where the anode and cathode are merely charge-transfer media, and the active masses experiencing the reduction and oxidation reactions are transported from the exterior of the cell [10]. That is, fuel cells are electrochemical mechanisms that continuously transform the chemical energy of an external source into electrical energy. The fundamental physical building block of a fuel cell is composed of an electrolyte layer in contact with a porous anode and a cathode on either side (see Figure 8.2). In a standard fuel cell, gaseous fuels are fed constantly to the anodic compartment and an oxidant, that is, oxygen, from air is supplied continuously to the cathodic compartment. The electrochemical reactions that occur at the electrodes generate an electric current.

A fuel cell, even though, containing parts and characteristics comparable to those of a typical battery is different in some respects. To be precise, the battery is an energy storage device, where the maximum energy accessible is established by the quantity of the chemical reactant stored within the battery itself; a fuel cell is an energy conversion mechanism that supposedly has the capability of producing electrical energy for as long as the fuel and the oxidant are supplied to the electrodes.

8.1.2 TYPES OF FUEL CELLS

Fuel cells can be classified into diverse types, where the most frequent categorization is by the type of electrolyte used in the cells, namely [6–9,11]:

- Polymer electrolyte fuel cell (PEFC)
- Alkaline fuel cell (AFC)
- Phosphoric acid fuel cell (PAFC)

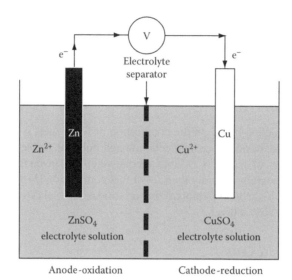

FIGURE 8.1   Graphic representation of the Daniel cell.

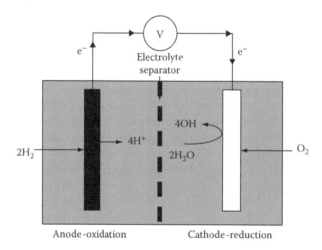

FIGURE 8.2   Schematic representation of a fuel cell.

- Molten carbonate fuel cell (MCFC)
- Solid oxide fuel cell (SOFC)

These fuel cells are listed in the order of their approximate operating temperature, ranging from 80°C for PEFCs, 100°C for AFCs, 200°C for PAFCs, 650°C for MCFCs, and 800°C–1000°C for SOFCs.

8.1.2.1   Polymer Electrolyte Fuel Cell

Figure 8.3 shows a schematic representation of a polymer electrolyte fuel cell (PEFC). In this instance, the oxidation reaction in the cathode is given by

$$H_2 \leftrightarrow 2H^+ + 2e^-$$

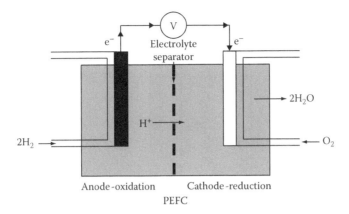

FIGURE 8.3 Schematic representation of a PEFC.

Besides, the corresponding reduction process in the cathode is represented as follows:

$$2H^+ + O_2 + 2e^- \leftrightarrow H_2O_2$$

$$2H^+ + H_2O_2 + 2e^- \leftrightarrow 2H_2O$$

The PEFC was first developed for the *Gemini* space vehicle by General Electric, USA. In this fuel cell type, the electrolyte is an ion-exchange membrane, specifically, a fluorinated sulfonic acid polymer or other similar solid polymer. In general, the polymer consists of a polytetrafluoroethylene (Teflon) backbone with a perfluorinated side chain that is terminated with a sulfonic acid group, which is an outstanding proton conductor. Hydration of the membrane yields dissociation and solvation of the proton of the acid group, since the solvated protons are mobile within the polymer. Subsequently, the only liquid necessary for the operation of this fuel cell type is water [7,8].

Another characteristic of the proton-conducting membrane is that it has low permeability to oxygen and hydrogen in the gas phase so that a high coulombic efficiency exists [7]. In addition, in this fuel cell type, the electrodes are normally formed on a thin layer on each side of a proton-conducting polymer membrane used as an electrolyte, and platinum catalysts are required for both the anode and the cathode for the proper operation of this fuel cell [9].

Water running in the membrane is decisive for an efficient performance; hence, the fuel cell has to operate under conditions where the byproduct, water, does not evaporate rapidly than it is generated since the membrane must be hydrated for to function properly. Thereafter, the operating temperature is usually less than 120°C.

8.1.2.2 Alkaline Fuel Cell

Figure 8.4 shows a schematic representation of an alkaline fuel cell (AFC). In this case, the oxidation reaction in the cathode is given by

$$H_2 + 2OH^- \leftrightarrow 2H_2O + 2e^-$$

Besides, the corresponding reduction process in the cathode is represented as follows:

$$H_2O + O_2 + 2e^- \leftrightarrow HO_2^- + OH^-$$

$$H_2O + HO_2^- + 2e^- \leftrightarrow 3OH^-$$

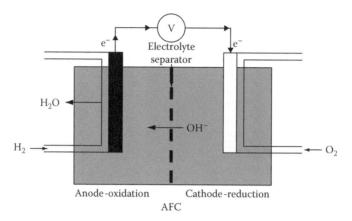

FIGURE 8.4   Schematic representation of an AFC.

The AFC type was originally created for the *Apollo* program, after that a modernized version has been developed and is even now in use to provide electrical power for shuttle missions. The electrolyte in this fuel cell is KOH, concentrated (85 wt %) for fuel cells operated at relatively high temperatures, that is, around 250°C, and less concentrated (35–50 wt %) for cells operated at lower temperatures, that is, less than 120°C [6,9,11]. In the construction of these fuel cells, the electrolyte is retained in a matrix, typically asbestos, and a wide range of catalysts, for example, Ni, Ag, metal oxides, and noble metals, can be used for both the hydrogen and the oxygen electrodes [8,9].

The fuel supply is limited to hydrogen; CO is a poison and CO_2 will react with the KOH to form K_2CO_3, thus altering the electrolyte [9].

8.1.2.3 Phosphoric Acid Fuel Cell

Figure 8.5 shows a schematic illustration of a phosphoric acid fuel cell (PAFC). This cell type is another fuel cell operating in acidic media, and the oxidation reaction in the anode is given by

$$H_2 \leftrightarrow 2H^+ + 2e^-$$

The corresponding reduction process in the cathode is represented as follows:

$$O_2 + 2H^+ + 2e^- \leftrightarrow H_2O$$

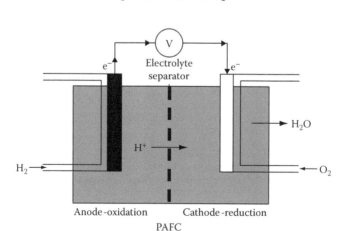

FIGURE 8.5   Schematic representation of a PAFC.

$$H_2O_2 + 2H^+ + 2e^- \leftrightarrow 2H_2O$$

This cell type works at about 200°C, since below 150°C, its conductivity is diminished, and above 220°C, the phosphoric acid is extremely volatile and tends to decompose [6].

The PAFC is based on an immobilized phosphoric acid electrolyte. The matrix universally used to retain the acid is silicon carbide, and the catalyst for both the anode and cathode is platinum [8]. The active layer of platinum catalyst on a carbon-black support and a polymer binder is backed by a carbon paper with 90% porosity, which is reduced to some extent by a Teflon binder [6,9].

In this fuel cell type, the water vapor pressure is noticeably reduced, since an acid is employed in the electrolyte; therefore, water management in the cell is easy [8].

8.1.2.4 Molten Carbonate Fuel Cell

Figure 8.6 shows a schematic illustration of a molten carbonate electrolyte fuel cell (MCFC).

Here, the fuel is a mixture of CO plus H_2 (syngas) obtained, for instance, by methane (CH_4) steaming. Therefore, the oxidation reaction in the anode is given by

$$CH_4 + 2H_2O \rightarrow CO_2 + 4H_2$$

followed by

$$H_2 + CO_3^{2-} \leftrightarrow CO_2 + H_2O + 2e^-$$

and the corresponding reduction process in the cathode is represented as follows:

$$O_2 + 2CO_2 + 4e^- \leftrightarrow 2CO_3^{2-}$$

The electrolyte in this fuel cell is generally a combination of alkali carbonates, which are retained in a ceramic matrix of $LiAlO_2$ [8]. This fuel cell type works at 600°C–700°C, where the alkali carbonates form a highly conductive molten salt with carbonate ions providing ionic conduction. At the high operating temperatures in the molten carbonate fuel cell, a metallic nickel anode and a nickel oxide cathode are adequate to promote the reaction [9]. Noble metals are not required.

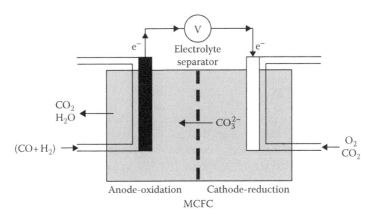

FIGURE 8.6 Schematic representation of an MCFC.

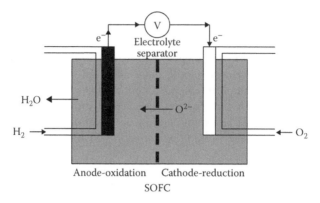

FIGURE 8.7   Schematic representation of an SOFC.

8.1.2.5   Solid Oxide Fuel Cell

The working principles behind a solid oxide fuel cell (SOFC) are schematically illustrated in Figure 8.7, where, similar to the other fuel cell types, the three key parts of an SOFC, a cathode, an anode, and an electrolyte, are shown. The electrolyte is, in a majority of cases, an oxygen-anion ceramic conductor, which is, as well, an electronic insulator [5]. In the SOFC the fuel can be methane (CH_4). Subsequently, in this case the oxidation reaction in the anode is given by

$$CH_4 + 4O^{2-} \rightarrow CO_2 + 2H_2O + 8e^-$$

or H_2 (Figure 8.7)

$$H_2 + O^{2-} \rightarrow H_2O + 2e^-$$

and the corresponding reduction process in the cathode is represented as follows:

$$O_2 + 4e^- \leftrightarrow 2O^{2-}$$

The SOFC works in a temperature range of 800°C–1000°C with O^{2-} conduction in the solid phase; the electrolyte in this fuel cell is a dense solid metal oxide, generally, Y_2O_3-stabilized with ZrO_2 [9]. The anode is usually composed of a porous cermet of Co or Ni catalyst on yttria-stabilized zirconia, that is, a Co-ZrO_2 or a Ni-ZrO_2 cermet [6]. Zirconia, in this case, acts to inhibit grain growth of the catalyst particles of cobalt or nickel and protects against thermal expansion [6]. The cathode is usually Sr-doped $LaMnO_3$, where the Sr dopant provides for oxygen transfer at the cathode–electrolyte interface; besides, a Mg- or a Sr-doped lanthanum chromate is used as the current collector and the intercell connection [8]. In Section 8.7, we describe in detail this fuel cell type.

8.2   SOLID ELECTROLYTES

Electrolytes, in most cases, are in a liquid or a molten state, and diverse carrier ions can migrate fluently in the bulk electrolyte. On the other hand, in various solids, ions can move rapidly with a low energy barrier; such solids demonstrate conductivities similar to those of molten and aqueous electrolytes and are recognized as solid electrolytes [12–37]. In this regard, the investigation of solid ionic conductors commenced in 1838, when Michael Faraday discovered that PbF_2 and A_2S are good electricity conductors [37].

One of the characteristic features of these materials is that only one kind of carrier ion migrates, since the other ions are required to maintain the rigid structure of the solid framework [12]. It is evident that both types of ions can be carriers of electric current in ionic solids. Nevertheless, since usually cations have smaller ionic radii than anions, and smaller ions should diffuse faster, cations normally must diffuse more rapidly in ionic solids than anions. For this reason, the greater part of the fast ionic conductors discovered are cationic conductors [37]. Another feature that affects the ionic diffusivity is the amount of the charge that the ion holds. When the charge on an ion is large, this ion is expected to be restricted to its crystallographic position by stronger electric attraction of the adjacent ions, which, because of the neutrality rule, are of opposite charge [37–39].

In Chapters 2 and 5, we have explained the structure and the transport mechanism of ions in solid electrolytes, respectively. In this chapter, we discuss their applications in electrochemical devices.

8.2.1 Defect Concentration in Ionic Compounds

In this section, we discuss the defect concentration of compounds that do not exhibit electronic conduction, such as pure NaCl.

In thermal equilibrium, some ionic crystals at a temperature above absolute zero enclose a certain number of Schottky pair defects, that is, anion and cation vacancies in the structure (see Section 5.7.1) [13]. Since the concentration of Schottky pair defects at equilibrium at an absolute temperature, T, obeys the mass action law, then [16]

$$C_{V_A} C_{V_C} = e^{-\frac{\Delta G_{\text{Schottky}}}{RT}} \tag{8.1}$$

where

C_{V_A} is the equilibrium concentration of anion vacancies
C_{V_C} is the total concentration of cation vacancies
$\Delta G_{\text{Schottky}}$ represents the Gibbs free energy of formation of the vacancy pair

Similarly, in thermal equilibrium, some ionic crystals at a temperature above absolute zero enclose a certain number of Frenkel pair defects, that is, anion and cation interstitials in the structure. Since the concentration of Frenkel pair defects at equilibrium at an absolute temperature, T, obeys the mass action law, then [16]

$$C_{I_C} C_{V_C} = e^{-\frac{\Delta G_{\text{Frenkel}}}{RT}} \tag{8.2}$$

where

C_{V_C} is the equilibrium concentration of cation vacancies
C_{I_C} is the total concentration of cation interstitials
$\Delta G_{\text{Frenkel}}$ represents the Gibbs free energy of formation of the interstitial pair

Generally, anion interstitials are rare, because the anionic radius is greater than the cationic radius. The rule of electrical neutrality in a material containing both Schottky and Frenkel defects requires that the positive and negative point defects must be balanced, that is

$$C_{V_C} = C_{V_A} + C_{I_C} \tag{8.3}$$

In the case of stoichiometric oxides, the equilibrium constant in the case of the Frenkel and Schottky mechanisms is given by Equations 5.60 and 5.61, which are equivalent to Equations 8.1 and 8.2.

8.2.2 Unipolar Ionic Conductivity in Solids

As described in Section 5.2, the charge drift in a generalized force is described by an atomic or molecular mobility, M_A, which is defined by

$$\bar{v}_A = M_A \bar{F}_A \tag{8.4}$$

where $\bar{v}_A$ is the average drift velocity. Where the only driving force is an electric field, the generalized force on the charged particle, A, is

$$\bar{F}_A = q_A \bar{E} \tag{8.5}$$

where $q_A = z_A F$ is the charge per mole of the moving particle, in which $F = 96,500$ [C/mol] is the Faraday number, and

$$\bar{E} = -\nabla \phi \tag{8.6}$$

Now, the flux is given by

$$\bar{J}_A = \bar{v}_A C_A = M_A C_A q_A \bar{E} = \sigma_A \bar{E} \tag{8.7}$$

where C_A is the concentration of the charged mobile species, A.

Equation 8.6 is an expression of Ohm's law, which for structures with defects can be expressed as follows [32,33]

$$\bar{J}_A = M_A C_A^N N q_A \bar{E} = \sigma_A \bar{E} \tag{8.8}$$

where C_A^N is the fractional occupancy of crystallographic equivalent lattice sites, N, per unit volume on which the charge carrier, A, moves. Consequently,

$$C_A = C_A^N N \tag{8.9}$$

is the charge carrier concentration. The mobility of a charge carrier, whose transport mechanism is diffusion, is given by the Nernst–Einstein relation

$$M_A = \frac{q_A D_A}{RT} \tag{8.10}$$

where D_A is the diffusivity describing the charge carrier transport.

Subsequently, from Equations 8.4 through 8.8

$$J_A = -D_A \frac{q_A C_A}{RT} \nabla \phi \tag{8.11}$$

and the ion current flowing per unit area is given by

$$I_A = q_A J_A = -D_A \frac{q_A^2 C_A}{RT} \nabla \phi \tag{8.12}$$

Now, we calculate a macroscopic expression for Ohm's law for an ionic conductor with only one moving ion, area, S, and length, l, and connected to an external circuit with the help of metallic electrodes, where a potential difference, V, is applied and a current I_e is detected (see Figure 8.8). In this case, the electronic current measured is given by (see Figure 8.8)

$$I_e = D_A \left(\frac{q_A^2 C_A}{RT} \right) \left(\frac{V}{l} \right) S \tag{8.13}$$

This electronic current is equivalent to J_A but in the opposite direction [16].

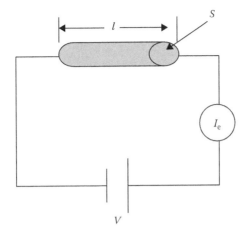

FIGURE 8.8 Ion conductor of area S and length l.

since

$$I_e = \rho_A V \tag{8.14}$$

Then

$$\rho_A = D_A \left(\frac{q_A^2 C_A}{RT} \right) \left(\frac{S}{l} \right) \tag{8.15}$$

where ρ_A the electrical conductance of the ion conductor specimen. Now,

$$\rho_A = \sigma_A \left(\frac{S}{l} \right) \tag{8.16}$$

where σ_A is the electrical conductivity of the ion-conducting material. Consequently, with the help of Equations 8.15 and 8.16, it is possible to calculate the conductivity of an ionic conductor with the help of macroscopic, current intensity–voltage (I–V) measurements, made with a circuit similar to that schematically presented in Figure 8.9, where

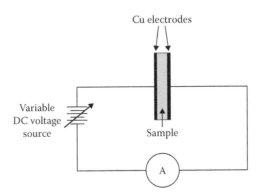

FIGURE 8.9 Schematic representation of the equipment to measure the I–V characteristic curve.

$$\sigma_A = D_A \left(\frac{q_A^2 C_A}{RT} \right) \tag{8.17}$$

is a characteristic of the tested material.

It is now necessary to describe the determination of conductivity in ionic solids. The circuit shown in Figure 8.9 is only a schematic representation. This type of measurement in high-impedance samples of ionic solids causes experimental problems that are avoided with the help of the so-called guard ring circuit, which prevents leakage currents affecting the measurement [16]. Besides, these circuits use an amplifier to deal with the high resistance of the sample.

The total charge current flow throughout an ionic conductor is composed of various charge carriers, such as electrons, holes, ions, and charge defects; then

$$I = \sum_i I_i \tag{8.18}$$

where the component currents will depend on the material properties. For example, proton conduction is ambipolar, that is, protons and electrons are transported simultaneously, and this effect is described separately in Section 8.2.5. Similarly, oxygen conduction is also ambipolar, that is, oxygen anions, O^{2-}, and electrons are transported simultaneously, and this effect is discussed in Section 8.2.6.

8.2.3 Examples of Unipolar Cationic Conductors

The majority of unipolar ionic conductors identified to date are polymorphic compounds with several phase transitions, where the phases have different ionic conductivities owing to modifications in the substructure of the mobile ions [28]. One of the first studied cationic conductors was α-AgI [21]. Silver iodide exhibits different polymorphic structures. AgI has a low-temperature phase, that is, β-AgI, which crystallizes in the hexagonal wurtzite structure type, and a high-temperature cubic phase, α-AgI, which shows a cubic CsCl structure type [20,22] (see Section 2.4.5).

Alpha silver iodide, the high-temperature phase of AgI, shows an extraordinarily high Ag^+ ionic conductivity. On the other hand, the conductivity at room temperature, that is, the conductivity of β-AgI is considerably lower [12,20]. In the α-AgI phase, there are only two Ag^+ ions distributed over the octahedral and tetrahedral positions of the cubic lattice, producing many vacancies that are accessible for Ag^+ ions to bypass through the structure [20]. Then, in order to operate these materials it is necessary to stabilize the high-temperature α phase at low temperatures [12]. Subsequently, a considerable amount of studies of ternary and quaternary compounds have been performed with the purpose of stabilizing fast ionic phases at low temperatures [29].

One largely investigated system in the field of silver ion conductivity is the RbI-AgI system, including the fast ionic conductor, $RbAg_4I_5$ [19,28,29]. Structural investigations of these compounds started 40 years ago and the field has been active during the last four decades. The mechanism of conduction and the favored conduction pathway in ionic conductors is important for the development of solid-state batteries [28].

New, optimized ion conductors, such as Ag_5Te_2Cl, demonstrate the characteristic performance of ion-conducting materials, specifically, polymorphism in combination with ion conduction in a high-temperature phase [28].

During the last 40 years, an outstanding Na^+ ion conductor has been developed, sodium β-alumina, or $Na_2O \cdot 11Al_2O_3$. This material has a hexagonal structure consisting of two hexagonal close-packed spinel-like blocks along the c-axis separated by a mirror plane with an oxygen density of 1/4 of the density of the spinel-like blocks, where these mirror planes contain mobile Na^+ ions [20]. Consequently, sodium β-alumina is a solid electrolyte with a layered structure where the Na^+ ions migrate along the two-dimensional conductive planes; consequently, the ionic conduction in this structure is to some extent restricted owing to the anisotropic conduction through the two-dimensional planes [12]. In order to eradicate this constraint, a new structure was proposed [23,24], which was designed with a suitable tunnel size for Na^+ migration in three dimensions. This novel polycrystalline material, that is, $Na_{1+x}Zr_2Si_xP_{3-x}O_{12}$, designed as a Na^+ super ionic conductor, or NASICON, shows acceptable stability in air [19].

It has been reported [25] that materials with the following unit cell composition, $Li_xLa_{1/3}Nb_{1-x}Ti_xO_3$ and $Li_xNd_{1/3}Nb_{1-x}Ti_xO_3$, exhibiting perovskite structure, are good lithium ion conductors. This is significant, since the solid-state rechargeable lithium battery is promising, owing to its high energy density and negative electrochemical potential. Another Li^+ conducting material specifically applied for the preparation of cathodes for lithium ion batteries is the spinel $LiMn_2O_4$, exhibiting orthorhombic structure [15]. Besides, lithium rare earth silicates, having a general composition, $LiMSiO_4$, where M = La, Nd, Sm, Eu, Gd, were reported as another type of Li^+-conducting solid electrolyte with high Li^+ ion conductivity [19]. In addition, the $LiMSiO_4$ Li^+ ion conductor exhibits an apatite-type crystalline structure [26,27]. Besides, the $Y_{0.8}Li_{0.6}PO_4 - Li_3PO_4$ mixed phase was confirmed to be predominantly a Li^+ ion–conducting solid electrolyte [19].

In addition, superionic conducting glasses, which are also named glassy electrolytes, are cationic conductor materials of importance because of their advantages such as isotropic conductivity, simplicity of preparation, superior thermal stability, and the accessible composition ranges which make them potential candidates for different applications. Among these materials, considerable attention has been given to silver ion–conducting glasses owing to their high-ionic conductivity [30].

8.2.4 ANIONIC CONDUCTORS

Both types of ions can be carriers of electric current in ionic solids; but the majority of fast ionic conductors discovered are cationic conductors, for the reasons previously explained.

Between, the existing anions only, F^- and Cl^- are found to be good carriers of electricity [37,40]. For example, rare-earth oxychlorides; between them, lanthanum oxychloride shows anionic conduction; however, its conductivity, is small; then to increase it, this material was doped with Ca^{2+} in the trivalent, La^{3+}, site of LaOCl in order to generate Cl^- vacancies [19,40].

8.2.5 PROTON CONDUCTORS

8.2.5.1 Introduction

In the 1960s, all the tools needed to treat hydrogen defects in oxides [41–43] were fundamentally developed by Wagner and collaborators. However, up to the end of the 1970s, real developments in the field did not take place. During the last years of this decade, some significant studies were carried out [44–48]. After that, it was realized that the introduction of defects in some perovskite structures determine the protonic conductivity of these materials. In this regard, Iwahara and collaborators [49,50], were the first who investigated protonic conductivity in $SrCeO_3$ and $BaCeO_3$ doped with trivalent cations such as Y, Yb, Gd, and Eu. These researchers identified these materials as good high-temperature proton conductors. Thereafter, extensive research has been carried out in this field (see Sections 5.7.3 through 5.7.7) [51–78].

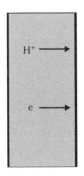

8.2.5.2 Conductivity in Proton Conductors

One possible general approach to calculate the conductivity in proton conductors is the phenomenological approach applying the formalism of nonequilibrium thermodynamics [13] (Section 5.7.6) to calculate the hydrogen diffusion coefficient in oxides [36].

FIGURE 8.10 Hydrogen transport in a mixed ion electron ionic conductor.

The flux density of protons is given by (see Figure 8.10) [77–80] the ambipolar equation (see Section 5.7.6)

$$J_{H^+} = -L_{HH} \frac{t_e}{2} \nabla\mu(H_2) \tag{8.19}$$

where t_e is the electron transference number given by

$$t_e = \frac{L_{ee}}{L_{HH} + L_{ee}} \tag{8.20}$$

in which L_{HH} and L_{ee} are the Onsager transport coefficients.

Now, from Ohm's law

$$I_{H^+} = \sigma_{H^+} \bar{E} \tag{8.21}$$

and from Equation 8.19, it is possible to get

$$J_{H^+} = -L_{HH} \cdot F \nabla\Phi \tag{8.22}$$

when $\nabla\mu_{H^+} = 0$.

Now,

$$I_{H^+} = F J_{H^+} \tag{8.23}$$

and

$$\bar{E} = -\nabla\Phi \tag{8.24}$$

Then

$$\sigma_{H^+} = F^2 L_{HH} \tag{8.25}$$

and

$$\sigma_{e^-} = F^2 L_{ee} \tag{8.26}$$

where σ_{H^+} and σ_e are the proton and electron conductivities, respectively. Thereafter, from Equation 8.20, 8.24, and 8.25

$$t_e = \frac{\sigma_e}{\sigma_{H^+} + \sigma_e} \tag{8.27}$$

is obtained, with which the electronic transference number can be calculated.

8.2.6 Oxide Conduction

8.2.6.1 Oxygen Conductors

Oxide ion conductors have been extensively investigated for their applications in fuel cells, oxygen sensors, oxygen pumps, and oxygen permeable membranes [81–108]. The ion conduction effect was discovered more than a century ago by Nernst in zirconia products [83,84]. To use zirconia, it is necessary to stabilize it. To do this, different oxides, such as CaO, MgO, Y_2O_3, and others, for example, adding from 16 to 26 wt % of CaO produces a cubic phase at all temperatures, are used [86] (see Section 2.4.3).

For oxide ion conductors, vacancy hopping is the major transport mechanism; consequently, the materials should contain oxygen vacancies to conduct. To obtain oxide conduction properties, a part of the Zr^{4+} must be substituted by another cation with a lower valence state, that is, Ca^{2+}, Sc^{3+}, Y^{3+}, or a rare-earth cation [84,86].

In a majority of ion conductor applications, high temperatures are required to achieve the relatively high oxygen fluxes required for efficient operation. Several questions need to be addressed with regard to the use of high operating temperatures, such as the high cost of materials, material stability and compatibility, and the thermal degradation of the electrolyte itself [82]. Diminishing the thickness of the solid electrolyte can raise the oxygen flux, and thus reduce the temperatures needed to attain a specified flux. However, there are restrictions to this approach [88]. It is, therefore, necessary to develop new materials that exhibit high oxide ion conductivities at relatively low temperatures, that is, below 800°C.

To reduce the working temperature, research efforts have been undertaken to search for new materials with conductivity values bigger than those of zirconia-based materials [82,84]. In this regard, ceria solid solutions have been extensively studied, since their conductivity is higher than those of zirconia-based phases. For example, the conductivity of a cerium gadolinium mixed oxide is about five times the value corresponding to yttrium zirconium oxide [82]. Samarium-doped ceria exhibits the highest ionic conductivity [89]. Besides, the high-temperature phase of bismuth oxide, that is, δ-Bi_2O_3, similar to stabilized zirconia, exhibits high oxide ion conductivity. The structure of δ-Bi_2O_3 is similar to the fluorite structure, with one-quarter of the anion sites vacant, and is formed at temperatures above 730°C. However, after doping with yttria, the bismuth oxide stabilizes the δ-phase down to 25°C [82].

Another group of materials that has displayed high oxide ion conductivity is based upon a layered bismuth perovskite-based structure, first reported by Aurivillius in 1949 [90–92]. The so-called Aurivillius phases are chemically expressed normally as $Bi_2A_{n-1}B_nO_{3n+3}$ [82], where A is a large 12-coordinated cation and B a small 6-coordinated cation. The structure is formed by n perovskite-like layers, $(A_{n-1}B_nO_{3n+1})^{2-}$, sandwiched between bismuth–oxygen fluorite-type sheets, $(Bi_2O_2)^{2+}$ [93,94].

Aurivillius phases have attractive ferroelectric properties [94]. However, in recent times, various researchers have investigated the oxide ion conductivity of these phases [82]. The first evidence of high oxide ion conductivity in an alternate $(Bi_2O_2)^{2+}$ layer/perovskite layer Aurivillius-type structure was published in 1988 [95]. The materials examined to date fall into two classes: those containing intrinsic oxygen vacancies, for example, $Bi_2VO_{5.5}$, and those with extrinsic oxygen vacancies, for instance, $Bi_2Sr_2Nb_2AlO_{11.5}$ [82].

Also, pyrochlore materials have been studied as solid electrolytes, since the electrical properties of pyrochlores can change after modification of their composition [87]. The stoichiometric composition of these oxides is $A_2B_2O_7$ (A^{3+} and B^{4+}), for example, $Ln_2B_2O_7$, where B = Ti, Zr, and Ru [20]. But, the general pyrochlore formula, $A_2B_2O_6O'_{1-\delta}$, shows the recognized capability of the pyrochlore structure to house an oxygen nonstoichiometry [84]. The pyrochlore structure can be described as an ordered, oxygen-deficient fluorite-type structure (see Section 2.3.6), where the cubic unit cell parameter of the pyrochlore is twice the fluorite one, that is, eight fluorite unit cells are required to completely portray the pyrochlore arrangement [84] (see Figure 8.10). To get a stoichiometric pyrochlore, 1/8 of the anions are removed systematically from the fluorite unit cell [20,87]. The fluorite unit cells contained in the pyrochlore structure can be separated in two groups, that is, group I and II; in the type I unit cell, the oxygen vacancy is located in the $\left(\frac{3}{4}, \frac{3}{4}, \frac{3}{4}\right)$ tetrahedral site of the fluorite cell, and in type II cubes, the oxygen vacancy is located in the $\left(\frac{1}{4}, \frac{1}{4}, \frac{1}{4}\right)$ tetrahedral site, where, in

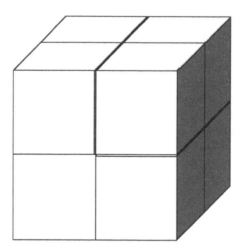

FIGURE 8.11 Schematic representation of the eight fluorite unit cells required to represent the pyrochlore structure.

the pyrochlore structure, types I and II are arranged in such a way that a type I cube shares faces only with type II cubes [84,87] (see Figure 8.11).

Ample studies on pyrochlore oxide electrolytes have been carried out, particularly on $Gd_2Ti_2O_7$- and $Gd_2Zr_2O_7$-based conductors, where the $Gd_2Ti_{1-x}Zr_xO_7$ solid solution is of great interest because the $x = 0$ member is an ionic insulator whereas the $x = 1$ end member is a good oxide ion conductor [96,97].

Another class of important materials with oxygen ion conduction are perovskites (see Section 2.4.4). Perovskite phases have been, initially, considered as possible electrode materials for SOFCs. The first report of the existence of oxide conductivity in a perovskite material was made using a calcium-doped lanthanum aluminate [98]. But the discovery of attractive oxygen conduction properties is rather recent [84].

In order to get high oxide ion conductivity, three criteria were defined [99]: low mean value of metal–oxygen bonding energy; an open structure, that is, high free volume; and a critical cation bottleneck, as large as possible, for O^{2-} migration. These criteria mean that both the A and B cation radius and valence state are determining factors [84]. In this sense, a perovskite doped on both the A- and B-sites, specifically, $La_{0.9}Sr_{0.1}Ga_{0.8}Mg_{0.2}O_{2.85}$, exhibiting a conductivity at 750°C close to 0.1 [S cm^{-1}], was synthesized [100,101]. These types of perovskites were studied, as well, with the following composition $(La_{0.9}Ln_{0.1})_{0.8}Sr_{0.1}Ga_{0.8}Mg_{0.2}O_{2.85}$, where Ln = Y, Nd, Sm, Gd, or Yb, and was shown that the conductivity diminishes in the following sequence, Nd > Sm > Gd > Yb > Y [102]. This change is an example of the effect of the crystal lattice symmetry on the oxygen mobility, since the conductivity diminishes linearly with the perovskite tolerance factor (see Section 2.4.4) from the ideal ion radii ratio for a cubic perovskite structure [84,103].

By doping both perovskite metals A and B, a different composition, $La_{1-x}Sr_xCo_{1-y}Fe_yO_{3-\delta}$, was obtained [104]. It was demonstrated that the $La_{1-x}Sr_xCo_{1-y}Fe_yO_{3-\delta}$ group of perovskites displays excellent oxygen permeation properties along with efficient oxygen reduction.

8.2.6.2 Conductivity in Oxygen Conductors

The simultaneous movement of ionic and electronic charge carriers under the driving force of a gradient in the electrochemical potential of oxygen facilitates transport of oxygen in the oxide bulk. The flux density of oxide anions is given (Figure 8.12) [77–79,109] by the ambipolar diffusion equation (see Section 5.7.6) [110,111]

$$J_{O^{2-}} = -L_{OO}\frac{t_e}{2}\nabla\mu(O_2)$$

(8.28)

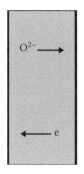

FIGURE 8.12 Oxygen transport in a mixed ion electron ionic conductor.

where t_e is the electron transference number, given by

$$t_e = \frac{L_{ee}}{4L_{OO} + L_{ee}} \tag{8.29}$$

Now, from Ohm's law

$$I_{O^{2-}} = \sigma_{O^{2-}} \bar{E} \tag{8.30}$$

Following a methodology similar to that applied in Section 8.2.5.2, it is possible to show that

$$\sigma_{O^{2-}} = 4F^2 L_{OO} \tag{8.31}$$

and

$$\sigma_{e^-} = F^2 L_{ee} \tag{8.32}$$

where $\sigma_{O^{2-}}$ and σ_e are the oxide anion and electron conductivities, respectively. Then, substituting Equations 8.31 and 8.32 in Equation 8.29, we get

$$t_e = \left(\frac{\sigma_e}{\sigma_{O^{2-}} + \sigma_e} \right) \tag{8.33}$$

which is another method to evaluate the electronic transference number.

8.2.7 ZEOLITE ELECTROLYTE

We have previously described the structure, adsorption, diffusion, and ion exchange properties of zeolites; we now describe zeolites as solid electrolytes [38,112]. During the 1950s and 1960s, the conduction mechanism in zeolites was ascribed to the charge-compensating extra-framework cations [113,114]. However, it was established later that for hydrated zeolites, proton or hydroxylic species in H_2O are the fundamental charge carriers [38,112,115–117]. Extra-framework cation conduction is important for dehydrated zeolites at relatively high temperatures [38].

In the case of hydrated zeolites at high filling factors, the zeolite secondary porosity is occupied by water molecules, since the secondary porosity is formed by the space between zeolite crystallites [118]. This water allows the electric connection between the zeolite crystallites, and then allows extended charge transport, that is, intra- and intercrystallite conductions [38,112,119].

In order to study the zeolite conduction mechanism, the electric current intensity versus voltage (I–V) characteristic curves of the hydrated and dehydrated Na-X zeolite (Si/Al = 1.25, provided by Laporte) in the form of cylindrical wafers compacted at a pressure of 250 MPa, obtaining wafers of different heights, h, specifically, $h = 1$, 2, and 3 mm and a diameter $d = 13$ mm, were measured [112]. The studied wafers (see Figure 8.9) were included between Cu electrodes with the help of a circuit comprised of a highly sensitive ammeter (1 nA of sensitivity), and a DC ramp voltage source, which allows to scan the voltage applied to the sample at a rate of 1.5 V/s. All the tested hydrated wafer samples were made to contact saturated water vapor at 300 K for 24 h to get a magnitude of water adsorption in the samples of about 10 mmol/g [112,119]. Figure 8.13 shows the I–V curves of a dehydrated and a hydrated sample [112]. It is evident that the obtained I–V curves, in the case of

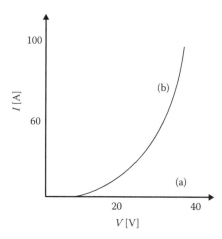

FIGURE 8.13   *I–V* characteristic curves of (a) dehydrated zeolite and (b) hydrated zeolite at 300 K.

the hydrated samples, which is a square voltage law, are normally related to space charge–limited current [32,120]. Space charge is a phenomenon that occurs in a semiconductor or an insulator, when the conducting material is pervaded with a net positive or negative charge, injected by one of the electrodes during the conduction process [32]. In this case, the current will be controlled by the space charge, because it will affect the charge carrier mobility in the bulk of the material. We analyze here a case of cationic injection by the Cu, cathode to the zeolite. For cation injection to Cu, electrodes (for other metals, the mechanism is similar), the proposed mechanism is [112]:

$$H_2O \leftrightarrow H^+ + OH^- \tag{8.34}$$

and

$$H^+ + Na\text{-}X \leftrightarrow Na^+ + H\text{-}X \tag{8.35}$$

That is, water dissociation and cationic exchange in the zeolite are the first processes. The second step (corroborated by measuring the Cu concentration in the surfaces of the wafers after the conduction test with the help of XRD-fluorescence spectrometry [112]) consists of the following processes:

$$2Cu + 2OH^- \leftrightarrow CuO_2 + H_2O + 2e^- \tag{8.36}$$

$$Cu_2O + 2OH^- + H_2O \leftrightarrow 2Cu(OH)_2 + 2e^- \tag{8.37}$$

The equation which describes the conduction for space charge–limited current in the case of a trap-free solid [32,120] is the Mott–Gurney equation [121]

$$J = \frac{9}{8}\varepsilon M \left(\frac{V^2}{L^3} \right) \tag{8.38}$$

where
  V is the applied voltage
  L is the sample thickness
  M is the charge-carrier drift mobility
  ε is the static dielectric constant in the solid

The Mott–Gurney equation is obtained as follows: the electric field E inside the conducting solid is given by the Poisson equation [32]

$$\frac{dE}{dx} = \frac{qp(x)}{\varepsilon}$$ (8.39)

where
  $p(x)$ is the density of injected charge
  q is its charge
  ε is the static dielectric constant in the solid

Now, the current density is given by

$$J = qMp(x)E(x)$$ (8.40)

where M is the charge-carrier drift mobility. Now, combining Equations 8.39 and 8.40, we get [32]:

$$(dE)^2 = \frac{2J}{\varepsilon M}dx$$ (8.41)

Thereafter, integrating Equation 8.41 with the condition

$$V = \int_0^L E(x)dx$$ (8.42)

Equation 8.38 is obtained.

The conditions of the experiment discussed here are different than the restrictions imposed to obtain the Mott–Gurney equation. However, at least qualitatively, Equation 8.38 can describe the space charge–limited current effect. To test this hypothesis, Equation 8.38 was experimentally tested, and it was shown that it is approximately satisfied (see Figures 8.13 and 8.14) [112].

The mobility in the present experimental arrangement is given by [120]

$$M = \frac{L^2}{\tau V_0}$$ (8.43)

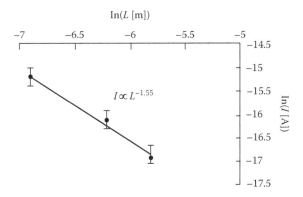

FIGURE 8.14 Plot of the dependence between the current, I [A], and the specimen thickness, L [m].

where

V_0 is the transition voltage from Ohm's law to the square voltage law

τ is the time constant of the capacitor, that is, the parallel plate capacitor, which constitutes the sample plus the electrode (see Figure 8.9)

Subsequently, the mobility was measured with the help of a pulse of amplitude, V_0, which allowed the calculation of τ [112]. The value of the mobility was $M = (3\pm1) \times 10^{-7}$ [m^2/Vs], which is very close to the values reported for H$^+$ and OH$^-$ in solution [112].

Therefore, hydrated zeolites conduct through intra- and intercrystalline cationic conduction where the material acts as a solid electrolyte, with the help of the water adsorbed in the primary and secondary porosities of the zeolite [112,119]. The conduction process is facilitated by oxidation–reduction reactions in the electrodes, initiated by OH$^-$ anions generated by an ion-exchange reaction between the protons generated by water dissociation and the Na$^+$ extra-framework cations included in the zeolite. These reactions cause charge injection, which controls the conduction process [112].

8.3 THERMODYNAMICS OF ELECTROCHEMICAL PROCESSES

One of the first quantitative studies of an electrochemical process, specifically, electrolysis, was carried out by Faraday in 1832 [122]. He obtained two laws that are defined as follows:

1. The weight of a substance generated by a cathode or anode reaction during an electrolytic process is directly proportional to the amount of electricity passed through the electrolytic cell.
2. The weights of diverse substances generated by the same amount of electricity are proportional to the equivalent weights of the substances.

Faraday's laws can be condensed in the following expression

$$w = \frac{ItM}{nF} \tag{8.44}$$

where

w is the amount of matter reacted, in grams
I is the current flowing, in amperes
t is the time elapsed during the whole electrolytic process
M is the molecular or atomic weight of the material being transformed
n is the number of electrons taking part in the electrolytic reaction
F is the Faraday number, which is defined as the charge of one equivalent of electrons

The fundamental thermodynamic equation for a reversible electrochemical transformation is given by [6,8,10]

$$\Delta G = \Delta H - T\Delta S \tag{8.45}$$

where

ΔG is the Gibbs free energy, change during the process
ΔH is the enthalpy or the energy released or absorbed by the reaction
ΔS is the entropy change
T is the absolute temperature

The terms ΔG, ΔH, and ΔS are state functions and depend only on the initial and final states of the reaction. Equation 8.45 can be written as follows

$$\Delta G^0 = \Delta H^0 - T\Delta S^0 \tag{8.46}$$

where the value of the function is for the material in the standard state, that is, at 298 K and for unit activity.

Since, ΔG corresponds to the net useful energy available from a given reaction, thereafter, in electrical terms, the net available electrical energy from a reaction in an electrochemical cell can be expressed as follows

$$\Delta G = -nFE \tag{8.47}$$

where
  n is the number of electrons transferred during the electrochemical reaction per mole of reactant
  F is the Faraday constant
  E is the voltage of the electrochemical cell

In other words, E is the electromotive force (EMF) of the reaction cell, where the voltage of the cell is unique for each reaction couple. Spontaneous processes have a negative free energy; consequently, an electrochemical process will have a positive EMF.

The van't Hoff isotherm identifies the free energy relationship for bulk chemical reactions as

$$\Delta G = \Delta G^0 + RT \ln\left(\frac{a_P}{a_R}\right) \tag{8.48}$$

where
  a_P is the activity of the products
  a_R is the activity of the reactants

Subsequently, substituting Equation 8.47 in Equation 8.48, we get the Nernst equation:

$$E_e = E^0 + \left(\frac{RT}{nF}\right)\ln\left(\frac{a_P}{a_R}\right) \tag{8.49}$$

where

$$E^0 = -\frac{\Delta G^0}{nF}$$

Equation 8.49 describes the equilibrium EMF of a concrete electrochemical cell in an open circuit regime, that is, when the current passed through an external circuit is zero.

8.4 KINETICS OF ELECTROCHEMICAL PROCESSES

8.4.1 OVERPOTENTIAL

Kinetic processes in an electrochemical reaction must be considered when current is extracted from a battery, that is, when the current passed through an external circuit is different from zero. In this case, while current is removed from the battery, the equilibrium voltage or open circuit voltage falls because of electrode polarization or overvoltage [6,8,10,66,123,124]. This electrode polarization effect occurs owing to kinetic limitations experienced by the reactions, and because of other processes taking place during the production of current flow.

Similar to chemical kinetics, the mechanism of electrochemical reactions regularly requires a series of physical, chemical, and electrochemical steps, comprising charge-transfer and charge-transport reactions. The velocity of these individual steps controls the kinetics of the electrode reactions and, thus, of the cell reaction. In this sense, three diverse kinetics effects for polarization must be taken into account:

- Activation polarization effect, which is associated with the kinetics of the electrochemical oxidation–reduction or charge-transfer reactions occurring at the electrode/electrolyte interfaces of the anode and the cathode.
- Ohmic polarization which is a phenomenon linked to the resistance of particular cell components and to the resistance as a result of contact problems among the cell components.
- Concentration polarization, which is a result of mass transport limitations throughout the cell function.

In this regard, the overpotential, η [V], is given by

$$\eta = E_e - E_P \tag{8.50}$$

where

E_e is the voltage measured in the case of an open circuit
E_P is the terminal voltage measured in a cell where a current, I, is flowing

The overvoltage, η, can be written as

$$\eta = \eta_A + \eta_\Omega + \eta_C + E_L \tag{8.51}$$

where

η_A is the activation overpotential owing to the kinetics of the electrode reactions
η_Ω is the overpotential caused by the ohmic resistance in the cell
η_C is the overpotential due to mass transport restrictions
E_L is the loss in voltage as a result of leaks across the electrolyte

8.4.2 Activation Polarization

8.4.2.1 Tafel Equation

The activation polarization takes place from kinetics impediments of the charge-transfer reaction occurring at the electrode/electrolyte interface; this form of kinetics is better understood applying the transition state theory.

The net current density, J, at an electrode is given by [10]

$$J = J_A - J_C = \frac{I_A}{S} - \frac{I_C}{S} \tag{8.52}$$

where

J_A and I_A are the anodic current density and the anodic current, respectively
J_C and I_C are the cathodic current density and the cathodic current, respectively
S is the electrode surface area

If the transition state theory is applied, the reaction takes a course comprising an activated complex, where the rate-limiting step is the dissociation of the activated complex [6,10,66,123,124]. Applying Equation 8.52 for a first-order reaction, the net current flow is given by the Butler–Volmer equation

$$J = J_0 e^{\frac{(1-\alpha)F\eta_A}{RT}} - J_0 e^{-\frac{\alpha F\eta_A}{RT}} \tag{8.53}$$

where

  J_0 is the exchange current density

  η_A is the overpotential or polarization or departure from equilibrium due to the kinetics of the electrode reactions, (see Equation 8.50)

  α is the transfer coefficient, which has a value which lies in the range $0 < \alpha < 1$ and is experimentally found to be $\alpha \approx 0.5$

When the overpotential is high and positive, corresponding to the anode during electrolysis

$$J = J_0 e^{\frac{(1-\alpha)F\eta_A}{RT}} \tag{8.54}$$

Consequently,

$$\ln\left(\frac{J}{J_0}\right) = (1-\alpha)\left(\frac{F}{RT}\right)\eta_A \tag{8.55}$$

When the overpotential is high and negative, corresponding to the cathode during electrolysis

$$J = -J_0 e^{-\frac{\alpha F\eta_A}{RT}} \tag{8.56}$$

Consequently,

$$\ln\left(-\frac{J}{J_0}\right) = -\alpha\left(\frac{F}{RT}\right)\eta_A \tag{8.57}$$

The activation polarization plot, named the Tafel plot, follows the equation

$$\eta_A = \pm\beta\log\frac{J}{J_0} \tag{8.58}$$

where

  β and J_0 are constants

  the slope β allows the calculation of the transfer coefficient, α

  J_0, estimated for $\eta_A = 0$, gives the exchange current density

To make Tafel plots (Figure 8.15) [124,125] subsequent to the subtraction of the ohmic losses (iR), the electrode overpotential, η, is plotted versus $\ln(J/J_0)$. For systems controlled by standard electrochemical kinetics, the slope at a high overpotential gives the anodic and the cathodic transference coefficients, α_A and α_C, respectively, and the intercept gives the exchange current density, J_0.

8.4.2.2 Calculation of the Transference Coefficient

Consider the following electrode reaction [126]

$$A^{n+} \rightarrow A^{(n+1)+} + e^- \tag{8.59}$$

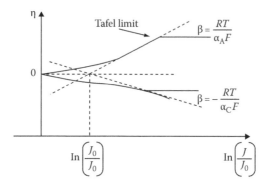

FIGURE 8.15 Tafel plots.

Where there is no transport control, it is possible to suppose that in the transition from A^{n+} to $A^{(n+1)+}$, an intermediary species (IS) is formed. This species has the property to share an electron during time, τ, with the electrode surface. It is then possible to describe the quantum state of the shared electron as follows [127]

$$|IS\rangle = C_1|I\rangle + C_2|II\rangle \tag{8.60}$$

where $|I\rangle$ and $|II\rangle$ are the quantum states of the electron in the ion and the surface states, respectively. Besides [127]

$$\delta = C_1 C_1^* \tag{8.61}$$

and

$$\alpha = C_2 C_2^* \tag{8.62}$$

are the occupation probabilities of the quantum states $|I\rangle$ and $|II\rangle$, where

$$\delta + \alpha = 1 \tag{8.63}$$

Therefore, the mean charge number of the IS is

$$\overline{q}_{IS} = (n + \delta) \tag{8.64}$$

where $\delta = 1 - \alpha$, since statistically a fraction, α, of the electron is given up to the electrode in a quantum sense. Now, it is possible to calculate the chemical potentials of the ion, A^+, in the bulk electrolyte phase and the IS as [127]

$$\mu_{A^{n+}} = \mu_{A^{n+}}^0 + RT \ln a_{A^+} \tag{8.65}$$

and

$$\mu_{IS} = \mu_{IS}^0 + RT \ln \gamma\theta \tag{8.66}$$

where
$a_{A^{n+}}$ is the activity coefficient of the ion, A^+, in the bulk electrolyte phase
θ is the IS surface concentration

γ is the IS activity coefficient

$\mu^0_{A^{n+}}$ and μ^0_{CAS} are the standard chemical potentials

We can now suppose that if the rate of transformation of the IS into A^{n+1} is slow compared to the rate of formation of the IS from A^{n+}, then the Nernst equation can be applied to obtain [127]

$$\mu_{IS} - \mu_{A^{n+}} = (\overline{q}_{IS} - q_{A^{n+}})F\phi \tag{8.67}$$

where

ϕ is the electrode potential

$q_{A^{n+}} = n$ is the charge number of the ion A^{n+}

The electrode potential, ϕ, is an overpotential generated by the electrode reaction; therefore, according to the notation followed here, it is possible to identify ϕ with η. Consequently, substituting Equations 8.65 and 8.66 in Equation 8.67 and making $\phi = \eta$, we get

$$\theta = K \frac{a_{A^{n+}}}{\gamma} e^{\frac{(1-\alpha)F\eta}{RT}} \tag{8.68}$$

where

$$K = e^{\frac{\mu^0_{A^{n+}} - \mu^0_{CAS}}{RT}} \tag{8.69}$$

Now, the oxidation rate can be expressed as follows [127]

$$\frac{d\theta}{dt} = -\frac{\theta}{\tau} \tag{8.70}$$

where τ, as previously stated, is the lifetime of the IS, and θ, expressed in moles per meter square, is the IS surface concentration. The current density is given by [127]

$$J = \frac{d\theta}{dt} \tag{8.71}$$

Then,

$$J = \frac{FKa_{A^{n+}}}{\tau\gamma} e^{\frac{(1-\alpha)F\eta}{RT}} = J_0 e^{\frac{(1-\alpha)F\eta}{RT}} \tag{8.72}$$

where Equation 8.72 is an expression of the Tafel equation in the oxidation case, where α is the transference coefficient. The exchange current is given by

$$J_0 = \frac{FKa_{A^{n+}}}{\tau\gamma} \propto \left[\frac{C/mol}{s(m^2/mol)} \right] = \left[\frac{A}{m^2} \right] \tag{8.73}$$

The reduction case can be treated similarly.

If we now suppose [127] that the system of electrons forming bonds with the electrode surface fulfills the Gibbs canonical ensemble [5], subsequently

$$\delta = \frac{e^{-\frac{E_1}{RT}}}{e^{-\frac{E_1}{RT}} + e^{-\frac{E_2}{RT}}} \tag{8.74}$$

and

$$\alpha = \frac{e^{-\frac{E_{\backslash 2}}{RT}}}{e^{-\frac{E_1}{RT}} + e^{-\frac{E_2}{RT}}} \tag{8.75}$$

where E_1 and E_2 are the energies of states $|I\rangle$ and $|II\rangle$. Consequently, if $E_1 \approx E_2$, the transference coefficient is $\alpha \approx 0.5$, which is the experimental value normally measured.

8.4.3 OHMIC POLARIZATION

Ohmic polarization takes place on account of resistance to the flow of ions and electrons in the battery. More precisely, ohmic polarization results from the resistance that arises as a result of the presence of such components in the battery as the electrolyte, electrodes, current collectors, and terminals. The overpotential generated is expressed by the term IR, in which R is the specific area resistance [6,8,66] and I is the flowing current. This type of polarization emerges and vanishes instantly, when the current flows and ceases, respectively. This is given by the Ohm's law relationship, $\eta_\Omega = IR$, between the current, I, and the overpotential, η_Ω, due to the ohmic resistance in the cell.

8.4.4 CONCENTRATION POLARIZATION

The transfer of reactants from the bulk solution to the electrode interface and in the reverse direction is an ordinary feature of all electrode reactions. As the oxidation–reduction reactions advance, the accessibility of the reactant species at the electrode/electrolyte interface changes. This is because of the concentration polarization effect, that is, η_C, which arises due to the limited mass transport capabilities of the reactant species toward and from the electrode surface, to substitute the reacted material to sustain the reaction [6,8,10,66,124]. This overpotential is usually established by the velocity of reactants flowing toward the electrolyte through the electrodes and the velocity of products flowing away from the electrolyte. The concentration overpotential, η_C, due to mass transport restrictions, can be expressed as

$$\eta_C = \frac{RT}{n} \ln\left(\frac{C}{C_0}\right)$$

where
  C is the concentration at the electrode surface
  C_0 is the concentration in the bulk of the solution

8.5 FUEL CELL EFFICIENCY

8.5.1 POLARIZATION CURVE

In a fuel cell, the most important parameter of its operation is the voltage output as a function of the electric current density withdrawn, or polarization curve (Figure 8.16), where all the terms incorporated in Equations 8.50 and 8.51 are included [8,66].

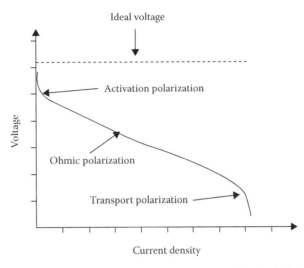

FIGURE 8.16 Schematic fuel cell polarization voltage [V] vs. current density [A/cm^2] curve.

Then, useful electric energy is attained from a fuel cell when current is extracted; however, during this process, the real cell potential is reduced from its equilibrium potential because of the irreversible losses previously explained.

8.5.2 THERMODYNAMIC EFFICIENCY OF A FUEL CELL

The thermal efficiency, E, of an energy conversion machine is described as the quantity of useful energy, E_U, generated relative to the change in stored chemical energy, ΔH, which is liberated when a fuel is reacted with an oxidant:

$$E = \frac{E_U}{\Delta H} \tag{8.76}$$

As previously defined in the ideal case of an electrochemical converter, such as a fuel cell,

$$E_U = \Delta G$$

Subsequently [128],

$$E = \frac{\Delta G}{\Delta H} = 1 - \frac{T\Delta S}{\Delta H} \tag{8.77}$$

where
    ΔS is the isothermal entropy variation of the reaction
    $T\Delta S$ is the reversible heat exchanged with the external environment

In the specific case of a fuel cell, if the fuel is hydrogen and the oxidant oxygen, the entire reaction in the cell is given by the following chemical equation [8]:

$$H_2(g) + \frac{1}{2}O_2(g) \rightarrow H_2O(l) \tag{8.78}$$

Consequently, the standard free energy change is given by

$$\Delta G_r^0 = G_{H_2O}^0 - G_{H_2}^0 - \frac{1}{2}G_{O_2}^0 = 237.1 \; [kJ/mol] \tag{8.79a}$$

and the standard enthalpy change is given by

$$\Delta H_r^0 = H_{H_2O}^0 - H_{H_2}^0 - \frac{1}{2}H_{O_2}^0 = 285.8 \; [kJ/mol] \tag{8.79b}$$

Thus, the thermal efficiency, E_{FC}^l, of an ideal fuel cell operating reversibly on pure hydrogen and oxygen at standard conditions would be

$$E_{FC}^l = \frac{237.1}{285.8} = 0.83 \quad or \quad E_{FC}^l = 83\% \tag{8.80}$$

8.5.3 ELECTROCHEMICAL EFFICIENCY OF A FUEL CELL

The efficiency of a standard fuel cell can be defined in terms of the relation between the operating cell voltage, E_P, and the ideal cell voltage, E_e [127].

$$E_{FC}^R = \frac{E_P}{E_e} = 1 - \frac{\eta}{E_e} \tag{8.81}$$

As previously described, the operating cell voltage is less than the ideal cell voltage on account of the losses associated with cell polarization, the ohmic loss, and leaks. The equilibrium potential, E_e, can be calculated from an understanding of the thermodynamics of the reaction in question. It is first necessary to calculate the change in Gibbs free energy, ΔG_r^e, for a reaction under specified conditions (Equation 8.78); the variation in the Gibbs free energy is given by

$$\Delta G_r^e = \Delta G_r^0 + RT \ln\left(\frac{P_{H_2} P_{O_2}^{\frac{1}{2}}}{P_{H_2O}}\right) \tag{8.82}$$

Using Equation 8.47, the EMF is calculated. The ideal voltage of a cell operating reversibly on pure hydrogen and oxygen at 1 atm pressure and 25°C is 1.229 V. In a real fuel cell, information about the equilibrium potential requires knowledge of the partial pressures of all the species involved in the reaction (see Equation 8.82). Then, the efficiency of a standard fuel cell at an operating voltage, E_P, is given by [8]:

$$E_{FC}^R = \frac{Real - Power}{Ideal - Power/0.83} = \frac{E_P J}{E_e J/0.83} = \frac{0.83 E_P}{E_e} = \frac{0.83}{1.229}E_P = 0.675 E_P$$

Since a fuel cell can be operated at different current densities, the related cell voltage controls the fuel cell efficiency, that is, diminishing the current density increases the cell voltage; thus the fuel cell efficiency increases.

8.5.4 Efficiency of an Internal Combustion Engine

An internal combustion engine transforms chemical energy into mechanical energy. The combustion of a hydrocarbon, that is, the source of chemical energy, is accompanied by an increase in temperature, since these chemical reactions are exothermic. In addition, as the reaction products are generally gases, the heat increases, and produces an expansion of the formed gases, which generate mechanical work by making the pistons in the internal combustion engine to run. In other cases, steam is generated to drive a steam cycle. The maximum efficiency of this engine is specified by the Carnot cycle thermal internal combustion engine efficiency, E_{ICE}^l, as follows

$$E_{ICE}^l = \frac{W_R}{(-\Delta H)} = 1 - \frac{T_2}{T_1} \tag{8.83}$$

where

W_R is the reversible work performed
ΔH is the enthalpy change of the reaction
T_1 and T_2 are the two absolute temperatures for the operation of the heat engine

The efficiency of an ICE is normally around 20% and, in general, the efficiencies of ICEs do not surpass 50% for the most efficient engines, for example, the steam turbines; thus, in general a fuel cell is more efficient than a heat engine [128].

8.6 ELECTROCHEMICAL IMPEDANCE SPECTROSCOPY

8.6.1 Impedance Analysis

With an alternating applied voltage, mathematically described with the help of the complex number representation, and using the following equation

$$V = V_0 e^{i\omega t} = V_0(\cos \omega t + i \sin \omega t)$$

where ω, the angular frequency, is expressed as $\omega = 2\pi f$, in which f is the frequency, it is possible to study the impedance behavior of an electrochemical cell, that is, perform an electrochemical impedance study [129–134]. The impedance, $\tilde{Z}(\omega)$, per se is the opposition that offers a circuit to the flow of an alternating current (AC) at a given frequency. It is measured as the complex ratio of the voltage and the current [129]:

$$\tilde{Z}(\omega) = \frac{\tilde{V}(\omega, t)}{\tilde{I}(\omega, t)} \tag{8.84}$$

Impedance spectroscopy is a helpful means for studying both the bulk transport properties of a material and the electrochemical reactions on its surface. The importance of impedance spectroscopy arises from the efficacy of the methodology in separating individual reaction-migration paces into a multistep process, since each reaction or migration step has, ideally, a single time constant related with it; consequently, each step can be separated in the frequency domain.

To carry out this type of study, a small AC amplitude voltage perturbation, $\Delta \tilde{V}(\omega, t)$, is applied, superimposed onto a DC bias voltage component, and the resulting alternating current response and its phase, $\Delta \tilde{I}(\omega, t)$, is measured [123,132]. Then, the electrochemical impedance of the system is thus defined as

$$\tilde{Z}(\omega) = \frac{\Delta\tilde{V}(\omega,t)}{\Delta\tilde{I}(\omega,t)}$$

Usually, the AC impedance experiments are performed over a broad range of frequencies, that is, from millihertz to megahertz frequencies. The interpretation of the resulting spectra is assisted by the similarity of the process to equivalent circuits, involving simple components such as resistors and capacitors. Generally, such equivalent circuits are not unique, and certainly there are a vast set of circuits that can represent any given impedance. Then, it is normal to choose a physically plausible circuit enclosing the smallest possible amount of components and, in a rather ad hoc way, give physical meaning to the derived parameters.

That is, in the specific case of electrochemical impedance spectroscopy (EIS), the steady, periodic linear response of a cell to a sinusoidal current or voltage perturbation is measured and analyzed in terms of gain and phase shift as a function of frequency, ω, where the results are expressed in terms of the impedance, $\tilde{Z}$. In this regard, the impedance response of an electrode or a battery is given by

$$\tilde{Z} = Z_r(\omega) + iZ_i(\omega)$$

where

Z$_r$ is the real part of the complex number which represents the impedance
$i = \sqrt{-1}$ is the imaginary unit
Z$_i$ is the imaginary part

The impedance can be represented in polar coordinates as follows

$$\tilde{Z} = |\tilde{Z}|e^{i\phi}$$

where

$$|\tilde{Z}| = \sqrt{Z_r^2 + Z_i^2} \quad \text{and} \quad \phi = \tan^{-1}\left(\frac{Z_i}{Z_r}\right)$$

The mathematical expressions which describe the impedance of some passive circuits are shown below, where a passive circuit is one that does not generate current or potential [129]. In this regard, the impedance response of simple passive circuit elements, such as a pure resistor with resistance R, a pure capacitor with capacitance C, and a pure inductor with inductance L, are given, respectively:

$$\tilde{Z}_{resistor} = R + i0$$

$$\tilde{Z}_{capacitor} = 0 + \frac{1}{i\omega C}$$

$$\tilde{Z}_{inductor} = 0 + i\omega L$$

8.6.2 Dielectric Spectroscopy and Impedance Spectroscopy

The measurements in an impedance spectroscopy of a simple electrolyte are normally obtained in the hertz to some megahertz frequency range with an impedance analyzer. For this purpose, impedance spectroscopy, as a methodology, has many similarities with dielectric spectroscopy (see Section 4.8.3).

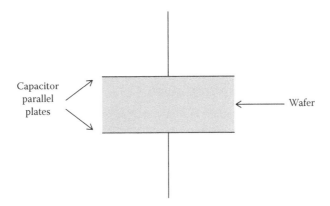

FIGURE 8.17 Parallel plate capacitor sample holder.

In impedance spectroscopy, the analyzer measures the response of the test sample, normally in the form of a cylindrical wafer with a radius of about 8–12 mm and a width of around 1–3 mm, which is included between the plates of a parallel capacitor sample holder (see Figure 8.17) [132]. The impedance analyzer measures the capacitance, C, and the conductance, G, of the capacitor. The capacitance is related as follows (see Chapter 4) (see Equation 4.41)

$$C = \frac{\varepsilon_0 \varepsilon' A}{d}$$

to ε', the real part of the complex dielectric constant of the wafer, and the conductance (G) is related as follows (see Equation 4.42)

$$G = \frac{\omega A \varepsilon_0 \left(\varepsilon_r'' + \dfrac{\sigma_0}{\omega \varepsilon_0} \right)}{d}$$

to σ_0, the conductivity of the wafer

where
   ε_r'' is the imaginary part of the complex dielectric constant of the wafer
   d is the wafer thickness
   A is the cross sectional area of the wafer
   ε_0 is the permittivity of free space

There are several models for the equivalent circuit of the electrolyte under an applied voltage, the simplest one is a resistor and a capacitor in parallel (see Figure 4.48). Applying this circuit, the real and imaginary parts of the complex impedance are obtained from the measured conductance and the capacitance using the relations [30,75]:

$$Z_r = \frac{G}{G^2 + \omega^2 C^2} \quad \text{and} \quad Z_i = \frac{\omega C}{G^2 + \omega^2 C^2} \tag{8.85}$$

In Sections 1.7 and 4.8.3, we have studied the dielectric relaxation phenomena and dielectric spectroscopy, respectively. In dielectric spectrometry, the methodology allowed us to measure the capacity and, consequently, the real part of the complex dielectric constant. The imaginary part of the complex dielectric constant was calculated, in this case, with the help of the Kramers–Kronig relations (see Section 1.7.6) whenever the conductivity is zero. The methodology explained in this section is more general, because it allows the measurement of capacitance and conductance, and with the help of Equation 8.85, the calculation of the impedance.

8.6.3 EQUIVALENT CIRCUITS FOR ELECTROCHEMICAL CELLS

One possible equivalent circuit of a battery is shown in Figure 8.18, in which C_{SC} is the capacitance of the electrical double layer, W the Warburg impedance for diffusion processes, R_I the internal resistance, and Z_A and Z_C the impedances of the electrode reactions [124,130].

In a simple case, the electrochemical reaction at the electrode–electrolyte interface of one of the electrodes of the battery can be represented by the so-called Randles circuit (Figure 8.19), which is composed of [129] a double layer capacitor formed by the charge separation at the electrode–electrolyte interface, in parallel to a polarization resistor and the Warburg impedance connected in series with a resistor, which represents the resistance of the electrolyte.

The polarization resistance is the slope of the current versus overpotential curve, that is, $dI/d\eta = 1/R_p$ (see Section 8.4). The impedance of this electrode reaction can be calculated with the help of the circuit (see Figure 8.19) whose impedance, if the Warburg resistance is not considered, is given by [132]:

$$\tilde{Z}(\omega) = R_s + \frac{R_p}{1 + i\omega R_p C_{dl}} = R_s + \frac{R_p}{1 + \omega^2 R_p^2 C_{dl}^2} - i\frac{\omega R_p^2 C_{dl}}{1 + \omega^2 R_p^2 C_{dl}^2} = Z_r + iZ_i \qquad (8.86)$$

The Warburg impedance, which is important at low frequencies, is related to the transport of the active species in the electrochemical reaction. The expression for the Warburg impedance in an infinite medium is given by [129,130]

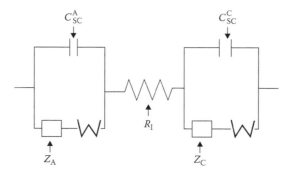

FIGURE 8.18 Equivalent circuit of a battery.

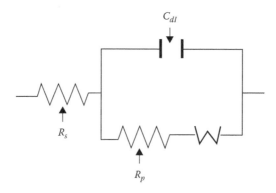

FIGURE 8.19 Electrochemical reaction at the electrode–electrolyte interface equivalent circuit.

$$\tilde{Z}_{\text{Warburg}}(\omega) = \left(\tau - i\frac{1}{\tau}\right)\sqrt{\omega} \tag{8.87}$$

where $\tau = C/\sqrt{D}$, in which, D, is the diffusion coefficient of the oxidizing or the reducing species and:

$$C = \frac{4RT}{\sqrt{2}(nF)^2 A\rho}$$

where

   A is the active electrode area
   ρ is the ionic concentration at the interface
   R, T, n, and F have their standard meanings

If in the equivalent circuit represented in Figure 8.19 the Warburg impedance is included, the whole impedance of the circuit is given by [131,132]

$$\tilde{Z}(\omega) = R_s + R_p\left(1 + \frac{\xi}{\sqrt{2\omega}}\right) - R_p^2\xi^2 C_{dl} - i\frac{R_p\xi}{\sqrt{2\omega}} \tag{8.88}$$

where

$$\xi = \frac{k_f}{\sqrt{D_O}} + \frac{k_b}{\sqrt{D_R}}$$

in which k_f and k_b are the forward and backward electron-transfer rate coefficients, respectively, and D_O and D_R are the diffusion coefficients of the oxidant and the reductant, respectively, for the reaction:

$$O + ne^- \leftrightarrow R$$

8.6.4 Methods for the Representation of Impedance Spectroscopy Data

In order to plot the experimental results of an AC impedance experiment, a methodology similar to the standard Argand diagram used by electrical engineers is applied, which shows positive imaginary quantities like inductances in its upper part, and negative imaginary quantities like capacitances in its lower part [129]. In this regard, in electrochemical applications, the impedance data is reported in the complex impedance plane. $-Z_i$ is plotted in the y-axis versus Z_r in the x-axis, both as parametric functions of the angular frequency, ω. This is the so-called Nyquist plot, where the impedance is represented as an imaginary number, explicitly as a vector of length, $|\tilde{Z}|$, making an angle, ϕ, with the, x-axis, that is, the axis where the real part of the imaginary number is represented (see Figure 8.20).

We now analyze the Nyquist plots corresponding to some circuits. In this regard, for a parallel network involving a resistance, R_p, and a capacitance, C_p, the impedance is given by

$$\frac{1}{\tilde{Z}_p} = \frac{1}{R_p} + i\omega C_p = \tilde{Y}_p$$

where $\tilde{Y}$ is the admittance. Consequently, the Nyquist plot in the case of the RC parallel circuit, representing a process with a small characteristic time, is a semicircle of radius, $R_p/2$, which meets

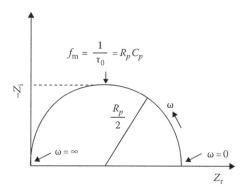

FIGURE 8.20 Plot of a RC parallel circuit (representing a process with a very small characteristic time).

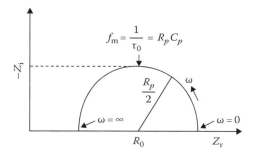

FIGURE 8.21 Plot of a RC parallel circuit (representing a process with not very small characteristic time).

the x-axis at $x = R$ for $\omega = 0$ and the origin, that is, $x = 0$ for $\omega = \infty$ (see Figure 8.19) [129]. The time constant of this simple circuit is defined as follows

$$\tau_0 = \frac{1}{f_m} = R_p C_p$$

(8.89)

where $f_m = \dfrac{\omega_m}{2\pi}$ is the frequency of the maximum of the semicircle. This relaxation time corresponds to the characteristic relaxation time of the electrochemical process under test.

For a process occurring with a not small characteristic time, the plot is a semicircle of radius, $R_p/2$, which meets the x-axis both at $x = R_0 + R_p/2$ for $\omega = 0$ and $x = R_0 - R_p/2$ for $\omega = \infty$ (see Figure 8.21) [75]. The time constant of this simple circuit is defined with the help of Equation 8.89, where $f_m = \omega_m/2\pi$ is also the frequency of the maximum of the semicircle. This relaxation time also corresponds to the characteristic relaxation time of the electrochemical process under test.

For a constant phase process, as a diffusion process, the plot is represented as a straight line with one slope (see Figure 8.22) [75]. This is evident from Equation 8.86, because when $-Z_i$ is plotted versus Z_r, the Warburg component is represented as a straight line with a unitary slope.

The entire diagram illustrating the performance of the electrode processes, corresponding to the simple equivalent circuit diagram of a battery which is shown in Figure 8.18, is presented in Figure 8.23 [6].

In the plot in Figure 8.23, it is evident that the ohmic factors are independent of frequency, after this, the ideal activation processes display a semicircular conduct with a frequency which is typical of the corresponding relaxation processes (see Equation 8.88 and Figures 8.20 and 8.21); finally, the concentration processes exhibit a diagonal conduct characteristic of diffusion processes (see Figure 8.22) often referred to as the Warburg behavior [124,129,130] (to see a real Nyquist plot related to an EIS test of a battery, see Section 8.9.1).

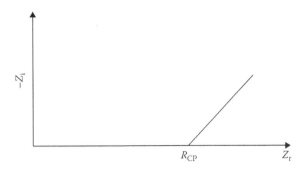

FIGURE 8.22 Plot of a constant phase process (very high characteristic time).

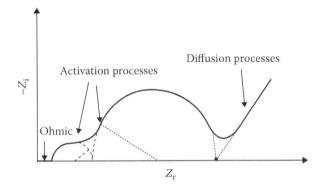

FIGURE 8.23 Plot related to the circuit shown in Figure 8.17.

To be more specific, we now refer to Equations 8.86 and 8.88, which represent the impedance of the equivalent circuit corresponding to the electrochemical reaction at the electrode–electrolyte interface. If in Equation 8.86 we calculate the limit for $\omega \to 0$, then the intercept of the plot in the real axis is $Z_r = R_s + R_p$; on the other hand, if the limit for $\omega \to \infty$ is calculated, then $Z_r = R_s$. Besides, at the frequency where a maximum of $Z_i (\omega)$ is detected, we have $R_p C_{dl} = 1/\omega_{max} = \tau$, where the time constant, τ, indicates how fast the electrochemical reaction is. Finally knowing $R_p C_{dl} = 1/\omega_{max}$, it is possible to calculate C_{dl} [129,130].

Including the Warburg impedance in Equation 8.88, it is evident that for $\omega \to 0$, we get $Z_r = R_s + R_p$ and for $\omega \to \infty$, $Z_r = R_s$; besides the intercept of the Warburg component in the real axis is $Z_r = R_s + R_p - R_p^2 \xi^2 C_{dl}$ [132].

In addition to the Nyquist representation, the Bode plot is as well applied for the description of impedance spectrometry data. In this case, the impedance data is represented in polar coordinates as $\widetilde{Z} = |\widetilde{Z}| e^{i\phi}$, and $|\widetilde{Z}| = \sqrt{Z_r^2 + Z_i^2}$ and $\phi = \tan^{-1} (Z_i/Z_r)$ are plotted versus frequency [129]. Related to this type of data description is the representation of impedance, where the real and imaginary parts of the impedance are plotted as a function of the frequency.

8.7 SUSTAINABLE ENERGY AND ENVIRONMENTAL SENSING TECHNOLOGY APPLICATIONS OF SOLID-STATE ELECTROCHEMISTRY

8.7.1 SOLID OXIDE FUEL CELL MATERIALS AND PERFORMANCE

8.7.1.1 Electrolyte

SOFCs are presently at the vanguard of research into a novel group of energy conversion systems because of their high efficiency, flexibility, and environmentally friendly nature. SOFCs transform the chemical energy of H_2 or a hydrocarbon fuel into electric power, the exhaust being water and carbon dioxide. It is necessary to remind that SOFCs works at high temperatures, that is, 800°C–1000°C. Besides, the cell can have two chambers or a one single chamber. We describe here the two-chamber fuel cells.

Currently, the standard SOFCs use yttria-stabilized zirconia (YSZ) (see Section 2.4.3) containing typically 8 mol% of YO_2 as the electrolyte, a ceramic metal (a cermet) composed of Ni plus YSZ as the anode, and lanthanum strontium manganite (LSM) perovskite ($La_{1-x}Sr_xMnO_{3-\delta}$) as the cathode material [5,7,9,135].

In an SOFC, the electrochemical reactions take place in the electrodes in the functional layer, that is, a zone within a distance of less than 10–20μm from the electrolyte surface [5,136–138]. The portion of the electrode beyond this width is principally a current collector structure, which has to be porous to permit the admission of gas to the functional layer where the oxidation and reduction reactions occur. Besides, the electrolyte has to be gas impermeable to avoid direct combination and combustion of the gases [137]. The essential parts of the SOFC, that is, the electrolyte, the anode, and the cathode, are made of ceramic materials produced with appropriate electrical conducting properties, chemical and structural stabilities, similar expansion coefficients, and negligible reactivity properties [135].

In Section 8.2, we have studied oxygen and proton conduction in solids; here we emphasize materials and performance of the cathode and the anode.

8.7.1.2 SOFC Cathode Materials and Performance

In the cathode of an SOFC, oxygen, supplied typically as air to one side of the oxygen permeable membrane, is reduced by the cathode to oxygen ions via the following overall half-cell reaction (see Figure 8.7):

$$4e^- + O_2 \rightarrow 2O^{2-} \tag{8.90}$$

The oxygen anions created are selectively transferred through the oxygen-permeable membrane to the anode, where they experience a similar half-cell reaction with a gaseous fuel, either H_2, syngas ($CO + H_2$), or a hydrocarbon, to produce H_2O and CO_2. For example, if the fuel is CH_4, the oxidation reaction in the anode is given by

$$CH_4 + 4O^{2-} - 8e^- \rightarrow CO_2 + 2H_2O$$

During this process, the electrons released and consumed at the anode and cathode produce some fraction of the reversible work of the reaction to the external circuit.

An SOFC cathode normally consists of a porous matrix cast onto an oxide ion-conducting electrolyte substrate (see Figure 8.24), where the cathode porosities are typically 25–40 vol% [66,123,137]. Besides, the cathode must be an electron conductor and catalytically active for the oxygen reduction reaction. However, because it is not an oxygen conductor, it must be porous with an optimized three-phase interface at which the reduction reaction takes place [33].

In the cathode, gaseous oxygen, O_2, diffuses through the gas phase included (see Figure 8.24) [123] inside the pores between the cathode and the electrolyte. Subsequently, it is reduced to O^{2-} somewhere within the matrix, (Equation 8.90), where the electron comes from the cathode, the oxygen from the gas phase, and the oxide anion goes through to the ion-conducting phase electrolyte (see Figure 8.24) [137].

In conclusion, the ideal cathode should be porous, electronically and ionically conducting, electrochemically active, and possess high surface areas. Given this, it is unusual for a single material to accomplish all of these functions, so a composite cathode, of which the electrocatalyst is one component, is often used [135].

Transition-metal oxides were first studied as SOFC cathodes owing to their excellent electrical conductivity, and as a comparatively inexpensive option to Pt, which during the 1960s was the

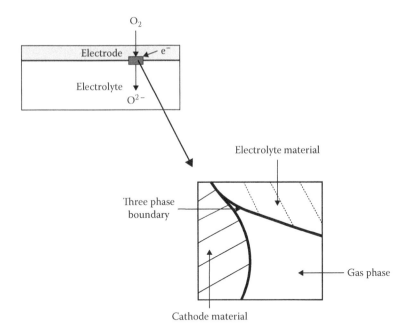

FIGURE 8.24 Electrode–electrolyte and the three-phase boundary.

only SOFC cathode material broadly studied. In this regard, one of the first materials of this kind studied was the $La_{1-x}Sr_xCoO_{3-\delta}$ perovskite, labeled LSC, which is currently one of the most studied mixed conductors. Later, different materials having a perovskite structure were studied; however $La_{1-x}Sr_xMnO_{3-\delta}$ (LSM) turned out to be the preferred material for SOFC cathodes. Because with the appropriate Sr content, that is, x, an almost exact thermal expansion match between LSM and YSZ can be attained [123]. Besides, Mn is commonly less reducible than other transition metals, for example, Co and Fe in a perovskite matrix. Therefore, LSM shows small or no chemical expansivity [139,140], which is an additional source of thermal–mechanical stress that can probably threaten the reliability of the electrode microstructure [123]. An additional benefit of LSM is that it is normally thermodynamically more stable than mixed conductors containing cobalt or iron [141]. Besides, it is moderately catalytically active for O_2 dissociation, and retains oxygen deficiency in the oxidizing atmosphere prevalent in the cathode chamber [142].

Another perovskite applied for the production of cathode materials is $ABO_{3-\delta}$, where the metal A are La and Sr, and the metal B are the transition metals Fe, Co, and Ni, for example, $La_{0.7}Sr_{0.3}Fe_{0.8}Ni_{0.2}O_{3-\delta}$; $La_{0.8}Sr_{0.2}Co_{0.8}Ni_{0.2}O_{3-\delta}$; and $SrCo_{0.8}Fe_{0.2}O_{3-\delta}$ [33].

8.7.1.3 SOFC Anode Materials and Performance

The fuel processing operation of an SOFC critically depends on the anode structure and composition, since the electrochemical reaction can only take place at the three-phase boundary. If there is a breakdown in connectivity in any one of the three phases, the reaction cannot take place. Besides, if ions from the electrolyte cannot reach the reaction site, if the gas-phase fuel molecules cannot reach the reaction site, or if electrons cannot be removed from the reaction site, this site cannot contribute to the performance of the cell [5].

The materials used for the anode construction have to be stable in the reducing atmosphere prevailing in the anode. Besides, these materials ought to be electronically conducting, and, finally, must have enough porosity to permit the diffusion [66,135,137]. In this region, the oxidation reaction takes place.

$$H_2 + O^{2-} \rightarrow H_2O + 2e^-$$ (8.91)

For this reaction to take place in the anode, gaseous hydogen, H_2, diffuses into the pores and is reduced somewhere within the matrix similarly as the reduction process schematically illustrated in Figure 8.24. The standard anode for an SOFC based on YSZ as the electrolyte consists in a NiO + YSZ composite, which in the reducing environment existing in the anode, the NiO reduces to elemental Ni, creating a porous structure with Ni deposited on the pore surface [5,33,137]. Elemental Ni is an excellent catalyst for the oxidation of H_2 in an SOFC anode, owing to the excellent catalytic properties of nickel for breaking hydrogen bonds, the low reactivity with other components, and fairly low cost [143].

Nevertheless, there are numerous problems when nickel is applied as an anode material, for example, its reactivity at high temperatures and in harsh environments, and the thermal expansion of nickel is significantly higher than that of YSZ. Besides, nickel can sinter at the cell operating temperature, ensuing in a reduction in the fuel electrode porosity [144]. These difficulties are reduced by forming a matrix of YSZ around the nickel particles, where the YSZ matrix inhibits sintering of the nickel particles, and reduces the thermal expansion coefficient of the fuel electrode, bringing it nearer to that of the electrolyte [136].

In SOFC research, recent reports of the direct electro-oxidation of hydrocarbon fuels have appeared [5]. In this regard, nickel is an outstanding catalyst for the cracking of hydrocarbons, but during the process, carbon is a build up in the catalyst surface. The carbon creation can produce blockage of gas channels, physical disintegration of the nickel structure, and fragmentation of the porous anode [5]. Therefore, nickel-based anodes are not appropriate to be used directly in dry natural gas without alteration of their catalytic properties. To solve this problem, ceria and other elements have been incorporated into the anode cermet. In a number of reports, Ni-ceria [145] or Cu-ceria were used [146]. It was found that they too have considerable limitations [137]. Therefore, mixed metal oxides have been studied as promising options for the preparation of SOFCs anodes, because they are possibly less likely to promote carbon production owing to greater accessibility of oxygen throughout the anode [144]. In this sense, throughout the past years, a lot of studies have been carried to investigate optional anode materials, especially the mixed conducting ceramics like ceria (CeO_2) (see Section 2.3.6) doped with gadolinium [137]. As ceria shows both ionic and some electronic conduction under reduced circumstances, this material can be used for anodes without forming a composite [147]. However, CeO_2 has a negative property, which is a variation in volume as a result of the partial reduction of Ce^{4+} to Ce^{3+} due to release of oxygen from the lattice. This effect can be reduced but not eliminated by partially doping CeO_2 with Gd. In this regard, as alternative materials for SOFC anodes, a mixture of [148] $La_{0.7}Sr_{0.3}Cr_{0.8}Mn_{0.2}O_{3-\delta}$, which is an electronic conductor; $Ce_{0.9}Gd_{0.1}O_{1.95}$, which is an ionic conducting oxide; and ≈ 4 wt % Ni, has been reported. In addition, titanium-doped YSZ has been broadly studied as an anode material for SOFCs [136].

Perovskite materials show outstanding thermal and mechanical stability at temperatures typical for the performance of SOFCs in contrast with the standard Ni/YSZ cermet, where nickel sintering and agglomeration are latent risks [144].

8.7.1.4 Interconnects

An additional component in a fuel cell is the interconnects or bipolar plates. This is a vital component in SOFC development, since it forms the connection between the anode of one cell and the cathode of the next in a stacked arrangement. That is, these components operate as connections between individual fuel cells in a fuel cell stack [128]. Then, the interconnects have to be electronically conductive and also possess good impermeability, chemical stability, and good mechanical properties since these components seal the gas chambers for the oxygen and fuel gas feed at either the anode or the cathode [66,137].

8.7.1.5 SOFC Fuel Processing

Fuel processing can be defined as the conversion of the raw primary fuel furnished to a fuel cell system into a fuel gas needed by the fuel cell [148,149]. SOFCs possess high fuel elasticity, because of the fact that O^{2-} anions are the species transported through the membrane, which permit SOFCs to work, in principle, on any combustible fuel [5]. The most important fuel for fuel cells is obviously hydrogen; however, as discussed in Section 6.12.2, hydrogen storage is a big drawback [1–4]. However, hydrogen storage in ammonia [4], alcohols [150,151], or hydrocarbons can also be an option [5]. Ammonia can be cracked and alcohols and hydrocarbons can be reformed into hydrogen-rich synthesis gases by several methods, such as partial oxidation, catalytic steam-reforming, and auto thermal reforming. Alcohols, specifically, methanol and ethanol, are suitable storage systems for hydrogen [8,150]. These liquids can with no trouble be transported using the existing infrastructure and have lesser restrictions regarding to safety than hydrogen storage. Methanol is normally made from natural gas and, for this reason, readily obtainable, even though, it is more expensive to produce than hydrogen. A direct feed system of methanol is in principle feasible for an SOFC; however, research endeavors to design a simple, efficient, compact, and inexpensive reformer is, at present, in progress [8]. It will be better for the fuel cell to operate with hydrogen.

Methanol steam-reforming is normally carried out in the range between 250°C–350°C, depending on the reforming process, and the catalyst usually applied is Cu/ZnO [151]. The chemical reactions taken into account in the steam-reforming of methanol, according to the literature, are the following ones [151–154]:

$$CH_3OH + H_2O \rightarrow CO_2 + 3H_2 \qquad (8.92a)$$

$$CH_3OH \rightarrow CO + 2H_2 \qquad (8.92b)$$

$$CO + H_2O \leftrightarrow CO_2 + H_2 \qquad (8.92c)$$

Reactions 8.92a and b are reversible and endothermic and occur with an increase in volume, suggesting that high methanol conversion can be attained at high temperatures and low pressures. Alternatively, the exothermic reaction 8.92c proceeds concurrently with methanol steam-reforming and without any volume change [155]. When carried out in a conventional reactor, these reactions generate a hydrogen-containing mixture; as a result, the obtained hydrogen needs purification before being fed to a fuel cell. Recent research on the methanol steam-reforming reaction has contemplated the use of a membrane reactor, in order to substitute the traditional reformer and its associated gas-cleaning unit. In this membrane reactor, methanol conversion, by means of steam-reforming and hydrogen purification, are combined, in order that the pure hydrogen outlet stream can be directly employed by the fuel cell [156].

Ethanol is a natural renewable product normally produced from biomass, which is an important factor for near-zero carbon dioxide (CO_2) emissions. It is accessible, easy to transport, ecofriendly, nontoxic, and it can be transformed by catalytic reactions into hydrogen, which is important for fuel cell fuel processing [151]. Section 9.8.10 discusses different processes for hydrogen production by the steam-reforming of ethanol.

In spite of the problems created by carbonaceous fossil fuels, at present, it is a reality that the global profusion of these fuels will guarantee that they continue to be an important energy resource for some years to come [149]. Natural gas, currently, is perhaps the most important fuel alternative for the many stationary fuel cells globally used [1,149].

Natural gas consists of a mixture of hydrocarbons of low boiling point, where methane is a component typically present in the maximum concentration, with lesser quantities of ethane, propane, other hydrocarbons, nitrogen, carbon dioxide, and traces of helium and hydrogen sulfide, where the whole composition fluctuates according to the source [157,158]. Biogas is, as well, a fuel, but it is obtained from biomass by an anaerobic digestion process, which generates a gas mostly composed of carbon dioxide and methane and is produced naturally in landfill sites or can be generated in an aerobic reactors [159]. In biogas production procedures, the composition of the output gas depends on the nature of the biomass used and the process applied; consequently, a significant amount of impurities are normally present in the gas, which has to be eliminated before passing it through a fuel cell system.

Steam-reforming is the catalytic reaction of hydrocarbon fuels with steam to obtain the hydrogen bound in the fuel and water [160]. Currently, about 96% of the hydrogen produced worldwide is generated by reforming hydrocarbons, principally using methane as the raw material. This is an established technology at the large scale [1,5,160]. During the catalytic methane steam-reforming, the relevant reactions are [161]

$$CH_4 + H_2O \leftrightarrow CO + 3H_2$$

and the water gas shift reaction is

$$CO + H_2O \leftrightarrow CO_2 + H_2$$

The process is carried out at relatively high temperatures in the range from 400°C to 875°C, at about 3 MPa pressure, in a Ni (12–20 wt %) catalyst supported in α-Al$_2$O$_3$.

8.7.2 POLYMER ELECTROLYTE FUEL CELLS

8.7.2.1 Electrolyte

PEFCs are significant energy conversion systems appropriate for use in numerous applications, in small portable systems and in bigger automotive and stationary applications [162–171]. Originally, polystyrene–sulfonic acid and sulfonated phenol–formaldehyde membranes were used to develop PEFCs. However, the effective operational life span of these polymers was inadequate, because of degradation produced in the working conditions of a fuel cell [162,165]. These problems were solved when a new polymer membrane based on a perfluorinated polymer with side chains terminating in sulfonic acid moieties, named Nafion, was developed by DuPont (see Figure 8.25) [168,169]. DuPont de Nemours Wilmington, DE, USA.

Currently, the most widely applied electrolyte in PEFCs is Nafion, manufactured by DuPont, Dow Chemical, Midland, MI, USA and other chemical companies. The Nafion polymer electrolyte is a good proton conductor. Besides, it has very low electron conductivity, and is gas impermeable in order to provide the necessary spatial separation between the anode oxidation and the cathode reduction reactions.

Nafion ionomers are produced by copolymerization of a perfluorinated vinyl ether comonomer with tetrafluoroethylene resulting in the chemical structure shown in Figure 8.25 [162,166]. This polymer and other related polymers consist of perfluorinated, hydrophobic, backbones that give chemical stability to the material. The material also contains sulfonated, hydrophilic, side groups that make hydration possible in the acidic regions, and also allow the transport of protons at low temperatures, since the higher limit of temperature is determined by the humidification of the membrane, since water is a sine qua non for conduction [166]. The material exhibits a proton conductivity of 0.1 S/cm at 80°C [162]. The membrane performance is then based on the hydrophilic character of the sulfonic acid groups, which allow proton transport when hydrated while the hydrophobic

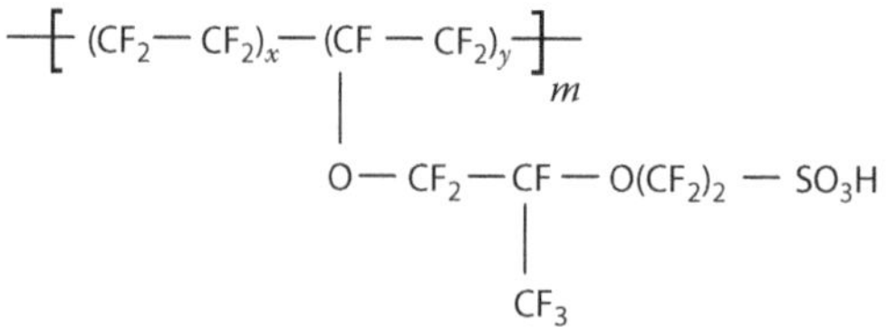

FIGURE 8.25 Nafion structure.

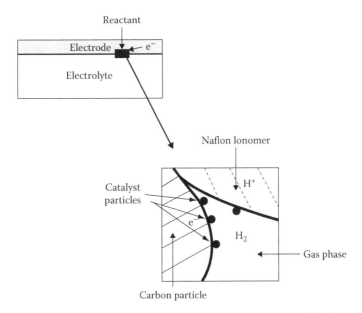

FIGURE 8.26 Scheme of the three-phase boundary zone for an electrode of a PEM fuel cell.

perfluorinated backbone provides morphological stability. Consequently, water management in the membrane is crucial for the operation of a PEFC.

8.7.2.2 Electrodes

Since the reactants in a fuel cell are normally gaseous products, the electrodes for the PEFCs are normally porous to guarantee the supply of the reactant (gases) to the active zones. As explained in the case of the SOFCs, the electrodes in the case of the PEFC normally consist of a porous matrix, since the electrochemical reactions take place in the three-phase boundary (see Figure 8.26), that is, at the interface between the electrode, the electrolyte, and the reactant gas [168,170].

The primary criteria for a good electrode is to provide a developed three-phase boundary between the gas supply, the catalyst particle, and the ionic conductor. Then the catalyst particles, which are normally expensive precious metals, such as Pt and Pt alloys, should be in direct contact with an electronic conductor to guarantee that the electrons are delivered to or removed from the reaction site. In this regard, the electronic conductivity is provided by a carbon support on top of which the catalyst particles are maintained [162].

8.7.3 ZEOLITES AS SOLID ELECTROLYTES IN BATTERIES

Aluminosilicate zeolites because of their structure, composition, and properties offer a superior ionic strength environment [172,173]. Even though these materials are electronic insulators, when hydrated, they are solid solutions of high ionic mobility, and when dehydrated exhibit fair ionic conductivity (see Section 8.2.7) [38,112,119,172]. The properties of aluminosilicate zeolites that are responsible for affecting the charge-transfer reactions in electrochemical systems are [172,174]:

- Zeolite size and shape selectivity due to its rigid pores and channels
- Ion-exchange properties
- Catalytic properties, specially acid catalysis
- Large water adsorption and desorption capacity

When zeolites are hydrated shows a notable ionic conductivity [112]. Consequently, since all electrode processes depend on the transport of charged species zeolites provide an excellent solid matrix for ionic conduction [172]. In 1965 [175], Freeman established the possibility of using zeolites in the development of a functional solid-state electrochemical system, that is, a battery where a zeolite, X, was used as the ionic host for the catholyte, specifically, Cu^2, Ag^+, or Hg^{2+}, and as the ionic separator in its sodium-exchanged form, that is, Na-X. Pressed pellets of Cu-X and Na-X were sandwiched between a gold current collector and a zinc anode. Then, the half-cell reactions are the oxidation of $Zn \rightarrow Zn^{2+} + 2e^-$ and the reduction of $Cu^{2+} + 2e^- \rightarrow Cu$, with type X providing a solid-state ionic path for cationic transport [175]. The electrochemical system obtained can be represented as follows: $(Au|Cu^{II}-X|Na-X|Zn)$.

The role of zeolites as solid electrolytes is linked to water adsorption in the zeolite secondary porosity (see Section 8.2.7), which can be evidenced in the use of natural or synthetic zeolites as solid electrolytes, since the cationic conduction of zeolites is highly enhanced by the adsorption of water in the secondary porosity [112].

Zeolite-based solid electrolytes have been used in the manufacture of Li secondary batteries with the zeolite acting as the host for Li ions [174,176]. These materials were applied as hosts for Zn^{II}, in $Zn-MnO_2$ batteries [177], and in Pb^{II} in lead acid batteries [178].

Zeolites have been also used in the production of fuel cell electrodes, fundamentally as catalyst supports for metals such as Ag, Pd, Rh, Pt, Ru, and Ni [174]. Besides, zeolite membranes have been used for the separation of the anode and the cathode sections of a fuel cell, because they allow the transfer of ions through it, for example, to carry an electric current in methanol fuel cells using a basic electrolyte such as carbonate, where the zeolite prevents the escape of methanol [174,179].

A Leclanché-type battery using a hydrated natural mordenite (sample MP, see Table 4.1) as the electrolyte has been prepared [180,181]. The battery was constructed as a pressed wafer of cylindrical geometry ($h = 1.5$ [mm] and $d = 11$ [mm]) where 0.4 g of the zeolite MP was included between a Zn metallic sheet and a mixture of graphite and MnO_2 powders to obtain the $(Zn(s)|MP|MnO_2(s)-C(graphite))$ system [180]. The obtained batteries were tested in open circuit and short circuit regimes. The open circuit voltage measured for the seven different batteries prepared was in the range of 1.70 ± 0.05 [V]. On the other hand, the short circuit current density measured fluctuated in the range of 69.23 [mA/cm²] $< J <$ 500 [mA/cm²] [180].

In order to undertake a large scale application of the obtained results, the zeolite MP was added to the electrolyte of Leclanché-type batteries, manufactured in the Yara Dry Cell Factory in Havana, Cuba [181]. The obtained batteries with the zeolite MP included in the electrolyte were tested under intermittent and continuous discharge procedures following the standard modus operandi of the Yara factory [181]. The results indicated that the batteries produced with the zeolite MP included in the electrolyte exhibited a better performance in comparison with the batteries produced following the standard technology [181].

Figure 8.27 shows an actual Nyquist plot (negative of the imaginary part of the complex impedance, Z_i versus the real part of the complex impedance Z_r) obtained by measuring the impedance of a Ni|Na-X|NiO₂ battery, which is composed of a wafer (with a diameter of 12 mm) in which were sandwiched films of Ni powder (0.5 g), Na-X zeolite powder (0.5 g), and NiO_2 powder (0.5 g) [132].

The measurements were obtained in the frequency range from 100 Hz to 1 MHz with the help of a Hewlett Packard impedance analyzer applying a constant bias voltage of 0 V and a modulating voltage of 500 mV [133]. Then, the real (Z_r) and imaginary (Z_i) parts of the complex impedance were calculated from the measured impedance.

8.7.4 SENSORS

Conductivity sensors are not strictly speaking solid-state electrochemical sensors; however, they are in some way related. The fact that a variation in electrical conductivity may be produced by the adsorption of a gas on the surface of semiconductor oxides has long been known [182]. Sensors of

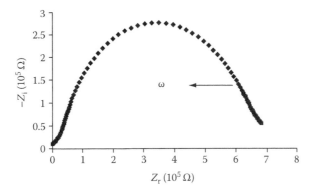

FIGURE 8.27 Nyquist plots (100 Hz to 1 MHz) of impedance spectroscopy of the Ni | Na-X | NiO$_2$ battery.

the conductivity type rely on the variation of electric conductivity of a film or a bulk material, whose conductivity is influenced by the existence in the surroundings of the sensor material of a concrete analyte [183]. One of the first materials tested was SnO$_2$. Currently, gas-sensing elements based on SnO$_2$ are commercially available [183].

Thin films, to attain enough sensitivity and response time, of oxide materials normally deposited on a substrate are typically used as gas sensors, owing to their surface conductivity variation following surface chemisorption [183,184]. Surface adsorption on a SnO$_2$ film deposited on alumina produces a sensitive and selective H$_2$S gas sensor [185]. In addition, a number of perovskite-type compounds are being used as gas sensor materials because of their thermal and chemical stabilities. BaTiO$_3$, for example, is used as sensor for CO$_2$ [183].

Solid-state devices based on stabilized zirconia electrolytes are of considerable significance in the energy industry. In this sense, the elevated ionic conductivity, and mechanical and chemical stabilities of zirconia have been applied in oxygen sensing, for prolonged periods, to monitor the operation of internal combustion engines in automobiles, to raise fuel efficiency and minimize emissions [186].

High-temperature stabilized NO$_x$ zirconia potentiometric sensors are also being utilized [187]. The electrochemical reactions on zirconia devices take place at the triple-phase boundary, that is, the junction between the electrode, electrolyte, and gas [186]. It has been reported that sensors composed of a WO$_3$ electrode, yttria-stabilized zirconia electrolyte, and Pt-loaded zeolite filters demonstrate high sensitivity toward NO$_x$, and are free from interferences from CO, propane, and ammonia, and are subject to minimal interferences from humidity and oxygen, at levels typically present in combustion environments [188]. In this sensor, a steady-state potential arises when the oxidation–reduction reaction [186,188]

$$2NO + 2O_2^- \leftrightarrow 2NO_2 + 4e^-$$

$$O_2 + 4e^- \leftrightarrow 2O^{2-}$$

takes place at the same time on the same electrode, where O$_2^-$ is the oxygen ion in the YSZ lattice. The measured potential is frequently identified as non-Nernstian or mixed-potential, as a consequence of the departure from the Nernstian relation. This type of potential arises when a nonequilibrium state is present, involving two or more electrochemical reactions.

Besides, potential-type sensors using battery systems, such as (Pt | MnO$_2$ | Cu), where Pt and Cu are the electrodes and MnO$_2$ as the solid electrolyte, have been developed [183]. The functionality of this system was verified for sensing water in a high humidity ambience [189]. Other battery systems consisting of nonsymmetrical electrodes in an oxide electrolyte have been applied in multigas sensing with several gas mixtures [190].

On the other hand, the theoretical performance of concentration electrochemical cells, based on perovskite materials with protonic and oxygen ion conduction properties, has been described as well [191]. Besides, a sensor for the detection of oxidizable gases that employs the production of a non-Nernstian electrode potential, using zirconia as the solid electrolyte, has been developed [192].

REFERENCES

1. G.W. Crabtree, M.S. Dresselhaus, and M.V. Buchanan, *Phys. Today*, 57, 39 (2004).
2. M. Jacoby, *Chem. Eng. News*, 83, 43 (2005).
3. W. Grochala and P.E. Edwards, *Chem. Rev.*, 104, 1283 (2004).
4. R. Roque-Malherbe, *Adsorption and Diffusion in Nanoporous Materials*, CRC Press, Boca Raton, FL, 2007.
5. S. McIntosh and R.J. Gorte, *Chem. Rev.*, 104, 4845 (2004).
6. M. Winter and R.J. Brodd, *Chem. Rev.*, 104, 4245 (2004).
7. C.-Y. Wang, *Chem. Rev.*, 104, 4727 (2004).
8. EG&G Services Parsons, Inc., *Fuel Cell Handbook* (6th edition), Science Applications International Corporation. Under Contract No. DE-AM26-99FT40575, US Department of Energy, Office of Fossil Energy, National Energy Technology Laboratory, Morgantown, WV, November, 2002.
9. K.V. Kordesch and G.R. Simader, *Chem. Rev.*, 95, 101 (1995).
10. P. Atkins, *Physical Chemistry* (6th edition), W.H. Freeman and Co., New York, 1998.
11. J. Larmine and D. Andrews, *Fuel Cell Systems Explained*, John Wiley & Sons, New York, 2000.
12. H. Aono, N. Imanaka, and G.-Y. Adachi, *Acc. Chem. Res.*, 27, 265 (1994).
13. M. Martin, in *Diffusion in Condensed Matter*, P. Heithans and J. Karger, (editors), Springer-Verlag, Berlin, 2005, p. 209.
14. C. Wagner and W. Schottky, *Z. Phys. Chem.*, B11, 163 (1930).
15. A.J. Jennings, *Annu. Rep. Prog. Chem., Sect. A*, 97, 475 (2001).
16. M.E. Glicksman, *Diffusion in Solids*, John Wiley & Sons, New York, 2000.
17. K.D. Kreuer, *Solid State Ionics*, 97, 1 (1997).
18. R. Oesten and R.A. Higgins, *Ionics*, 1, 427 (1995).
19. G.-Y. Adachi, N. Imanaka, and S. Tamura, *Chem. Rev.*, 102, 2405 (2002).
20. C.N.R. Rao and J. Gopalakrishnan, *New Directions in Solid State Chemistry* (2nd edition), Cambridge University Press, Cambridge, UK, 1997.
21. C. Tubant and E. Lorenz, *Z. Phys. Chem.*, 87, 513 (1914).
22. C.E. Housecroft and A.G. Sharpe, *Inorganic Chemistry* (2nd edition), Pearson Education Limited, Essex, UK, 2005.
23. H.Y.-P. Hong, *Muter. Res. BUZZ.*, 11, 183 (1976).
24. J.B. Goodenough, H.Y-P. Hong, and J.A. Kafalas, *Mater. Res. Bull.*, 11, 203 (1976).
25. L. Latie, G. Villeneuve, D. Conte, and G.L. Flem, *J. Solid State Chem.*, 51, 293 (1984).
26. M. Sato, Y. Kono, and K. Uematsu, *Chem. Lett.*, 1425 (1994).
27. M. Sato, Y. Kono, H. Ueda, K. Uematsu, and K. Toda, *Solid State Ionics*, 83, 249 (1996).
28. T. Nilges, S. Nilges, A. Pfitzner, T. Doert, and P. Böttcher, *Chem. Mater.*, 16, 806 (2004).
29. K. Funke, *Prog. Inorg. Chem.*, 11, 345 (1976).
30. N. Baskaran, *J. App. Phys.*, 92, 825 (2002).
31. S. Hull, D.A. Keen, D.S. Sivia, and P. Berastegui, *J. Solid State Chem.*, 165, 363 (2002).
32. K.C. Kao, *Dielectric Phenomena in Solids*, Elsevier, Amsterdam, the Netherlands, 2004.
33. J.B. Goodenough, in *Mixed Ionic Conducting Perovskites for Advanced Energy Systems*, N. Orlovskaya and N. Browning, (editors), NATO Science Series Vol. 183, Kluwer Academic Publishers, Dordrecht, the Netherlands, 2004, p. 1.
34. P. Kofstad, *Non-Stoichiometry, Diffusion and Electrical Conductivity of Binary Metal Oxides*, Wiley, New York, 1972.
35. T. Norby, M. Wideroe, R. Glöckner, and Y. Larring, *Dalton Trans.*, 3012 (2004).
36. S. Nieto, R. Polanco, and R. Roque-Malherbe, *J. Phys. Chem. C*, 111, 2809 (2007).
37. P. Padmakumar and S. Yashonath, *J. Chem. Sci.*, 118, 135 (2006).
38. R. Roque-Malherbe, in *Handbook of Surfaces and Interfaces of Materials*, Vol. 2, H.S. Nalwa, (editor), Academic Press, New York, Chapter 13, 2001, p. 509.
39. R. Roque-Malherbe, L. Lemes-Fernandez, L., Lopez-Colado, C. de las Pozas, and A. Montes-Caraballal, in *Natural Zeolites '93 Conference Volume International Committee on Natural Zeolites*, D.W. Ming and F.A. Mumpton, (editors), Brockport, NY, 1995, p. 299.

40. N. Imanaka, K. Okamoto, and G. Adachi, *Chem. Lett.*, 130 (2001).
41. K.-D. Kreuer, *Chem. Mater.*, 8, 610 (1996).
42. S. Stotz and C. Wagner, *Ber. Bunsenges. Phys. Chem.*, 70, 781 (1967).
43. C. Wagner, *Ber. Bunsenges. Phys. Chem.*, 72, 778 (1968).
44. T.S. Elleman, L.R. Zumwalt, and K. Verghese, *Proc. Top. Meet. Technol. Controlled Nucl. Fusion*, 3, 763 (1978).
45. R.M. Roberts, T.S. Elleman, H. Palmour III, and K. Verghese, *J. Am. Ceram. Soc.*, 62, 495 (1979).
46. J.B. Bates, J.C. Wang, and R.A. Perkins, *Phys. Rev. B*, 19, 4130 (1979).
47. J.C. Cathcart, R.A. Perkins, J.B. Bates, and L.C. Manley, *J. Appl. Phys.*, 50, 4110 (1979).
48. S.K. Mohapatra, S.K. Tiku, and F.A. Kröger, *J. Am. Ceram. Soc.*, 62, 50 (1979).
49. H. Iwahara, K. Uchida, and K. Ogaki, *J. Electrochem. Soc.*, 135, 529 (1988).
50. H. Iwahara, H. Uchida, and K. Morimoto, *J. Electrochem. Soc.*, 137, 462 (1990).
51. N. Bonanos, B. Ellis, K.S. Knight, and M.N. Mahmood, *Solid State Ionics*, 35, 189 (1989).
52. N. Bonanos, *Solid State Ionics*, 53–56, 967 (1992).
53. N. Bonanos, *J. Phys. Chem. Solids*, 54, 867 (1993).
54. C.Y. Jones, J. Wu, L.-P. Li, and S.M. Haile, *J. App. Phys.*, 97, 114908 (2005).
55. D. Shima and S.M. Haile, *Solid State Ionics*, 97, 443 (1997).
56. T. Yajima and H. Iwahara, *Solid State Ionics*, 50, 281 (1992).
57. K. Liang and A. Nowick, *Solid State Ionics*, 61, 77 (1993).
58. T. Norby, *Solid State Ionics*, 40–41, 857 (1990).
59. N. Bonanos, K.S. Knight, and B. Ellis, *Solid State Ionics*, 79, 161 (1995).
60. C.N.R. Rao, J. Gopalakrishnan, and K. Vidyasagar, *Indian J. Chem. Sect.*, 23A, 265 (1984).
61. D.M. Smyth, *Annu. Rev. Mater. Sci.*, 15, 329 (1985).
62. D.M. Smyth, in *Properties and Applications of Perovskite-Type Oxides*, L. Tejuca and J.L.G. Fierro, (editors), Marcel Dekker, New York, 1993, p. 47.
63. J. Wu, L.-P. Li, W.T.P. Espinosa, and S.M. Haile, *J. Mater. Res.*, 19, 2366 (2004).
64. J. Wu, R.A. Davies, M.S. Islam, and S.M. Haile, *Chem. Mater.*, 18, 846 (2005).
65. L. Li, A. Li, and E. Iglesia, *Stud. Surf. Sci. Catal.*, 136, 357 (2001).
66. S.M. Haile, *Acta Materialia*, 51, 5981 (2003).
67. H. Iwahara, *Solid State Ionics*, 86–88, 9 (1996).
68. H. Iwahara, *Proceedings in 18th Risø International Symposium on Material Science*, F.W. Poulsen, N. Bonanos, S. Linderoth, M. Mogensen, and B. Zachau-Christiansen, (editors), Risø National Labs., Roskilde, Denmark, 1996, p. 13.
69. F. Kroger and V. Vink, *Solid State Phys.*, 3, 307 (1956).
70. H. Iwahara, H. Uchida, and N. Maeda, *J. Power Sources*, 7, 293 (1982).
71. T. Norby and P. Kofstad, *Solid State Ionics*, 20, 169 (1986).
72. W. Münch, G. Seifert, K.-D. Kreuer, and J. Maier, *Solid State Ionics*, 86–88, 647 (1996).
73. M. Cherry, M.S. Islam, J.D. Gale, and C.R.A. Catlow, *J. Phys. Chem.*, 99, 14614 (1995).
74. M.S. Islam, *J. Mater. Chem.*, 10, 1027 (2000).
75. J. Wu, PhD dissertation, California Institute of Technology, Pasadena, CA, 2005.
76. T. Scherban and A.S. Nowick, *Solid State Ionics*, 35, 189 (1989).
77. T. Norby and R. Hausgrud, in *Non-porous Inorganic Membranes*, A.F. Sammells and M.V. Mundschau, (editors), Wiley-VCH Verlag GmbH & Co., Weinheim, Germany, 2006, p. 1.
78. L. Li. and E. Iglesia, *Chem. Eng. Sci.*, 58, 1977 (2003).
79. H.J. Bouwmeester and A.J. Burggraaf, in *Fundamentals of Inorganic Membrane Science and Technology*, A.J. Burggraaf, (editor), Elsevier, Amsterdam, the Netherlands, 1996, p. 435.
80. I. Prigogine, *Thermodynamics of Irreversible Process*, John Wiley & Sons, New York, 1967.
81. T. Kudo and K. Fueki, *Solid State Ionics*, VCH Publishers, New York, 1990.
82. K.R. Kendall, C. Navas, J.K. Thomas, and H.-C. zur Loye, *Chem. Mater.*, 8, 642 (1996).
83. T.H. Etsell and S.N. Flengas, *Chem. Rev.*, 70, 339 (1970).
84. J.C. Boivin and G. Mairesse, *Chem. Mater.*, 10, 2870 (1998).
85. M. Fernandez-Garcia, A. Martinez-Arias, J.C. Hanson, and J.A. Rodríguez, *Chem. Rev.*, 104, 4063 (2004).
86. D.R. Askeland, *The Science and Engineering of Materials* (3rd edition), PWS Publishing Company, Boston, MA, 1994.
87. J.B. Thomson, A.R. Armstrong, and P.G. Bruce, *J. Am. Chem. Soc.*, 118, 11129 (1996).
88. H. Deng, M. Zhou, and B. Abeles, *Solid State Ionics*, 74, 75 (1994).
89. H. Inana and H. Tagawa, *Solid State Ionics*, 83, 1 (1996).
90. B. Aurivillius, *Ark. Kemi.*, 1, 463 (1949).

91. B. Aurivillius, *Ark. Kemi.*, 1, 499 (1949).

92. B. Aurivillius, *Ark. Kemi.*, 2, 519 (1949).

93. A.D. Rae, J.G. Thompson, and R.L. Withers, *Acta Crystallog. B*, B48, 418 (1992).

94. N.C. Hyatt and K.S. Knight, in *Synthesis, Properties and Crystal Chemistry of Perovskite-Based Materilas*, W. Wong-Ng, A. Goyal, R. Guo, and A.S. Bhalla, (editors), Proceedings of the 106th Annual Meeting of the American Ceramic Society, American Ceramic Society, Westerville, OH, 2005, p. 151.

95. F. Abraham, M.F. Debreuille-Gresse, G. Mairesse, and G. Nowogrocki, *Solid State Ionics*, 28–30, 529 (1988).

96. S.A. Kramer and H.L. Tuller, *Solid State Ionics*, 28–30, 465 (1995).

97. H.L. Tuller, *Solid State Ionics*, 94, 63 (1997).

98. T. Takahashi and H. Iwahara, *Energy Convers.*, 11, 105 (1971).

99. R.L. Cook and A.F. Sammells, *Solid State Ionics*, 45, 311 (1991).

100. M. Feng and J.B. Goodenough, *J. Eur., Solid State Inorg. Chem.*, 31, 663 (1994).

101. T. Ishihara, H. Matsuda, and Y. Takita, *J. Am. Chem. Soc.*, 116, 3801 (1994).

102. T. Ishihara, H. Matsuda, and Y. Takita, *Solid State Ionics*, 79, 147 (1995).

103. M.A. Peña and J.L.G. Fierro, *Chem. Rev.*, 101, 1981 (2001).

104. Y. Teraoka, T. Nobunaga, and N. Yamazoe, *Chem. Lett.*, 503 (1988).

105. C. Chen, Z. Zhang, G. Jiang, C. Fan, W. Liu, and H.J.M. Bouwmeester, *Chem. Mater.*, 13, 2797 (2001).

106. L. Yang, L. Tan, X.H. Gu, W.Q. Jin, L.X. Zhang, and N.P. Xu, *Ind. Eng. Chem. Res.*, 42, 299 (2003).

107. S.G. Li, W.Q. Jin, P. Huang, N.P. Xu, J. Shi, Z.C. Hu, E.A. Payzant, and Y.H. Ma, *AIChE J.*, 45, 276 (1999).

108. L. Yang, X.H. Gu, L. Tan, W.Q. Jin, L.X. Zhang, and N.P. Xu, *Ind. Eng. Chem. Res.*, 41, 4273 (2001).

109. M. Anca-Dragan, PhD dissertation, Department of Mathematic, Informatics and Natural Sciences, Institute of Physical Chemistry, RWTH-Aachen, Germany, 2006.

110. C. Wagner, *Z. Phys. Chem. B.*, 21, 25 (1933).

111. C. Wagner, *Progr. Solid State Chem.*, 10, 3 (1975).

112. O. Vigil, J. Fundora, H. Villavicencio, M. Hernández, and R. Roque-Malherbe, *J. Mat. Sci. Lett.*, 11, 1825 (1992).

113. I.R. Beattie and A. Dyer, *Trans. Faraday Soc.*, 53, 61 (1957).

114. D.N. Stamires, *J. Chem. Phys.*, 36, 3184 (1962).

115. K.E. Simonens and E.S. Skou, in *Solid State Protonic Conductors II*, J.B. Goodenough, J. Jensen, and M. Kleitz, (editors), Odense University Press, Denmark, 1983, p. 155.

116. E. Krogh-Andersen, J.G. Krogh-Andersen, E. Skou, and S. Yde-Andersen, *Solid State Ionics*, 18–19, 1180 (1986).

117. N. Knudsen, E. Krogh-Andersen, J.G. Krogh-Andersen, and E. Skou, *Solid State Ionics*, 28–30, 627 (1988); 35, 51 (1989).

118. R. Roque-Malherbe, *Mic. Mes. Mat.*, 41, 227 (2000).

119. M. Hernández-Vélez and R. Roque-Malherbe, *J. Mat. Sci. Lett.*, 14, 1112 (1995).

120. M.A. Lampert and P. Mark, *Current Injection in Solids*, Academic Press, New York, 1970.

121. N.F. Mott and R.W. Gurney, *Electronic Processes in Ionic Crystals*, Dover Publications Inc. New York, 1940.

122. L. Pauling, *General Chemistry*, Dover Publications Inc., New York, 1988.

123. S.B. Adler, *Chem. Rev.*, 104, 4791 (2004).

124. A.J. Bard and L.R. Faulkner, *Electrochemical Methods* (2nd edition), John Wiley & Sons, New York, 2001.

125. V.S. Bagotsky, (editor), *Fundamentals of Electrochemistry*, John Wiley & Sons, New York, 2005.

126. B.E. Conway, *J. Electrochem. Soc.*, 124, 410C (1977).

127. R. Roque-Malherbe and R. Pascual, *Revista CNIC*, 11, 18 (1980); *Chemical Abstracts*, Vol. 95 No. 140775t.

128. L. Carrette, K.A. Friedrich, and U. Stimming, *Fuel Cells*, 1, 5 (2005).

129. M.E. Orazem and B. Tribollet, *Electrochemical Impedance Spectroscopy*, John Wiley & Sons, New York, 2008.

130. R.G. Linford and S. Hackwoodt, *Chem. Rev.*, 81, 327 (1981).

131. H. Hillebrandt and M. Tanaka, *J. Phys. Chem. B*, 105, 4270 (2001).

132. S.-M. Park and J.-S. Yoo, *Anal. Chem.*, 75, 455 A, (2003).

133. S. Nieto, M. Correa and R. Roque-Malherbe (in preparation).

134. E. Barsoukov and R. Mac Donald, (editors), *Impedance Spectroscopy, Theory, Experiments and Applications* (2nd edition), Wiley-Interscience, New York, 2005.

135. N.Q. Minh, *J. Am. Cer. Soc.*, 78, 563 (1993).

136. M. Juhl, S. Primdahl, C. Manon, and M. Mogensen, *J. Power Sources*, 61, 183 (1996).
137. M.M. González Cuenca, PhD dissertation, Twente University Press, Enschede, the Netherlands, 2002.
138. F. Dogan, T. Susuki, P. Jasinski and H.U. Anderson, in *Synthesis Properties and Crystal Chemistry of Perovskite Based Materials*, W. Wong-Ng, A. Goyal, R. Guo, and S. Bhalla (editors), Proceedings of the 106th Annual Meeting of the American Ceramic Society, ACS, Westerville, OH, 2005, p. 39.
139. A. Atkinson and T. Ramos, *Solid State Ionics*, 129, 259 (2000).
140. S.B. Adler, *J. Am. Ceram. Soc.*, 84, 2147 (2001).
141. J. Mizusaki, H. Tagawa, K. Naraya, and T. Sasamoto, *Solid State Ionics*, 49, 111 (1991).
142. Y. Takeda, R. Kanno, M. Noda, Y. Tomida, and O. Yamamoto, *J. Electrochem. Soc.: Electrochem. Sci. Technol.*, 134, 2656 (1987).
143. B.C.H. Steele, *Nature*, 414, 345 (2001).
144. S.W. Tao and J.T.S. Irvine, in *Mixed Ionic Electronic Conducting Perovskites for Advanced Energy Systems*, N. Orlouskaya and N. Browning (editors), NATO Science Series, vol. 173, Kluwer Academic Publishers, Dordrecht, the Netherlands, 2004, p. 87.
145. S. Park, J.M. Vohs, and R.J. Gorte, *Nature*, 404, 265 (2000).
146. S.D. Park, J.M. Vohs, and R.J. Gorte, *Nature*, 404, 625 (2000).
147. O.A. Marina, C. Bagger, S. Primdahl, and M. Mogensen, *Solid Oxide Fuel Cell*, 123, 199 (1999).
148. J. Liu, B.D. Madsen, Z. Ji, and S.A. Barnett, *Electrochem. Solid State Lett.*, 5, A122 (2002).
149. D. Hart, *J. Power Sources*, 86, 23 (2000).
150. A. Siddle, K.D. Pointon, R.W. Judd, and S.L. Jones, *Fuel Processing for Fuel Cells: A Status Review and Assessment of Prospects*, ETSU F/03/00252/REPURN031644, Advantica Ltd., Loughborough, UK, 2003.
151. A. Haryanto, S. Fernando, N. Murali, and S. Adhikari, *Energ. Fuel.*, 19, 2098 (2005).
152. B. Emonts, J.B. Hansen, S.L. Jörgensen, B. Höhlein, and R. Peters, *J. Power Sources*, 71, 288 (1998).
153. F. Gallucci, L. Paturzo, and A. Basile, *Ind. Eng. Chem. Res.*, 43, 2420 (2004).
154. B.A. Peppley, J.C. Amphlett, L.M. Kearns, and R.F. Mann, *Appl. Catal. A: Gen.*, 189, 21 (1999).
155. B.A. Peppley, J.C. Amphlett, L.M. Kearns, and R.F. Mann, *Appl. Catal. A: Gen.*, 189, 31 (1999).
156. X. Zhang, H. Hu, Y. Zhu, and S. Zhu, *Ind. Eng. Chem. Res.*, 45, 7997 (2006).
157. R. Roque-Malherbe, L. Lemes, C. de las Pozas, L. Lopez, and A. Montes, in *Natural Zeolites 93 Conference Volume*, D.W. Ming and F.A. Mumpton, (editors), International Committee on Natural Zeolites, Brockport, NY, 1995, p. 299.
158. R. Roque-Malherbe, L. Lemes, M. Autie, and O. Herrera, *8th International Zeolite Conference Extended Abstracts*, J.C. Jansen, L. Moscou, and M.F.M. Post, (editors), International Zeolite Association, Amsterdam, the Netherlands, 1988, p. 137.
159. E. Sanchez and R. Roque-Malherbe, *Biotechnol. Lett.*, 9, 671 (1987).
160. J.N. Armor, *App. Catalysis A. General*, 176, 159 (1999).
161. J.M. Thomas and W.J. Thomas, *Principle and Practice of Heterogeneous Catalysis*, VCH Publishers, New York, 1997.
162. V. Rao, K.A. Friedrich, and U. Stimming, in *Handbook of Membrane Separations: Chemical, Pharmaceutical, and Biotechnological Applications*, A.K. Pabby, A.N. Sastre, and S.S. Rizvi, (editors), CRC Press Boca Raton, FL, 2008, p. 759.
163. P. Costamagna and S. Srinivasan, *J. Power Sources*, 102, 242 (2001); 102, 253 (2001).
164. V. Metha and J.S. Cooper, *J. Power Sources*, 114, 32 (2003).
165. M. Doyle and G. Rajendran, *Handbook of Fuel Cells: Fundamentals, Technology and Applications*, V. Vielstich, H.A. Gasteiger, and A. Lamm, (editors), John Wiley & Sons, New York, 2003, p. 647.
166. K.A. Mauritz and R.B. Moore, *Chem. Rev.*, 104, 4535 (2004).
167. S. Gottesfeld and T. Zawodzinski, in *Advances in Electrochemical Science and Engineering*, R.C. Alkire, H. Gerischer, D.M. Kolb, and C.W. Tobias, (editors), Wiley-VCH, New York, 1997, p. 197.
168. M.A. Hickner, H. Ghassemi, Y.S. Kim, B.R. Einsla, and J.E. McGrath, *Chem. Rev.*, 104, 4587 (2004).
169. Q. Li, R. He, J.O. Jensen, and N.J. Bjerrum, *Chem. Mater.*, 15, 4896 (2003).
170. A.Z. Weber and J. Newman, *Chem. Rev.*, 104, 4679 (2004).
171. K.C. Lauzze and D.J. Chmielewski, *Ind. Eng. Chem. Res.*, 45, 4661 (2006).
172. D. Rolison, *Chem. Rev.*, 90, 867 (1990).
173. D. Rolison, *Stud. Surf. Sci. Catal.*, 85, 543 (1994).
174. A. Walcarius, in *Handbook of Zeolite Science and Technology*, S.M. Auerbach, K.A. Carrado, and P.K. Dutta, (editors), Marcel Dekker New York, 2003, p. 721.
175. D.C. Freeman Jr., US Patent 3,186,875, (1965).
176. S. Slane and M. Solomon, *J. Power Sources*, 55, 7 (1995).
177. W. Yang and H. Yang, *Dianchi*, 21, 3, (1991).

178. Furukawa Battery Co., Ltd., Japan Patent No. 58,012,263 (1983).
179. G.J. Bratton, T. Naylor, and A.C.C. Tseung, World Patent No. 9,852,243 (1998).
180. M. Hernandez-Velez, A. Blanco, R. Roque-Malherbe, H. Villavicencio, F. Fernandez, A. Berazain, and J.M. Albella, *Bol. Soc. Esp. Ceram. Vidrio*, 34, 409 (1995).
181. H. Villavicencio, M. Hernandez-Velez, O. Vigil-Galan, A. Rodriguez-Rivere, R. Roque-Malherbe, M. Betancourt, and A. Diaz, in *Zeolites '91 Memoirs of the 3rd International Conference on the Occurrence, Properties and Utilization of Natural Zeolites*, Havana, Cuba, April 9–11, 1991, G. Rodríguez and J.A. González, (editors), International Conference Center Press, Havana, Cuba, Vol. 1, 1992, p. 220.
182. W.H. Brattain and J. Bardeen, *Bell Systems. Tech. J.*, 32, 1 (1953).
183. J. Janata, M. Josowicz, P. Vanysek, and D.M. DeVaney, *Anal. Chem.*, 70, 179 (1998).
184. T. Arakawa, in *Properties and Applications of Perovskite Type Oxides*, L.G. Tejuca and J.L.G. Fierro, (editors), Marcel Dekker, New York, 1993, p. 361.
185. D.J. Yoo, J. Tamaki, S.J. Park, N. Miura, and N. Yamazoe, *J. Mater. Sci. Lett.*, 14, 1391 (1995).
186. J.-C. Yang and P.K. Dutta, *J. Phys. Chem. C*, 111, 8307 (2007).
187. N.F. Szabo and P.K. Dutta, *Sens. Actuators, B*, 88, 168 (2003).
188 J.-C. Yang and P.K. Dutta, *Sensors and Actuators* 125, 30 (2007).
189. K. Miyazaki, C.N. Xu, and M. Hieda, *J. Electrochem. Soc.*, 141, L35 (1994).
190. D.Y. Wang *J. Phys. Chem. Solids*, 55, 1471 (1994).
191. J.R. Frade, *Solid State Ionics*, 78, 87 (1995).
192. G. Baier, V. Schuele, and A. Vogel, *Appl. Phys. A*, A57, 51 (1993).

9 Heterogeneous Catalysis and Surface Reactions

9.1 INTRODUCTION

The definition of a catalyst, as per the International Union of Pure and Applied Chemistry (IUPAC), is that "a catalyst is a substance that increases the rate of a reaction without modifying the overall standard Gibbs energy change in the reaction" [1]. The chemical process of increase of the reaction rate is called catalysis and the catalyst is both a reactant and a product of the reaction. That is, the catalyst is restored after each catalytic act. Besides, the catalyst does not influence the final equilibrium composition after the cessation of the reaction.

Catalysis can be classified as homogeneous catalysis, in which only one phase is involved, and heterogeneous catalysis, the case of interest here, in which the reaction occurs at or near an interface between phases [1].

The catalytic effect was not noticed by the alchemists. The first heterogeneous catalytic reaction, that is, the dehydration of ethanol in active clay, was studied by Priestley in 1778; later Van-Marum, in 1796, was the first to use metallic catalysts for the dehydrogenation of ethanol [2]. In 1813, Thenard revealed that ammonia is decomposed into nitrogen and hydrogen when passed over different burning metals. In 1823, Dulong found that the activity of different metals such as iron, copper, silver, gold, and platinum for decomposing ammonia decreased in the order given. In 1814, Kirchoff reported that acids aid the hydrolysis of starch to glucose; in 1817, H. Davy and E. Davy reported the oxidation of hydrogen by air over platinum and later Faraday studied why platinum facilitates oxidation reactions [2]. However, the term catalysis has its origins in the work of Berzelius, who, in 1836, examined some previous results on chemical change in both homogeneous and heterogeneous systems. In the further development of the science of catalysis, Nerst, Kirchoff, Ostwald, Sabater, Langmuir, and others were, as well, involved.

Since the start of the development of the science and technology of catalysis, it has become an important operation in the chemical industry. In this regard, catalytic technologies are extremely significant for the economic development and expansion of the chemical industry. Heterogeneous catalysts offer numerous inherent benefits over their homogeneous counterparts, such as simplicity of product separation and catalyst reuse.

In this chapter, the basic principles of heterogeneous catalysis and surface reactions, and chemical, sustainable energy, and pollution abatement applications of heterogeneous catalysts are described [3–5].

9.2 GENERAL PROPERTIES OF CATALYSTS

A catalyst acts by making available a new route with a lesser activation energy, and because of this the reaction progresses more rapidly and equilibrium is attained more rapidly [1]. If the catalyst is included in other phases or it is, by itself, another phase, then, it is a heterogeneous catalyst. In this case, the reaction generally takes place at the surface, where normally the reactants are adsorbed on the surface prior to the occurrence of the catalytic process. In this regard, a lot of catalysts rely on adsorption, since the adsorption process can modify the electronic structure of the surface and can promote or inhibit the activity of the catalyst.

The main characteristics of a good catalyst are activity and selectivity, including reproducibility, thermal and mechanical stability, and the capacity for easy regeneration [6–8].

Catalyst activity is related to the nature, the number, the strength, and the spatial arrangement of the chemical bonds that are momentarily created between the reactants and the surface, which relies on the composition, structure, and morphology of the solid catalyst [9]. In this regard, catalyst activity greatly depends on the active component, or components included in the catalyst composition. A catalyst is composed of a major active component, the proportion of which surpasses that of other components, and secondary components, which are included to improve catalyst activity, and which are called additives or, sometimes, promoters or modifiers [2,9].

The promoter is a substance that by itself does not have catalytic properties; however, when it is added to the catalyst, it improves its properties. The promoter can be related to the stabilization of the catalyst structure or to the modification of the chemical properties of the catalyst surface [2].

The term impurity is reserved for trace quantities of other components over which the investigator or the manufacturer has little control. The reaction rate under specified conditions of temperature, composition, and pressure is used to measure the catalytic activity of the solid, as described later.

Selectivity, on the other hand, means high yield of a particular product, that is, the capacity of the catalyst to conduct the conversion of the reactants in one particular path. Very often a reactant or a set of reactants can, at the same time, experience some parallel reactions, giving different, specific products, which react further in consecutive reactions to yield diverse secondary products. Consequently, high selectivity indicates a high yield of a desired product, while blocking out unwanted competitive and consecutive reactions [1,4].

The activity of a catalyst is mainly restricted to a portion of the catalyst's surface called active sites; thereafter to reduce the activity of a catalyst, it is normally necessary only a little amount of substances named poisons [2].

Stability indicates that the catalyst properties will not noticeably change during the catalytic process. Sources of instability are change in the grain size distribution via attrition during contact with other grains in the reactor, particle agglomeration with loss of surface area, loss of mechanical strength because of structural damage, and phase and chemical changes of the catalyst during operation [8]. On the other hand, coke formation, poisoning of active sites, and loss of active components, all lead to the unsteadiness of the catalyst. Besides, catalysts which include a material that is photosensitive, for example, TiO_2 (titania), can exhibit increased chemical instability when exposed to sunlight.

Catalysts can be regenerated, that is, the performance of deactivated catalysts can be improved by regeneration. That is, when the catalyst activity and/or selectivity is reduced during operation, a particular treatment allows the proper activity and/or selectivity of the catalysts to be restored [7]. Deactivation is produced by inhibition, fouling, or sintering, and all of these can be reversed, by the removal of poisons, or fouling agents, like coke, or by the re-dispersion of the active species [8]. The reproducibility of a catalyst is related to the consistency of its properties in different sets of production lots [7].

In Chapters 4 and 6, several methods of characterization of solids that are normally used for catalyst testing were described. In particular, the parameters which characterize the surface morphology of a porous catalyst are the same that characterize a porous adsorbent, that is, the specific surface area, S [m^2/g], the micropore volume, W^{MP} [cm^3/g], the sum of the micropore and mesopore volumes, that is, the pore volume, W [cm^3/g], and the pore size distribution (PSD), $\Delta V_p / \Delta D_p$ (see Chapter 6).

Heterogeneous catalysts are solid materials that sometimes consist of the bulk material itself, for example, acid zeolite catalysts [10] or fused catalysts [11]. Or in other cases of an active component or components deposited, as a rule, on a highly developed area support, for example, silica, alumina, carbon or in some cases a zeolite. The function of the support is to enhance the catalyst properties, for example, the stability of the active component or components, or in some cases to be even included in the catalytic reaction, for example, by providing acidic sites in bifunctional zeolite catalysts [10].

In a broad sense, an acid site can be defined as a site on which a base is chemically adsorbed. Conversely, a basic site is a site on which an acid is chemically adsorbed. Specifically, a Brönsted acid site has a propensity to give a proton, and a Brönsted base has the tendency to receive a proton. Additionally, a Lewis acid site is capable of taking an electron pair and a Lewis basic site is capable of providing an electron pair. These processes can be studied by following the color modifications of indicators, and by using infrared (IR) and nuclear magnetic resonance (NMR) spectroscopies, and calorimetry of adsorption of the probe molecules (see Chapter 4).

Another type of a catalytic site is the so-called redox site, which can be divided in two categories [8]:

1. Sites that include atoms, which in the presence of the reaction products are oxidized or reduced.
2. Sites that are capable of creating a charge-transfer complex by interaction with an electron acceptor or an electron donor.

These sites can be studied by electron paramagnetic resonance (EPR), ultraviolet (UV) spectrometry, x-ray photoelectron spectroscopy (XPS), and other methods. Besides, a quantitative study of redox sites can be carried out with the help of a volumetric or gravimetric study of the adsorption of the oxidizing or the reducing molecules.

Another common distinctive feature of catalysts is the shape of the catalyst pellets, which can be cylinders, rings, spheres, monoliths, and other forms. The shape has an effect in the empty space of a catalytic reactor, and, consequently, will affect mass transport as well mechanical strength.

In Chapter 2, we have studied the structure and the morphology of materials, and in Chapter 3 some of the methods of materials syntheses were studied. Among the materials included were some of the catalysts and supports, which are described here. These chapters should be referred to for the structure and syntheses methods, and this chapter focuses on catalysts and the modification of materials to obtain catalysts.

9.3 CRYSTALLINE AND ORDERED NANOPOROUS HETEROGENEOUS CATALYSTS

Zeolites and related materials, pillared clays, and mesoporous molecular sieves (MMSs) are the most important acid catalysts. In this section, the catalytic properties of these materials are explained [12–22].

9.3.1 Acid Zeolite Catalysts: Brönsted Type

Ammonium zeolites are transformed into acid zeolites by the decomposition of ammonium cations: the zeolite is modified to its acid form by exchanging sodium or any other charge-balancing cation present in the zeolite for ammonium, as follows [18]:

$$Na\text{-Zeolite} + NH_4^+ \leftrightarrow NH_4\text{-Zeolite} + Na^+$$

Then, the zeolite is heated in an air flow to obtain the acid form of the zeolite, as follows:

$$NH_4\text{-Zeolite} \xrightarrow{\text{Heat}} H\text{-Zeolite} + NH_3$$

This process should be followed by the ultrastabilization of the acid zeolite. This procedure is one of the basic operations in the industrial production of acid catalysts, consisting of controlled dealumination produced by thermal treatment in a water vapor atmosphere, which increases the thermal stability of the zeolite [19].

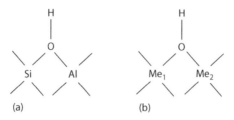

FIGURE 9.1 (a) Bridged OH group for aluminosilicate zeolites and (b) bridged OH group in the general case.

In an acid zeolite, the proton is connected to the oxygen atom bonded to the neighboring silicon and aluminum atoms (see Figure 9.1a). The obtained acid site structure, which is called the bridged OH group, is responsible for the Brönsted acidity of zeolites. The bridged OH group is illustrated in Figure 9.1b, where Me_1 and Me_2 are Si, Al, P, Ge, Ga, Fe, B, be Cr, V, Zn, Zr, Co, Mn, or any other metal [14]. The atoms, Me_1 and Me_2, in this structure are in such a combination that the electroneutrality principle is fulfilled. For example, if Me_1 is P(V), then Me_2 can be Co(II). In addition to the OH acid sites shown in Figure 9.1, it is possible to find terminal silanol groups in zeolites, that is, Si-OH. This group is not acidic.

The most important OH groups are those shown in Figure 9.1a, and, consequently, the most frequently studied. Therefore, the OH bridged groups of aluminosilicate zeolites have been widely studied by microcalorimetry, IR spectrometry, thermoprogrammed desorption, NMR, and other methods (see Chapters 4 and 6). The general conclusion about the acid strength of the bridged OH groups is that the acidity of this site is related to the ease with which the proton is released, that is, if the interaction between the proton and the oxygen is weak, the site is more acidic [17,21]. If we consider that this interaction is merely electrostatic, the previous statement means that the interaction depends on the negative charge of oxygen and the positive charge of hydrogen [17,21]. This charge is a function of the chemical composition of the neighborhood of the OH bridge sites, according to the rule of average electronegativity, or the Sanderson principle of equalization of the electronegativity. The Sanderson electronegativity equalization principle states that in a molecule composed of atoms with different electronegativities, the electrons will be restructured in such a way that they will be equally attracted to the nuclei in the bond. Consequently, the average molecule electronegativity is postulated to be the geometric mean of the compound atoms of the molecule under consideration [22]. Then, a rise in the electronegativity in the vicinity of the OH group induces an electronic density transfer from the less electronegative atom, hydrogen, to the most electronegative, oxygen; subsequently, an increase in acidity occurs fact accompanied by an increase in the length of the OH bond [17,21]. The deprotonation energy is, as well, related to the zeolite geometry, due to the long-range ordered structure of the zeolites and the zeolite composition [20].

To increase the acidity of the acid zeolites, these materials are dealuminated. The dealumination process can be done by a hydrothermal treatment (steaming), acid leaching, or by treatment in flowing $SiCl_4$ at 200°C–300°C or using hexafluorsilicates [18]. These treatments produce high-silica zeolites. The dealumination treatment also creates mesopores by extraction of Al from the zeolite lattice, thereby causing a partial collapse of the framework. One example of a dealuminated acid catalyst is the USY zeolite, that is, ultrastable zeolite Y, which is widely used in catalytic cracking [20]. Other zeolites are synthesized with high silica/alumina ratios [15,16]. For example, the majority of the members of the ZSM family of molecular sieves and some of the members of the SSZ family are high-silica zeolites. It is common that in the zeolite deamination process, dealumination occurs to some extent and extra-framework Al (EFAL) species can be generated. It is well known that the calcination of the NH_4^+ form of zeolites can lead to dealumination, as described later [18].

9.3.2 BIFUNCTIONAL ZEOLITE CATALYSTS

The thermal reduction of zeolites exchanged with metals is the method used for the preparation of bifunctional catalysts for hydrocarbon conversion. The bifunctional zeolite catalysts are composed of both acid sites and metal clusters. The preparation methods of these catalysts encompass three steps: ion exchange, calcination, and reduction. The ion-exchange process is carried out with aqueous solutions of salts or, more commonly, metal complexes that will be incorporated into the zeolite cavities and channels. For the case of metals with high catalytic activity, such as Pt, Pd, and Rh, the complexed ions, $[Pt(NH_3)_4]^{2+}$, $[Pd(NH_3)_4]^{2+}$, and $[Rh(NH_3)_5(H_2O)]^{3+}$ in aqueous solution, are exchanged with sodium zeolite [24]. After the ion exchange, the produced solid is calcined in a flow of air or oxygen to remove water and the ligands of the exchanged cations. Finally, the solid produced is reduced in a hydrogen flow, by passing a stream of a mixture of about 4 wt % (or in some cases more than 4 wt %) H_2 plus about 96 wt % of Ar or N_2. The reduction process is described by

$$M^{n+} + \frac{n}{2}H_2 \rightarrow M^0 + nH^+$$

which produces metallic clusters and Brönsted acid sites.

9.3.3 ACID ZEOLITE CATALYSTS: LEWIS TYPE

In addition to the Brönsted acidity in zeolites, these materials also have Lewis acidity. According to Lewis, an acid is an electron pair acceptor, a broader definition than that given by Brönsted; since a proton is a particular case of an electron pair acceptor, then the definition of Lewis covers practically all acid–base processes, whereas the definition of Brönsted represents only particular types [25].

In zeolite-based catalysts, the Lewis acidity is related to the existence of extra-framework Al (EFAL) species formed during the zeolite dealumination process [18]. It occurs frequently in zeolite activation, for example, during the calcination process

$$NH_4\text{-Zeolite} \xrightarrow{\text{Heat}} H\text{-Zeolite} + NH_3$$

dealumination takes place to a certain degree, and EFAL species are generated. These EFAL species produce Lewis acid sites, as determined by pyridine adsorption, and also increase the acidity of the framework Brönsted sites due to a polarization effect, thereby increasing the catalytic activity of medium- and large-pore zeolites [20]. The extra-framework Al species acting as Lewis acids perform an important function in cracking bottoms, although they have a tendency to generate more gases and coke [25].

9.3.4 BASIC ZEOLITE CATALYSTS

The catalytic activity of basic zeolites was first reported in the early 1970s. It was stated that the side-chain alkylation of toluene was effectively catalyzed by alkali, ion-exchanged X- and Y-type zeolites, which is a typical base-catalyzed reaction [23]. The most significant basic sites in zeolites are the framework negative oxygen sites, which are Lewis basic sites, whose strengths are correlated with the fractional negative charge of the oxygen, satisfying the intermediate Sanderson electronegativity, as was formerly argued with respect to Brönsted acidity [6].

A bonding model of a basic zeolite [23], as shown in Figure 9.2, where the configurations (Figure 9.2a and b) are in resonance, was proposed. In this arrangement (Figure 9.2a), the extra-framework cation forms a covalent bond with the framework oxygen, while, in that shown in Figure 9.2b, the cation forms an ionic bond with the negatively charged zeolite lattice.

While the electronegativity of the cation increases and approaches that of the oxygen, the contribution of the configuration (Figure 9.2a) increases to reduce the net charges on the lattice, and,

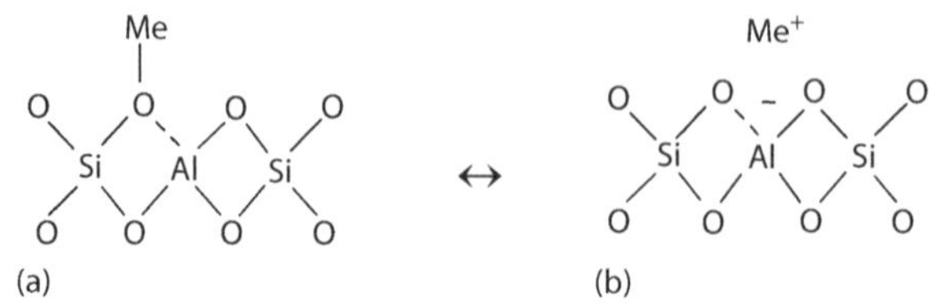

FIGURE 9.2   Schematic bonding model of a basic zeolite.

subsequently, decreases the basicity [6,26]. The basic strength of the zeolite, exchanged with an alkaline cation, increases in the following order Li-Z < Na-Z < K-Z < Rb-Z < Cs-Z [23,26].

To produce basic zeolites, two methods have been developed [18]: one line of attack is to ion exchange with alkali metal ions, and the other is to impregnate the zeolite pores with fine particles that can act as bases themselves. The first method, that is, ion exchange, generates moderately weak basic sites, while the latter results in strong basic sites [23].

9.3.5   CATALYSTS OBTAINED BY THE ISOMORPHOUS SUBSTITUTION OF Ti IN ZEOLITES

The isomorphous substitution of T atoms by other elements produces novel hybrid atom molecular sieves with interesting properties. In the early 1980s, the synthesis of a zeolite material where titanium was included in the MFI framework of silicalite, that is, in the aluminum-free form of ZSM-5, was reported. The name given to the obtained material was titanium silicate (TS-1) [27]. This material was synthesized in a tetrapropylammonium hydroxide (TPAOH) system substantially free of metal cations. A material containing low levels (up to about 2.5 atom %) of titanium substituted into the tetrahedral positions of the MFI framework of silicalite was obtained [28]. TS-1 has been shown to be a very good oxidation catalyst, mainly in combination with a peroxide, and is currently in commercial use. It is used in epoxidations and related reactions. TS-1, additionally an active and selective catalyst, is the first genuine Ti-containing microporous crystalline material.

9.3.6   PILLARED CLAYS

Zeolites become ineffective when the size of the reactants taking part in the catalyzed reaction are higher than the dimensions of the zeolites pores. In this case, the approach should be to overcome such a limitation by preparing catalysts with larger pores. In this regard, an additional group of materials with large pores, whose frameworks are composed of layered structures with pillars in the inter-lamellar region, was developed. These materials are the so-called pillared clays (PILCS) [29–31] (See Section 2.5.4).

In the early history of acid catalysis, clays had a significant utility. The acid leaching of natural smectite clays produced amorphous silica–alumina [Si–Al] materials with large surface areas and active acid sites, which were used as acid catalysts [31].

The intercalation of alkylammonium cations in clays, specifically tetraalkylammonium, for the first time, was carried out by Barrer and MacLeod. These researchers prepared pillared materials by exchanging the alkali and alkaline earth cations present in a montmorillonite clay with tetraalkylammonium [29]. Nevertheless, the obtained material was thermally unstable and, consequently, had no viable application in catalysis.

The idea of maintaining the clays permanently expanded by intercalating strong inorganic pillars was advanced independently by Brindley and Semples [32] and Vaughan, Lussier, and Maggee [33]. These investigators used oxyhydroxyaluminum cations as the pillaring agent [32,33]. These two researches led to a global attention for PILCS, which has a similarity with zeolites, since both materials are microporous. However, in the case of PILCS, it was, at least theoretically possible, to obtain large distances between the clay sheets [31].

Clay minerals are hydrous aluminum phyllosilicates made of sheets or layers constituted of tetrahedra and octahedra [34,35] (see Section 2.5.4). This mineral type includes the following groups: kaolinite, smectite, illite, and chlorite [35]. If the PILCS obtained with the help of a clay expansion is to be used as a molecular sieve, then, the pillared material ought to have the these attributes: uniform spacing between the pillars, suitable gallery heights, and layer rigidity [31,36]. Among the diverse layered phases, those belonging to the smectite group are suited to satisfy the above-mentioned conditions [12].

In the case of a smectite, each layer comprises two sublayers of tetrahedra with an inserted octahedral layer, where between the layers an interlayer space is formed in which the exchangeable cations are located (see Figure 9.3). That is, the smectite structure can be described on the basis of layers containing two sublayers of silica tetrahedra squeezed into a layer of an octahedra of Al^{3+} or Mg^{2+}, that is, a 2:1 layered clay [34]. The replacement of some of the Al^{3+} with Mg^{2+} or Li^+, or the isomorphous replacement of tetrahedral Si^{4+} with Al^{3+}, results in a certain amount of total negative charge on the layer, compensated in turn by the presence of hydrated cations in the interlayer region (see Figure 9.3).

At first view, the pillaring mechanism is straightforward since it consists of the exchange of the charge-compensating cations of the clay by a cationic oligomer, that is, P_n^{v+} made from n cations bound by oxo- or hydroxo- bridges with a total charge, $v+$ [31]. The common method for expanding clays consists of exchanging the cations in the interlamellar position, to be precise, Na^+, K^+, and Ca^{2+}, with larger inorganic oligomeric hydroxyl metal cations, formed by the hydrolysis of metal salts of Al, Zr, Ga, Cr, Si, Ti, Fe, and mixtures of them [36]. That is, if M is the symbol for the clay, the following exchange occurs in the gallery or interlayer space [31]:

Step 1: $MNa_v + P_n^{v+} \leftrightarrow MP_n + vNa^+$

where the oligomer results from the following condensation reaction:

Step 2: $nP^{Z+} \leftrightarrow P_n^{v+} + (nZ - v)H^+ + xH_2O$

After these stages, the treated sample undergoes a meticulous thermal treatment. During this treatment, the dehydration and dehydroxylation of the exchanged material take place, the stable metal oxide clusters then develop, which carry out the layer separation, generating a two-dimensional interlayer space with an aperture, which, if the process is correctly performed, can be larger than 10 Å [12,36] (see Figure 2.25). The obtained materials are stable up to 625–725 K [31].

The creation of Brönsted acid sites in a dioctahedral clay such as the beidellite, or a trioctahedral smectite, such as the saponite, is carried out using the same methodology previously explained for zeolites:

$$Na\text{-}PILCS + NH_4^+ \leftrightarrow NH_4\text{-}PILCS + Na^+$$

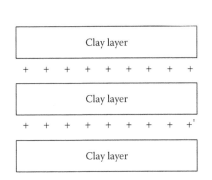

FIGURE 9.3   Schematic representation of a clay.

After this, the sample is heated in an air flow to obtain the acid form of the PILCS:

$$NH_4\text{-PILCS} \xrightarrow{\text{Heat}} H\text{-PILCS} + NH_3$$

In addition to the Brönsted acid sites produced by the above-explained methodology, in PILCS ion exchanged with multivalent cations and partially dehydrated, Lewis acid sites are produced.

The main interest in developing acidic PILCS was associated with their applications as cracking catalysts and in the synthesis of fine chemicals. Certainly, the prospect of making PILCS wherein the huge gas-oil and other molecules can diffuse and meet the active sites was a motivation for the development of these catalysts [12].

9.3.7 Mesoporous Molecular Sieves

Regardless of the large amount of work dedicated to zeolites and related materials, the dimensions and easy accessibility of pores in these materials are confined to the microporous scale [37]. This limits the uses of zeolites and related materials to catalyzing reactions where only small molecules take part. As a result, during the recent years, major efforts have been focused on obtaining materials that have larger pore sizes [16]. As a significant result of these endeavors, in 1992, researchers at the Mobil Corporation discovered the M41S family of mesoporous molecular sieves (MMSs) [38–42]. In Section 2.5.2, the structure of the family of MMS, and in Section 3.5, the syntheses methods were described. Here, we consider the modifications that must be carried out in these materials to obtain the catalysts.

The materials belonging to the MMSs family show very high surface areas with regular pore size dimensions [37,40,41]. These properties, which are not catalytic by themselves, are of high value for the manufacture of catalytically active phases [12,43–54]. Subsequently, the MMS, MCM-41, that can be a good support for catalytically active metallic phases, was proposed. It was shown that the MCM-41 impregnated with Cr has good catalytic activity for olefin oligomerization [40]. In addition to the attributes of MMS as catalyst supports, it is possible to generate Brönsted acid sites on the surface of the mesoporous structures; this allows novel options for the manufacture of acid and bifunctional catalysts [12]. The use of MCM-41 as an acid catalyst for Friedel–Crafts alkylation and acylations [43–45,48–51] have been reported, where the advantages of MCM-41 were manifested [40].

On the other hand, if Brönsted acidity is generated, it is possible to generate basicity by exchanging the protons with alkaline ions [12]. Specifically, Na-MCM-41 and Cs-MCM-41 catalysts exhibit satisfactory performance in base catalysis [40]. Besides, there exists the option of setting up transition metals in the MMS walls with the purpose of developing catalytic redox properties that will be effective in selective oxidation.

The isomorphous substitution of T atoms by other elements is capable of producing a novel hybrid atom molecular sieve with interesting properties; a typical example is titanium-containing ZSM-5 zeolite (TS-1) [27]. Titanium-substituted MCM-41 has been synthesized in an acidic system by using either ionic surfactants (CTA$^+$) or primary amine (DDA) as the template [46], and it has been applied as a catalyst [12].

9.4 AMORPHOUS, POROUS HETEROGENEOUS CATALYSTS AND SUPPORTS

9.4.1 Amorphous Acid Silica–Alumina

The catalytic activity of amorphous silica–alumina ([Si–Al]) in reactions via carbonium ions is due to the existence of Brönsted acid sites on their surface. Consequently, amorphous [Si–Al] acid catalysts provide acid sites and transport to the active sites easily. As a result, amorphous [Si–Al] acid catalysts have been widely operated as cracking catalysts. Acid zeolites have been successfully applied as cracking catalysts. However, in some industrial applications of acid catalysts, for example, in the cracking of hydrocarbons of high molecular weight, zeolites are not useful, since large pores are required in order to assist the internal diffusion of reactants and products. Then, in some cases amorphous acid [Si–Al] has to be applied instead of zeolites.

The method generally used for the synthesis of amorphous [Si–Al] is by the appropriate combination of sodium silicate and sodium aluminate, as explained for the synthesis of aluminosilicate zeolites in Section 3.4.1. But, in this case, the gel is not hydrothermally treated in an autoclave in order to crystallize a zeolite. It is instead thermally dried (see Section 2.7.1) to obtain an amorphous [Si–Al]. The produced amorphous material is then exchanged with NH_4^+, using a method similar to that previously explained for zeolites:

$$Na\text{-}[Si - Al] + NH_4^+ \leftrightarrow NH_4\text{-}[Si - Al] + Na^+$$

Afterward, it is calcined to get the acid catalyst as follows:

$$NH_4\text{-}[Si - Al] \xrightarrow{\text{Heat}} H\text{-}[Si - Al] + NH_3$$

[Si–Al] synthesis and zeolite synthesis have similarities; advances in the synthesis methods of high surface area amorphous [Si–Al] have been attained by the use of TPAOH in an alkali-free reaction mixture, such as the precursor gel for the synthesis of a ZSM-5 zeolite [15].

To conclude this section, it is necessary to state that besides their application in catalytic cracking, amorphous silica–alumina acid catalysts have been applied in other hydrocarbon transformations, such as isomerization of olefins, paraffins, and alkyl aromatics, the alkylation of aromatics with alcohols and olefins, and in olefin oligomerization [55].

9.4.2 METALLIC CATALYSTS SUPPORTED ON AMORPHOUS MATERIALS

Many practical catalysts consist of one or several active components, deposited on a high surface area support, whose purpose is the dispersion of the catalytically active component or components and their stabilization against sintering. In some cases, the support is passive; in other cases, the support is not inert; it is also part of the catalyst, combining oth functions, that is, the support and the active catalytic phases [56].

We focus here on supported metallic catalysts, since metals are widely used as catalysts. However, the metal catalyst is often expensive. Then, it is usually applied in a finely dispersed form on a high surface area support, since in this circumstance, a large fraction of the metal atoms are exposed to the reactant molecules.

Metals frequently used as catalysts are Fe, Ru, Pt, Pd, Ni, Ag, Cu, W, Mn, and Cr and some of their alloys and intermetallic compounds, such as Pt-Ir, Pt-Re, and Pt-Sn [5]. These metals are applied as catalysts because of their ability to chemisorb atoms, given an important function of these metals is to atomize molecules, such as H_2, O_2, N_2, and CO, and supply the produced atoms to other reactants and reaction intermediates [3]. The heat of chemisorption in transition metals increases from right to left in the periodic table. Consequently, since the catalytic activity of metallic catalysts is connected with their ability to chemisorb atoms, the catalytic activity should increase from right to left [4]. A Balandin volcano plot (see Figure 2.7) [3] indicates a peak of maximum catalytic activity for metals located in the middle of the periodic table. This effect occurs because of the action of two competing effects. On the one hand, the increase of the catalytic activity with the heat of chemisorption, and on the other the increase of the time of residence of a molecule on the surface because of the increase of the adsorption energy, decrease the catalytic activity since the desorption of these molecules is necessary to liberate the active sites and continue the catalytic process. As a result of the action of both effects, the catalytic activity has a peak (see Figure 2.7).

In heterogeneous catalysis, transition metal nanoparticles are supported on different substrates and are utilized as catalysts for different reactions [57], such as hydrogenations and enantioselective synthesis of organic compounds [58], oxidations and epoxidations [59], and reduction and decomposition [57].

Among the supports that have been used in the preparation of supported transition metal nanoparticles are carbon, silica, alumina, titanium dioxide, and polymeric supports [57], and the most frequently used support is alumina [56]. These supports normally produce an effect on the catalytic activity of the metallic nanoparticles supported on the amorphous material [60]. In Chapter 3, different methods for the preparation of metallic catalysts supported on amorphous solids were described [61–71].

9.5 PHOTOCATALYSTS

9.5.1 INTRODUCTION

The term photocatalysis was introduced during the 1930s. It is a change in the rate of chemical reactions under the action of light in the presence of photocatalysts that absorb light quanta and are involved in the chemical transformations of the reaction participants [72,73]. The study of catalysts functioning by photoinduced electron transfer at molecule–semiconductor interfaces is a relatively new branch of catalysis [72–91]. To obtain a good photocatalyst, many structural parameters are important, such as particle size, crystalline quality, morphology, specific surface area, and the surface state, among others.

Photocatalytic oxidation and reduction have received considerable attention during the last few years [75–78,88]. Specifically, the photocatalytic oxidation and reduction of organic compounds in water has become very important. The net process involves oxidizing or reducing the organic compound to an intermediate stage containing oxygen or to carbon dioxide, water, and a mineral acid if a heteroatom such as nitrogen or chlorine is present.

Photoexcitation of a semiconductor irradiated with UV-light photons, with an energy which matches the photocatalyst band gap energy, yields electron–hole pairs (see Figures 2.10 and 2.11 in Section 2.3.3). The mechanism of the photocatalytic decomposition of organic compounds is supposed to follow these steps: under UV-light illumination, absorption of photons creates an electron–hole pair, if the photon energy is higher than the band gap; thereafter, the pairs migrate to the surface and are trapped by OH surface groups and other surface sites, forming hydroxyl, $^{\bullet}OH^{-}$, hydrosuperoxide, $^{\bullet}H_2O$, and superoxide, $^{\bullet}O_2^{-}$, radicals, and finally these free radicals cause the oxidation of the organic compounds [75,78,88]. This process can be used in various areas such as elimination of odor from drinking water, degradation of oil spills in surface water systems, and degradation of harmful organic contaminants such as herbicides, pesticides, and refractive dyes.

These processes take place on photocatalytic active sites. These are surface sites where chemical transformations take place after transition of the photocatalytic active center into the active state via a photophysical process [72,73].

9.5.2 TITANIUM OXIDE

The photoactivation of TiO_2 has received considerable attention from scientists and engineers during the last years. This has produced some outstanding developments, such as [86] the application of TiO_2 for the creation of solar cells, as a means for environmental cleanup, and its antimicrobial applications.

There are three common TiO_2 polymorphs in nature, which, in the order of abundance, are rutile, anatase, and brookite, an additional synthetic phase called $TiO_2(B)$, and some high-pressure polymorphs [87]. In photocatalysis, the rutile and anatase polymorphs are primarily used [72,73]. However, the anatase structure is regarded to be the most efficient among the TiO_2 polymorphs, even though the physicochemical background of this has not been understood properly. Both the rutile and anatase crystal structures are in distorted octahedron classes. In rutile (see Figure 2.9), a slight distortion from orthorhombic structure occurs, where the unit cell is stretched beyond a cubic shape. In anatase, the distortion of the cubic lattice is more significant, and thus the resulting symmetry is less orthorhombic [86].

The n-type semiconductor properties of TiO_2-based materials are basic to realize the photocatalytic functions [87]. Both rutile and anatase phases are semiconductors with a band gap of 3.10 eV for rutile and 3.23 eV for anatase [75]. With its band gap between 3.1–3.2 eV, pure TiO_2 has a photothreshold that includes less than 10% of the total active zone of the solar spectrum. However, by doping TiO_2 with particular impurities, this onset energy for photoactivation can be decreased, probably even rising the photoactivity [86]. The doping of TiO_2 has been carried out fundamentally with heterocations, or C and/or N [86,87].

9.5.3 OTHER PHOTOCATALYSTS

Even though, TiO_2 exhibits excellent photocatalytic activities in VOC and other organic compound decomposition, it is desirable to study new photocatalysts with different structures that exhibit higher activity. In this regard, in recent times, a group of novel photocatalysts made of highly donor-doped (110) layered perovskites, for example, $La_2Ti_2O_7$, $Sr_2Nb_2O_7$, La_4-$CaTi_5O_{17}$, and $Ca_2Nb_2O_7$, which were found to be much more efficient than bulk-type TiO_2, have been developed [89,90]. Layered perovskite means that there exists interspacing between perovskite slabs consisting of a TiO_6 or a NbO_6 octahedron [77]. This higher activity resulted from the highly donor-doped electronic structure, which leads to a narrower depletion region and a more facile separation of charge carriers.

Selective photocatalysis can be achieved through the use of proper molecular sieves, such as the titanosilicate ETS-10, an alternative approach. The photocatalytic activity of ETS-10 is due to the presence of photoexcitable Ti–O–Ti chains and the three-dimensional interconnected pore system of large 12-membered ring channels in its structure, which endow the material with excellent diffusion properties [83]. This regular channel system has a direct effect in determining the shape selectivity of the degradation process [84].

Another approach is the introduction of some transition metals, such as Cr, V, Fe, Cu, Mn, Co, Ni, Mo, and La into the synthesis mixture of the MCM-41, MMS with a Si/Me ratio of 80 [85]. It was then demonstrated that the presence of the transition metal salts in the gel during the hydrothermal synthesis process hinders the action of the template, which results in MCM-41 pores that are not well formed. These materials were then loaded with TiO_2, via the solgel method, and the activity of the TiO2/TM-MCM-41 catalysts in the degradation of the 4-chlorophenol reaction was tested in the presence of UV and/or visible light. It was shown that although some metals are deleterious, others can improve the performance of photocatalysts and even enable them to utilize visible light [85].

A wide range of other semiconductors and materials have been tested for photocatalytic activity. In general, they have been found to be less active than titanium dioxide; significant work has been carried out in V_2O_5, Fe_2O_3, ZnO, ZnS, CdS, Pt/CdS, ZnTe, $ZrTiO_4$, MoS_2, SnO_2,Sb_2O_4, Sn/SbO_2, CeO_2, WO_3, and Nb_2O_5 [79].

9.6 KINETICS OF SURFACE REACTIONS

9.6.1 STEPS IN A HETEROGENEOUS CATALYTIC REACTION

The first stage in all heterogeneous catalytic reactions is the adsorption of reacting species on the active phase of the catalyst. During this step, the intramolecular bonds of the reacting species are broken or weakened. Afterward, the adsorbed species react with each other or with the gas phase species. This process generally occurs in consecutive steps until the desired product is reached, which desorbs after this. During this process, the function of the catalyst is to decrease the activation energy of the rate-limiting reaction stage. Heterogeneous catalytic reactions are Explained with the help of two mechanisms: the Langmuir–Hinshelwood (L–H) or the Eley–Rideal (E–R) mechanisms [92,93]. In the first mechanism, the product generation takes place by means of a reaction of two or more adsorbed species followed by desorption of the product, where the adsorbed species are in thermal equilibrium with the catalyst surface (see Figure 9.6). Consequently, the reaction is initiated by the thermal energy provided by the surface. Within the framework of the Eley–Rideal mechanism, a new chemical bond is shaped by a straight collision among a gas phase molecule or atom with an adsorbed species (see Figure 9.7). Then, after being formed, the product desorbs instantly. In this case, if the reaction is activated, the energy required to overcome the barrier emerges from the translational and/or internal energy of the impinging molecule/atom.

The whole heterogeneous catalytic process, where the reactant and products are fluids proceed, in general, through the following steps:

1. Diffusion of the reactant molecule or molecules to the solid catalyst
2. Adsorption of at least one of the reactants
3. Chemical reaction between the molecules adsorbed on neighboring locations
4. Desorption of products
5. Diffusion of products into the bulk of the fluid phase

If we analyze all the steps taking part in the process, the final result will be very complicated. The diffusion, adsorption, and desorption processes are fast enough in comparison with the chemical reaction. Therefore, the adsorption process is in equilibrium during the catalytic reaction, since it is a fast process, and as a result, we can use an adsorption isotherm, for example, the Langmuir isotherm to calculate the amount of reactant in the surface.

The chemical reaction step is normally composed of various steps, and a broad diversity of rate laws and reaction mechanisms are relevant for surface-catalyzed reactions. However, if the simple assumption that the chemical reaction consists of a sole unimolecular or bimolecular elementary reaction or a rate determining simple reaction followed by one or more fast steps, is made, then the reaction kinetics can be mathematically treated [92].

9.6.2 Reaction Rate

We can express the heterogeneous catalytic chemical reaction as follows

$$\sum_i \upsilon_{A_i} A_i = \upsilon_A A + \upsilon_B B + \cdots + \upsilon_C C + \upsilon_D D + \cdots$$

where
A, B,… are the reactants
C, D,… are the products
υ_{A_i} is negative for the reactants and positive for the products

Subsequently, the conversion rate, J_C, in particular conditions of composition, pressure, and temperature is given by

$$J_C = \frac{1}{\upsilon_B} \left(\frac{dn_B}{dt} \right) \tag{9.1}$$

where
υ_B is the stoichiometric coefficient of the species B in the overall reaction
n_B are the moles of B
t is time

This parameter is a measure of the catalytic activity of the solid for the conversion of any species, B, under these conditions. The quantity of the catalyst to which the reaction rate is referred may be expressed by the mass (W), volume (V), or the surface area (S) of the solid catalyst. Then, the obtained rates are [8]

- Mass-specific rate: $r_W = \dfrac{J}{W}$ in [mol·s^{-1}·kg^{-1}]
- Volume-specific rate: $r_V = \dfrac{J}{V}$ in [mol·s^{-1}·m^{-3}]
- Specific-area rate: $r_A = \dfrac{J}{A}$ in [mol·s^{-1}·m^{-2}]

Additionally, the rate referred to the number of catalytic sites is known as the turnover rate, v_t, or turnover frequency (TOF) [94]. The TOF is defined as the number of molecules reacting per active site in unit time, that is, it is the number of revolutions of a catalytic cycle per unit time [95]

$$v_t = \frac{1}{v_B N} \frac{dN_B}{dt} \tag{9.2}$$

where
  N is the number of active sites
  $N_B = n_B N_A$
  N_A is the Avogadro number

Now, the rate is given for a gas phase reaction by [4]

$$r_x = kF(P_A, P_B, \ldots)$$

where

$$k = A_0 \exp\left(-\frac{E_a^{app}}{RT}\right)$$

is the apparent rate coefficient where A_0 is the pre-exponential factor and E_a^{app} is the apparent activation energy measured in the kinetic experiment. Besides, $F(P_A, P_B, \ldots)$ is a pressure related term, where P_A, $P_B, \ldots$ are the partial pressures of the reactants, which depends on the reaction mechanism.

9.6.3 Unimolecular Decomposition

A unimolecular decomposition can be expressed by the following equation:

$$A(g) \leftrightarrow B(g) + C(g)$$

An example of a reaction of unimolecular decomposition is the decomposition of NH_3 on metal surfaces is

$$2NH_3 \leftrightarrow N_2 + 3H_2$$

that can be very important in the hydrogen economy, if ammonia is used as the hydrogen carrier in storage systems [37,96].

We now consider that the elementary reaction will be the unimolecular surface reaction expressed by the following equation

$$A_a \rightarrow B_a + C_a, \text{ rate constant: } k$$

where A_a, B_a, and C_a are adsorbed species. Consequently, the specific-area rate

$$r_A = \frac{J}{A} = k\theta_A \,,$$

(9.3)

where k [mol· m^{-2}·s^{-1}] is the rate constant and

$$\theta_A = \frac{K_A P_A}{1 + K_A P_A + K_B P_B + K_C P_C}$$

(9.4)

We now suppose that the products are weakly adsorbed. Then,

$$1 + K_A P_A \gg K_B P_B + K_C P_C$$

Therefore,

$$r_A = k \frac{K_A P_A}{1 + K_A P_A}$$

(9.5)

This means that the decomposition occurs uniformly across the surface, where the products are weakly bound and rapidly desorbed; consequently, the rate-determining step is the surface decomposition step. This type of reaction shows two rate-limiting laws corresponding to the two extreme behaviors of the Langmuir isotherm. That is, at low pressure, θ_A is small and proportional to the pressure, and the rate becomes first order in A(g):

$$r_A = kK_A P_A$$

(9.6a)

On the other hand, at high pressures, θ_A is almost equal to one, and the reaction is of zero-order rate:

$$r_A = k$$

(9.6b)

It is a well-known experimental fact that the rate constant, k, obeys the Arrhenius law

$$k = A \exp\left(-\frac{E_{int}}{RT}\right)$$

(9.7)

while the temperature dependence of the equilibrium constant complies with the following relationship:

$$RT \ln K_A = -\Delta G_a = -\Delta H_a + T\Delta S_a$$

That is,

$$K_A = B \exp\left(-\frac{\Delta H_a}{RT}\right)$$

(9.8)

Consequently, since for low pressures $r_A = kK_A P_A = k_{obs} P_A$, the effective or observed activation energy for the process at low pressures is given by (see Figure 9.4)

$$r_A = kK_A P_A = k_{obs} P_A = A \exp\left(-\frac{E_{obs}}{RT}\right)P_A = C \exp\left[-\left(\frac{E_{int} + \Delta H_a}{RT}\right)\right]P_A$$

(9.9)

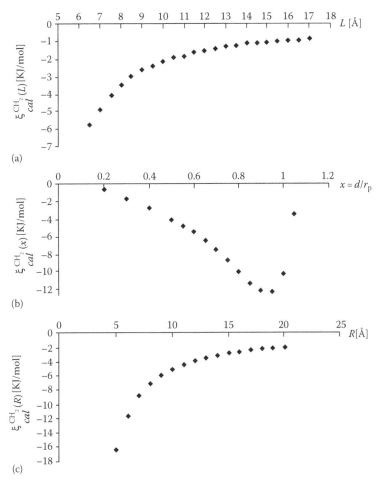

FIGURE 9.4 Graphic representation of Equations 9.18a through c.

Then, the effective activation energy for the reaction at a low pressure is [2]

$$E_{obs} = E_{int} + \Delta H_a \tag{9.10}$$

where the true or intrinsic activation energy, E_{int}, is decreased by the heat of adsorption, ΔH_a, of the reactant. On the other hand, at a high pressure, $r_A = k$ applies, and the activation energy is equal to the intrinsic activation energy.

$$r_A = k = A \exp\left(-\frac{E_{int}}{RT}\right) \tag{9.11}$$

9.6.4 Calculation of the Adsorption Enthalpy of *n*-Paraffins in Nanoporous Crystalline and Ordered Acid Catalysts, and Its Relation with the Activation Energy of the Monomolecular Catalytic Cracking Reaction

9.6.4.1 Introduction

As an example of a unimolecular decomposition reaction, we study the monomolecular catalytic cracking reaction of *n*-paraffins in high-silica acid zeolites or other crystalline or ordered acid porous materials, in this section [97–102].

Catalytic cracking is a chemical reaction for the conversion of vacuum distillates and residues into olefinic gas, high-octane gasoline, and diesel oil. This process is performed by cracking a vaporized feed over a solid acid catalyst. The acid forms of zeolites, PILCS, and MMSs have been applied in catalytic cracking and have been investigated for their applications in these reactions [12,20]. Although, it is generally accepted that Brönsted acid centers are implicated in these reactions, efforts to relate the number of protons and their acid strength to a particular catalytic reaction have repeatedly led to incoherent results, mainly in the case when the catalytic properties of different structures are evaluated [102,103].

In the case of alkane cracking, dispersion forces between the alkane molecules and the siliceous walls of the zeolites and perhaps other nanoporous crystalline and ordered materials are possibly the most important interactions for stabilizing adsorption in the cavities, since the proton affinity of alkanes is low [104] and the electrostatic interactions between the alkane and the adsorbent are negligible [97].

In this section, some ideas developed by Haag [100], Gorte [101,104], and others [98,99,102] employing the slit [105], cylindrical [106], and spherical pore models [107] to develop a model for the description of the channels and cavities of zeolites and other nanoporous acid catalysts [97] and the united-atom model to describe the n-alkanes of m carbons are described [108].

With the model developed, we describe alkane adsorption in nanoporous crystalline and ordered acid catalysts with the intention of calculating three mathematical equations, one each for the geometry of the pore system, for describing the adsorption enthalpy and its relation with the activation energy, and for the monomolecular cracking of n-paraffins [97]. This methodology can also be applied to other unimolecular reactions.

9.6.4.2   Unimolecular Catalytic Cracking

The reaction of hydrocarbon cracking operates via two mechanisms: monomolecular and bimolecular [98,99]. In the monomolecular cracking of an alkane molecule, the hydrocarbon is protonated to form a high-energy transition state that can look like a tightly coordinated nonclassical pentacoordinated carbonium ion [98,99]. Dehydrogenation or cracking of the carbonium ion produces the formation of hydrogen or a paraffin and a carbenium ion that can be desorbed as an alkene or can react further [98]. Therefore, for the unimolecular case, the reaction rate at low pressures is

$$r_A = kK_A P_A = k_{obs} P_A$$

Subsequently, the effective or observed activation energy for the process can be expressed as follows:

$$r_A = kK_A P_A = k_{obs} P_A = A \exp\left(-\frac{E_{obs}}{RT}\right) P_A = C \exp\left[-\left(\frac{E_{int} + \Delta H_a}{RT}\right)\right] P_A \qquad (9.12)$$

9.6.4.3   Calculation of the Adsorption Enthalpy

In the case of high–acid silica zeolites, Haag has experimentally shown [100] that for monomolecular cracking the apparent or observed activation energy, E_{obs}, is the sum of the heat of adsorption, ΔH_a, of the alkane and the intrinsic activation energy, E_{int}

$$E_{obs} = E_{int} + \Delta H_a \qquad (9.13)$$

The fulfillment of Equation 9.13, within experimental error, is obvious from the analysis of the data reported in Ref. 100 (see Figure 9.5). As a result, sorption effects are responsible for the changes in the reaction rates for monomolecular paraffin cracking.

As described in Section 6.5.3

$$\Delta G^{\text{ads}} = U_0 + P_{\text{a}} \tag{9.14}$$

where U_0 and P_0, are the adsorbate–adsorbent and adsorbate–adsorbate interaction energies respectively.

The author and collaborators have shown that aromatic hydrocarbons during adsorption in acid zeolites, at about 400 K, behave as a gas in an external force field [109]. Therefore, it is logical to suppose that during the cracking process, the reacting molecules behave as an ideal gas in an external field, given that cracking occurs at relatively low pressures and to some extent at high temperatures. Subsequently, Equation 9.14 is applicable. Then, since the entropy change during adsorption is small:

$$\Delta G^{\text{ads}} \approx \Delta H^{\text{ads}} \approx U_0 + P_{\text{a}} \tag{9.15}$$

For the adsorption of alkanes, the dispersion energy dominates [104]; as a result, in the case of the adsorption of n-alkanes, these hydrocarbons can be approximately described with a united-atom model, where the CH_2 and CH_3 groups are considered as single interaction centers with parameters similar to those that characterize CH_4. Therefore, the n-alkane of m carbons in the framework of this approximation is considered as a linear set of m CH_2 groups, if we ignore the effect of one of the H atoms in the CH_3 terminal groups [108]. Within the framework of this model, the alkane–solid interactions are described by the Lennard–Jones potential, and the alkane–alkane interactions between two united atoms is described, as well, by the Lennard–Jones potential [108]. The whole n-paraffin interacts with a single acid group, as the OH groups in highly siliceous acid zeolites are at a distance of about 10–20 Å depending of the Si/Al relation, a distance larger than the length of the hydrocarbon chain [109].

Applying the united-atom model for an n-alkane, the total average potential for the different geometries of the catalyst pore systems is given by [97]

$$\xi_T(L) = m\xi^{CH_2}(L) + \xi_{AB}^{C_mH_{2m+2}} \tag{9.16a}$$

$$\xi_T(r_p) = m\xi^{CH_2}(r_p) + \xi_{AB}^{C_mH_{2m+2}} \tag{9.16b}$$

$$\xi_T(R) = m\xi^{CH_2}(R) + \xi_{AB}^{C_mH_{2m+2}} \tag{9.16c}$$

where

m is the carbon number

$\xi^{CH_2}(L)$, $\xi^{CH_2}(r_p)$, and $\xi^{CH_2}(R)$ are the average potentials for the action of the dispersion and repulsion forces with a CH_2 group for the three pore geometries (see Equations 6.30b, 6.32b, and 6.34)

$\xi_{AB}^{C_mH_{2m+2}}$ is the alkane interaction with the acid site, which does not depend on the pore diameter and the pore geometry

In view of the fact that the cracking process takes place at relatively low pressures and somewhat high temperatures [97], we do not consider the adsorbate–adsorbate interaction. The following equations define $\xi_{\text{cal}}^{CH_2}(L)$, $\xi_{\text{cal}}^{CH_2}(r_p)$, and $\xi_{\text{cal}}^{CH_2}(R)$ [97]:

$$\xi_{\text{cal}}^{CH_2}(L) = \left(\frac{N_{\text{AS}}A_{\text{AS}}}{2\sigma^4(L-2d)}\right)\left(\frac{\sigma^4}{3(L-d)^3} - \frac{\sigma^{10}}{9(L-d)^9} - \frac{\sigma^4}{3d^3} + \frac{\sigma^4}{9d^9}\right) \tag{9.17a}$$

TABLE 9.1
Physical Properties of Methane and the Oxide Ion

Atomic Species	Polarizability, α [10^{-24} cm^3]	Magnetic Susceptibility, χ [10^{-29} cm^3]	Diameter, d [Å]	Surface Density, N_S [10^{18} atm/m^2]
Methane (CH$_4$)	2.59 [110]	2.89 [110]	3.0 [97]	6.0 [97]
Oxide ion	2.50 [110]	1.30 [110]	2.8 [107]	13.1 [107]

$$\xi_{cal}^{CH_2}(r_p) = \frac{3}{4}\pi N_A \left(\frac{N_{AS}A_{AS}}{d^4}\right) \left(\sum_{k=0}^{\infty}\left[\frac{1}{2k+1}\left(1-\frac{d}{r_p}\right)^{2k}\left\{\frac{21}{32}\alpha_k\left(\frac{d}{r_p}\right)^{10}-\beta_k\left(\frac{d}{r_p}\right)^{4}\right\}\right]\right) \quad (9.17b)$$

$$\xi_{cal}^{CH_2}(R) = \frac{6(N_l\varepsilon_{aA}^*)R^3}{(R-d)^3}\left[-\left(\frac{d}{R}\right)^6\left(\frac{1}{12}T_1+\frac{1}{8}T_2\right)+\left(\frac{d}{R}\right)^{12}\left(\frac{1}{90}T_3+\frac{1}{80}T_4\right)\right] \quad (9.17c)$$

Table 9.1 shows a set of values for the parameters α, χ, d, and N_s for CH$_4$ as the adsorbate and the oxide ion (a zeolite, for example) as the adsorbent [97,107,110].

Afterward, with the help of the Kirkwood–Muller equation (see Section 6.9), it is possible to obtain the following equations to numerically describe $\xi_{cal}^{CH_2}(L)$, $\xi_{cal}^{CH_2}(r_p)$, and $\xi_{cal}^{CH_2}(R)$ [97]

$$\xi_{cal}^{CH_2}(L) = \left(\frac{23.21\times10^3}{(L-0.60)}\right)\left(\frac{1.85\times10^{-3}}{(L-0.30)^3}-\frac{2.54\times10^{-7}}{(L-0.30)^9}-0.050\right) \quad (9.18a)$$

$$\xi_{cal}^{CH_2}(r_p) = 29.65\times10^3\left[\sum_{k=0}^{\infty}\left[\frac{1}{2k+1}\left(1-\frac{d}{r_p}\right)^{2k}\left\{\frac{21}{32}\alpha_k\left(\frac{d}{r_p}\right)^{10}-\beta_k\left(\frac{d}{r_p}\right)^{4}\right\}\right]\right] \quad (9.18b)$$

$$\xi_{cal}^{CH_2}(R) = \frac{4.1R^5}{(R-0.3)^3}\left[-\left(\frac{0.3}{R}\right)^6\left(\frac{1}{12}T_1+\frac{1}{8}T_2\right)+\left(\frac{0.3}{R}\right)^{12}\left(\frac{1}{90}T_3+\frac{1}{80}T_4\right)\right] \quad (9.18c)$$

where L, r_p, and R are in nanometers and for $\xi_{cal}^{CH_2}(\rho)$, $\rho = L$, r_p, or R, which is given in kilojoules per moles. To graphically represent Equation 9.18, these equations are plotted in Figure 9.4. The calculations were performed with a Scientific Notebook program [111].

Finally, it is evident that for the different geometries of the pore system

$$\xi_T(\rho) \approx \Delta H^{ads} \quad (9.19)$$

where $\rho = L$, r_p, or R.

9.6.4.4 Calculation of the Activation Energy

Within the framework of the transition state theory [112,113], the observed activation energy, E_{obs}, for a monomolecular catalytic process in the heterogeneous case is $E_{obs} = E_0 + \Delta H_{ads}$ [act. complex], where E_0 is the energy of the reaction without a catalyst and ΔH_{ads}[act. complex] is the adsorption enthalpy of the activated complex [114]. In the monomolecular cracking of n-alkanes catalyzed by an acid zeolite, the activated complex is the adsorbed alkanium ion, that is, the protonated alkane ($[CR_2H_3{}^+]$-ZO^-) [98,99]. Therefore,

$$E_{obs} = E_0 + \left(\Delta H_{ads}[\text{carbocation}]\right) \tag{9.20}$$

and then combining Equation 9.20 with Equation 9.16 [97], we get

$$E_{obs} \approx E_0 + m\xi^{CH_2} + \xi_{AB}^{C_mH_{2m+2}} \tag{9.21}$$

where the expression for ξ^{CH_2} depends on the pore geometry. Equation 9.21 can be used for the description of the experimental data of adsorption and cracking of n-paraffins in the H-ZSM-5 zeolite [100] reported in Table 9.1. If we consider that $E_0 = E_{int}$, we are able to fit the following equation (see Figure 9.5) [97]:

$$y = ax + b \tag{9.22}$$

where

$y = E_{obs}$
$x = m$
$a = \xi^{CH_2}(\rho)$, where $\rho = r_p$
$b = E_0 + \xi_{AB}^{C_mH_{2m+2}}$

The outcome of the fitting process was very good, and the calculated values for the parameters a and b were $a = -12.75\,\text{kJ/mol}$ and $b = 196.5\,\text{kJ/mol}$. If we now made an average of the values reported for $[\Delta H_{ads} - m \cdot a]$, we get

$$\overline{\xi}_{AB}^{C_mH_{2m+2}} \approx <[\Delta H_{ads} - m \cdot a]> \approx -3 \pm 5\,[\text{kJ/mol}]$$

where $\overline{\xi}_{AB}^{C_mH_{2m+2}}$ is the average acid–base interaction and $\sigma = \pm 5\,\text{kJ/mol}$ is the standard deviation. The acid–base interaction term can be calculated as

$$\xi_{AB}^{C_mH_{2m+2}} \approx \Delta H_{ads} - m\left(\xi^{CH_2}(\rho)\right) = \Delta H_{ads} - m \cdot a \approx 0$$

which numerically corroborates the findings of other authors, confirming that the interaction of the alkane with the acid site is negligible [104].

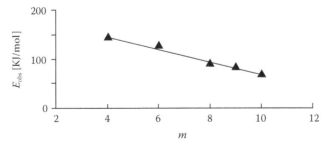

FIGURE 9.5 Fitting of Equation 9.21 with the data of the observed activation energy for the cracking of C_4–C_{10} alkanes in H-ZSM-5 vs. the number of carbon atoms (m) reported in Ref. [100].

9.6.4.5 Numerical Evaluation of the Model

From Equations 9.19 through 9.21, it can be shown that [97]

$$\xi_T(\rho) = m\xi^{CH_2}(\rho) + \xi_{AB}^{C_m H_{2m+2}} \tag{9.23}$$

and we get

$$\xi_{exp}^{CH_2}(\rho) \approx \frac{\Delta H_a - \xi_{AB}^{C_m H_{2m+2}}}{m} \tag{9.24}$$

Then, using data reported in the literature [99,104] and considering that $\xi_{AB}^{C_m H_{2m+2}} \approx 0$ [kJ/mol] [104], Tables 9.2 and 9.3 show the experimental values for $\xi_{exp}^{CH_2}$ calculated with Equation 9.24, and the theoretical values for $\xi_{cal}^{CH_2}(x)$ and $\xi_{cal}^{CH_2}(R)$ calculated with the help of Equations 9.18b and c in terms of the variables $x = d/r_p$ and R, using the Scientific Notebook program [111].

The results obtained with the slit pore model are not reported since the calculated values do not agree with the experiment [97]. This is an expected outcome, since the zeolite geometry can be modeled with the cylindrical pore or the spherical pore geometries, but not with the slit pore geometry.

The results obtained with the cylindrical pore geometry (Table 9.2) are in reasonable agreement with the reported experimental data [97]. However, for the H-Y zeolite, the cylindrical pore model did not provide a good result, since the pore system of the zeolite Y resembles a three-dimensional cylindrical system [115]. The appropriate model for the zeolite Y is the spherical geometry pore [107]; in this regard, the results reported in Table 9.3 shows that only the zeolite Y is properly described with the spherical geometry pore model [97].

TABLE 9.2
Experimental ($-\xi_{exp}^{CH_2}$) and Calculated ($-\xi_{calc}^{CH_2}$) Values for the Cylindrical Pore Model for H-ZSM-5, H-MOR, H-Beta, H-USY, and H-MCM-41

Zeolite	$-\xi_{exp}^{CH_2}$ [kJ/mol]	$-\xi_{calc}^{CH_2}(r_p)$ [kJ/mol]	$d = 2r_p$ [Å]
H-ZSM-5 [99]	14.3	12.5	6.2
H-MOR [99]	11.5	11.0	7.2
H-USY [99]	8.3	8.0	8.4
H-ZSM-5 [104]	12.7	12.5	6.2
H-Beta [104]	10.7	11.0	7.2
H-Y [104]	7.8	8.0	7.2
H-MCM41	—	1.6	20

TABLE 9.3
Experimental ($-\xi_{exp}^{CH_2}$) and Calculated ($-\xi_{calc}^{CH_2}$) Values for the Spherical Pore Model for H-ZSM-5, H-MOR, H-Beta, and H-USY

Zeolite	$-\xi_{exp}^{CH_2}$ [kJ/mol]	$-\xi_{calc}^{CH_2}(R)$ [kJ/mol]	R [Å]
H-ZSM-5 [99]	14.3	18,700	3.2
H-MOR [99]	11.5	1164	3.6
H-USY [99]	8.3	7.9	7.5
H-ZSM-5 [104]	12.7	18,700	3.2
H-Beta [104]	10.7	1164	3.6
H-Y [104]	7.8	7.9	7.5

The results reported in this Section 9.6.4 clearly shows that the pore geometry is very important for the understanding of how the reaction takes place as was previously stated by Derouane and collaborators [116,117].

9.6.5 Bimolecular Reaction

A bimolecular catalytic reaction can be expressed by the following equation:

$$A(g) + B(g) \leftrightarrow Products$$

An example of a bimolecular reaction is the synthesis of NH_3 on promoted iron catalysts:

$$N_2 + 3H_2 \leftrightarrow 2NH_3$$

which is an important industrial reaction.

9.6.5.1 Langmuir–Hinshelwood Mechanism

In the Langmuir–Hinshelwood (L–H) mechanism for surface-catalyzed reactions, the reaction takes place between two surface-adsorbed species [4,5]. As a substitute for concentration, we use surface coverage, and the rate is expressed in this term. We consider that the elementary reaction in the L–H mechanism is the bimolecular surface reaction expressed by the following equations:

$$A_a + B_a \rightarrow C_a, \text{ rate constant: } k_{AB}$$

$$(A)_a + (B)_a \rightarrow (C)_a + (D)_a, \text{ rate constant: } k_{AB}$$

where $(A)_a$, $(B)_a$, $(C)_a$, and $(D)_a$ are the adsorbed species (see Figure 9.6).

As a result, the reaction rate is

$$r_{AB} = k_{AB}\theta_A\theta_B$$

where, k_{AB} is the rate constant and

$$\theta_A = \frac{K_A P_A}{1 + K_A P_A + K_B P_B} \tag{9.25a}$$

and

$$\theta_B = \frac{K_B P_B}{1 + K_A P_A + K_B P_B} \tag{9.25b}$$

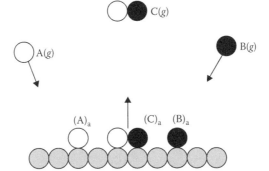

FIGURE 9.6 Schematic representation of the Langmuir–Hinshelwood mechanism.

If both reactants comply with the Langmuir adsorption model without dissociation, in which K_A and K_B are the equilibrium constants and P_A and P_B are the partial pressures, we can write the reaction rate as follows:

$$r_{AB} = k_{AB} \frac{K_A K_B P_A P_B}{(1 + K_A P_A + K_B P_B)^2}$$

We now suppose that A is strongly adsorbed and B is poorly adsorbed, that is, $K_A > K_B$, then

$$r_{AB} = k_{AB} \frac{K_A K_B P_A P_B}{(K_A P_A)^2} = k_{AB} \frac{K_B P_B}{K_A P_A}$$

If we suppose that A and B are poorly adsorbed, that is, $1 + K_A P_A + K_B P_B \approx 1$, then

$$r_{AB} = k_{AB} K_A K_B P_A P_B \tag{9.26}$$

For low pressures, we get the same dependence, that is

$$r_{AB} = k_{AB} K_A K_B P_A P_B$$

The rate constant, k_{AB}, obeys the Arrhenius law:

$$k_{AB} = A \exp\left(-\frac{E_{act}}{RT}\right) \tag{9.27}$$

9.6.5.2 Eley–Rideal Mechanism

In the Eley–Rideal (E-R) mechanism, the reaction takes place between a surface-adsorbed species and a gaseous reactant. That is, in this mechanism, only one of the reactants is bound to the surface [4,5]. Consequently, in the E–R mechanism, a gas phase molecule hits the adsorbed molecule and the reaction continues as follows

$$(A)_a + B(g) \rightarrow C(g), \text{ rate constant: } k_A$$

where

(A)$_a$ is an adsorbed reactant species
B(g) is the gaseous reactant
C(g) is the product (see Figure 9.7)

The reaction rate is then a function of the fraction of the concentration of the adsorbed molecules on the surface and the pressure of the gas molecule, as follows

$$r_A = k_A \theta_A P_B, \tag{9.28}$$

where k_A is the rate constant and

$$\theta_A = \frac{K_A P_A}{1 + K_A P_A} \tag{9.29}$$

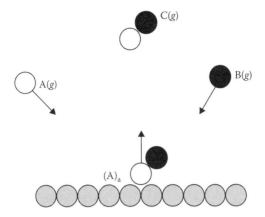

FIGURE 9.7 Schematic representation of the Eley–Rideal mechanism.

where the adsorbed reactant, A, complies with the Langmuir adsorption model without dissociation, where K_A is the equilibrium constant and P_A is the partial pressure. We can write the reaction rate as follows:

$$r_A = k_A \frac{K_A P_A P_B}{1 + K_A P_A}$$

If we now suppose that the pressure of both reactants is very high, then the surface is practically totally covered, $\theta_A \approx 1$, and the rate reduces to

$$r_A = k_A P_B \tag{9.30}$$

Therefore, the rate-determining step in this case is the rate at which the gas molecules, B(g), hit the catalyst surface. If we suppose that the pressure of A is low, then $1 + K_A P_A \approx 1$ and

$$r_A = k_A K_A P_A P_B \tag{9.31}$$

Here, the adsorption turns out to be the rate-determining step. In the E–R mechanism, generally the rate constant, k_A, obeys the Arrhenius law [4]:

$$k_A = A \exp\left(-\frac{E_{act}}{RT}\right) \tag{9.32}$$

9.6.6 COMPOSITE MECHANISM REACTIONS

A reaction that includes more than one elementary reaction is supposed to take place by a composite mechanism. There are two major evidences for a composite mechanism [118]:

1. The kinetic equation for the reaction does not match its stoichiometry.
2. There is experimental confirmation, direct and indirect, for intermediates of such a nature that it is needed to assume that more than one elementary reaction is involved.

There are various kinds of composite mechanisms, for instance:

1. Reactions occurring in parallel, such as

$$A \rightarrow Y$$

$$A \rightarrow Z$$

are named parallel or simultaneous reactions.

When there are concurrent reactions there is occasionally competition, as in the scheme:

$$A + B \rightarrow Y$$

$$A + C \rightarrow Z$$

where B and C compete with one another for A.

2. Reactions occurring in forward and reverse directions, such as

$$A + B \leftrightarrow Z$$

are named opposing reactions.

3. Reactions taking place in sequence, such as

$$A \rightarrow X \rightarrow Y \rightarrow Z$$

are recognized as consecutive reactions.

4. Reactions are supposed to show feedback if a substance formed in one step influences the rate of a preceding step. For instance, in the following schematic representation

$$A \rightarrow X \rightarrow Y \rightarrow Z$$

the intermediate, Y, could catalyze the reaction $A \rightarrow X$, that is, a positive feedback, or it can inhibit this reaction, that is, a negative feedback.

5. Chain reactions are composite reactions that occasionally comprise a cycle of elementary reactions, such that specific reaction intermediates consumed in one step are regenerated in another. If such a cycle is replicated more than one time, the reaction is identified as a chain reaction. The reaction intermediates can be atoms, molecules, free radicals, or ions. There are diverse types of chain reactions comprising straight-chain reactions, branching-chain reactions, energetic branching-chain reactions, and degenerate branching-chain reactions.

9.7 EXAMPLES OF SURFACE REACTIONS

9.7.1 REACTION BETWEEN NITRIC OXIDE AND THE SURFACE OF IRON

We now study the reaction between nitric oxide (NO) and iron (Fe) [119–121]. The whole reaction, which is a combination of a surface reaction and an oxidation process, is expressed as follow [121]:

$$8NO + 3Fe \rightarrow Fe_3O_4 + 4N_2O$$

$$4N_2O + 3Fe \rightarrow Fe_3O_4 + 4N_2$$

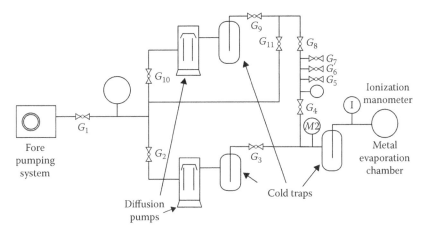

FIGURE 9.8 Diagram of the high-vacuum system.

FIGURE 9.9 High-vacuum system and mass spectrometer photography.

The reaction was studied in a laboratory assembled high-vacuum system (see Figures 9.8 and 9.9), consisting of two vacuum circuits, one for the evacuation of the evaporation chamber and the other for the evacuation of the gas introduction system [119,122,123]. Figure 9.8 shows the fore pump, diffusion pumps, cold traps, ionization manometer, and the metal evaporation chamber [119]. The evaporation was carried out at $P = 10^{-6}$ Torr, with the help of a thread helicoidal filament made of a wolfram (W) wire of 0.5 mm diameter and 10 cm length [119]. The whole vacuum system was coupled with a mass spectrometer, Hitachi RMU-6D, in order to follow the reaction kinetics [119,122,123] (see Figure 9.9). The procedures followed in order to study the reaction were as follows:

1. Fe was evaporated in the center of a 5 L spherical Pyrex glass container, connected to the vacuum system (shown in Figure 9.8), whose walls were covered with a fresh iron surface (this container was enclosed in a furnace to control the reaction temperature). Then, NO from the sample introduction system was introduced into the evaporation chamber and the reaction followed at different temperatures, with the help of the mass spectrometer [119].
2. The second procedure was using a quartz container, connected to the vacuum, the sample introduction system, and the mass spectrometer. Included in a furnace was an iron powder

that was poured in the bottom, then, NO from the sample introduction system was introduced into the evaporation chamber and the reaction followed with the help of the mass spectrometer, at different temperatures [121].

The gas reactants and products of the reaction were studied with a mass spectrometer and the solid reaction of the oxidation of iron was studied with Mössbauer spectrometry with electron diffraction. The Mössbauer study of the oxidized iron powder was carried out in a constant acceleration equipment [121]. The electron diffraction study of the oxidized iron film, evaporated over a carbon covered transmission electron microscopy sample holder and introduced into the 5 L spherical Pyrex glass container, where the Fe evaporation takes place, was carried out with the help of an Hitachi 100 C transmission electron microscope [119].

The methodology used to follow the reaction in the gas phase consisted of taking out gas samples from the container where the reaction was taking place through the connections with the mass spectrometer and measuring a mass spectrum in the mass/charge (m/e) range from 20 to 60 [m/e] (see Figure 9.10).

In the reaction chamber, Ar, a noble gas which does not chemically react with the Fe, is introduced as the reference gas before the introduction of NO. The introduced Ar allows us to define the parameters (see Figure 9.10)

$$\alpha_X = \frac{h_{X^+}}{h_{Ar^+}}$$

where $X \equiv NO^+$, NO_2^+, and N_2^+

In Figure 9.11, the time progress of the parameter is plotted

$$\alpha_X - \alpha_{X_0} = \frac{h_X}{h_{Ar^+}} - \frac{h_{X_0}}{h_{Ar_0^+}}$$

versus time, where this parameter measures the amount of reactant, that is, NO, and reaction products, N_2O and N_2, evolved during the reaction between NO and Fe [119,121].

The data reported in Figure 9.11 shows that the decomposition of NO during its reaction with Fe fulfills the following consecutive reaction scheme [121]

$$2NO \rightarrow N_2O \rightarrow N_2$$

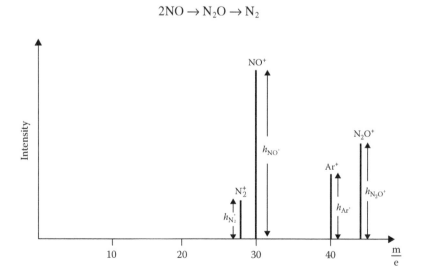

FIGURE 9.10 Standard mass spectrum.

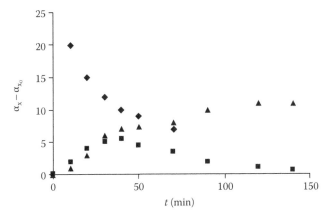

FIGURE 9.11 Time evolution of $\alpha_X - \alpha_{X_0}$ vs. time of the reactants and reaction products for the reaction between NO and Fe, where the symbol ♦ describes $\alpha_{N_2} - \alpha_{N_{2_0}}$, the sign ▲ expresses $\alpha_{NO} - \alpha_{NO_0}$, and the character ■ denotes $\alpha_{N_2O} - \alpha_{N_2O_0}$.

and the oxygen lost during the decomposition reaction reacts with the Fe to form Fe_3O_4. The reaction mechanism of this reaction for the gas-phase components is as follows [121]:

$$2NO(g) + \text{Adsorption site} \leftrightarrow 2(NO)_a \rightarrow (O)_a + (N_2O)_a \rightarrow (N_2)_a + (O)_a$$

$$(NO_2)_a \leftrightarrow NO_2(g)$$

$$(N_2)_a \leftrightarrow N_2(g)$$

On the other hand, the formation of the solid-phase components formation follows the following scheme [121]:

$$4(O)_a + 3Fe \rightarrow Fe_3O_4(s)$$

As shown by the Mössbauer spectrum (see Figure 9.12), where are reported the position of the lines for iron and all the possible iron oxides, that is, FeO, Fe_2O_3, and Fe_3O_4 and was shown that the only oxide present was Fe_3O_4 [121].

The oxidation of the evaporated iron thin films after the action of NO at 527 K was established with the help of electron diffraction. In Figure 9.13a and b, the electron diffraction patterns of the solid phase reactant, Fe, and the product of the action of NO over Fe are reported [119].

Figure 9.13a shows the electron diffraction pattern of the Fe thin film before the reaction, and Figure 9.13b shows the thin film after the reaction with NO at 527 K [119]. The determination the cell parameters of the solid phases present in the films with the help of the electron diffraction patterns allows to identify the iron oxide formed during the reaction, that is, Fe_2O_3 [119].

9.7.2 REACTION BETWEEN CARBON MONOXIDE AND THE SURFACE OF NICKEL

We now study the reaction between carbon monoxide (CO) and nickel (Ni) [122,123]. The reaction was tested in the high-vacuum systems described to study the reaction between NO and Fe (see Figures 9.8 and 9.9). The evaporation of Ni was also carried out at $P = 10^{-6}$ Torr, with the help of a filament made of a wolfram (W) wire of 0.5 mm diameter and 10 cm length [122], as was previously described [119]. The whole vacuum system was attached to a mass spectrometer to analyze the reaction products [122,123] (see Figure 9.9). The methodology to test the reaction was as follows: Ni was

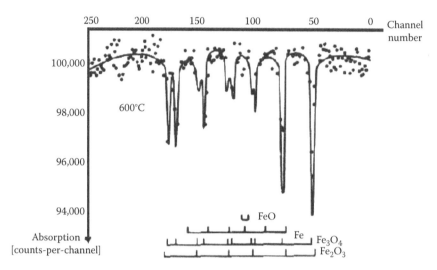

FIGURE 9.12 Mössbauer spectrum of the solid-phase products of the reaction between NO and Fe.

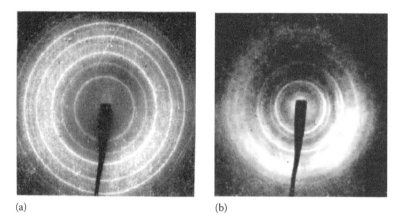

(a)   (b)

FIGURE 9.13 Electron diffraction patterns of the solid-phase reactant and product of the reaction between NO and Fe: (a) initial Fe thin film, and (b) thin film reacted with NO at 527 K.

evaporated in the center of a 5 L spherical Pyrex glass container connected to the vacuum system (shown in Figure 9.8), whose walls were covered with a fresh nickel surface. The container was included in a heating system to manipulate the reaction temperature. CO was then introduced into the evaporation chamber and the reaction was carried out at 343 K and 35 Torr pressure. The reaction products were analyzed with the mass spectrometer by taking gas samples from the container [122].

The reaction products detected in the gas phase were $Ni(CO)_4$ and some gaseous molecular species showing a mass–charge relation 56 and 84 [m/e]. After a careful examination of all the possible reaction products, it was realized that the 56 and 84 mass number peaks, under the existing experimental conditions, can only be assigned to the charged molecular species $[(CO)_2]^+$ and $[(CO)_2]^+$, respectively [122].

In order to evaluate the probable existence of the related neutral molecular species, theoretical calculations were carried out on two molecular CO dimer (see Figure 9.14a) and trimer (see Figure 9.14b) structures, applying the CNDO/SM calculation method [123].

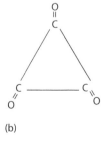

FIGURE 9.14 Proposed CO (a) dimer and (b) trimer molecular species.

The obtained results indicate that the species are stable but highly reactive, because they are strongly polarized in the ground state [123]. This fact can explain the nonoccurrence of these molecular species under normal conditions.

The existence of the carbon monoxide dimer is currently accepted as a weakly bound molecular complex [CO · · · CO] [124]. Besides, it has been found that weakly bound molecular clusters of carbon monoxide dimers and cyclic trimers, confined in zeolite cavities and channels, are stabilized under the zeolite framework cationic field [125]. These structures are similar in geometric configuration to those reported in Figure 9.14; however these are not covalently bonded, but are only weakly bound.

9.8   PACKED BED PLUG-FLOW CATALYTIC REACTOR

9.8.1   Laboratory Scale Reactor

The laboratory-sized catalytic reactor is a straight tubular tube packed with a catalyst to be tested, and through which the reactant gas flows. The granular particles of the catalyst should be of large sizes, because if the particle size is small, a pressure drop is generated inside the reactor. A good practice is to construct the reactor following the following approximate relation $d_R/d_P \geq 10$, where d_R is the reactor diameter and d_P is the particle size. For the laboratory test of a material, $d_R/d_P \approx 10$ is a good choice. In order to maintain the proportions of the reactor dimensions, the following rule can be applied $L/d_R \leq 10$, where L is the reactor length and d_R is the reactor diameter. For the laboratory test of an ion-exchange material, $D/d_R \approx 10$ is a good option. These options are only very approximate design criteria, which are exclusively justified for laboratory catalytic tests.

For example, a microreactor can be constructed with a tube of about 0.5–1 cm in diameter and 5–10 cm in length; then, in this case it will contain only a few grams of the catalyst. When these microreactors work in a continuous form, the isothermal conditions are obtained by controlling the reaction conversion to low values; then, the reactor works in the so-called differential mode, where the gas passes through the catalysts in a plug-flow regime. The plug-flow regime signifies that the fluid velocity profile is plug shaped, and is uniform at all radial positions, as explained in Section 6.11. In this case, the packed bed plug-flow catalytic reactor is packed randomly with particles of a solid catalyst, clean or just regenerated. The catalytic reaction process is supposed to be very fast relative to the convection and diffusion effects. The plug-flow catalytic reactor has a cross sectional area, A, column length, L, and catalyst mass, W (see Figure 9.15).

This reactor operates in a steady-state regime and a volumetric flow rate passing through is

$$F = \frac{\Delta V}{\Delta t} = \frac{\text{Volume}}{\text{Time}}$$

of reactants with an initial concentration C_X^0 [mass/volume] The volume of the empty bed is $V_B = \varepsilon V$, where V is the bed volume and ε is the fraction of free volume in the bed. The interstitial fluid velocity, u, is defined as

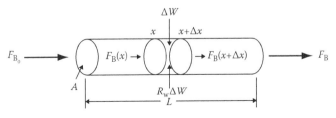

FIGURE 9.15   Schematic representation of a packed bed plug-flow catalytic reactor.

$$u = \frac{F}{A}$$

In addition, the contact or residence time of the fluid passing though the reactor, τ, is calculated with equation:

$$\tau = \frac{V_B}{F}$$

9.8.2 Equations Governing the Plug-Flow Packed Bed Reactor

The mass balance equation, expressed in moles, for the catalytic reactor is given by (see Figure 9.16)
Input − Output + Production = Accumulation
where, in a steady state, the accumulation is zero.

The production of B in a catalytic reaction is (see Equation 9.1)

$$WR_W = W\vartheta_B r_W \qquad (9.33)$$

measured in moles per second of reactant B, where ϑ_B is the stoichiometric coefficient of B in the reaction and W is the mass [kg] of the catalyst in the reactor. Now, it is evident that

$$R_W = \vartheta_B r_W \qquad (9.34)$$

where $R_B = \vartheta_B r_W$ is in

$$\frac{\text{mol/s}}{\text{kg of catalyst}} = \frac{\text{mol}}{\text{s} \cdot \text{kg}}$$

The molar balance equation for a packed bed-flow catalytic reactor is given by [126]

$$F_B(x) - F_B(x + \Delta x) + R_W \Delta W = 0 \qquad (9.35)$$

where $F_B(x) = C_B(x)F$. Then dividing Equation 9.35 by ΔW and taking the limit for $\Delta W = \rho_W A \Delta x \rightarrow 0$, we get the differential equation for the molar balance for a packed bed-flow reactor:

$$\frac{dF_B}{dW} = R_W \qquad (9.36)$$

Now, defining the conversion, X_B, of the reactant, B as [8]

$$X_B = \frac{C_{B_0} - C_B}{C_{B_0}} = \frac{\text{Moles B (reacted)}}{\text{Moles B (introduced)}}$$

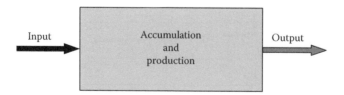

FIGURE 9.16 Mass balance in a reactor.

it is possible to write

$$F_B = F_{B_0}(1 - X_B)$$

or in a differential form

$$dF_B = -F_{B_0} dX_B$$

Consequently,

$$\frac{dX_B}{d\left(\dfrac{W}{F_{B_0}}\right)} = \frac{dX_B}{dT} = -R_W \qquad (9.37)$$

where

$$T = \frac{W}{F_{B_0}}$$

is the space time [8], related to the residence time as follows

$$T = \frac{W}{F_{B_0}} = \frac{\rho_W V}{F C_{B_0}} = \frac{\rho_W}{\varepsilon C_{B_0}}\left(\frac{V_B}{F}\right) = \frac{\rho_W}{\varepsilon C_{B_0}} \tau$$

where ρ_W is the apparent density of the packed-bed catalytic reactor.

9.8.3 Solution of the Governing Equation for the First-Order Chemical Reaction

A first-order chemical reaction can be represented as follows

$$A \rightarrow B$$

or

$$A \rightarrow B + C$$

Since a single reactant is flowing through the reactor, the governing equation can be expressed as follows:

$$\frac{dX}{d\left(\dfrac{W}{F_0}\right)} = \frac{dX}{dT} = -R_W$$

For the first-order reaction, we have

$$-R_W = kC_A = k(1 - X)$$

since

$$X = \frac{C_{A_0} - C_A}{C_{A_0}}$$

Consequently,

$$\frac{dX}{d\left(\dfrac{W}{F_0}\right)} = k(1-X)$$

Therefore,

$$\frac{dX}{(1-X)} = k d\left(\frac{W}{F_0}\right) \tag{9.38}$$

Now integrating Equation 9.38

$$\int_0^X \frac{dX}{(1-X)} = k \int_0^{\frac{W}{F_0}} d\left(\frac{W}{F_0}\right)$$

we get

$$\ln\left(\frac{1}{1-X}\right) = k\left(\frac{W}{F_0}\right) \tag{9.39}$$

9.8.4 Steps in a Catalytic Reaction in a Packed-Bed Reactor

The global process by which a heterogeneous catalytic reaction takes place in a packed-bed reactor can be described in a succession of steps, where the rate of the reaction is equivalent to the rate of the slowest stage in the whole mechanism of the reaction. The stages of a catalytic reaction in a packed-bed reactor are (see Figure 9.17) [126,127]

1. Diffusion of the reactants from the bulk fluid to the external surface of the catalyst pellet
2. Diffusion of the reactants from the pore opening to the catalyst surface
3. Adsorption on the surface
4. Reaction on the surface
5. Desorption of products

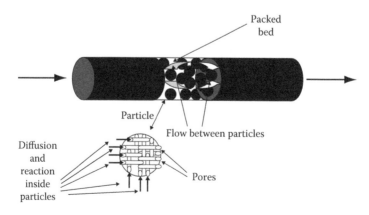

FIGURE 9.17 Schematic representation of a packed bed reactor.

6. Diffusion of the products from the catalyst surface to the pore entrance
7. Diffusion of the products from the external surface of the catalyst pellet to the bulk fluid

We can have two extreme cases:

1. When the mass transport steps, that is, stages, 1, 2, 6, and 7, are very fast compared with the reaction phases, that is, steps, 3, 4, and 5, then the diffusion does not affect the global rate of the reaction. These types of processes are characterized by the fact that the concentration of the reactants and products in the vicinity of the active catalytic site is equivalent to the concentration in the bulk fluid [126]. Besides, these processes occur at low temperatures [5].
2. When the diffusion steps, that is, stages 1, 2, 6, and 7, are very slow compared with the reaction steps, that is, steps 3, 4, and 5, the diffusion affects the overall rate of the reaction [126]. These types of processes are characterized by the fact that the overall rate of the reaction changes when the flow conditions change. These processes occur at high temperatures [5].

9.9 CHEMICAL, SUSTAINABLE ENERGY, AND POLLUTION ABATEMENT APPLICATIONS OF HETEROGENEOUS CATALYSTS

9.9.1 AMMONIA SYNTHESIS

Ammonia is the second largest synthetic commodity manufactured by the chemical industry, with the global production capacity exceeding 140 million metric tons. Haber, in 1909, demonstrated that ammonia can be produced at a high pressure by the reaction

$$3H_2(g) + N_2(g) \leftrightarrow NH_3 \qquad -\Delta H_{773K} = 109 \text{ [kJ/mol]}$$

This process is favored by a high pressure and a low temperature.

The promoted iron catalyst to accelerate this reaction was discovered by Bosch, Mittasch, and coworkers, in 1909. Consequently, the industrial process for the production of ammonia is named the Haber–Bosch process. In this process, ammonia is formed by the reaction between N_2 and H_2 using a Fe_3O_4 (magnetite) catalyst promoted with Al_2O_3, CaO, K_2O, and other oxides.

The active Fe is formed from the magnetite through a reduction produced by the reactant mixture; both Al_2O_3 and CaO are structural promoters which preserve the high surface area of the active iron catalyst [5]. The K influences the activity per unit area of the Fe by enhancing of the velocity of dissociative nitrogen chemisorption by increments of the adsorption energy [129].

It has been established that the molecules implicated in the ammonia synthesis reaction undergo the following adsorption–desorption steps [127,129]:

$$H_2(g) \leftrightarrow 2H_{ad}$$

$$N_2(g) \leftrightarrow N_{2,ad} \leftrightarrow 2N_{ad}$$

$$N_{ad} + H_{ad} \leftrightarrow NH_{ad}$$

$$NH_{ad} + H_{ad} \leftrightarrow NH_{2,ad}$$

$$NH_{2,ad} + H_{ad} \leftrightarrow NH_{3,ad} \leftrightarrow NH_3(g)$$

From the aforesaid mechanisms, it is evident that both H_2 and N_2 experience a dissociation during the adsorption process on the surface of the catalyst, N_2 more slowly than H_2. An Auger spectroscopy study showed that the density of adsorbed atomic nitrogen drops very fast at higher pressures of H_2, signifying that adsorbed N is involved in the reaction and that the dissociative adsorption of N_2 is the rate-determining step [127]. A similar conclusion was obtained by a study of the kinetics of the overall reaction described by the Temkin–Pyshev mechanism.

9.9.2 CATALYTIC CRACKING OF HYDROCARBONS

Synthetic zeolites have three major commercial applications as detergents, in separation and adsorption processes, and as catalysts. Use as detergents is the major application. However, from an economic point of view, catalysis is the most important application, and fluid cracking catalysts (FCC) are the most important one. Oil refining is one of the most important industries worldwide. It is composed of different processes, such as distillation, desulfurization, catalytic reforming, dewaxing, and cracking [10]. Industrially, the cracking process is carried out FCC facilities, in which the conversion of residual hydrocarbon material vacuum-distillates and residues into olefinic gases, high octane gasoline, and diesel products. This process employs an acidic catalyst, and the reaction is carried out at high temperatures, normally 480°C–550°C at about 0.2–0.3 MPa [10].

The catalysts used for cracking before the 1960s were amorphous [Si–Al] catalysts. The replacement of these catalysts by faujasite zeolites was a big step forward in the oil refining industry, which led to an increase in the production of gasoline [20]. The acid catalyst, currently used in FCC units, is generally composed of 5–40 wt % of 1–5 μm crystals of the H-Y zeolite included in a porous particle composed of an active matrix, which in turn is composed of amorphous alumina, silica, or [Si–Al] and a binder. The porous particle allows the diffusion of the reactants and products of the cracking reaction to and from the micropores of the zeolite [10].

9.9.3 DECOMPOSITION OF AMMONIA FOR HYDROGEN PRODUCTION AND OTHER APPLICATIONS

The decomposition of ammonia

$$NH_3 \leftrightarrow \frac{1}{2}N_2 + \frac{3}{2}H_2$$

is an endothermic process, where $\Delta H = 46$ [kJ/mol]. Metals, such as Al, Fe, Re, Rh, Ni, Pt, Ru, W, and Ir, are active as catalysts for ammonia decomposition [129]. The decomposition of ammonia on tungsten, for example, has been widely studied [128]. It has been found that the reaction is zero order in these studies. That is, the reaction rate is independent of the pressures of ammonia, hydrogen, and nitrogen [130]. This zero-order kinetics signifies that the catalytic sites are saturated during the reaction with ammonia molecules. However, it has been shown [131] that when ammonia contacts the tungsten surface, hydrogen formation occurred even at 150°C, but no nitrogen molecules were produced. Therefore, at normal reaction temperatures, specifically 600°C or more, it is unlikely that the surface is saturated by adsorbed molecular ammonia; conversely, it is more likely to suppose that the catalyst surface is completely covered by chemisorbed nitrogen, and we can consider that nitrogen desorption is the rate-determining step [128].

Currently, the general agreement is that the recombinative desorption of N_2 is the rate-limiting step [128]. To obtain an optimum ammonia decomposition catalyst, a low recombinative desorption temperature for N_2 is necessary. Then, a comprehensive understanding of the reaction of ammonia decomposition over metallic catalysts depends significantly upon understanding the interaction of the reaction products (hydrogen and nitrogen) with the metal catalyst in question. Currently, Pt and Ir, which offer comparable temperatures for recombinative nitrogen desorption with acceptable activities, are among the best catalysts for this reaction [128].

One possible application of the ammonia decomposition reaction can be for hydrogen generation [37,96]. The major concern with the use of hydrogen, on a large scale, for transportation is that it is highly unstable [133]. Hence, it is possible to get more stable hydrogen by cracking NH_3 [96] previously adsorbed on silica [132].

9.9.4 Fischer–Tropsch Synthesis

In 1925, Franz Fischer and Hans Tropsch at the Kaiser-Wilhelm Institute of Coal Research in Mülheim in the Ruhr, Germany, patented a methodology to manufacture liquid hydrocarbons from carbon monoxide gas (CO) and hydrogen (H_2) with metal catalysts. The hydrocarbons produced in the process are primarily composed of liquid alkanes, that is, paraffins and the other products are olefins, alcohols and solid paraffins, that is, waxes. The required gas mixture of carbon monoxide and hydrogen, which is known as the synthesis gas, or syngas, is produced with the help of a reaction of coke or coal with water steam and oxygen at temperatures of about 900°C. Then, coal gasification and Fischer–Tropsch hydrocarbon synthesis jointly results in a two-stage sequence of reactions, which allow the manufacture of liquid fuels like diesel and petrol from the solid combustible coal [128]. The Fisher–Tropsch process is normally catalyzed by both Fe and Co at pressures from 1 to 6 MPa and temperatures from 200°C to 300°C, producing fuels of high quality, owing to a very low aromaticity and absence of sulfur [134].

The Fisher–Tropsch plants build in Germany before World War II and during World War II produced about 16,000 barrels (1 barrel = 0.159 m³) per day of liquid fuels from coal, employing a Co catalyst in fixed-bed reactors [5]. However, during the 1950s, the Fisher–Tropsch process turned out to be uneconomical as a consequence of the abundant supply of crude oil. Nevertheless, currently considerable attention is being paid to develop alternatives of the Fisher–Tropsch process to generate liquid fuels from natural gas, biomass, oil sands, oil shales, and coal [134].

The Fisher–Tropsch catalytic process is carried out with a metallic catalyst [128], by a reductive oligomerization of carbon monoxide in the presence of hydrogen [135]:

$$nCO + mH_2 \leftrightarrow C_x H_y O_z$$

This reaction yields paraffins, olefins, and oxygenated products such as alcohols, aldehydes, ketones, acids, and esters. The probability of formation decreases in the following order [5]: paraffins > olefins > oxygenated products. The reaction consists in a surface polymerization reaction where the reactants, CO and H_2, adsorb and dissociate at the surface of the catalyst and react to form a chain initiator (CH_3), the methylene (CH_2) monomer, and water. The hydrocarbons are formed by CH_2 insertion into metal-alkyl bonds and subsequent dehydrogenation or hydrogenation to an olefin or paraffin, respectively [134]. Then, the oligomerization in the Fisher–Tropsch reaction can be described as a chain reaction where the monomer is $-CH_2-$ and the reaction entails the individual process of propagation of the chain followed by termination, obeying the Anderson–Schultz–Flory chain-length distribution.

In order to explain the mechanism of the Fisher–Tropsch reaction, some authors derived the Langmuir–Hinshelwood or Eley–Rideal types of rate expressions for the reactant consumption, where in the majority of cases the rate-determining step is supposed to be the formation of the building block or monomer, methylene [134].

9.9.5 Water–Gas Shift Reaction for Hydrogen Production and Other Applications

The water–gas shift reaction:

$$CO + H_2O \leftrightarrow CO_2 + H_2 \qquad \Delta H^0 = -41 \text{ [kJ/mol]}$$

is a reversible, exothermic reaction, thermodynamically not propitious at elevated temperatures.

Normally, the industrial procedure to produce the water–gas shift reaction proceeds in two stages using a $CuO/ZnO/Al_2O_3$ catalyst for the low temperature shift, and a Fe_2O_3 catalyst structurally promoted with CrO for the high temperature shift, where the hematite, Fe_2O_3, is reduced to magnetite, Fe_3O_4, during the water–gas shift reaction, and then the magnetite is the active phase [136].

9.9.6 ETHANOL DEHYDRATION

At present, roughly 80% of the current global energy needs comes from fossil fuels. Besides, oil is used as a raw material for the production of several chemical products. Ethanol (C_2H_5OH), a natural product obtained from biomass, is, on the one hand, a renewable source of energy that would be an important factor for near-zero carbon dioxide (CO_2) emissions, on the other hand, it is the basis for a C_2 chemistry, that is, a raw material for the production of different chemical products [19,21,137–147]. Besides, ethanol is accessible, can be easily transported, biodegradable, has low toxicity, and can be transformed by catalytic reactions [137].

Synthetic zeolites are the most important materials used currently in industry for catalyst preparation. However, natural zeolites are not contemplated in catalyst manufacturing because of the impurities present in the natural raw materials; nevertheless, in some reactions, such as the isomerization of hydrocarbons, this contamination does not affect the catalytic transformation; therefore, acid natural zeolites can be used for this purpose [19]. Furthermore, acid clinoptilolites were tested for catalytic cracking with success [19,21,137–143]. We have shown [19,21,138–143] that the acid clinoptilolite, used as catalyst in the reaction of ethanol dehydration, exhibits high selectivity for ethylene production due to steric restrictions imposed on the formation of diethyl ether. The scheme of the ethanol dehydration reaction is shown in Figure 9.18 [145].

Table 9.4 shows the results corresponding to the reaction of ethanol dehydration in a flow reactor at atmospheric pressure and at temperatures of 403 and 453 K, in comparison with several synthetic zeolites, where the reaction is described with the help of the following parameters: $X_T(\%)$, total conversion; $S_O(\%)$, selectivity to olefin; and $S_E(\%)$, selectivity to ether [21]. It is necessary to state that the olefin was ethylene and the ether was diethyl ether [21,138–143].

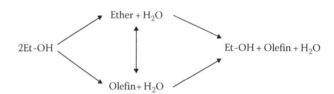

FIGURE 9.18 Mechanism of ethanol dehydration.

TABLE 9.4
Ethanol Dehydration in a Flow Reactor at Atmospheric Pressure

T [K]	403	403	403	453	453	453
Sample	X_T [%]	S_O [%]	S_E [%]	X_T [%]	S_O [%]	S_E [%]
H-HEU	8.1	90	10	95	96	4
H-MFI	8.6	6	94	90	89	11
H-MOR	51	77	23	100	100	0
H-LTL	57	7	93	100	100	0
H-FAU	38	0	100	100	100	0

TABLE 9.5
Characteristics of the Studied Zeolite Frameworks

Framework Type	Channel System	Pore	Pore Diameter [Å]
HEU	Bidimensional	10 MR	7.2×4.4
		8 MR	4.0×4.4
		8 MR	4.1×4.7
MFI	Bidimensional	10 MR	5.1×5.8
		10 MR	5.2×5.4
MOR	Monodimensional	12 MR	6.7×7.0
		8 MR	2.9×5.7
LTL	Monodimensional with lobes	12 MR	7.1
		Lobe	7.5
FAU	Tridimensional with cages	12 MR	7.4
		Cage	13.5

The reported catalysts were prepared as follows: the natural zeolite rock HC (see Table 4.1 in Chapter 4), the Na-ZSM-5 zeolite (Si/Al = 25) (provided by Costa-Novella and Nefiodov), and the Na-MOR (Si/Al = 5.5), K-LTL (Si/Al = 3.0), and the Na-Y (Si/Al = 2.1) zeolites provided by Union Carbide were exchanged with ammonium and then deammoniated by repeated exchange with a 3 M ammonium chloride solution at 373 K for 2 h followed by calcinations at 673 K for 4 h (see Section 9.3.1) to obtain the H-HEU, H-MFI, H-MOR, H-LTL, and H-FAU samples [21].

From Table 9.4, it is evident that the natural and MFI zeolites have lower yields than the zeolites with larger pores (see Table 9.5); besides, the MFI zeolite has a higher selectivity to ether and the natural zeolite to ethylene [21]. It is evident that the zeolite with higher selectivity to ethylene at a low temperature is the natural zeolite, and that at the higher temperature, the selectivity is toward the olefin for all zeolites [21].

A kinetic study was carried out in an impulse microreactor [19,140,141] (see Section 9.8) for a catalyst prepared with the natural zeolite CMT-C (see Table 4.1). The methodology to obtain the catalyst H-CMT-C is as follows: first, the NH$_4$-CMT-C is obtained, and second, it is activated at 450°C using a flow of 30 L/h of air saturated with water vapor at room temperature (27°C) for 4 h for ultrastabilization [140,141]. During this study, it was shown that the catalyst is rapidly deactivated during the reaction. However, if water is added to the feed, the catalyst deactivation is clearly reduced [140,141].

Another kinetic study was as well carried out in an impulse microreactor [19,140,141] (see Section 9.8) for a catalyst prepared to obtain the NH$_4$-CMT-C (see Table 4.1) sample by applying the methodology previously described, using the natural zeolite rock labeled CMT-C as the raw material; however to obtain the catalyst, the H-CMT-C sample was activated at 450°C using a flow of 30 L/h of air saturated with water vapor at room temperature (27°C) for 4 h for ultrastabilization [140,141]. During this study, it was shown that the catalyst is rapidly deactivated during the reaction; however, if water is added to the feed, the catalyst deactivation is clearly reduced; it was as well shown that the catalyst stability was improved by the addition of water in the feed [140,141].

Figure 9.19 shows the ethylene molar yield (%) of the ethanol dehydration reaction versus the ethanol feed (mol) for a mixture of ethanol (70%) + H$_2$O (30%), supplied with a spatial velocity of 5 h^{-1} to an isothermic fixed-bed flow reactor consisting of 20 cm^3 of the bed filled with the H-CMT-C sample (0.2–0.6 mm particle diameter) at 718 K [19,140].

It is obvious (see Figure 9.19) that the catalyst is stabilized by water addition to the feed. It is well known that the mechanism of formation of diethyl ether is of the Eley–Rydeal type [148] (see Section 9.6), and the mechanism of formation of ethylene is of the Langmuir–Hinshelwood type [145]. Figure 9.20 shows the fitting of the experimental data obtained during the kinetic study

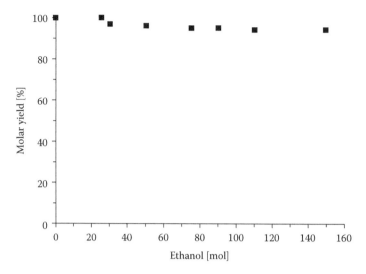

FIGURE 9.19 Ethylene molar yield [%] in the catalytic dehydration of ethanol vs. ethanol (70%) plus H_2O (30%) feed [mol] at 718 K.

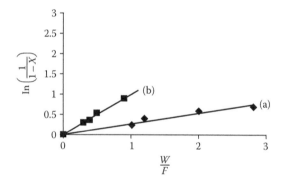

FIGURE 9.20 Fitting of the experimental data of the kinetic study in an impulse microreactor of the ethanol dehydration reaction for a feed of ethanol (60%) plus H_2O (40%) at (a) 673 K and (b) 723 K.

of the ethanol dehydration reaction by feeding pulses of a mixture of ethanol (60%) plus H_2O (40%) at 673–723 K [140,141].

It is clear from Figure 9.20 that the experimental data is accurately fitted to the first-order kinetic equation for both reaction temperatures [140,141].

9.9.7 Oxidation of CO

One of the best understood reactions is the catalytic oxidation of carbon monoxide on Pt or Pd catalysts:

$$2CO + O_2 \leftrightarrow 2CO_2$$

Particularly, the formation of CO_2 by the oxidation of CO over platinum metal catalysts is the reaction that has been studied most extensively in the past years and whose elementary steps appear to be best understood. This reaction is an important process in automobile catalytic converters [92] and its mechanism proceeds as follows [129]:

$$O_2(g) + \text{Adsorption site} \rightarrow (CO)_a$$

$$CO(g) + \text{Adsorption site} \rightarrow (CO)_a$$

$$(CO)_a + (O_2)_a \rightarrow (CO_2)_a + 2 \text{ Adsorption sites}$$

where $(CO)_a$, $(O_2)_a$, and $(CO_2)_a$ are adsorbed species in specific adsorption sites in the catalyst surface. That is, the O_2 molecule is adsorbed on the metallic (Pt or Pd) surface where the molecule is dissociated, while the CO molecule is strongly chemisorbed on the metallic surface. Then, the reaction takes place by means of a Langmuir–Hinshelwood mechanism, which has been confirmed by experiment [129].

9.9.8 Water Treatment by Heterogeneous Photocatalysis

Heterogeneous photocatalysis using titanium dioxide as the photocatalyst is a valuable methodology for the decomposition of several contaminants in air, in water, or on solid surfaces. The process was explained in Section 9.5. Here, more details about the primary events occurring on an UV-illuminated TiO_2 catalyst in relation to the photodegradation of organic pollutants are given. That is [81], during the first step of this process, the absorption of the photon at $\lambda < 385$ nm occurs, followed by pair generation (see Figure 2.10); then, the holes produced in the valence band oxidize the organic contaminants to produce organic radical cations, or oxidize chemisorbed OH^- or H_2O to produce $^\bullet OH$ radicals. Along with this, the conduction band electrons reduce O_2 to yield $^\bullet O_2^-$; besides, other oxidants can also be reduced by the conduction band electrons.

When TiO_2 is used as the photocatalyst, the photodegradation processes are nonselective, and, thus, only a given component from a mixture can be possibly degraded. Therefore, the preparation of selective photocatalysts is an interesting goal, since it opens new potential fields of application where nonselective, TiO_2 catalysts cannot be used; this can be the case for separation processes and for the selective elimination of pollutant molecules from a mixture [83,84] (see Section 9.5).

9.9.9 Other Sources of Activation of a Photocatalyst. Mechanical Activation

Another source for the activation of a photocatalyst, for example, TiO_2, is ultrasonic irradiation. When this process is applied in water, which is used as the medium for the propagation of the ultrasonic waves, it causes the formation and collapse of microscopic bubbles, which provoke local high temperature and pressure [149]. Consequently, the ultrasonic irradiation of water containing a photocatalyst can be applied as an energy source to generate electron–hole pairs within the photocatalyst, and this effect can be applied to decompose organic compounds. The degradation of phenol in water was carried out by ultrasonic irradiation with the help of a composite of particles of TiO_2 and activated carbon as the catalysts [150].

The author has previously worked in tribochemistry [151], and is aware of the fact that the action of milling in a tribochemical process, as in a sonochemical process, generates high pressure and temperature. Therefore, milling can be used as an energy source for the photocatalyst, for example, TiO_2 particles, to generate electron–hole pairs. Consequently, we studied the degradation of phenol in water by the mechanical activation of rutile, consisting of ball-milling TiO_2 (rutile) particles in the presence of a 100 mg/L phenol solution [152], that is, phenol in water solutions (100 mg/L) were ball milled for 0, 12, 18, 24, 48, and 72 h with or without TiO_2 (rutile) powder, and the reaction products examined with UV spectrometry and gas chromatography coupled with mass spectrometry (GC/MS). It was shown that in the case when rutile was not incorporated in the process, phenol was not decomposed, but when rutile was integrated in the reaction process, phenol was decomposed (see Figure 9.21) [152].

In this regard a new process for the elimination of organic pollutants in aqueous solutions is proposed [152]. This methodology is in essence catalytic, analogous to that occurring during the activation of a photocatalyst with light or the radiation of a sonocatalyst with sound.

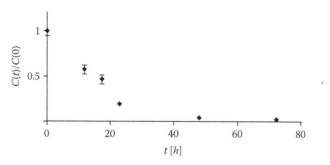

FIGURE 9.21 Kinetics of the ball milling process with rutile incorporated in the reactor.

9.9.10 Hydrogen Production by Photocatalytic Water Splitting

Hydrogen, unlike fossil fuels, burns without emitting environmental contaminants. Hydrogen is believed to be one of the possible energy carriers of the future, because it can have a significant function in reducing the current environmental pollution produced by fossil fuels [37,137].

A huge problem in the implementation of the hydrogen economy is the production of hydrogen, which currently is carried out with the help of fossil fuels. One possible source of hydrogen is water. For example, during electrolysis, electric energy is used to cause a nonspontaneous chemical reaction to decompose water to form hydrogen and oxygen gas as follows:

$$2H_2O \rightarrow 2H_2(g) + O_2(g) \qquad \Delta G_0 = 474.4 \text{ [kJ/mol]}$$

Water splitting by electrolysis is an oxidation–reduction process where the oxidation reaction occurs at the anode and the reduction reaction at the cathode (see Section 2.3.3).

The anodic reaction is $2H_2O \rightarrow O_2 + 4H^+ + 4e^-$

The cathodic reaction is $4H^+ + 4e^- \rightarrow 2H_2(g)$

Another technique to decompose water is with the help of a semiconductor photocatalyst (see Figure 2.11) [78]. For example, Pt-loaded TiO_2 had been known to decompose water to O_2 and H_2 by UV-light irradiation only in the gas phase in the presence of H_2O vapor but not in the liquid phase [78].

In addition to TiO_2, heterogeneous photocatalysts such as ZrO_3, $NaTaO_3$, $BaTa_2O_6$, $SrTa_2O_6$, $K_4Nb_6O_{17}$, and $K_3TaSi_2O_{13}$ can decompose water to evolve O_2 and H_2 without loading any cocatalysts under UV irradiation [77]. The activities of the majority of these photocatalysts are augmented by loading a cocatalyst, such as NiO_x, as a proton reduction site. In this regard, high donor-doped layered perovskites loaded with NiO have demonstrated photocatalytic ability for water splitting under UV irradiation. These layered structures with a basic composition $A_mB_mO_{3m+2}$, where $m \equiv 4$ or 5, $A \equiv$ Ca or Sr, and $B \equiv$ Nb or Ti, showed elevated quantum yields and a stoichiometric evolution of H_2 and O_2 [78]. On the other hand, $SrTiO_3$, $KTaO_3$, Ta_2O_5, Rb_4-Nb_6O_{17}, $K_2La_2Ti_3O_{10}$, $Rb_2La_2Ti_3O_{10}$, $Cs_2La_2Ti_3O_{10}$, $CsLa_2Ti_2NbO_{10}$, $Na_2Ti_6O_{13}$, and $BaTi_4O_9$ can also evolve O_2 and H_2 with the aid of cocatalysts, such as Pt, Rh, and RuO_2 [77].

9.9.10.1 Solar Water Splitting with Quantum Boost

The process for hydrogen production by solar water splitting with quantum boost has been developed by Science Applications International Corporation (SAIC), San Diego, California, and the Florida Solar Energy Center at the University of Central Florida (FSEC-UCF), Cocoa Beach, Florida with the collaboration of the author's group (see Figure 9.22) [153].

The proposed approach is based on splitting water by applying a two-step high-temperature cycle, which includes one step utilizing solar (visible and/or near-UV) photons according to the following scheme [153]:

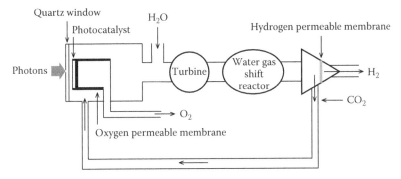

FIGURE 9.22   Diagram of water splitting with the quantum boost reactor.

$$CO_2 + heat/h\upsilon \rightarrow CO + \frac{1}{2}O_2 \qquad \Delta H^0_{298} = 283 \ [kJ/mol]$$

$$CO + H_2O \rightarrow CO_2 + H_2 \qquad \Delta H^0_{298} = -41 \ [kJ/mol]$$

where, in the first reaction, not only solar heat energy but also the quantum boost exercised by the solar, near-UV and visible photons is considered, which conduct CO_2 thermal dissociation at the temperature range of 1500–2500 K, easily attainable with solar concentrators. The second reaction is the well-known water–gas shift reaction.

To carry out this process, a directly irradiated solar reactor, consisting of a quartz window, the oxygen-permeable ceramic membrane arrangement coated with a photocatalyst (which is exposed to concentrated solar radiation), a turbine, a water–gas shift reactor, and a unit for hydrogen separation has to be used (see Figure 9.22). The ceramic oxygen-permeable membrane accelerates the process of CO_2 decomposition by catalytically shifting the reaction equilibrium and preventing the reverse reaction by a withdrawal of oxygen with the membrane [153]. The function of the photocatalyst is to enhance the kinetics of CO_2 dissociation via an interaction with the active sites, leading to weakening of C=O bonds [153]. The hot, gaseous stream leaving the photocatalyst/membrane arrangement section of the reactor is quenched with water, producing steam that can be used for generating electric power in the turbine. After cooling the effluent to 700 K, the CO–water vapor mixture enters a water–gas shift reactor to produce hydrogen and CO_2 for recycling. Finally, a ceramic hydrogen-permeable membrane purifies the $CO_2 + H_2$ mixture to obtain pure hydrogen and carbon dioxide to feed it back to the first step of the cycle in the next round of the process [153].

9.9.11   Hydrogen Production by Steam-Reforming of Ethanol

We are capable of producing hydrogen from different sources, for example, coal, natural gas, liquefied petroleum gas, propane, methane, gasoline, light diesel, dry biomass, biomass-derived liquid fuels, such as methanol, biodiesel, and from water. Among the liquid sources to produce hydrogen, ethanol is an excellent aspirant.

The steam-reforming of natural gas to produce hydrogen is the more energy effective methodology currently [154]. However, the steam-reforming of ethanol is an excellent alternative, where the aim of the procedure is to generate a large amount of hydrogen and carbon dioxide by breaking the ethanol molecule in the presence of steam over an appropriate catalyst [137,155–160]. The overall steam-reforming reaction of ethanol is represented as follows [155]:

$$C_2H_5OH + 3H_2O \rightarrow 2CO_2 + 6H_2 \qquad \Delta H^0_{298} = 347.4 \ [kJ/mol]$$

However, numerous reaction pathways takes place in the ethanol steam-reforming process, depending on the catalysts used [137]. These pathways include dehydration to ethylene and water, followed by ethylene polymerization to form coke [156,157]; decomposition or cracking to methane followed by steam-reforming [156,158]; dehydrogenation to acetaldehyde [156] followed by decarboxylation or steam-reforming of C_2H_4O; decomposition into acetone, CH_3COCH_3, followed by steam reforming [159]; and steam reforming of ethanol to syngas [156]. These processes are accompanied by the water–gas shift reaction, coking, and other reactions.

Tremendous progress is needed to obtain a highly selective catalyst to develop a process for hydrogen generation by ethanol steaming, which is a sustainable solution for the hydrogen production problem.

9.9.12　Porous Polymers as Catalysts

Cation- and anion-exchange resins are widely applied as catalysts, when reactions can be carried out at temperatures lower than 423 K [161–171]. Cation-exchange resins in acid forms have Brönsted acid sites and when exchanged with metallic cations contain Lewis acid sites, while anion exchange resins forms base species to carry out base-catalyzed reactions [169]. The principles previously described in this chapter for these sites in other catalysts, can be applied for the catalytic action of these polymers.

The pioneer work in this field was carried out on polystyrene-supported acid catalysts [161]. Thereafter, several works on the use of sulfonic, strong acidic cation exchangers as acid catalysts were reported for alkylation, hydration, etherification, esterification, cleavage of ether bonds, dehydration, and aldol condensation [162,168–171]. Besides, industrial applications of these materials were evaluated with reactions related to the chemistry of alkenes, that is, alkylation, isomerization, oligomerization, and acylation. [163,169]. Also, Nafion, an acid resin which has an acid strength equivalent to concentrated sulfuric acid, can be applied as an acid catalyst. It is used for the alkylation of aromatics with olefins in the liquid or gas phases and other reactions; however, due to its low surface area, the Nafion resin has relatively low catalytic activity in gas-phase reactions or liquid-phase processes where a nonpolar reactant or solvent is employed [166].

On the other hand, base-catalyzed reactions have been as well studied. In this regard, the aldol condensation of butyraldehyde with formaldehyde was investigated at 313–353 K in water and methanolic solutions, over anion-exchange resins, for example, gel-type acrylic-divinylbenzene or macroporous styrene-divinylbenzene polymers, both having weak basic properties [169]. It was shown that the gel-type acrylic-based catalysts have very high aldol selectivity [172]. In addition, the cross-aldol condensation reaction of terpenoid aldehyde with acetone, or benzaldehyde with saturated and α-β-unsaturated aldehyde, was successfully performed using anion-exchange resins as the solid base catalysts for the synthesis of 5,5-dimethyl-3-styryl-2-cyclohexen-1-one and other compounds [169,173].

REFERENCES

1. A.D. McNaught and A. Wilkinson, *IUPAC Compendium of Chemical Terminology* (2nd edition), Royal Society of Chemistry, Cambridge, UK, 1997.
2. Ya. Guerasimov, V. Dreving, E. Eriomin, A. Kiseliov, V. Lebedev, G. Pacehnkov, and Ashliguin, *Curso de Fisica Quimica, Tomo II*, Editorial MIR, Moscow, 1971.
3. D.V. Shriver and P.W. Atkins, *Inorganic Chemistry* (3rd edition), W.H. Freeman & Co., New York, 1999.
4. G.A. Somorjai, *Introduction to Surface Chemistry and Catalysis*, John Wiley & Sons, New York, 1994.
5. J.M. Thomas and W.J. Thomas, *Principle and Practice of Heterogeneous Catalysis*, VCH Publishers, New York, 1997.
6. H.G. Karge, M. Hunger, and H.K. Beyer, in *Catalysis and Zeolites. Fundamentals and Applications*, J. Weitkamp and L. Puppe, (editors), Springer-Verlag, Berlin, 1999, p. 198.
7. J.F. LePage, in *Preparation of Solid Catalysts*, G. Ertl, H. Knozinger, and J. Weitkamp, (editors), Wiley-VCH, Weinheim, Germany, 1999, p. 3.

8. J. Haber, J.H. Block, L. Berlnek, R. Burch, J.B. Butt, B. Delmon, G.F. Froment et al., *Pure Appl. Chem.*, 63, 1227 (1991).
9. B. Delmon, in *Preparation of Solid Catalysts*, G. Ertl, H. Knozinger, and J. Weitkamp, (editors), Wiley-VCH, Weinheim, Germany, 1999, p. 541.
10. P.M.M. Blauwhoff, J.W. Gosselink, E.P. Kieffer, S.T. Sie, and W.H.J. Stork, *Catalysis and Zeolites. Fundamentals and Applications*, Springer-Verlag, Berlin, 1999, p. 437.
11. R. Schlögl, in *Preparation of Solid Catalysts*, G. Ertl, H. Knozinger, and J. Weitkamp, (editors), Wiley-VCH, Weinheim, Germany, 1999. p. 11.
12. A. Corma, *Chem. Rev.*, 97, 2373 (1997).
13. F. Marquez-Linares and R. Roque-Malherbe, *Facets-IUMRS J.*, 2, 14 (2003); 3, 8 (2004).
14. J.A. Martens and P.A. Jacobs, *Catalysis and Zeolites. Fundamentals and Applications*, Springer-Verlag, Berlin, 1999, p. 53.
15. C.S. Cundy and P.A. Cox, *Chem. Rev.*, 103, 663 (2003).
16. G.J.A.A. Soler-Illia, C. Sanchez, B. Lebeau, and J. Patarin, *Chem. Rev.*, 102, 4093 (2002).
17. C. delas Pozas, C. Díaz-Aguila, E. Reguera-Ruiz, and R. Roque-Malherbe, *J. Solid State Chem.*, 93, 215 (1991).
18. G.H. Kühl, in *Catalysis and Zeolites. Fundamentals and Applications*, J. Weitkamp and L. Puppe, (editors), Springer-Verlag, Berlin, 1999, p. 81.
19. R. Roque-Malherbe, in *Handbook of Surfaces and Interfaces of Materials*, Vol. 5, H.S. Nalwa, (editor), Academic Press, New York, 2001, p. 495.
20. A. Corma, *Chem. Rev.*, 95, 559 (1995).
21. C. de las Pozas, R. López-Cordero, J.A. González-Morales, N. Travieso, and R. Roque-Malherbe, *J. Mol. Catal.*, 83, 145 (1993).
22. R.T. Sanderson, *Chemical Bonds and Bond Energy*, Academic Press, New York, 1976.
23. H. Hattori, *Chem. Rev.*, 95, 537 (1995).
24. W.M.H. Sachtler, in *Preparation of Solid Catalysts*, G. Ertl, H. Knozinger, and J. Weitkamp, (editors), Wiley-VCH, New York, 1997, p. 388.
25. A. Corma and H. Garcia, *Chem. Rev.*, 103, 4307 (2003).
26. D. Barthomeuf, *Stud. Surf. Sci. Catal.*, 65, 157 (1991).
27. B. Notari, *Stud. Surf. Sci. Catal.*, 37, 413 (1988).
28. H. Robson, (editor), *Verified Synthesis of Zeolitic Materials*, Elsevier, Amsterdam, the Netherlands, 2001.
29. R.M. Barrer and D.M. Mcleod, *Trans. Faraday Soc.*, 50, 980 (1954).
30. P. Grange, *J. Chem. Phys.*, 87, 1547 (1990).
31. J.J. Fripiat, in *Preparation of Solid Catalysts*, G. Ertl, H. Knozinger, and J. Weitkamp, (editors), Wiley-VCH, New York, 1997, p. 284.
32. G.W. Brindley and R.E. Sempels, *Clay Miner.*, 12, 299 (1977).
33. D.E. Vaughan, R.J. Lussier, and J.S. Magee, US Patent 4,176,090 (1979).
34. H.H. Murray, *Applied Clay Mineralogy*, Elsevier Science & Technology Books, Amsterdam, the Netherlands, 2007.
35. D. M. Moore and R. C. Reynolds, *X-Ray Diffraction and the Identification and Analysis of Clay Minerals*, Oxford University Press, Oxford, UK, 1997.
36. D.E. Vaughan, *Catal. Today Pillared Clays*, 2, 187 (1988).
37. R. Roque-Malherbe, *Adsorption and Diffusion of Gases in Nanoporous Materials*, CRC Press-Taylor & Francis, Boca Raton, FL, 2007.
38. C.T. Kresge, M.E. Leonowicz, W.J. Roth, J.C. Vartuli, and J.S. Beck, *Nature*, 359, 710 (1992).
39. J.S. Beck, J.C. Vartuli, W.J. Roth, M.E. Leonowicz, C.T. Kresge, K.D. Schmitt, C.T-W. Chu et al., *J. Am. Chem. Soc.*, 114, 10834 (1992).
40. X.S. Zhao, G.Q. Lu, and G.J. Millar, *Ind. Eng. Chem. Res.*, 35, 2075 (1996).
41. T.J. Barton, L.M. Bull, G. Klemperer, D.A. Loy, B. McEnaney, M. Misono, P.A. Monson et al., *Chem. Mater.*, 11, 2633 (1999).
42. M.E. Davies, *Nature*, 417, 813 (2002).
43. R.K. Kloetstra and H. van Bekkum, *J. Chem. Res. Symp.*, 1, 26 (1995).
44. E. Armengol, M.L. Cano, A. Corma, H. Garcia, and M.T. Navarro, *J. Chem. Soc. Chem. Comm.*, 519 (1995).
45. H. Van Bekkum, A.J. Hoefnagel, M.A. Van Koten, E.A. Gunnewegh, A.H.G. Vogt, and H.W. Kouwenhoven, *Stud. Surf. Sci. Catal.*, 83, 379 (1994).
46. P.T. Tanev, M. Chibwe, and T.J. Pinnavaia, *Nature*, 368, 321 (1994).
47. A. Corma, A. Martinez, V. Martinez-Soria, and J.B. Monton, *J. Catal.*, 153, 25 (1995).

48. E. Armengol, A. Corma, H. Garcia, and J. Primo, *Appl. Cat. A.*, 126, 391, (1995).
49. E. Armengol, A. Corma, H. Garcia, and J. Primo, *Appl. Cat. A.*, 129, 411, (1997).
50. E.A. Gunnewegh, S.S. Gopie, and H. Van Bekkum. *J. Mol. Cat.* A., 106, 151, (1996).
51. K.R. Kloestra and H. VanBekkum, *J. Chem. Res.*, 1, 26 (1995).
52. K. Wilson and J.H. Clark, *Pure Appl. Chem.*, 72, 1313 (2000).
53. M.J. Climent, A. Corma, S. Iborra, M.T. Navarro, and J. Primo, *J. Catal.*, 161, 786, (1996).
54. M.J. Climent, A. Corma, S. Iborra, S. Miquel, J. Primo, and F. Ray, *J. Catal.*, 183, 76 (1999).
55. A.T. Aguayo, J.M. Arandes, A. Romero, and J. Bilbao, *Ind. Eng. Chem. Res.*, 26, 2403 (1987).
56. M. Che, O. Clause, and Ch. Marcilly, in *Preparation of Solid Catalysts*, G. Ertl, H. Knozinger, and J. Weitkamp, (editors), Wiley-VCH, Weinheim, Germany, 1999, p. 315.
57. C. Burda, X. Chen, R. Narayanan, and M.A. El-Sayed, *Chem. Rev.*, 105, 1025 (2005).
58. B.F.G. Johnson, *Top. Catal.*, 24, 147 (2003).
59. M. Haruta, *Chem. Rec.*, 3, 75 (2003).
60. A. Wieckowski, E.R. Savinova, and C.G. Vayenas, *Catal. Electrocatal. Nano. Surf.*, 847 (2003).
61. B.C. Gates, in *Preparation of Solid Catalysts*, G. Ertl, H. Knozinger, and J. Weitkamp, (editors), Wiley-VCH, Weinheim, Germany, 1999, p. 371.
62. B.L. Cushing, V.L. Kolesnichenko, and C.J. O'Connor, *Chem. Rev.*, 104, 3893 (2004).
63. R. Duchateau, *Chem. Rev.*, 102, 3525 (2002).
64. R.K. Iler, *The Chemistry of Silica*, John Wiley & Sons, New York, 1979.
65. X.S. Zhao and G.Q. Lu, *J. Phys. Chem. B.*, 102, 1556 (1998).
66. E.P. Plueddemann, *Silane Coupling Agents* (2nd edition), Plenum Press, New York, 1991.
67. M.G. Voronkov, S.V. Kirpichenko, A.T. Abrosimova, A.I. Albanov, V.V. Keiko, and V.I. Lavrentyev, *J. Organomet. Chem.*, 326, 159 (1987).
68. K. Flodström, V. Alfredsson, and N. Källrot, *J. Am. Chem. Soc.*, 125, 4402 (2003).
69. C. Louis and M. Che, in *Preparation of Solid Catalysts*, G. Ertl, H. Knozinger, and J. Weitkamp, (editors), Wiley-VCH, Weinheim, Germany, 1999, p. 341.
70. Y. Iwasawa, in *Preparation of Solid Catalysts*, G. Ertl, H. Knozinger, and J. Weitkamp, (editors), Wiley-VCH, Weinheim, Germany, 1999, p. 427.
71. J.W. Geus and A.J. van Dillen, in *Preparation of Solid Catalysts*, G. Ertl, H. Knozinger, and J. Weitkamp, (editors), Wiley-VCH, Weinheim, Germany, 1999, p. 460.
72. S.E. Braslavsky and K.N. Houk, *Pure Appl. Chem.*, 60, 1055 (1988).
73. J.W. Verhoven, *Pure Appl. Chem.*, 68, 2223 (1996).
74. J.C. Yu, J. Lin, D. Lo, and S.K. La, *Langmuir*, 16, 7304 (2000).
75. A.L. Linsebigler, G. Lu, J.T. Yates Jr., *Chem. Rev.*, 95, 735 (1995).
76. M. Calatayud, P. Mori-Sanchez, A. Beltran, A. Martın-Pendas, E. Francisco, J. Andres, and J.M. Recio, *Phys. Rev. B*, 64, 184113 (2001).
77. M.A. Peña and J.L.G. Fierro, *Chem. Rev.*, 101, 1981 (2001).
78. M. Yagi and M. Kaneko, *Chem. Rev.*, 101, 21 (2001).
79. D.M. Blake, NREL/TP-430-22197, January, 1997.
80. S. Ghosh-Mukerji, H. Haick, M. Schvartzman, and Y. Paz, *J. Am. Chem. Soc.*, 123, 10776 (2001).
81. C. Chen, P. Lei, H. Ji, W. Ma, and J. Zhao, *Environ. Sci. Technol.*, 38, 329 (2004).
82. Y.V. Kolenko, B.R. Churagulov, M. Kunst, L. Mazerolles, and C. Colbeau-Justin, *Appl. Catal. B-Environ.*, 54, 51 (2004).
83. F.X. Llabres, P. Calza, C. Lamberti, C. Prestipino, A. Damin, S. Bordiga, E. Pelizzetti, and A. Zecchina, *J. Amer. Chem. Soc.*, 125, 2264 (2003).
84. P. Calza, C. Paze, E. Pelizzetti, and A. Zecchina, *Chem. Comm.*, 2130, (2001).
85. E. Reddy, B. Sun, and P. Smirniotis, *J. Phys. Chem. B*, 108, 19198 (2004).
86. T.L. Thompson and J.T. Yates Jr., *Chem. Rev.*, 106, 4428 (2006).
87. M. Fernandez-Garcıa, A. Martınez-Arias, J.C. Hanson, and J.A. Rodríguez, *Chem. Rev.*, 104, 4063 (2004).
88. C. Cheng, P. Lei, H. Ji, W. Ma, J. Chao, H. Hidaka, and N. Serpone, *Env. Sci. Tech.*, 38, 329 (2004).
89. H.G. Kim, D.W. Hwang, J. Kim, Y.G. Kim, and J.S. Lee, *Chem. Comm.*, 1077 (1999).
90. D.W. Hwang, H.G. Kim, J. Kim, K.Y. Cha, Y.G. Kim, and J.S. Lee, *J. Catal.*, 193, 40 (2000).
91. C.D. Lindstrom and X.-Y. Zhu, *Chem. Rev.*, 106, 4281 (2006).
92. I.N. Levine, *Physical Chemistry* (5th edition), McGraw-Hill Higher Education, New York, 2002.
93. P.W. Atkins, *Physical Chemistry* (6th edition), W.H. Freeman & Co., New York, 1998.
94. M. Boudart, *Chem. Rev.*, 95, 661 (1995).

95. M. Boudart and G. Djega-Mariadassou, *Kinetics of Heterogeneous Catalytic Reactions*, Princeton University Press, Princeton, NJ, 1984.
96. R. Roque-Malherbe, F. Marquez, W. del Valle, and M. Thommes, *J. Nanosci Nanotech.*, 8, 5993 (2008).
97. R. Roque-Malherbe and F. Diaz-Castro, *J. Mol. Catal. A.*, 280, 194 (2008).
98. H.H. Kung, B.A. Williams, S.M. Babitz, J.T. Miller, and R.Q. Snurr, *Catal. Today*, 52, 91 (1999).
99. S.M. Babitz, B.A. Williams, J.T. Miller, R.Q. Snurr, W.O. Haag, and H.H. Kung, *App. Catal. A.*, 179, 71 (1999).
100. W.O. Haag, *Stud. Surf. Sci. Catal.*, 84B, 1375 (1994).
101. R.J. Gorte and D. White, *Mic. Mes. Mat.*, 35–36, 447 (2000).
102. S. Kotrel, M.P. Rosynek, and J.H. Lunsford, *J. Phys. Chem. B*, 103, 818 (1999).
103. E.G. Derouane, *J. Mol. Catal., A*, 134, 29 (1998).
104. L. Yang, K. Trafford, O. Kresnawahjuesa, J. Sepa, R.J. Gorte, and D. White, *J. Phys. Chem. B*, 105, 1935 (2001).
105. G. Horvath and K. Kawazoe, *J. Chem. Eng. Jpn.*, 16, 470 (1983).
106. A. Saito and H.C. Foley, *AIChE J.*, 37, 429 (1991).
107. L.S. Cheng and R.T. Yang, *Chem. Eng. Sci.*, 49, 2599 (1994).
108. B. Smit and J. IIja-Siepman, *J. Phys. Chem.*, 98, 8442 (1994).
109. R. Roque-Malherbe, R. Wendelbo, A. Mifsud, and A. Corma, *J. Phys. Chem.*, 99, 14064 (1995).
110. D.R. Lide, (editor), *Handbook of Chemistry and Physics* (83rd edition), CRC Prees, Boca Raton, FL, 2002–2003.
111. D.W. Hardy and C.L. Walker, *Doing Mathematics with Scientific Workplace and Scientific Notebook*, MacKisham Software Inc., Poulsbo, WA, 2003.
112. S. Glasstone, K.J. Laidler, and H. Eyring, *The Theory of Rate Processes*, Mc-Graw-Hill, New York, 1964.
113. E.M. Sevick, A.T. Bell, and D.N. Theodorou, *J. Chem. Phys.*, 98, 3196 (1993).
114. R.A. van Santen, *J. Mol. Catal. A. Chem.*, 107, 5 (1996).
115. Ch. Baerlocher, W.M. Meier, and D.H. Olson, *Atlas of Zeolite Framework Types*, Elsevier, Amsterdam, the Netherlands, 2001.
116. E.G. Derouane, J.M. Andre, and A. Lucas, *J. Catal.*, 110, 58 (1988).
117. E.G. Derouane, in *Zeolites Micropoorous Solids: Synthesis, Structure and Reactivity*, E.G. Derouane, F. Lemos, C. Naccache, and F. Ramos-Riveiro, (editors), Kluwer Academic Publications, Dordrecht, the Netherlands, 1992. p. 511.
118. K.J. Laidler, E.T. Denisov, J.T. Herron, J. Villermaux, J.A. Kerr, A.J. Bard, D.L. Baulch et al., *Pure Appl. Chem.*, 68, 149 (1996).
119. R. Roque-Malherbe and J. Buttner, *Revista CNIC*, 6, 1 (1975); Chemical Abstracts Vol. 84 No. 93517w.
120. R. Roque-Malherbe, A.A. Zhujovitskii, and B.S. Bokstein, *Revista CNIC*, 8, 41 (1977); Chemical Abstracts Vol. 93 No. 13799n.
121. R. Roque-Malherbe, B.S. Bokstein, and A.A. Zhujovitskii, *Revista de Metalurgia CENIM*, 15, 287 (1979); Chemical Abstracts Vol. 93 No. 13843x.
122. R. Roque-Malherbe, R. Sanchez, and A. Rosado, *Revista CNIC*, 8, 87 (1977); Chemical Abstracts Vol. 90 No. 161499k.
123. L. Montero, R. Roque-Malherbe, J. Fernandez-Bertran, and A. Rosado, *J. Mol. Struct.*, 85, 393 (1981).
124. L.A. Surin, D.N. Fourzikov, T.F. Giesen, S. Schlemmer, G. Winnewisser, V.A. Panfilov, B.S. Dumesh, G.W.M. Vissers, and A. van der Avoird, *J. Phys. Chem. A*, 11, 12238 (2007).
125. B.S. Shete, V.S. Kamble, N.M. Guptaand, and V.B. Kartha, *J. Phys. Chem. B*, 102, 5581 (1998).
126. H.S. Fogler, *Elements of Chemical Reaction Engineering* (3rd edition), Prentice Hall, Upper Saddle River, NJ, 2002.
127. G. Ertl, *Pure Appl. Chem.*, 52, 2052 (1980).
128. A.K. Santra, B.K. Min, C.W. Yi, K. Luo, T.V. Choudhary, and D.W. Goodman, *J. Phys. Chem. B*, 106, 340 (2002).
129. K. Tamura, *Pure Appl. Chem.*, 52, 2067 (1980).
130. C.N. Hinsheiwood and R.E. Burk, *J. Chem. Soc.*, 127, 1105 (1925).
131. W. Frankenburger and A. Holder, *Trans. Faraday Soc.*, 28, 229 (1932).
132. B. Civarelli, E. Garrone and P. Ugliengo, *Langmuir*, 15, 5829 (1999).
133. A. Raissi, *Hydrogen, Fuel Cells, and Infrastructure Technologies*, DoE, FY, 2002, Progress Report.
134. G.P. van der Laan, PhD thesis, University of Groningen, the Netherlands, 1999.
135. M. Roper, in *Catalysis is C_1-Chemistry*, W. Keim, (editor), D. Reidel Publishing Co., Dordrecht, the Netherlands, 1983, p. 41.

136. M.S. Wainwright, in *Preparation of Solid Catalysts*, G. Ertl, H. Knozinger, and J. Weitkamp, (editors), Wiley-VCH, Weinheim, Germany, 1999, p. 28.

137. A. Haryanto, S. Fernando, N. Murali, and S. Adhikari, *Energ. Fuel.*, 19, 2098 (2005).

138. R. Roque-Malherbe, C. de las Pozas, G. Rodriguez-Fuentes, and G.M. Plavnik, *Revista Cubana de Fisica*, 4, 135 (1984); Chemical Abstracts Vol. 103 No. 221602v.

139. R. Roque-Malherbe, G. Rodriguez, C. Hernandez, A. Suzarte, and R. Lopez-Cordero. *Revista Cubana de Fisica*, 3, 141 (1983); Chemical Abstracts Vol. 102 No. 23833n.

140. G. Rodriguez, PhD thesis, National Center for Scientific Research, Havana, Cuba, 1987.

141. G. Rodriguez, C. Lariot, R. Roque-Malherbe, R. Lopez, J.A. Gonzales, and J.J. García, in *Actas Simposio Iberoamericano de Catálisis*, Merida, Venezuela, 1986, p. 1287.

142. R. Roque-Malherbe and G. Rodriguez, Patent Certificate No. 21742, ONIITEM, Vedado, Havana, Cuba (1987).

143. R. Roque-Malherbe and G. Rodriguez, Patent Certificate No. 22083, ONIITEM, Vedado, Havana, Cuba (1992).

144. E.G. Deroaune, J.B. Nagy, J.H.C. vant Hoof, B.P. Spekman, J.C. Vedrine, and C. Naccache, *J. Catal.*, 53, 40 (1978).

145. P.A. Jacobs, *Carboniogenic Activity of Zeolites*, Elsevier, Amsterdam, the Netherlands, 1977.

146. T. Isao, S. Masahiro, M. Inabe, and K. Murata, *Catal. Lett.*, 105, 249 (2005).

147. S. Arenamnart and W. Trakarnpruk, *Int. J. App. Sci. Eng.*, 4, 21 (2006).

148. P.B. Venuto and P.S. Landis, *Adv. Catal.*, 18, 259 (1968).

149. Y.G. Adewuyi, *Ind. Eng. Chem. Res.*, 40, 4681 (2001).

150. M. Kubo, H. Fukuda, X. J. Chua, and T. Yonemoto, *Ind. Eng. Chem. Res.*, 46, 699 (2007).

151. R. Roque-Malherbe, J. Oñate, and J. Fernandez-Bertran, *Solid State Ionics*, 34, 193 (1991); J.O. Martinez, C. Diaz-Aguila, E.R Reguera, J. Fernandez, and R. Roque-Malherbe, *Hyperfine Interact.*, 73, 371 (1992); J. Fernandez-Bertran and R. Roque-Malherbe, *React. Solids*, 8, 141 (1989).

152. M. Cotto, A. Emiliano, S. Nieto, J. Duconge, and R. Roque-Malherbe, *J. Colloid Interface Sci.* (in press) (2009).

153. R. Taylor, C. Mullich, A. Raissi, and R. Roque-Malherbe, Solar high-temperature water-splitting cycle with quantum boost, Solicitation Number DE-PS36-03GO093007, to the Department of Energy Program: Hydrogen Production Using High Temperature Thermochemical Water Splitting Cycles, SAIC-FSEC-UT, 2004.

154. J.N. Armor, *Appl. Catal. A*, 176, 159 (1999).

155. S. Velu, N. Satoh, C.S. Gopinath, and K. Suzuki, *Catal. Lett.*, 82, 145 (2002).

156. G.A. Deluga, J.R. Salge, L.D. Schmidt, and X.E. Verykios, *Science*, 303, 993 (2004).

157. S. Cavallaro, *Energ. Fuel.*, 14, 1195 (2000).

158. A.N. Fatsikostas and X.E. Verykios, *J. Catal.*, 225, 439 (2004).

159. J. Llorca, J.P.R. de la Piscina, J. Sales, and N. Homs, *Chem. Commun.*, 641 (2001).

160. F. Mariño, M. Boveri, G. Baronetti, and M. Laborde, *Int. J. Hydrogen Energ.*, 29, 67 (2004).

161. H. Widdecke, *Br. Polym. J.*, 16, 188 (1984).

162. A. Chakrabarti and M.M. Sharma, *React. Polym.*, 20, 1 (1993).

163. M.M. Sharma, *React. Funct. Polym.*, 26, 3 (1995).

164. P.C. Chopade and M.M. Sharma, *React. Funct. Polym.*, 28, 253 (1996).

165. S.-K. Ihm, J.-H. Ahn, and Y.-D. Jo, *Ind. Eng. Chem. Res.*, 35, 2946 (1996).

166. Q. Sun, M.A. Harmer, and W.E. Farneth, *Ind. Eng. Chem. Res.*, 36, 5541 (1997).

167. M.A. Harder and Q. Sun, *App. Catal. A. Gen.*, 221, 45 (2001).

168. E.L. du Toit, MEng (Chemical) thesis, University of Pretoria, South Africa, 2003.

169. G. Gelbard, *Ind. Eng. Chem. Res.*, 44, 8468 (2005).

170. B.C. Gates and L.N. Johanson, *J. Catal.*, 14, 69 (1969).

171. B.C. Gates, *Catalytic Chemistry*, John Wiley & Sons, New York, 1992.

172. V. Serra-Holm, T. Salmi, P. Maki-Arvela, E. Paatero, and L.P. Lindfors, *Org. Process Res. Dev.*, 5, 368 (2001).

173. H. Naka, Y. Kaneda, and T. Kurata, *J. Oleo Sci.*, 50, 813 (2001).

10 Membranes

10.1 INTRODUCTION

Membranes have been employed for the treatment of a diversity of fluids ranging from gases, wastewater, seawater, milk, yeast suspensions, and others [1–22]. Membrane development started with liquid-phase separation processes. In this regard, Nollet carried out the first documented membrane experiment in 1748, using a section of a pig's bladder as the membrane. As a result of this experiment, he introduced the term "osmosis" [10]. However, the manufacture of asymmetric cellulose acetate membranes in the early 1960s by Loeb and Sourirajan is normally accepted as the first milestone for membrane technology [2]. However, the earliest large-scale gas separation membrane procedure was carried out in the mid-1940s by the U.S. government to separate the different isotopic compositions of UF_6 for nuclear fuel enrichment [6]. Nevertheless, commercially important gas separation membrane processes were introduced only in late 1970s and early 1980s [10].

Membranes owe their popularity largely to the following advantages [9]: separations can be performed continuously under mild conditions, with low energy consumption, and using no additives. However, some disadvantages include fouling, short life span, and low selectivity or flux.

Organic membranes have been commercially distributed for many years. However, in recent times, there is an increasing attention for extremely selective inorganic membranes owing to their exceptional stability characteristics that are unsurpassed by organic membranes, specifically, chemical stability in corrosive atmospheres and high temperature tolerance [7]. Inorganic membranes are employed in gas separation [8–10], catalytic reactors [12,20], gasification of coal, water decomposition, and other applications [9,13,15,16].

The synthesis and characterization of permselective membranes are new areas of activities in materials science [1,12,20]. Nevertheless, to use these membranes commercially, much work remains to be done [12,20]. It is necessary to increase membrane permeance, to resolve brittleness problems, to be capable of scaling-up the process, and to increase the membrane area per unit volume [12,20].

In this chapter, gas–solid systems, with an emphasize on inorganic permeable materials, to produce dense and porous membranes for chemical, sustainable energy, and pollution abatement applications, are considered. However, since the most important membranes currently in use are the polymeric porous membranes, then these are discussed at the end of the chapter.

10.2 DEFINITIONS AND NOMENCLATURE

10.2.1 SOME DEFINITIONS

A membrane is a barrier that allows selective mass transport between two phases. It is selective since various components are able to pass through the membrane more efficiently than others. This makes membranes an appropriate means to separate a mixture of components. That is, a membrane is a permselective barrier between two phases that can be permeated owing to a driving force, such as pressure, concentration, or electric field gradient [18,19]. The phases on either side of the membrane can be liquid or gaseous.

Considering the basic material used for membrane production, membranes are categorized as organic, which are normally made of polymers, or inorganic, which includes membranes made of glass, metal, and ceramic [6–10].

Considering the microstructure of membranes, they can be categorized as porous, which allow transport through their pores, or dense, which permit transport through the bulk of the material [19]. Porous membranes are classified as microporous, mesoporous, and macroporous (see Section 6.2).

With regard to their morphology, membranes can be classified as symmetric, which possess a homogeneous structure, or asymmetric, which consist of two or more sheets with dissimilar characteristics [19]. Symmetric membranes are prepared with a single solid phase, and asymmetric membranes are produced with two ends composed of different solid phases or more slides composed of different solid phases [8]. That is, symmetrical membranes have a uniform structure, and asymmetric membranes are composed of a number of layers with different structures. Hence, membranes consisting of different layers of different materials are called composite membranes [10]. Asymmetric membranes are normally designed to guarantee the mechanical stability of the whole membrane, that is, the morphology is arranged in a composite form, comprising a highly permeable, usually porous support, and a narrow coat of the microporous or dense separating material [19]. For example, macroporous and mesoporous membranes are used as support for the synthesis of asymmetric membranes with a microporous or dense, end thin film, that is, the macro- and/or mesopores of the support are covered with films of a micropore or dense materials, where the support gives mechanical strength while the end layer is intended to carry out selective separations [8,13,19].

Different methods have been used to deposit microporous thin films, including solgel, pyrolysis, and deposition techniques [20]. Porous inorganic membranes are made of alumina, silica, carbon, zeolites, and other materials [8]. They are generally prepared by the slip coating method, the ceramic technique, or the solgel method (Section 3.7). In addition, dense membranes are prepared with metals, oxides, and other materials (Chapter 2).

10.2.2 Membrane Unit

Figure 10.1 [10] schematically shows the parts that constitute a membrane unit.

The two sections of the membrane unit are called the feed side or upstream side, and permeate side or downstream side. Besides, the flow resulting after permeation is named the residue or retentate flow. Finally, on the permeate side the exit flow is termed the permeate flow [6].

10.2.3 Permeance and Permeability

When pressure is the driving force of the process, the gaseous molecules will be transported from the high-pressure side to the low-pressure side of the membrane (see Figure 10.2 [18,19]). The permeation cell (Figure 10.2) is coupled with two (or more) pressure transducers to measure the reject pressure, P_1, and the permeate pressure, P_2, and a mass flow meter (F) to measure the flux (J) passing through the membrane [18,19].

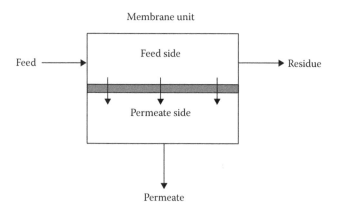

FIGURE 10.1 Membrane unit in the case of a gas separation process.

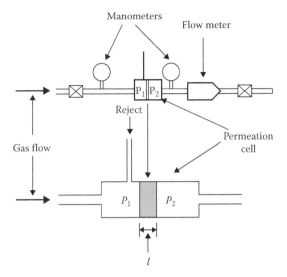

FIGURE 10.2 Schematic diagram of a permeation test facility.

Consequently, for single component gases, which are nondissociated during the process, and a linear pressure drop across the porous media, this transport process follows Darcy' law [16,19]:

$$J = B\left(\frac{\Delta P}{l}\right) = \Pi \Delta P \tag{10.1a}$$

$$J = \frac{Q}{V_m A} \tag{10.1b}$$

$$\Pi = \frac{B}{l} \tag{10.1c}$$

where
  A is the effective membrane area
  B is the permeability [mol/m·s·Pa]
  J is the molar gas flow [mol/m²·s]
  l is the membrane thickness [m]
  $\Delta P = P_1 - P_2$ is the transmembrane pressure [Pa]
  Π is the gas permeance [mol/m²·s·Pa]
  Q is the gas filtrate flux [m³/s]
  V_m is the molar volume of the flowing gas [m³/mol]

since for an ideal gas

$$V_m = \frac{V}{n} = \frac{RT}{P}$$

10.2.4 SELECTIVITY

Selectivity, often represented as the permeance ratio of two gases, is the capability of the membrane to separate a given gas mixture into its components [7]. Selectivity is a measure of the membrane separation efficiency, that is, the selectivity factor, $\alpha_{\frac{A}{B}}$, of two components, A and B, in a mixture and is defined as follows [9]:

$$\alpha_{\frac{A}{B}} = \left[\frac{\left(\dfrac{y_A}{y_B} \right)}{\left(\dfrac{x_A}{x_B} \right)} \right] \tag{10.2}$$

where

y_A and y_B are the fractions of components, A and B, in the permeate
x_A and x_B are the fractions of the components, A and B, in the feed

10.3 PERMEABILITY IN DENSE MEMBRANES

As explained in Chapter 5, the transport mechanism in dense crystalline materials is generally made up of incessant displacements of mobile atoms because of the so-called vacancy or interstitial mechanisms. In this sense, the solution–diffusion mechanism is the most commonly used physical model to describe gas transport through dense membranes. The solution–diffusion separation mechanism is based on both solubility and mobility of one species in an effective solid barrier [23–25]. This mechanism can be described as follows: first, a gas molecule is adsorbed, and in some cases dissociated, on the surface of one side of the membrane, it then dissolves in the membrane material, and thereafter diffuses through the membrane. Finally, in some cases it is associated and desorbs, and in other cases, it only desorbs on the other side of the membrane. For example, for hydrogen transport through a dense metal such as Pd, the H_2 molecule has to split up after adsorption, and, thereafter, recombine after diffusing through the membrane on the other side (see Section 5.6.1).

In the solution–diffusion separation mechanism, the permeating species dissolves in the membrane material and then diffuses responding to the chemical potential gradient. Consequently, the equation governing the solute flux is [24,25]

$$J_s = D_s k_s \frac{(C_1 - C_2)}{l} \tag{10.3}$$

where

J_s is the solute flux
D_s is the diffusivity of the solute in the membrane
C_1 is the solute concentration in the reject side
k_s is the solute distribution coefficient
C_2 is the solute concentration in the permeate side

10.3.1 HYDROGEN TRANSPORT IN METALLIC DENSE MEMBRANES

Metallic membranes for hydrogen separation can be of many types, such as pure metals Pd, V, Ta, Nb, and Ti; binary alloys of Pd, with Cu, Ag, and Y; Pd alloyed with Ni, Au, Ce, and Fe; and complex alloys of Pd alloyed with more than one metal [3]. Body-centered cubic metals, for example, Nb and V, have higher permeability than face-centered cubic metals, for instance, Pd and Ni [26–29]. Even though Nb, V, and Ta possess a permeability greater than that of Pd, these metals develop oxide layers and are complicated to be used as hydrogen separation membranes [29]. Especially, the Pd and Pd-based membranes have in recent times obtained renovated consideration on account of the prospects of a generalized use of hydrogen as a fuel in the future [26]. We emphasize on these types of membranes in this chapter.

Hydrogen transport through Pd and Pd-based alloys comprises the next steps [30,31]. The H_2 molecules during adsorption are dissociated on top of the metal surface, giving a proton to the interstitial sites and an electron to the metal conduction band (see Section 2.4.2). The second step is the diffusion of atomic H, since the proton will be surrounded by an electron cloud [32], through the bulk of the metal. Finally, an associative desorption process of H_2 molecules occurs from the metal surface at the other end of the membrane.

Hydrogen separation through a Pd-based film takes place via a solution–diffusion mechanism. Then, the equation for the hydrogen flux is written in terms of Sievert's law [33,34] (see Figure 5.11)

$$J_{H_2} = B_{H_2} \frac{P_1^{\frac{1}{2}} - P_2^{\frac{1}{2}}}{l} \tag{10.4}$$

where
J_{H_2} is the hydrogen flux
B_{H_2} is the hydrogen permeability
l is the membrane thickness
P_1 and P_2 are the reject and permeate pressures

Permeability is an essential property of the materials that constitute the membrane and is independent of membrane thickness [23]. Additionally, permeability can be described as the product of the diffusion coefficient and the solubility constant (see Equation 10.3) and is temperature dependent. Permeability can then be represented by the following Arrhenius-type expression (see Section 5.6.1)

$$B = K_B e^{-\frac{E_a^B}{RT}} \tag{10.5}$$

where
K_B is the pre-exponential factor
E_a^B is the activation energy

and the working temperatures of the Pd and Pd-based membranes are in the range of 300°C–600°C [6]. In a general case [34,35]

$$J_{H_2} = B_{H_2} \frac{P_1^n - P_2^n}{l} \tag{10.6}$$

where n is the pressure exponent, that is, the exponential dependence of hydrogen flux on the pressure, which for nonmetals (see Section 10.3.2) deviates from the value corresponding to Sievert's law, that is, $n = 1/2$.

To conclude this section, it is necessary to state that Pd and Pd-based membranes are currently the membranes with the highest hydrogen permeability and selectivity. However, the cost, availability, their mechanical and thermal stabilities, poisoning, and carbon deposition problems have made the large-scale industrial application of these dense metal membranes difficult, even when prepared in a composite configuration [26,29,33–37].

10.3.2 HYDROGEN PERMEATION IN OXIDE CERAMIC MEMBRANES

Proton-conducting materials [38–47], analogous to oxygen conductors but with stationary oxygen anions, can show mixed protonic–electronic conductivity, without considerable oxygen transport in hydrogen or water atmospheres [40,41]. These materials have not been widely studied in comparison to oxygen conductors, but they can be used to separate H_2 from gas flows. Consequently, some mixed proton–electron conductor oxides were identified that can be applied as hydrogen-permeable dense ceramic materials (see Sections 5.7.6 and 8.2.5).

High-temperature hydrogen permeation driven by a partial hydrogen pressure difference across membranes is considered now as hydrogen exchange at the interfaces between the gas and solid phases, that is, hydrogen molecules are oxidized to protons at the end of the membrane which is making contact with the high-pressure hydrogen gas phase. Thereafter, a diffusion of protons and electrons is produced in the bulk of the material, where the electrical current ensuing from the transport of protons is balanced by an electronic current within the membrane. When the protons are transported to the low-pressure hydrogen end of the membrane, the protons are reconverted to hydrogen molecules. That is, hydrogen transport through ceramic membranes is based on the solution–diffusion mechanism in the dense ceramic material. However, in this case, ambipolar diffusion is present. In Sections 5.7.6 and 8.2.5, these cases of proton transport and conduction were described.

In this regard, hydrogen flux by proton transport in a dense oxide membrane, in the short circuit case, is described by the ambipolar diffusion expression (see Figure 10.3a) [40,48,49]

$$\bar{J}_{H^+} = -\frac{1}{2F^2}(\sigma_{H^+} t_e)\nabla \mu_{H_2} \tag{10.7}$$

where

$$t_e = \frac{\sigma_e}{\sigma_{H^+} + \sigma_e} \tag{10.8}$$

is the electronic transference number. If we have a membrane where the transport is driven by a protonic–electronic ambipolar charge, it is obvious that the driving force for permeation is the hydrogen partial pressure gradient. Then, in a steady-state situation, we can integrate the flux density expression (Equation 10.7) over the thickness of the membrane, maintaining the flux density constant. Consequently, the flux through the membrane is (see Figure 10.3a)

$$J_{H^+} = -\frac{RT}{2F^2 l} \int_{P_{H_2}^I}^{P_{H_2}^2} \sigma_{H^+} t_e \, d \ln P_{H_2}(g) \tag{10.9}$$

To get an explicit solution of Equation 10.9, by integration, it is necessary to make some assumptions [39]. In this regard, if the membrane material shows a high electronic conduction

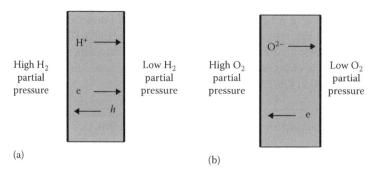

(a) (b)

FIGURE 10.3 (a) Hydrogen and (b) oxygen permeation through dense oxide ceramic membranes.

$$t_e \approx 1 \tag{10.10}$$

and σ_{H^+} can be written with the help of the following expression

$$\sigma_{H^+} = \sigma_{H^+}^0 (P_{H_2})^n \tag{10.11}$$

where $\sigma_{H^+}^0$ and n are constants. Then, Equation 10.9 can be integrated if we know both constants.

10.3.3 PERMEATION IN DENSE OXIDE MEMBRANES

As described in Section 8.2.6, along with YSZ, mixed oxygen-ion, and electron-conducting oxides with a perovskite-type structure, the so-called Aurivillius phase and pyrochlore materials are fundamentally used for the production of a variety of high-temperature electrochemical devices [50–58].

When membranes prepared with the mixed conducting oxides are located in an oxygen partial pressure, P_{O_2}, gradient at elevated temperatures, a spontaneous flow of oxygen molecules from the high-pressure end to the low end of the membrane results. During this high-temperature process, oxygen molecules are reduced to oxide anions at the end of the membrane, which faces the high oxygen pressure. After this, the oxide anions are transported by the vacancy mechanism through the oxide membrane to the low oxygen pressure end of the membrane, where the oxide anions are reconverted to oxygen molecules (see Figure 10.3b). This process is a chemical reaction that proceeds at a finite rate, involving the chemisorption of oxygen as a molecule at the surface, the splitting of the oxygen bond to form oxygen atoms, the charge transfer, and the incorporation of oxygen anions into the oxide [58]. Then, oxygen permeation driven by an oxygen partial pressure difference across membranes is the result of oxygen exchange at the interfaces between the gas and solid phases and counterdiffusion of oxide ions and electrons in the bulk of the material, where the electrical current resulting from the transport of oxide ions is balanced by an electronic current within the membrane material, so that neither the electrode nor the external circuitry is necessary to produce the oxygen permeation (see Figure 10.3b). It is accepted that for moderately thick membranes, the oxygen permeation is restricted by the counterdiffusion of oxide ions and electrons in the volume of the membrane. Consequently, a reduction in the thickness of membrane results in a proportional increase in the flux. However, if the thickness is decreased at some point, then the oxygen exchange at the interface of the gas and solid phases turns out to be rate limiting, so that an additional decrease in the membrane's thickness cannot result in a substantial increase in the flux [51].

When oxygen is transported through a membrane in a steady state, there is no net charge current and the flux of oxygen anions, O^{2-}, can be written using the ambipolar diffusion expression, as follows [49]

$$\bar{J}_{O^{2-}} = -\frac{1}{8F^2}\left(\frac{\sigma_{O^{2-}}\sigma_e}{\sigma_{O^{2-}} + \sigma_e}\right)\nabla\mu_{O_2} \tag{10.12}$$

since in this case, we have a system with an oxygen anion–electronic ambipolar charge transport. Then, if we have a membrane where the transport takes place by this type of process, it is evident that the driving force for permeation is the oxygen partial pressure gradient. Therefore, in a steady-state situation, we can integrate the flux density (Equation 10.12) over the thickness of the membrane maintaining the flux density constant; then, the flux through the membrane is given by [15,48,49,57]:

$$J_{O_2} = -\frac{RT}{16F^2 l}\int_{P_{H_2}^1}^{P_{H_2}^2}\frac{\sigma_{O^{2-}}\sigma_e}{\sigma_{O^{2-}} + \sigma_e}\,d\ln P_{O_2}(g) \tag{10.13}$$

To integrate Equation 10.13, the ambipolar conductivity is defined as [58]

$$\sigma_{ambipol} = \frac{\sigma_{O^{2-}} \sigma_e}{\sigma_{O^{2-}} + \sigma_e} \qquad (10.14)$$

which can also be written with the help of the following empirical expression

$$\sigma_{ambipol} = \sigma^0_{ambipol} (P_{O_2})^n \qquad (10.15)$$

where
$\sigma^0_{ambipol}$ is a constant
n is the order of the ambipolar conductivity

Then, if we know both constants, it is possible to integrate Equation 10.13 to get an explicit expression for the oxygen flux.

10.4 PERMEATION IN POROUS MEMBRANES

10.4.1 INTRODUCTION

According to the classification system proposed by the International Union of Applied Chemistry (IUPAC), pores are divided into three groups on the basis of their size, that is, macropores (more than 50 nm), mesopores (from 2 to 50 nm), and micropores (less than 2 nm) [59]. Porous membranes are grouped as macroporous, mesoporous, and microporous. Particularly, porous inorganic membranes are made of carbon, alumina, titania, zirconia, silica, silicon carbide, silicon nitrides, zeolites, cordierite, and other materials [15,19]. They are generally prepared by the slip coating method, the ceramic technique, the solgel method, pyrolysis, and deposition techniques [8–10]. The materials used for inorganic porous membrane preparation experience phase transformations, structural changes, and sintering at high temperature; as a result, the maximum temperature at which porous inorganic membranes can be used ranges from 400°C to 1000°C depending on the material used [11].

10.4.2 TRANSPORT MECHANISMS IN POROUS MEMBRANES

The classification of pores is also based on the difference in the types of molecular interactions controlling adsorption in the different groups. That is, in micropores, the overlapping surface forces of opposing pore walls are predominant; in mesopores, surface forces and capillary forces are significant, while for the macropores, the contribution of the pore walls to surface forces is very small [15,19].

There are four well-known types of diffusion [19,30,31,60–70]: gaseous or molecular diffusion [61], Knudsen diffusion [62–64], liquid diffusion [60], and atomic diffusion in solids [30,31]. If the diffusion process in a pore system occurs at sufficiently high temperatures, as is usually the case in applications, we will fundamentally have three regimes with different diffusivities according to the pore diameter (see Figure 5.28 [71]). For macropores, generally, collisions between the molecules take place much often than collisions with the wall. Therefore, molecular diffusion is the main mechanism [19]. As the dimension of the pores diminish, the number of collisions with the wall rises. Subsequently, Knudsen diffusion takes over and the mobility begins to depend on the dimensions of the pore [19]. At still lesser pore sizes, that is, in the range of 2 nm or less, when the pore diameter becomes comparable to the size of the molecules, the molecules will constantly undergo interactions with the pore surface; diffusion in micropores typically takes place in the configurational diffusion regime [60].

10.4.3 VISCOUS AND KNUDSEN FLOWS

To distinguish between the viscous and Knudsen gas-phase flows (Section 5.8.2), the ratio between the mean free path, λ, and the characteristic length of the flow geometry, L, is defined as

$$\frac{\lambda}{L} = K_n \tag{10.16}$$

which is commonly named the Knudsen number, K_n [64].

A viscous flow takes place when the effects of viscosity become significant. This type of flow can be categorized as either a laminar flow or a turbulent flow [72]. A laminar flow is one with no considerable mixing of neighboring fluid particles, apart from molecular motion. In a turbulent flow, the quantities which typify the flow exhibit a random variation with the time and space coordinates. The quantity used to predict the type of flow regime is the Reynolds number (Re), which is a dimensionless parameter defined as [72]

$$Re = \frac{vL}{\eta} \tag{10.17}$$

where
  L is the characteristic length, for example, pipe diameter, that is, the length of the flow field
  v is the flow velocity
  η is the dynamic viscosity

If $Re < 2000$, then the flow is laminar [69].

10.4.4 DARCY'S LAW FOR VISCOUS FLOW

In the simplest situation of a flow through a straight cylindrical pore, Darcy's law, based on the Hagen–Poiseuille equation, describes the process with the expression [73]

$$J_v = \left(\frac{r^2}{8\eta V_m}\right)\left(\frac{\Delta P}{l}\right) \tag{10.18}$$

where r is the pore radius.

Darcy's law for laminar flow in a real macroporous membrane is described by the following equation [16]

$$J_v = B_v\left(\frac{\Delta P}{l}\right) \tag{10.19}$$

where

$$B_v = \frac{k}{\eta V_m} \tag{10.20}$$

in which
  k is the permeation factor [m^2]
  η is the dynamic viscosity of the gas [Pa s]

For the description of this flow, the Carman–Kozeny expression [16] can be applied, since the Hagen–Poiseuille equation is not valid, given that usually inorganic macroporous and mesoporous membranes are prepared by the sinterization of packed quasispherical particles, which develop a random pore structure [19]. In this case, the Carman–Kozeny factor for a membrane formulated with pressed spherical particles is [74]

$$k = \frac{\varepsilon d_v^2}{16C} \tag{10.21}$$

where

$$\varepsilon = 1 - \frac{\rho_A}{\rho_R} \tag{10.22}$$

Therefore [74],

$$k \approx \frac{\varepsilon d_v^2}{77} \tag{10.23}$$

in which
   $C = 4.8 \pm 0.3$ is the Carman–Kozeny constant
   d_v is the membrane pore diameter
   ε is the membrane porosity
   ρ_A is the apparent membrane density [g/cm^3]
   ρ_R is the real membrane density [g/cm^3] [18,19]

The "dusty gas model" (DGM) developed by Mason and collaborators also accounts for the viscous mechanisms in real porous systems [75]. Within the framework of this model, permeability for the viscous flow is given by the following expression [75,76]

$$B_v = \frac{\varepsilon d_p^2}{8\tau \eta V_m} \tag{10.24}$$

where
   ε is the porosity
   τ is the tortuosity
   d_p is the average pore diameter of the porous medium

The viscous diffusion mechanism is also valid for transport process in the liquid phase. Then, if we have a liquid filtration process through a porous (i.e., macroporous or mesoporous) membrane, the following form of the Carman–Kozeny equation can be used [9]

$$J_v = \left(\frac{\varepsilon^2 \rho}{K\eta S^2 (1-\varepsilon)^2} \right) \left(\frac{\Delta P}{l} \right) \tag{10.25}$$

in which
   ε is the porosity
   S is the pore area
   K is a constant
   ρ is the molar density that can be applied

10.4.5 DARCY'S LAW FOR KNUDSEN FLOW

As the membrane pore widths diminish, or the mean free path of the molecules rise, the permeating particles tend to collide more with the pore walls than among themselves [60,64]. In this
case, the Knudsen flow regime is established; therefore, the molar gas flow, J, for the Knudsen
flow, in a straight cylindrical mesopore of length, l, and trans-pore pressure, $\Delta P = P_1 - P_2$, is
given by [64]

$$J_K = D_K\left(\frac{\Delta P/kT}{l}\right) \tag{10.26}$$

where [19] (see Section 5.8.4)

$$D_K^* = \frac{d_P}{2}\left(\frac{\pi kT}{2M}\right)^{1/2} \tag{10.27}$$

expresses the diffusivity of a Knudsen gas. For a real mesoporous membrane, which is shaped
by a complex pore network, the expression for the permeation flux across the membrane is
given by [20]

$$J_K = \left(\frac{G}{(2MkT)^{1/2}}\right)\left(\frac{\Delta P}{l}\right) \tag{10.28}$$

in which G is a geometrical factor. If M is expressed in molar units, then

$$J_K = \left(\frac{G}{(2MRT)^{1/2}}\right)\left(\frac{\Delta P}{l}\right) \tag{10.29}$$

the geometrical factor G can be calculated with the help of a simple model. Within the framework
of this model, it is assumed [9,70] that the diffusivity in a porous material, D, can be related to the
diffusivity inside a straight cylindrical pore, D_K, with a diameter equal to the mean pore diameter
of the pore network by a simple factor, (ε/τ), that is,

$$D = \left(\frac{\varepsilon}{\tau}\right)(D_K) \tag{10.30}$$

where ε is the porosity, which takes into consideration the fact that transport takes place only
throughout the pore and not through the solid matrix. The other effects are grouped together into a
parameter called the tortuosity factor, τ. Consequently,

$$G = \frac{d_P\varepsilon(\pi)^{1/2}}{2\tau} \tag{10.31}$$

10.4.6 TRANSPORT IN ZEOLITE MEMBRANES

In zeolites, the rate of molecular diffusion depends on the position of charge-compensating cations
in the pore network and the structure of the framework [77–81]. Since mass transport in microporous media takes place in an adsorbed phase [82,83], this transport can be envisaged as activated
molecular hopping between fixed sites [60,82,84] (for more details, see Section 5.9.1).

10.4.7   ZEOLITE-BASED MEMBRANES

Recent developments in the synthesis of membranes prepared from zeolites have led to considerable attention in these type of membranes [17,19,22,85–97]. The pore sizes of zeolites are of molecular dimensions, and these materials have the capability of separating mixtures of molecules by molecular sieving, if we have a continuous zeolite membrane [95]. Besides, zeolites are relatively stable in high-pressure and high-temperature environments, and show good chemical stability. These properties allow zeolite membranes to be used in conditions that would be too demanding for polymeric membranes. In addition, the extremely ordered pore network in zeolite membranes makes selective separation of species with no significant dissimilarities in adsorption and transport properties possible [8,19,92]. Additionally, other zeolite-based separation methods are expensive and energy consuming, and steady-state separation processes based on zeolite membranes are economically and environmentally better options [13,22]. This line of reasoning justifies the tremendous research devoted to the development of zeolite membranes for new applications [8,13,17,19,22,85–95], particularly for gas and vapor separations [97,98].

The task of developing syntheses methods for a new zeolite membrane is complicated. The first attempts to prepare zeolite membranes took place in the late 1980s [96]. Even though the early efforts did not lead to the synthesis of high-quality zeolite membranes, throughout the early 1990s, numerous researchers obtained excellent MFI-type zeolite membranes with fine permeation and separation properties [13,95]. Zeolites membranes are prepared over a support made of stainless steel, alumina, silicon carbide, mullite, zirconia, titania, or other porous supports, both as flat disks and tubes [22,96]. For the preparation of zeolite membranes, two methodologies can be generally applied [13,22]: the in situ membrane growth and the seeded growth techniques. The in situ membrane growth methodology is carried out by placing an appropriate support in contact with a precursor solution or gel in an autoclave; afterward, a zeolite layer is produced on the support under hydrothermal conditions following rules of the hydrothermal synthesis of zeolites (see Section 3.4) [13,18]. The other method consists of seeding the support in order to improve the nucleation of the crystals on the support surface [13]. Usually, the complexity in zeolite membrane synthesis increases with growing aluminum content in the zeolite; therefore, the synthesis of A-type membranes is difficult. However, a highly hydrophilic A-type zeolite membrane has been synthesized under hydrothermal conditions on the outer surface of a porous α-alumina tube [7,22,88].

10.4.8   PERMEATION FLOW IN ZEOLITE MEMBRANES

Transport and adsorption processes in microporous materials have been, during the last years, a topic of significant research activity [19,92–94,98]. Section 5.9.2 dealt with the phenomenological description of diffusion in zeolites. Applying this methodology to a membrane, it is possible to express the flux of a gas through the zeolite membrane in isothermal conditions as follows [19,70,92–94]

$$J = -Mq\frac{\partial\mu}{\partial z}$$

where
    M is the mobility
    q is the concentration of the gas in the zeolite micropores
    μ is the chemical potential

Then,

$$J = -MRT\frac{\partial \ln P}{\partial \ln q}\frac{\partial q}{\partial z} \tag{10.32}$$

Consequently,

$$J = -D_0 \Phi \frac{\partial q}{\partial z} \qquad (10.33)$$

in which

D_0 is the intrinsic or corrected diffusion coefficient

Φ is the thermodynamic factor

Under steady-state conditions, Equation 10.33 can be integrated to give the steady-state flux [92]

$$\int_0^l J \mathrm{d}z = -\int_{q_1}^{q_2} D_0 \Phi \, \mathrm{d}q \qquad (10.34)$$

where q_1 and q_2 are the adsorbate concentrations in the zeolite at $z = 0$ and $z = l$ and (see Section 5.9.2)

$$D_A = D_0 \Phi$$

is the Fickean or transport diffusion coefficient [71] which depends on q, the adsorbate concentration in the pores. The calculation of the adsorbate concentration can be carried out, for example, with the help of the following adsorption isotherm equation

$$\frac{q}{q^m} = \frac{K_0 P}{1 + K_0 P}$$

where $K_0 = K_0(T)$ and q^m is the adsorbate concentration which saturates the zeolite pores and cavities (see Section 6.7.3). Now, applying the Barrer and Jost model [99], which is valid for a Langmuir system, it is possible to get an equation equivalent to Equation 5.113, which gives a procedure to calculate the corrected or intrinsic diffusion coefficient [19,70,92–94]

$$D_0 = D_A (1 - \theta)$$

where $\theta = \dfrac{q_a}{q^m}$ is the fractional saturation of the adsorbent. Subsequently, the integration of Equation 10.34 gives [92–94]

$$J = \frac{q^m}{l} D_0 \left(\frac{1 + K_0 P^1}{1 + K_0 P^2} \right) \qquad (10.35)$$

where P^1 and P^2 are the gas pressures at $z = 0$ and $z = l$.

As described in Section 5.9.1, we can calculate the self-diffusion coefficients for localized and mobile adsorption sites with the help of Equations 5.104 and 5.105 [82]. Now, it is possible to consider that [70]

$$D_0 \approx D^*$$

and that the total flux through the membrane is given by [94]

$$J = J_L + J_M \qquad (10.36)$$

where

$$J_{L} = \left(\frac{\nu l_0^2 q^m \exp\left(-\dfrac{E_g}{RT}\right)}{l} \right) \left(\frac{1 + K_0 P^1}{1 + K_0 P^2} \right) \tag{10.37a}$$

and

$$J_{M} = \left(\frac{\dfrac{1}{2}\left(\dfrac{RT}{\pi M}\right) l_0 q^m \exp\left(-\dfrac{E_g}{RT}\right)}{l} \right) \left(\frac{1 + K_0 p^1}{1 + K_0 p^2} \right) \tag{10.37b}$$

are the fluxes related with the localized (J_L) and mobile (J_M) adsorption sites. Finally, it is possible to calculate the permeance of a zeolite membrane as follows:

$$\Pi = \frac{J}{\Delta P} = \frac{\left\{ \left[\dfrac{\dfrac{1}{2}\left(\dfrac{RT}{\pi M}\right) l_0 q^m \exp\left(-\dfrac{E_g}{RT}\right)}{l} \right] + \left(\dfrac{\nu l_0^2 q^m \exp\left(-\dfrac{E_g}{RT}\right)}{l} \right) \right\} \left[\dfrac{1 + K_0 P^1}{1 + K_0 P^2} \right]}{\Delta P}$$

10.5 ZEOLITE-BASED CERAMIC POROUS MEMBRANE

10.5.1 CARBON DIOXIDE PERMEATION IN A ZEOLITE-BASED CERAMIC POROUS MEMBRANE

Porous zeolite-based membranes are prepared using a ceramic technique (see Section 3.7.1) by thermal transformation of natural clinoptilolite at 700°C–800°C [18,19]. In the membranes thus produced, configurational diffusion is not feasible, because during the thermal treatment of the zeolite to obtain the ceramic membrane, the clinoptilolite framework collapsed [100]. Also, the particles used for the sintering process to make the tested membranes have a size of 220–500 µm; subsequently, in these membranes only macropores are present [18]. In addition, adsorption and capillary condensation on the surface of the membrane will be weak, owing to the relatively high temperature (300 K) and rather low pressures (0.2–1.4 MPa); therefore, in this kind of permeation, only Knudsen or gaseous flow can take place [18,19]. In order to carry out a permeation test of these membranes, the permeability [B] and permeance [P] of H_2 was measured using Darcy's law [18,19] (see Section 10.2.3).

Figures 10.4 and 10.5 show the experimental results of CO_2 permeation in some of these membranes [18]. In Table 10.1, the permeance and permeability experimental results are reported [18]. In Table 10.2, the estimated pore diameters of the membranes are reported [18].

The mean free path, λ, of the CO_2 molecules at the temperatures and pressures of the permeation experiment are by far smaller than the membrane pore size, d_v, that is, $d_v \gg \lambda$. Then, Knudsen flow is not possible since the determining process is gaseous laminar flow through the membrane pores [18]. It is therefore feasible to apply Darcy's law for gaseous laminar flow (Equations 10.19 through 10.23).

In addition, for the description of this flow it is necessary to employ the Carman–Kozeny relation (Equation 10.19), since the Hagen–Poiseuille equation is not suitable, as the membranes were obtained by the sinterization of packed quasispherical particles [18].

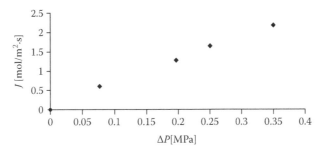

FIGURE 10.4 Permeation test of CO_2 in a membrane prepared by thermal treatment of a clinoptilolite powder of 500 mm particle diameter at 700°C for 2 h.

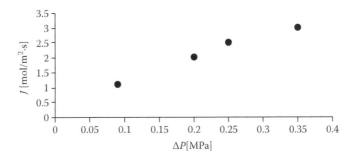

FIGURE 10.5 Permeation study of CO_2 in a membrane produced by thermal treatment of a clinoptilolite powder of 220 μm particle diameter at 800°C for 1 h.

TABLE 10.1
Carbon Dioxide Permeability, B, and Permeance, Π, in the Studied Membranes

Sample d_p [μm]	Sample Treatment Temp. [°C]	Sample Treatment Time [h]	$B \times 10^8$ [mol m^{-1} s^{-1} Pa]	$\Pi \times 10^6$ [mol m^{-2} s^{-1} Pa]
220	700	2	0.5	1.8
500	700	2	1.7	6.3
220	800	1	0.7	1.7
500	800	1	2.4	8.9

TABLE 10.2
Estimated Membrane Pore Diameters, $_v$

Sample $_p$ [μm]	Sample Treatment Temp. [°C]	Sample Treatment Time [h]	$_v$ [μm]
220	700	2	35
500	700	2	79
220	800	1	36
500	800	1	84

TABLE 10.3
Carbon Dioxide Permeation, k, and Membrane Average Pore Diameter, d_v

Sample d_p [μm]	Sample Treatment Temp. [°C]	Sample Treatment Time [h]	$k \times 10^{12}$ [m²]	d_v [μm]
220	700	2	3.4	32
500	700	2	11.6	58
220	800	1	5.1	36
500	800	1	16.5	84

Then, Equations 10.19 through 10.23 are used to measure the membrane pore diameter (d_v) (see Table 10.3). The results coincide fairly well with the values previously estimated (see Table 10.2) [18].

10.5.2 In Situ Synthesis of an AlPO₄–5 Zeolite over a Ceramic Porous Membrane

Aluminosilicate zeolites are usually synthesized under hydrothermal conditions from solutions containing sodium hydroxide, sodium silicate, and sodium aluminate (see Section 3.4.1) [101–103]; it is also possible to grow aluminosilicate zeolite crystals using aluminosilicates as raw materials, under hydrothermal conditions in a solution containing sodium hydroxide [104].

The in situ membrane growth technique cannot be applied using the zeolite-based ceramic porous membrane as support, under hydrothermal conditions in a solution containing sodium hydroxide. The high pH conditions will cause membrane amorphization and lead to final dissolution. Therefore, we tried to synthesize an aluminophosphate zeolite such as AlPO₄-5 [105] over a zeolite porous ceramic membrane. For the synthesis of the AlPO₄-5–zeolite-based porous membrane composite, the in situ membrane growth technique [7,13,22] was chosen. Then, the support, that is, the zeolite-based porous ceramic membrane, was placed in contact with the synthesis mixture and, subsequently, subjected to a hydrothermal synthesis process [18]. The batch preparation was as follows [106]:

1. 7 g H_2O plus 3.84 g of H_3PO_4 were properly mixed.
2. (1) plus 2.07 g $(C_2H_5)_3N$, where $(C_2H_5)_3N$ was added in drops and mixed.
3. (2) plus 5.23 g $Al(C_3H_7O)_3$, added in small amounts at 0°C with intense stirring, then stir the mixture at room temperature for 2 h.
4. 0.83 g HF plus 89.2 g H_2O were properly mixed.
5. (3) plus (4) were stirred for 2 h.

Finally, the obtained gel was poured on the membrane in a 150 mL Teflon-lined steel autoclave and was treated for 24, 48, and 72 h at 180°C without agitation [18].

Figure 10.6 shows the XRD pattern of the zeolite-based ceramic membrane produced by sintering a clinoptilolite powder ($d_v \approx 500$ μm) at 800°C for 2 h, covered with an AlPO₄-5 molecular sieve produced with the methodology previously described.

It is evident that the ceramic membrane, which is represented in the XRD pattern (see Figure 10.6) by the amorphous component of the XRD profile, was covered by the AlPO₄-5 molecular sieve, since the crystalline component of the obtained XRD pattern fairly well coincides with the standards reported in the literature [107]. Consequently, the porous support was successfully coated with a zeolite layer, which was shaped by the hydrothermal process as previously described. Thus, a composite membrane, that is, an AlPO₄-5 molecular sieve thin film zeolite-based ceramic, was produced.

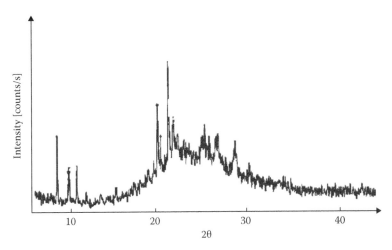

FIGURE 10.6 XRD profile of the porous ceramic membrane covered with a $AlPO_4$-5 molecular sieve.

10.6 CHEMICAL, SUSTAINABLE ENERGY, AND POLLUTION ABATEMENT APPLICATIONS OF INORGANIC MEMBRANES

10.6.1 HYDROGEN AND OXYGEN SEPARATIONS

We have studied the proton and oxide anion transport, conduction, and permeation in metals, dense oxide ceramics and, also briefly, in polymers (see also Chapters 2, 5, and 8). This section describes the application of proton and oxide permeation in these materials in hydrogen and oxygen separations.

10.6.1.1 Hydrogen Separations

Roughly, hydrogen separation membranes can be classified in four classes, based on the used materials: polymer, metallic, carbon, and ceramic membranes [6].

Hydrogen separation with polymer membranes is now an established technology; however, these membranes have inadequate mechanical strength, comparatively high sensitivity to swelling and compaction, and are vulnerable to be damaged by specific chemicals, all of which make polymeric membranes less attractive than inorganic membranes for hydrogen separation [3].

With respect to inorganic membranes, the Pd-membrane is possibly the oldest membrane used in hydrogen separation processes [26]. Even today, Pd and Pd-based membranes are receiving increasing attention from the separation and reaction research communities, with studies being carried out on the preparation of thin metal membranes supported on porous substrates [26–29]. In this regard, Pd and Pd-based membranes can be used to generate hydrogen for practical objectives with a purity of up to 99.99% [3,29]. However, it is necessary to recognize that the cost is possibly the main obstacle for the production of Pd membranes, and recent studies have been focused on thin metallic membranes [27]. Thin membranes, normally supported on porous ceramic, glass, or stainless steel membranes, would decrease the cost of materials as well as increase the hydrogen flux [3]. Two of the supports that are extensively applied are porous glass and porous ceramic [37]. However, fitting ceramics to metal do not have enough mechanical stability; consequently, stainless steel might be applied as a support material because of its mechanical durability, its thermal expansion coefficient, close to that of Pd, and its ease of gas sealing [4].

With respect to carbon membranes, the molecular sieving carbon membranes, produced as unsupported flat, capillary tubes, or hollow fibers membranes, and supported membranes on a macroporous material are good in terms of separation properties as well as reasonable flux and stabilities, but are not yet commercially available at a sufficiently large scale, because of brittleness and cost among other drawbacks [3,6].

Dense ceramic membranes for hydrogen purification has been explored during the last years; if the working temperature of the hydrogen separation method is adequate, hydrogen-permeable ceramic membranes can be useful in hydrogen separation applications. The selectivity of these membranes is high. However, high temperatures, near 900°C, are necessary to attain suitable fluxes [47]; besides, the chemical stability in gas flows containing specific molecules, for example, CO_2 and SH_2, is a problem [45]. Consequently, these materials are still in an early stage of development and are not applied as is the case for Pd-based membranes [6].

10.6.1.2 Oxygen Separations

Oxygen is the fourth largest chemical produced worldwide by industry [108]. The separation of air components or oxygen enrichment has progressed significantly during the past 10 years. At present, oxygen is generally produced by cryogenic distillation or pressure swing adsorption. These are relatively expensive processes; instead, ceramic membrane methods are economically more promising.

Oxygen-enriched air generated by membranes can be applied in diverse areas, including chemical and related industries, medicine, metallurgy, food packaging, and other fields [2]. Accordingly, the manufacture of oxygen by air separation is of large significance in both environmental and industrial processes. Besides, the application of pure oxygen instead of air as the oxidant in different industrial processes would diminish the cost outstandingly as the separation of nitrogen from subsequent product streams turns out to be needless [52]. However, oxygen separation using membranes is yet not really developed. This is because the majority of the industrial oxygen applications needs a purity higher than 90%, which is without difficulty attainable by adsorption or cryogenic technologies, but not by membranes [2].

Oxide-based perovskite-type mixed ionic conductors, with different compositions, can be prepared as high oxygen permeation fluxes ceramic membranes [57]. For example, the perovskite-based oxides with the composition $La_{1-x}Sr_xCo_{1-y}Fe_yO_{3-\delta}$ have appreciably high oxygen permeation fluxes at high temperatures [51,53,109–111]. As well, the perovskite oxides with the following composition $SrCo_{0.95-x}Fe_xZr_{0.05}O_{3-\delta}$ [54,112,113] exhibit high oxygen permeation properties in combination with improved chemical and thermal stabilities [54].

In most case studies, disk-shaped membranes with a restricted membrane area were generally used for oxygen permeation, because of the fact that they are easily prepared using the standard static-pressing method [114]. A manifold planar stack can be used to increase the membrane area to plant scale [115]. However, several problems such as sealing, connection, and pressure resistance have to be confronted [52]. Therefore, these dense inorganic membranes are being manufactured in tubular configurations [2] to overcome these difficulties [115,116]. However, their small surface-to-volume ratios and the wide membrane walls make these configurations not advantageous for industrial uses [52]. Instead, hollow-fiber membranes show many benefits over planar and tubular membranes, for example, superior surface area-to-volume ratio, and simplicity of high-temperature sealing [117].

10.6.2 Catalytic Membrane Reactors

Advancements in the field of catalytic membrane reactors and separation membranes for high-temperature applications turned out to be practical only recently with the development of high-temperature-resistant membranes [1,12,118–139]. In general, membrane reactors are used to shift the equilibrium in thermodynamically limited reactions [119–121]. Consequently, with the synthesis of new, highly selective inorganic membranes, the option of using these membranes in the acceleration of different reactions became a reality; therefore, a great quantity of research worldwide are being currently carried out on these types of reactors [12,20,118,119].

The first applications of membranes in catalysis [120] principally involved dehydrogenation reactions, where the function of the membrane was simply hydrogen removal. These studies were fundamentally performed on palladium and palladium alloys and confirmed the existence of membranes that can permeate H_2 with high selectivity [12,20,118,119]. However, problems arise when pure palladium is used because of the α- to β-phase transition in hydrogen-containing environments; then, it is necessary to stabilize the membrane by alloying Pd with other elements, for example, a commercial alloy of 23–25 wt % Ag is produced by Johnson Matthey, Hatton Garden, London, UK [121].

Currently, several types of membrane reactors are under investigation for dehydrogenation reactions, for example, the dehydrogenation of propane to propene [122,123], or of ethylbenzene to styrene [124]. In addition, the dehydrogenation of H_2S has been studied in membrane reactors [125].

Hydrogen can be manufactured by steam reforming and shift conversion of natural gas or other hydrocarbons. In conventional steam reformers, high conversions of natural gas are obtained at reformer outlet temperatures of around 850°C–900°C. However, pure hydrogen can be produced at appreciably lower temperatures by incorporating a membrane in the reactor that selectively takes away hydrogen during conversion [2]. Consequently, reforming reactions have been studied in membrane reactors. Among these reactions, the most important is the steam reforming of various hydrocarbons [121], especially methane steam reforming, which is the major source of hydrogen in the world [126].

Besides, syngas can be produced from methane steams reforming and by the partial oxidation of methane. Nevertheless, both processes have thermodynamic constraints. In this regard, membrane reactors affect equilibrium conversion, owing to the selective removal of one of the products from the reaction zone [127,128]. Specifically, a palladium membrane enables only hydrogen to permeate through it, changing the conversions toward values higher than those obtained at thermodynamic equilibrium [127,128]. Significant efforts have been made in studying the water–gas shift reaction in membrane reactors [129,130].

In addition to the Pd-based membranes, microporous silica membranes for hydrogen permeation [8] can be produced by a special type of chemical vapor deposition [140] named chemical vapor infiltration (CVI) [141]. A large amount of studies have been carried out on silica membranes made by CVI for hydrogen separation purposes [8,121]. CVI [141] is another form of chemical vapor deposition (CVD) [140] (see Section 3.7.3). CVD involves deposition onto a surface, while CVI implies deposition within a porous material [141]. Both methods use almost similar equipment [140] and precursors (see Figure 3.19); however, each one functions using different operation parameters, that is, flow rates, pressures, furnace temperatures, and other parameters.

Hydrogen-permselective silica membranes [8,10] can be, as well, synthesized by a particular application of solgel techniques [142,143]. Additionally, hydrogen permselective asymmetric membranes composed of a dense ceramic of a proton-conducting perovskite over a porous support have been developed [40]. However, some difficulties with respect to their stability is possible in certain reactive environments [121].

10.7 EXAMPLES OF POLYMERIC MEMBRANES

Polymeric materials are the most extensively applied for membrane preparation in science and industry [144–150]. Polymers for the preparation of membranes are applied, fundamentally, for gas separation [146], reverse osmosis seawater desalination [147,148], microfiltration, ultrafiltration, and nanofiltration [149,150].

Different methods for the preparation of porous polymeric membranes have been developed, for example, the track-etch methodology, the expanded-film procedure, and the phase separation process. The track-etch process consists of the following steps: a polycarbonate (Figure 10.7) film is radiated by charged fission particles; thereafter, the film is passed trough an etching solution, which attacks the sites where the polymer was broken by the irradiation process, forming sensitized nucleation tracks, and where pores are formed [144]. Another methodology for the preparation of porous polymeric membranes is the expanded-film process, where crystalline polymers are used, for example, polypropylene. To create the pores, the material is extruded close to its melting point

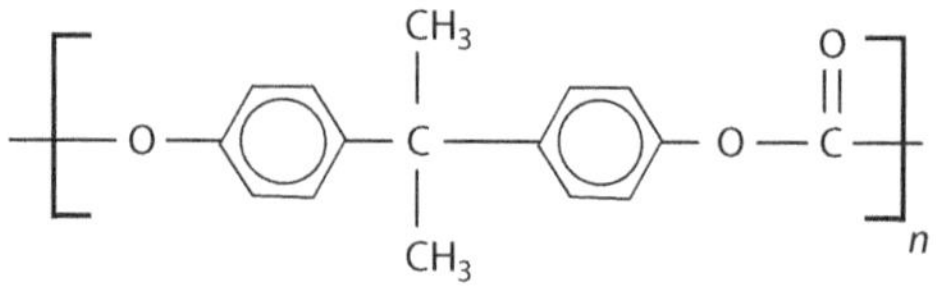

FIGURE 10.7 Polycarbonate.

temperature by applying a fast draw-down rate. The extruded material is cooled, annealed, and stretched up to three times its initial length [10]. This stretching process generates slit-like pores in the range between 20 and 250 nm [144]. The phase separation, phase inversion, or polymer precipitation process is a methodology for the production of membranes that involves the precipitation of a polymer solution into a polymer-rich solid phase that forms the membrane and a polymer-poor liquid phase that forms the membrane pores or void spaces [145]. This process of precipitation can be accomplished in different forms, that is, cooling precipitation, precipitation by immersion in a nonsolvent, and by solvent evaporation precipitation [10,145].

Particularly, the nonsolvent immersion, that is, the Loeb–Sourirajan preparation method is an important methodology. In this method, a polymer solution is cast into a film and the polymer precipitated by immersion into water [10,144]. The nonsolvent (water) quickly precipitates the polymer on the surface of the cast film, producing an extremely thin, dense-skin layer of the membrane [10,144]. The polymer under the skin layer precipitates gradually, ensuing in a more porous polymer sublayer [145]. Following polymer precipitation, the membrane is usually annealed in order to improve solute rejection [10,144].

Polymeric gas separation membranes are produced normally as composites with a selective skin layer made of, for example, poly(dimethyl)siloxane, on a support structure composed of, for instance, polypropylene [146].

In a reverse osmosis system, pressurized water containing dissolved salts contact the feed side of the membrane; water depleted of salt is withdrawn as a low-pressure permeate. Then, these membranes are permeable to water but impermeable to salt ions. That is, if a pressure (4–7 MPa for seawater desalination) is applied to the highly concentrated solution, the flow of the liquid can be forced to occur in the reverse in the direction of osmosis. Desalination using reverse osmosis membranes has become very useful for producing freshwater from brackish water and seawater, since the method requires low capital and operating costs compared to other processes like multistage flash [11,148]. Membrane materials for reverse osmosis range from polysulfone (see Figure 10.8) and polyethersulfone, to cellulose acetate, cellulose diacetate, and aromatic polyamide (see Figure 10.9) [11].

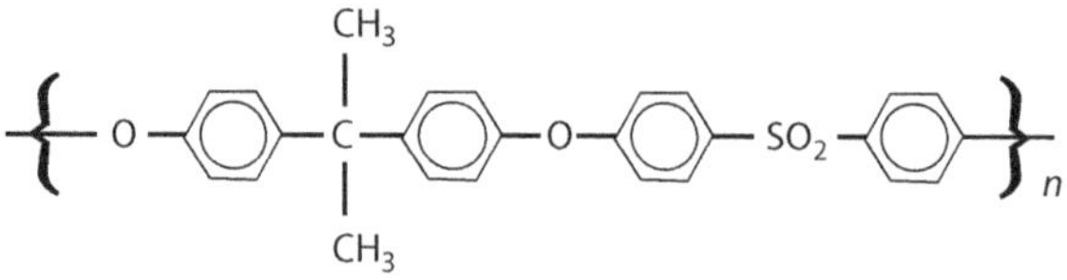

FIGURE 10.8 Polysulfone.

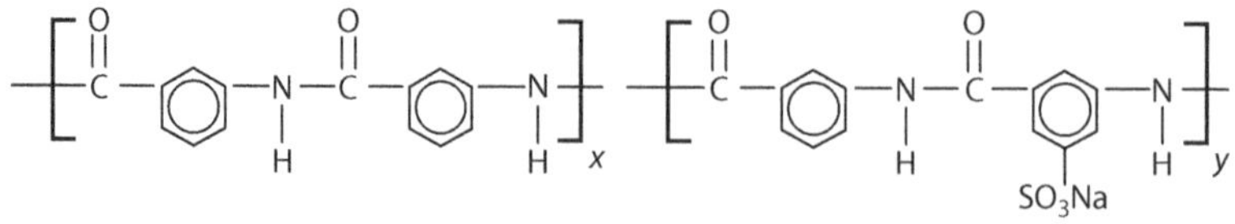

FIGURE 10.9 Aromatic polyamide.

In micro- and ultrafiltrations, the mode of separation is by sieving through fine pores, where microfiltration membranes filter colloidal particles and bacteria from 0.1 to 10 mm, and ultrafiltration membranes filter dissolved macromolecules. Usually, a polymer membrane, for example, cellulose nitrate, polyacrilonytrile, polysulfone, polycarbonate, polyethylene, polypropylene, polytretrafluoroethylene, polyamide, and polyvinylchloride, permits the passage of specific constituents of a feed stream as a permeate flow through its pores, while other, usually larger components of the feed stream are rejected by the membrane from the permeate flow and incorporated in the retentate flow [10,148,149].

During reverse osmosis and ultrafiltration membrane concentration, polarization and fouling are the phenomena responsible for limiting the permeate flux during a cyclic operation (i.e., permeation followed by cleaning). That is, membrane lifetimes and permeate (i.e., pure water) fluxes are primarily affected by the phenomena of concentration polarization (i.e., solute build up) and fouling (e.g., microbial adhesion, gel layer formation, and solute adhesion) at the membrane surface [11].

REFERENCES

1. H.P. Hsieh, *Inorganic Membranes for Separation, and Reaction, Membrane Science and Technology Series 3*, Elsevier, Amsterdam, the Netherlands, 1996.
2. E. Drioli and M. Romano, *Ind. Eng. Chem. Res.*, 40, 1277 (2001).
3. S. Adhikari and S. Fernando, *Ind. Eng. Chem. Res.*, 45, 875 (2006).
4. M.E. Ayturk, E.E. Engwall, and Y.H. Ma, *Ind. Eng. Chem. Res.*, 46, 4295 (2007).
5. J. Ghassemzadeh, L. Xu, T.T. Tsotsis, and M. Sahimi, *J. Phys. Chem. B*, 104, 3892 (2000).
6. S.C.A. Kluiters, *Status Review on Membrane Systems for Hydrogen Separation*, Intermediate Report, 5th Research Framework, European Union Project MIGREYD, Contract No. NNE5-2001-670, ECN-C-04-102, 2004.
7. M. Tsapatsis, PhD dissertation, California Institute of Technology, Pasadena, CA, 1994.
8. S. Morooka and K. Kusakabe, *MRS Bull.*, March, 25 (1999).
9. M. Mulder, *Basic Principles of Membrane Technology*, Kluwer Academic Publishers, Dordrecht, the Netherlands, 1996.
10. R.W. Baker, *Membrane Technology and Applications*, John Wiley & Sons, New York, 2004.
11. M.F. Goosen, S.S. Sablani, and R. Roque-Malherbe, in *Handbook of Membrane Separations: Chemical, Pharmaceutical, and Biotechnological Applications*, A.K. Pabby, A.N. Sastre, and S.S. Rizvi, (editors), CRC Press, Boca Raton, FL, 2008, p. 325.
12. G. Saracco, H.W.J.P. Neomagus, G.F. Versteeg, and W.P.M. Swaaij, *Chem. Eng. Sci.*, 54, 1997 (1999).
13. N. Sankar and M. Tsapatsis, in *Handbook of Zeolite Science, and Technology*, S. Auerbach, K.A. Carrado, and P.K. Dutta, (editors), Marcel Dekker, New York, 2003, p. 867.
14. I.F.J. Vankelecom, *Chem. Rev.*, 102, 3779 (2002).
15. C. Guizard and P. Amblard, in *Handbook of Membrane Separations: Chemical, Pharmaceutical, and Biotechnological Applications*, A.K. Pabby, A.N. Sastre, and S.S. Rizvi, (editors), CRC Press, Boca Raton, FL, 2008, p. 139.
16. S. Mauran, L. Rigaud, and O. Coudeville, *Transport Porous Med.*, 43, 355 (2001).
17. R. Krishna, in *Handbook of Zeolite Science and Technology*, S. Auerbach, K.A. Carrado, and P.K. Dutta, (editors), Marcel Dekker, New York, 2003, p. 1105.
18. R. Roque-Malherbe, W. del Valle, F. Marquez, J. Duconge, and M.F.A. Goosen, *Sep. Sci. Tech.*, 41, 73 (2006).
19. R. Roque-Malherbe, *Adsorption and Diffusion in Nanoporous Materials*, CRC Press, Boca Raton, FL, 2007.
20. G. Saracco and V. Specchia, *Catal. Rev. Sci. Eng.*, 36, 305 (1994).
21. H. Strathmann, L. Giorno, and E. Drioli, *An Introduction to Membrane Science and Technology*, Institute of Membrane Technology, University of Calabria, Italy, 2006.
22. M. Arruebo, R. Mallada, and M.P. Pina, in *Handbook of Membrane Separations: Chemical, Pharmaceutical, and Biotechnological Applications*, A.K. Pabby, A.N. Sastre, and S.S. Rizvi, (editors), CRC Press, Boca Raton, FL, 2008, p. 269.
23. W.J. Koros and G.K. Fleming, *J. Membr. Sci.*, 83, 1 (1993).
24. H.K. Lonsdale, U. Merten, and R.L. Riley, *J. Appl. Polym. Sci.*, 9, 1341 (1965).

25. J.M. Benito, A. Conesa, and M.A. Rodríguez, *Bol. Soc. Esp. Ceram. Vid.*, 43, 829 (2004).

26. H. Gao, Y.S. Lin, Y. Li, and B. Zhang, *Ind. Eng. Chem. Res.*, 43, 6920 (2004).

27. Y.S. Lin, *Sep. Purif. Technol.*, 25, 39 (1995).

28. S.N. Paglieri and J.D. Way, *Sep. Purif. Methods*, 31, 1 (2002).

29. S. Uemiya, *Top. Catal.*, 29, 79 (2004).

30. H. Mehrer, in *Diffusion in Condensed Matter*, P. Heithans and J. Karger, (editors), Springer-Verlag, Berlin, 2005, p. 3.

31. M.E. Glicksman, *Diffusion in Solids*, John Wiley & Sons, New York, 2000.

32. K.-D. Kreuer, *Chem. Mater.*, 8, 610 (1996).

33. R. Dittmeyer, V. Hollein, and K. Daub, *J. Mol. Catal. A*, 173, 35 (2001).

34. H.D. Tong, F.C. Gielens, J.G.E. Gardeniers, H.V. Jansen, C.J.M. van Rijn, M.C. Elwenspoek, and W. Nijdam, *Ind. Eng. Chem. Res.*, 43, 4182 (2004).

35. F. Roa and J.D. Way, *Ind. Eng. Chem. Res.*, 42, 5827 (2003).

36. D.J. Edlund and J.M. McCarthy, *J. Membr. Sci.*, 107, 147 (1995).

37. S. Uemiya, *Sep. Purif. Methods*, 28, 51 (1999).

38. H.A. Meinema, R.W.J. Dirrix, H.W. Brinkman, R.A. Terpstra, J. Jekerle, and P.H. Kösters, *INTERCERAM*, 54, 86 (2005).

39. T. Norby and R. Hausgrud, in *Non-porous Inorganic Membranes*, A.F. Sammells and M.V. Mundschau, (editors), Wiley-VCH Verlag GmbH & Co., Weiheim, Germany, 2006, p. 1.

40. L. Li, A. Li, and E. Iglesia, *Stud. Surf. Sci. Catal.*, 136, 357 (2001).

41. H. Iwahara, K. Uchida, and K. Ogaki, *J. Electrochem. Soc.*, 135, 529 (1988).

42. N. Bonanos, *Solid State Ionics*, 53–56, 967 (1992).

43. J. Wu, R.A. Davies, M.S. Islam, and S.M. Haile, *Chem. Mater.*, 17, 846 (2005).

44. T. Norby, M. Wideroe, R. Glöckner, and Y. Larring, *Dalton Trans.*, 3012 (2004).

45. J. Wu, PhD thesis, California Institute of Technology, Pasadena, CA, 2005.

46. R. Oesten and R.A. Higgins, *Ionics*, 1, 427 (1995).

47. S. Nieto, R. Polanco, and R. Roque-Malherbe, *J. Phys. Chem. C*, 111, 2809 (2007).

48. C. Wagner and W. Schottky, *Z. Phys. Chem.*, 21, 25 (1933).

49. C. Wagner, *Prog. Solid State Chem.*, 10, 3 (1975).

50. J.C. Boivin and G. Mairesse, *Chem. Mater.*, 10, 2870 (1998).

51. C. Chen, Z. Zhang, G. Jiang, C. Fan, W. Liu, and H.J.M. Bouwmeester, *Chem. Mater.*, 13, 2797 (2001).

52. X. Tan, Y. Liu, and K. Li, *Ind. Eng. Chem. Res.*, 44, 61 (2005).

53. X. Qi, Y.S. Lin, and S.L. Swartz, *Ind. Eng. Chem. Res.*, 39, 646 (2000).

54. L. Yang, L. Tan, X.H. Gu, W.Q. Jin, L.X. Zhang, and N.P. Xu, *Ind. Eng. Chem. Res.*, 42, 2299 (2003).

55. V.L. Kozhevnikov, I.A. Leonidov, J.A. Bahteeva, M.V. Patrakeev, E.B. Mitberg, and K.R. Poeppelmeier, *Chem. Mater.*, 16, 5014 (2004).

56. X. Chang, C. Zhang, Z. Wu, W. Jin, and N. Xu, *Ind. Eng. Chem. Res.*, 45, 2824 (2006).

57. H.J.M. Bouwmeester and A.J. Burggraaf, in *Fundamentals of Inorganic Membrane Science and Technology*, A.J. Burggraaf and L. Cot, (editors), Elsevier, Amsterdam, the Netherlands, 1996, p. 435.

58. M. Anca-Dragan, PhD dissertation, Department of Mathematic, Informatics and Natural Sciences, Institute of Physical Chemistry, RWTH-Aachen, Germany, 2006.

59. K.S.W. Sing, D.H. Everett, R.A.W. Haul, L. Moscou, R.A. Pirotti, J. Rouquerol, and T. Siemieniewska, *Pure App. Chem.*, 57, 603 (1985).

60. J. Xiao and J. Wei, *Chem. Eng. Sci.*, 47, 1123 (1992).

61. R. Reif, *Fundamentals of Statistical and Thermal Physics*, McGraw-Hill, Boston, MA, 1965.

62. C.N. Satterfield, *Heterogeneous Catalysis in Practice*, McGraw-Hill, New York, 1980.

63. M.R. Wang and Z.X. Li, *Phys. Rev. E*, 68, 046704 (2003).

64. J.-G. Choi, D.D. Do, and H.D. Do, *Ind. Eng. Chem. Res.*, 40, 4005 (2001).

65. R.W. Barber and D.R. Emerson, in *Advances in Fluid Mechanics IV*, M. Rahman, R. Verhoeven, and C.A. Brebbia, (editors), WIT Press, Southampton, UK, 2002, p. 207.

66. S.A. Schaaf and P.L. Chambre, *Flow of Rarefied Gases*, Princeton University Press, Princeton, NJ, 1961.

67. J. Karger, S. Vasenkow, and S.M. Auerbach, in *Handbook of Zeolite Science and Technology*, S. Auerbach, K.A. Carrado, and P.K. Dutta, (editors), Marcel Dekker, New York, 2003, p. 341.

68. A. Kapoor, R.T. Yang, and C. Wong, *Catal. Rev. Sci. Eng.*, 31, 129 (1989).

69. R.J.R. Ulhorn, K. Keizer, and A.J. Burggraaf, *J. Membr. Sci.*, 66, 271 (1992).

70. J. Karger and D.M. Ruthven, *Diffusion in Zeolites and Other Microporous Solids*, John Wiley & Sons, New York, 1992.

71. M.F.M. Post, *Stud. Surf. Sci. Catal.*, 58, 391 (1991).
72. M.C. Potter and D.C. Wiggert, *Mechanics of Fluids* (3rd edition), Brooks/Cole-Thomson Learning, Pacific Groove, CA, 2002.
73. R.B. Bird, W.E. Stewart, and E.N. Lightfoot, *Transport Phenomena* (2nd edition), John Wiley & Sons, New York, 2002.
74. P. Carman, *Flow of Gases Through Porous Media*, Butterworth, London, UK, 1956.
75. E.A. Mason, and A.P Malinauskas, *Gas Transport in Porous Media. The Dusty Gas Model*, Elsevier, Amsterdam, the Netherlands, 1983.
76. V. Papavassiliou, C. Lee, J. Nestlerode, and M.P. Harold, *Ind. Eng. Chem. Res.*, 36, 4954 (1997).
77. W.J. Mortier, *Compilation of Extraframework Sites in Zeolites*; Butterworth, London, UK, 1982.
78. Ch. Baerlocher, W.M. Meier, and D.M. Olson, *Atlas of Zeolite Framework Types* (5th edition), Elsevier, Amsterdam, the Netherlands, 2001.
79. A. Corma, *Chem. Rev.*, 97, 2373 (1997).
80. N.Y. Chen, T.F. Degnan Jr., and C.M. Smith, *Molecular Transport and Reaction in Zeolites*, VCH, New York, 1994.
81. R. Snurr and J. Karger, *J. Phys. Chem. B*, 101, 6469 (1997).
82. R. Roque-Malherbe, R. Wendelbo, A. Mifsud, and A. Corma, *J. Phys. Chem.*, 99, 14064 (1995).
83. R. Roque-Malherbe, *Mic. Mes. Mat.*, 41, 227 (2000).
84. J. de la Cruz, C. Rodriguez, and R. Roque-Malherbe, *Surf. Sci.*, 209, 215 (1989).
85. M. Matsukata and E. Kikuchi, *Bull. Chem. Soc. Jpn.*, 70, 2341 (1997).
86. J. Coronas and J. Santamaria, *Sep. Purif. Methods*, 28, 127 (1999).
87. K.H. Bennett, K.D. Cook, J.L. Falconer, and R.D. Noble, *Anal. Chem.*, 71, 1016 (1999).
88. K. Aoki, K. Kusakabe, and S. Morooka, *Ind. Eng. Chem. Res.*, 39, 2245 (2000).
89. X. Lin, J.L. Falconer, and R.D. Noble, *Chem. Mater.*, 10, 3716 (1998).
90. C.D. Baertsch, H.H. Funke, J.L. Falconer, and R.D. Noble, *J. Phys. Chem.*, 100, 7676 (1996).
91. I. Kumakiri, T. Yamaguchi, and S.-I. Nakao, *Ind. Eng. Chem. Res.*, 38, 4682 (1999).
92. D.S. Sholl, *Ind. Eng. Chem. Res.*, 39, 3737 (2000).
93. A.J. Burggraaf, *J. Memb. Sci.*, 155, 45 (1999).
94. L.J.P. van den Broeke, W.J.W. Bakker, F. Kateijn, and J.A. Moulijn, *Chem. Eng. Sci.*, 54, 245 (1999).
95. J.C. Poshusta, V.A. Tuan, J.L. Falconer, and R.D. Noble, *Ind. Eng. Chem. Res.*, 37, 3925 (1998).
96. H. Suzuki, US Patent, 4,699,892 (1987).
97. A. Tavolaro and E. Drioli, *Adv. Mater.*, 11, 975 (1999).
98. J. Ghassemzadeh, L. Xu, T.T. Tsotsis, and M. Sahimi, *J. Phys. Chem. B*, 104, 3892 (2000).
99. R.M. Barrer and W. Jost, *Trans. Faraday Soc.*, 45, 928 (1949).
100. M. Hernández-Vélez, O. Raymond, A. Alvarado, A. Jacas, and R. Roque-Malherbe, *J. Mat. Sci. Lett.*, 14, 1653 (1995).
101. R.M. Barrer, *Hydrothermal Chemistry of Zeolites*, Academic Press, London, UK, 1982.
102. D.W. Breck, *Zeolite Molecular Sieves*, Wiley, New York (1974).
103. C.S. Cundy and P.A. Cox, *Chem. Rev.*, 103, 663 (2003).
104. C. de las Pozas, D. Díaz-Quintanilla, J. Pérez-Pariente, R. Roque-Malherbe, and M. Magi, *Zeolitas*, 9, 33 (1989).
105. R. Roque-Malherbe, R. López-Cordero, J.A. González-Morales, J. Onate, and M. Carreras, *Zeolites*, 13, 481 (1993).
106. H. Robson, (editor), *Verified Synthesis of Zeolitic Materials* (2nd edition), Elsevier, Amsterdam, the Netherlands, 2001.
107. M.M.J. Treacy and J.B. Higgins, *Collection of Simulated XRD Powder Patterns for Zeolites* (4th edition), Elsevier, Amsterdam, the Netherlands, 2001.
108. J.N. Armor, in *Materials Chemistry and Emerging Discipline*, L.V. Interrante, L.A. Casper, and A.B. Ellis, (editors), American Chemical Society, Washington, DC, 1995, p. 321.
109. Y. Teraoka, H.M. Zhang, S. Furukawa, and Yamazoe, *Chem. Lett.*, 1743 (1985).
110. L.W. Tai, M.M. Nasrallah, H.U. Anderson, D.M. Sparlin and S.R. Sehlin, *Solid State Ionics*, 76, 259 (1995).
111. Y. Teraoka, T. Nobunaga, and N. Yamazoe, *Chem. Lett.*, 503 (1988).
112. L. Qiu, T.H. Lee, L.-M. Liu, Y.L. Yang, and A.J. Jacobson, *Solid State Ionics*, 76, 321 (1995).
113. B. Ma, U. Balachandran, J.-H. Park, and C.U. Segre, *Solid State Ionics*, 83, 65 (1996).
114. Y. Zeng, Y.A. Lin, and S.L. Swartz, *J. Membr. Sci.*, 150, 87 (1998).
115. P.N. Dyer, R.E. Richards, S.L. Russek, and D.M. Taylor, *Solid State Ionics*, 134, 21 (2000); S. Li, W. Jin, P. Huang, N. Xu, J. Shi, and Y.S. Lin, *J. Membr. Sci.*, 166, 51 (2000).

116. Y. Lu, A.G. Dixon, W.R. Moser, Y.H. Ma, and U. Balachandran, *J. Membr. Sci.*, 170, 27 (2000).

117. J. Luyten, A. Buekenhoudt, W. Adriansens, J. Cooymans, H. Weyten, F. Servaes, and R. Leysen, *Solid State Ionics*, 135, 637 (2000).

118. J. Zaman and A. Chakma, *J. Membr. Sci.*, 92, 1 (1994).

119. J.N. Armor, *Catal. Today*, 25, 199 (1995).

120. V.M. Gryaznov, *Platinum Metals Rev.*, 30, 68 (1986).

121. A. Nijmeijer, PhD dissertation, University of Twente, Enschede, the Netherlands, 1999.

122. Y. Yildirim, E. Gobina, and R. Hughes, *J. Membr. Sci.*, 135, 107 (1997).

123. J.P. Collins, R.W. Schwartz, R. Sehgal, T.L. Ward, C.J. Brinker, G.P. Hagen, and C.A. Udovich, *Ind. Eng. Chem. Res.*, 35, 4398 (1996).

124. J.C.S. Wu, T.E. Gerdes, J.L. Pszczolkowski, R.R. Bhave, P.K.T. Liu, and E.S. Martin, *Sep. Sci. Tech.*, 25, 1489 (1990).

125. T. Kameyama, M. Dokiya, M. Fujihige, H. Yokokawa, and K. Fukuda, *Int. J. Hydrogen Energ.*, 8, 5 (1983).

126. T. Johansen, K.S. Raghuraman, and L.A. Hacket, *Hydrocarb. Process.*, 8, 119 (1992).

127. L. Paturzo and A. Basile, *Ind. Eng. Chem. Res.*, 41, 1703 (2002).

128. E. Kikuchi and Y. Chen, *Natural Gas Conversion V*, Vol. 119, Elsevier Science, New York, 1998, p. 441.

129. E. Kikuchi, S. Uemiya, N. Sato, H. Inoue, H. Ando, and T. Matsuda, *Chem. Lett.*, 489 (1989).

130. S. Uemiya, N. Sato, H. Ando, and E. Eikuchi, *Ind. Eng. Chem. Res.*, 30, 585 (1991).

131. J.P. Collins and J.D. Way, *Ind. Eng. Chem. Res.*, 32, 3006 (1993).

132. G.R. Gavalas, *Ceramic Membranes for Hydrogen Production from Coal*, Annual Technical Report, DE-FG26-00NT40817, California Institute of Technology, Pasadena, CA, March 18, 2003.

133. S.N. Paglieri and J.D. Way, *Sep. Purif. Methods*, 31, 1 (2002).

134. E. Wicke and G.H. Nernst, *Ber. Buns. Phys. Chem.*, 68, 224 (1964).

135. A. Kulprathipanja, G.O. Alptekin, J.L. Falconer, and J.D. Way, *Ind. Eng. Chem. Res.*, 43, 4188 (2004).

136. A.G. Knapton and A.G. Platinum, *Metals. Rev.*, 21, 44 (1977).

137. G.L. Holleck, *J. Phys. Chem.*, 74, 1370 (1970).

138. S. Maestas and T.B. Flanagan, *J. Phys. Chem.*, 77, 850 (1973).

139. J.E. Philpott, *Platinum Metals Rev.*, 29, 12 (1985).

140. G. Cicala, G. Bruno, and P. Capezutto, in *Handbook of Surfaces and Interfaces of Materials*, Vol. 1, H.S. Nalwa, (editor), Academic Press, New York, 2001, p. 509.

141. T.M. Besmann, *Processing Science for Chemical Vapor Infiltration, High Temperature Materials Laboratory Industry/Government Briefing*, Oak Ridge National Laboratory, Oak Ridge, TN, 1990.

142. A.C. Pierre and G.M. Pajonk, *Chem. Rev.*, 102, 4243 (2002).

143. B.L. Cushing, V. L. Kolesnichenko, and C.J. O'Connor, *Chem. Rev.*, 104, 3893 (2004).

144. R.W. Baker, in *Membrane Separation Systems. Recent Developments and Future Directions*, Vol. II, R.W. Baker, E.L. Cussler, W. Eykamp, W.J. Koros, R.L. Riley, and H. Strathmann, (editors), SciTech Publishing Inc., Raleigh, NC, 1991, p. 100.

145. H. Strathmann, in *Handbook of Industrial Membrane Technology*, M. Porter, (editor), Noyes Publications, Park Ridge, NJ, 1990, p. 1.

146. M.-B. Haag, in *Handbook of Membrane Separations: Chemical, Pharmaceutical, and Biotechnological Applications*, A.K. Pabby, A.N. Sastre, and S.S. Rizvi, (editors), CRC Press, Boca Raton, FL, 2008, p. 65.

147. R. Rautenbach and R. Albrecth, *Membrane Processes*, John Wiley & Sons, New York, 1991.

148. R. Droste, *Theory and Practice of Water and Wastewater Treatment*, John Wiley & Sons, New York, 1997.

149. L.J. Zeman and A.L. Zydney, *Microfiltration and Ultrafiltration: Principles and Applications*, CRC Press, Boca Raton, FL, 1996.

150. S. Sridar and B. Smitha in *Handbook of Membrane Separations: Chemical, Pharmaceutical, and Biotechnological Applications*, A.K. Pabby, A.N. Sastre, and S.S. Rizvi, (editors), CRC Press, Boca Raton, FL, 2008, p. 1101.

Index